Palladacycles

Edited by
Jairton Dupont and
Michel Pfeffer

Further Reading

Hashmi, A. S. K., Toste, D. F. (eds.)

Modern Gold Catalyzed Synthesis

450 pages
2009
Hardcover
ISBN: 978-3-527-31952-7

Stepnicka, P. (ed.)

Ferrocenes
Ligands, Materials and Biomolecules

672 pages
2008
Hardcover
ISBN: 978-0-470-03585-6

Tolman, W. B. (ed.)

Activation of Small Molecules
Organometallic and Bioinorganic Perspectives

382 pages with 147 figures and 24 tables
2006
Hardcover
ISBN: 978-3-527-31312-9

Handbook of C-H Transformations

Applications in Organic Synthesis

688 pages in 2 volumes with 43 figures and 81 tables
2005
Hardcover
ISBN: 978-3-527-31074-6

Knochel, P. (ed.)

Handbook of Functionalized Organometallics
Applications in Synthesis

690 pages in 2 volumes with 824 figures and 1 tables
2005
Hardcover
ISBN: 978-3-527-31131-6

Evans, P. A. (ed.)

Modern Rhodium-Catalyzed Organic Reactions

496 pages with 336 figures and 102 tables
Hardcover
ISBN: 978-3-527-30683-1

Tamaru, Y. (ed.)

Modern Organonickel Chemistry

346 pages with 297 figures and 30 tables
2005
Hardcover
ISBN: 978-3-527-30796-8

Palladacycles

Synthesis, Characterization and Applications

Edited by
Jairton Dupont and Michel Pfeffer

WILEY-VCH
WILEY-VCH Verlag GmbH & Co. KGaA

The Editors

Prof. Dr. Jairton Dupont
UFRGS, Institute of Chemistry
Laboratory of Molecular Catalysis
Av. Bento Goncalves
9500 Porto Alegre 91501-970 RS
Brasil

Dr. Michel Pfeffer
Université Louis Pasteur UMR 7177
Laboratoire Synthèses Métallo-Induites
4, rue Blaise Pascal
67070 Strasbourg
France

Library of Congress Card No.: applied for

British Library Cataloguing-in-Publication Data
A catalogue record for this book is available from the British Library.

Bibliographic information published by the Deutsche Nationalbibliothek
The Deutsche Nationalbibliothek lists this publication in the Deutsche Nationalbibliografie; detailed bibliographic data are available on the Internet at <http://dnb.d-nb.de>.

Typesetting SNP Best-set Typesetter Ltd., Hong Kong
Printing Strauss GmbH, Mörlenbach
Binding Litges & Dopf GmbH, Heppenheim
Cover Design Grafik-Design Schulz, Fußgönheim

Printed in the Federal Republic of Germany
Printed on acid-free paper

ISBN: 978-3-527-31781-3

Contents

Palladacycles: Synthesis, Characterization and Applications. Edited by Jairton Dupont and Michel Pfeffer

ISBN: 978-3-527-31781-3

List of Contributors

Martin Albrecht
University of Fribourg
Department of Chemistry
Chemin du Musée 9
CH-1700 Fribourg
Switzerland

Diego A. Alonso
Universidad de Alicante
Facultad de Ciencias
Departamento de Química Orgánica
Apdo. 99
03080 Alicante
Spain

Robin B. Bedford
University of Bristol
School of Chemistry
Cantock's Close
Bristol BS8 1TS
UK

Duncan W. Bruce
Université Louis Pasteur
Institut de Physique et Chimie des Matériaux de Strasbourg
CNRS UMR 7504
23 rue du Loess BP 43
67034 Strasbourg
Cedex 2
France

Jean-Pierre Djukic
Université Louis Pasteur
Institut de Chimie
CNRS UMR 7177
4, Rue Blaise Pascal
67000 Strasbourg
France

Bertrand Donnio
Université Louis Pasteur
Institut de Physique et Chimie des Matériaux de Strasbourg
CNRS UMR 7504
23 rue du Loess BP 43
67034 Strasbourg
Cedex 2
France

Jairton Dupont
UFRGS, Institute of Chemistry
Laboratory of Molecular Catalysis
Av. Bento Goncalves
9500 Porto Alegre 91501-970 RS
Brasil

Robertus J. M. Klein Gebbink
Utrecht University
Faculty of Science
Chemical Biology and Organic Chemistry
Padualaan 8
3584 C.H. Utrecht
The Netherlands

Palladacycles: Synthesis, Characterization and Applications. Edited by Jairton Dupont and Michel Pfeffer

ISBN: 978-3-527-31781-3

John A. Gladysz
Texas A8M University
Department of Chemistry
P.O. Box 30012
College Station, Texas
77842-3012, USA

Ties J. Korstanje
Utrecht University
Faculty of Science
Chemical Biology and Organic Chemistry
Padualaan 8 3584 C.H. Utrecht
The Netherlands

Gerard van Koten
Utrecht University
Faculty of Science
Chemical Biology and Organic Chemistry
Padualaan 8 3584 C.H. Utrecht
The Netherlands

Mario R. Meneghetti
Universidade Federal de Alagoas
Instituto de Química e Biotecnologia
Av. Lourival de Melo Mota s/n
5 7072-970 Maceió–AL
Brazil

David Morales-Morales
Universidad Nacional Autonoma de México
Instituto de Quimica
Circuito Exterior S/N. Ciudad Universitaria
Coyoacan. C.P. 04510
México D.F.

Carmen Nájera
Universidad de Alicante
Facultad de Ciencias
Departamento de Química Orgánica
Apdo. 99
03080 Alicante
Spain

Francesco Neve
Università della Calabria
Dipartimento di Chimica
Cubo 14/C
Ponte P. Bucci
87030 Arcavacata di Rende
Italy

Mª Teresa Pereira
Universidad de Santiago de Compostela
Facultad de Química
Departamento de Química Inorgánica
Avenida das Ciencias S/N
15782 Santiago de Compostella
Spain

Michel Pfeffer
Université Louis Pasteur UMR 7177
Laboratoire de Synthèses Metallo-Induites
4, rue Blaise Pascal
67070 Strasbourg
France

Niels J. M. Pijnenburg
Utrecht University
Faculty of Science
Chemical Biology and Organic Chemistry
Padualaan 8
3584 C.H. Utrecht
The Netherlands

Alexander D. Ryabov
Carnegie Mellon University
Department of Chemistry
4400 Fifth Avenue
Pittsburgh
Pennsylvania 15213
USA

John Spencer
Reader in Medicinal Chemistry
School of Science
University of Greenwich at
Medway
Chatham Maritime
Kent
ME4 4TB
UK

Esteban P. Urriolabeitia
Instituto de Ciencia de Materiales de
tragón (CSIC-Zaragoza University)
Department of Organometallic
Compounds
Pedro Cerouna 12, Ciudad
Universitaria
50009 Zaragoza
Spain

José M. Vila
Universidad de Santiago de
Compostela
Facultad de Química
Departamento de Química Inorgánica
Avenida das Ciencias, s/n
15782 Santiago de Compostella
Spain

1
Introduction

David Morales-Morales

1.1
Introduction

Since their discovery in the mid-1960s palladacycle compounds have represented a very interesting topic of research [1]–first identified as important intermediates in palladium mediated organic synthesis [2] and more recently due to their unique physical properties, these compounds have experienced a renaissance that has been fundamental in the recent development of homogeneous catalysis. This is particularly true in the case of C—C cross-coupling reactions [3]. In general, these compounds can be synthesized in a very facile manner, making it possible to modulate both their steric and electronic properties or even include chiral motifs in their structures to enable them for potential applications in enantioselective transformations as chiral auxiliaries [4]. Other important areas where palladacycles have found recent applications include their use as mesogenic [5] and photoluminescent agents [5h, 6] as well as biological applications for cancer treatment (bio-organometallic chemistry) [7]. Consequently, the present chapter covers some general concepts regarding palladacycle compounds such as a general definition, a brief historical overview, a proposal of a general classification based on some excellent recent reviews and, finally, a brief description of the future outlook for these very interesting species.

1.2
Definition

In general, a palladacycle (Figure 1.1) can be defined as any palladium compound containing one palladium–carbon bond intramolecularly stabilized by one or two neutral donor atoms (Y), where the organic moiety acts as a C-anionic four-electron donor ligand or as a C-anionic six-electron donor ligand.

Palladacycles: Synthesis, Characterization and Applications. Edited by Jairton Dupont and Michel Pfeffer

ISBN: 978-3-527-31781-3

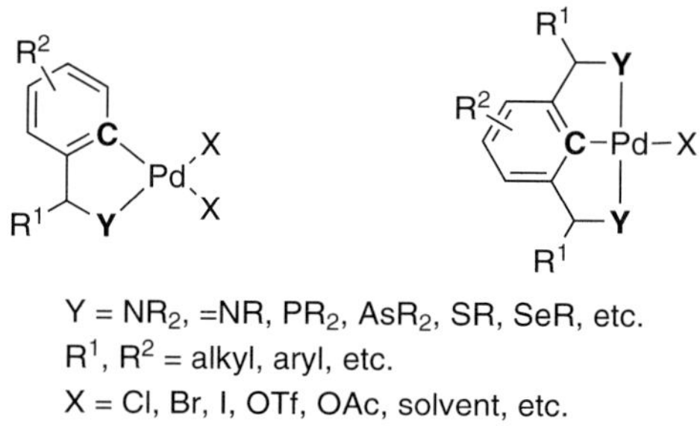

Figure 1.1 Structural definition of a palladacycle.

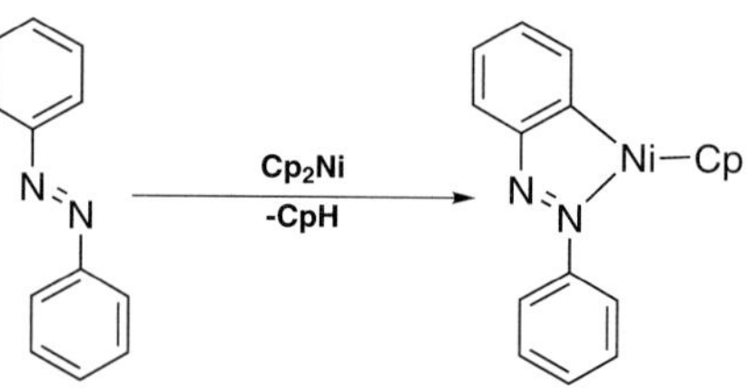

(1)

Scheme 1.1

1.3 Historical Overview

Historically, there are probably three different events that have defined the development of the chemistry of palladacycles, one being the discovery of the cyclometallation reaction in 1963 by Kleinman and Dubeck [8] when they reacted azobenzene with $NiCp_2$ to obtain a five-membered metallacycle **(1)** (Scheme 1.1). The structure originally proposed by Kleinman and Dubeck considered the nickel center to be coordinated η^2 to the N=N π-bond **(2)** [8]. This chemistry was soon extended to other group-10 transition metals. Thus, between 1965 [9] and 1968 [10] Cope, Siekman and Friedrich carried out analogous reactions of azobenzene and *N,N*-dimethylbenzylamines, this time using $PdCl_2$ or Li_2PdCl_4, to afford the first isolated, well-characterized palladacycles (Scheme 1.2).

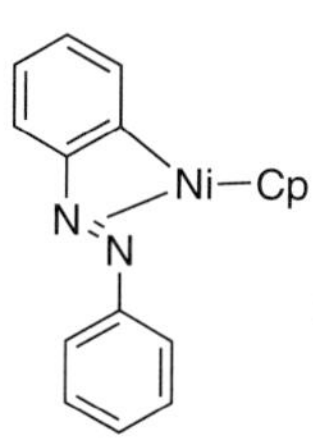

(2)

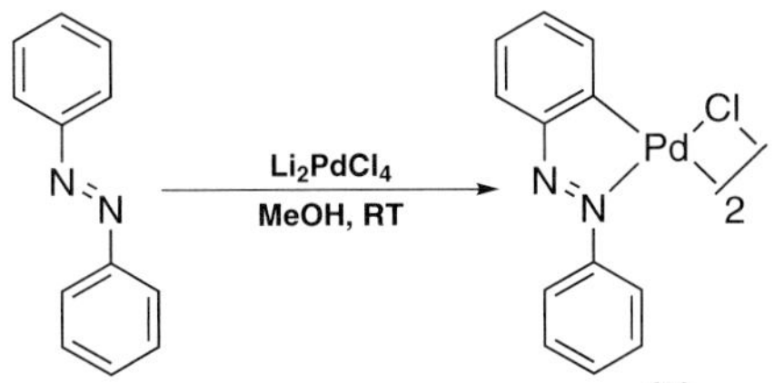

(3)

Scheme 1.2

The physical properties these compounds exhibited, in particular the high thermal stability in the solid state, led to the third and probably most important fact, which was the introduction by Herrmann *et al.* in 1995 of the cyclopalladated tri-*o*-tolyl-phosphine complex **(4)** as catalyst precursor for palladium-catalyzed Heck and other cross-coupling reactions [11]. This raised high expectations for this class of compounds, as these species could activate more economic substrates than those applied thus far (aryl iodides or aryl triflates), such as aryl chlorides, hence potentially enabling the industrial application of these cross-coupling reactions mediated by palladacycle catalysts [12]. Since then, palladacycles have been ubiquitous in catalytic transformations, playing important roles as catalyst precursors or active intermediates in cascade transformations leading to complex molecular architectures and so forth [2, 3].

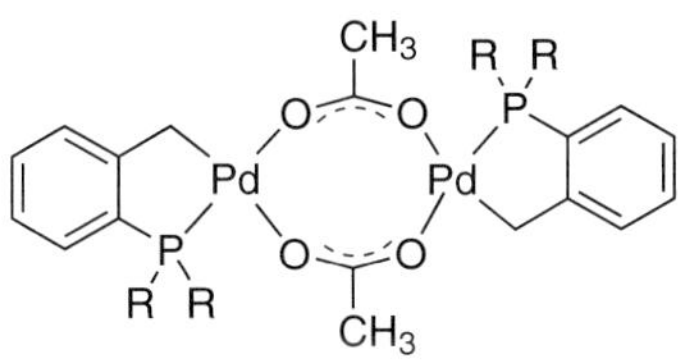

R = *o*-Tol

(4)

1.4 Classification of Palladacycles (Types)

According to the established definition, palladacycles can be divided into two different classes based on the organic fragment: anionic four-electron (CY) or six-electron donor (YCY) complexes [1t, 1w, 1x].

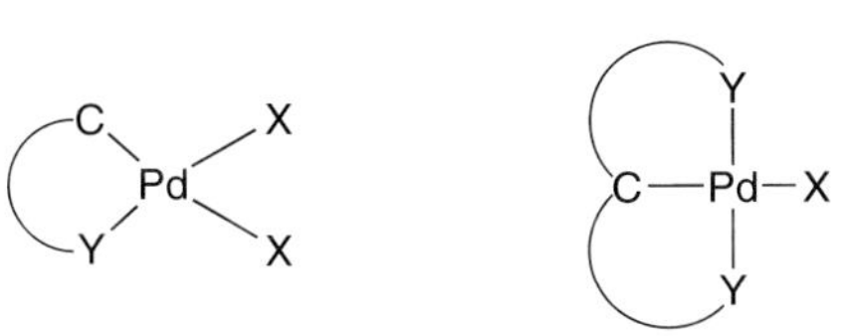

CY **YCY**

Hence, palladacycles of the type CY usually exist as halogen **(5)** or acetate **(6)** bridged dimers (Scheme 1.3) [1w, 13a], as two geometric isomers, *cisoid* and *transoid* conformations.

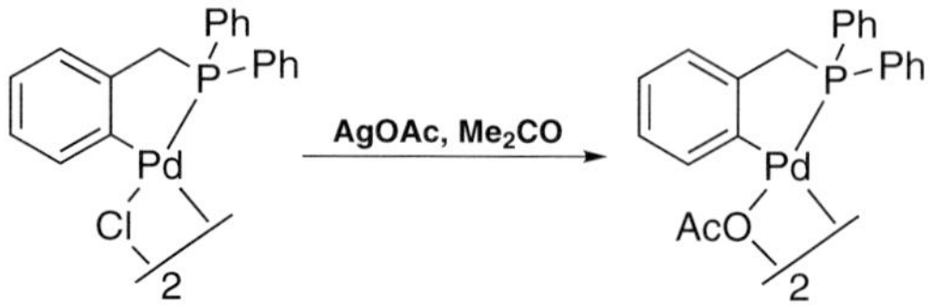

(5) **(6)**

Scheme 1.3

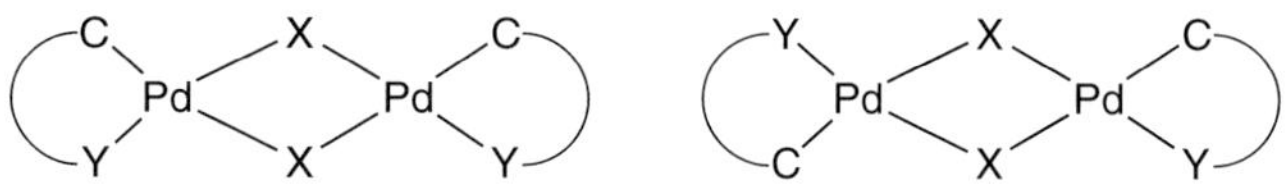

cisoid-palladacycle *transoid*-palladacycle

Additionally, CY species can be divided into neutral, cationic **(7)** [14] or anionic **(8)** [15]; the neutral species can be found as monomers **(9)** [16], dimers **(10)** [10] or bis-cyclopalladated **(11)** [17] complexes, depending on the nature of the other ligands X.

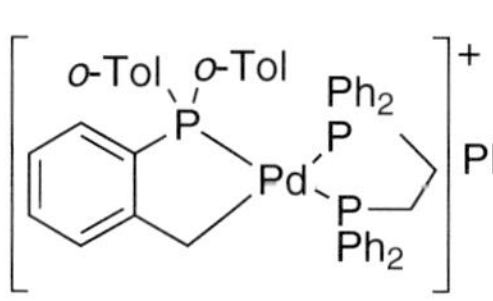

(7)

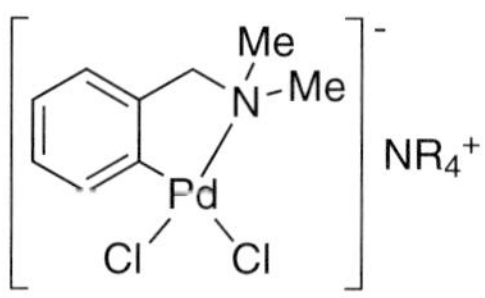

(8)

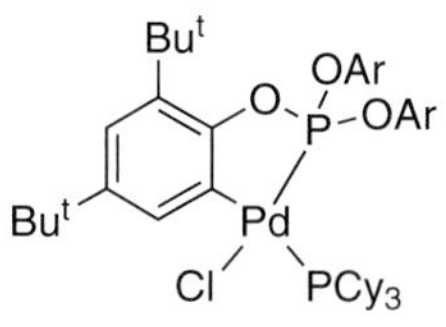

Ar= C_6H_3-2,4-Bu^t_2

(9)

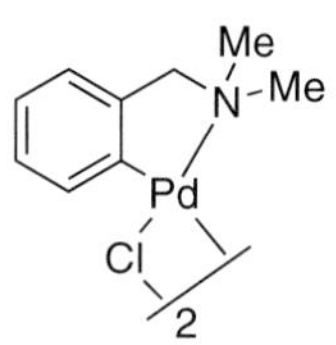

(10)

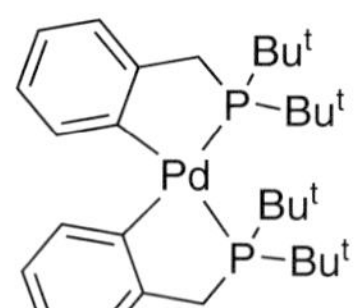

(11)

The position of the C—H bond to be activated with respect to the donor atom Y, as well as the hybridization of the carbon atom in the C—H bond being metallated, undoubtedly influences the ease of cyclometallation, and although formal energetic considerations regarding the strength of aromatic and aliphatic C—H bonds have been performed [13b, 18], these data are of little utility due to the complex combination of various factors determining the metallation process. However, from analyses of the available experimental results, it can be concluded that for the vast majority of known complexes the metallated carbon is usually an aromatic sp^2 carbon [10, 15–17] (species **8–11**) and less commonly an sp^3 aliphatic **(12)** [19], benzylic **(13)** [11, 20]) or sp^2 vinylic **(14)** [21] carbon.

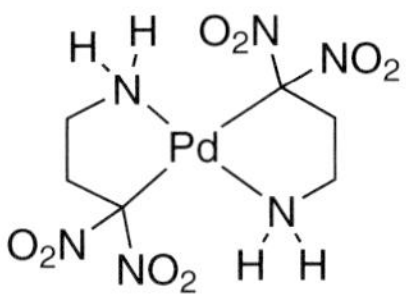

(12)

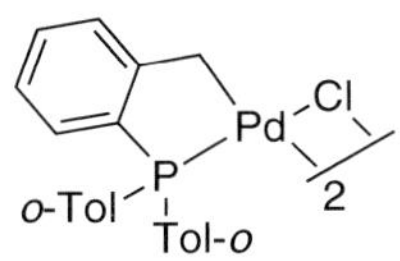

(13)

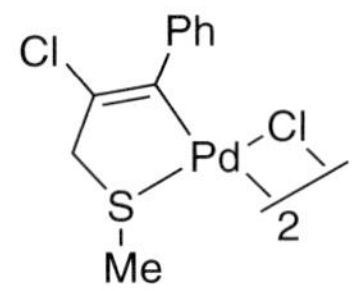

(14)

On the other hand, the position of the C—H bond with respect to the Y donor atom determines the size of the palladacycle. Thus, although CY-type metallated rings can vary from 3 to 11 members, the most common palladacycles are usually five- or six-membered rings. Palladacycles of three and four members are usually unstable, as are those larger than six members, which generally undergo facile reductive elimination [1u, 2b, 22]; consequently, examples of well-characterized compounds of this kind are rare. The structures of some isolated, well-characterized palladacycles are shown here of three **(15)** [23], four **(16)** [24], five **(17)** [25], six **(18)** [26], seven **(19)** [27], eight **(20)** [28], nine **(21)** [29] and ten **(22)** [29] members.

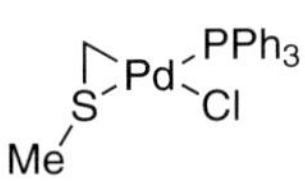

(15)

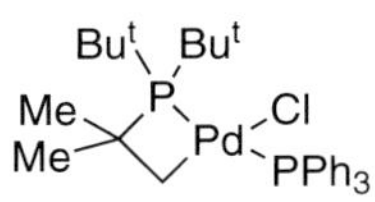

(16)

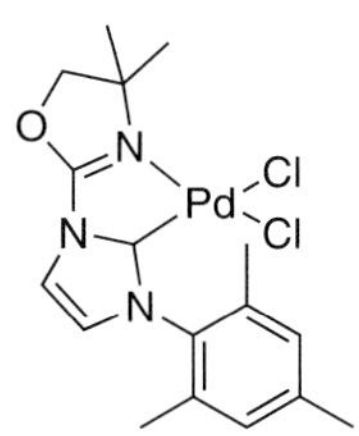

(17)

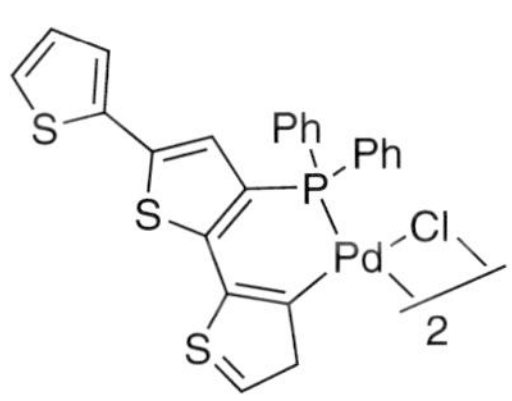

(18)

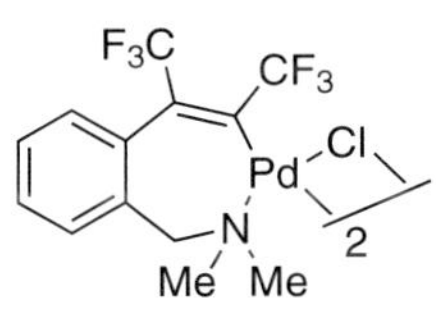

(19)

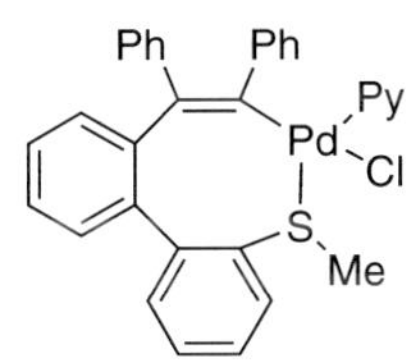

(20)

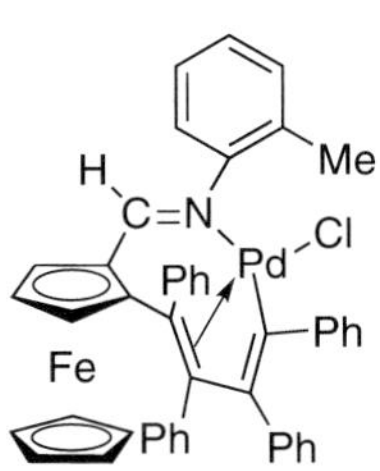

(21)

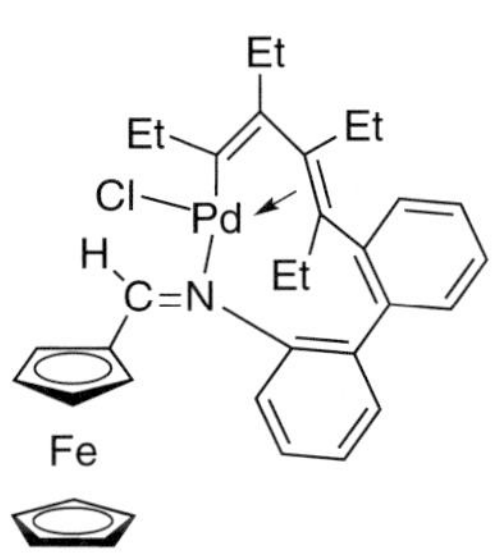

(22)

The above discussion is also valid for YCY palladacycles or pincer-type complexes [1o, 1r, 1s, 30]. The most common arrangement found for these species is that having two equivalent five-membered rings **(23)** [31]. In addition, recently, unsymmetrical mixed five- and symmetric six-membered **(24)** [32] and six-membered complexes **(25)** [33] have been isolated and characterized.

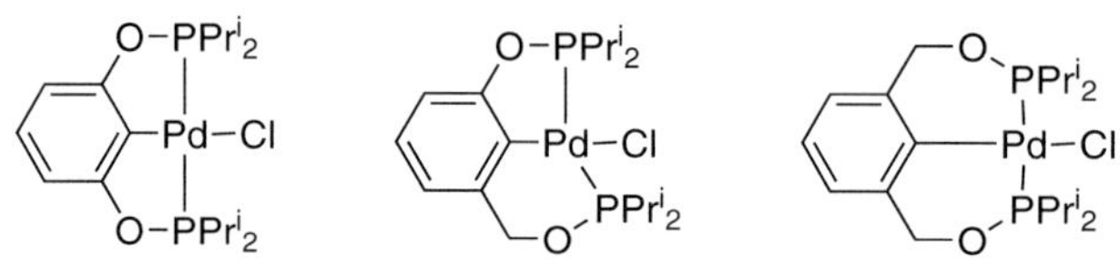

(23) **(24)** **(25)**

On the other hand, the donor atoms (Y), the other important part of palladacycles, can theoretically influence the palladation process by the basicity and the coordination ability of the donor atom. However, studies carried out with phosphines differing in the nature of their substituents at the phosphorus atoms revealed that these factors are relatively insignificant [34]. Thus, complexes derived from numerous phosphines can be synthesized by similar synthetic methods – even YCY symmetric five-membered palladium compounds containing the $P(C_6F_5)_2$ fragment **(26, 27)** [35], were synthesized in a very facile manner via a C–H activation process (Scheme 1.4). Conversely, the analogous YCY compound derived from the fluorinated thioether $^-SC_6F_5$ **(28)** has not yet been synthesized (Scheme 1.5) [36]; this is probably being due to the low availability of the electron pair in the sulfur. These results clearly call for more detailed studies to shed more light on the potential effect of the Y donor atom in the cyclometallation process.

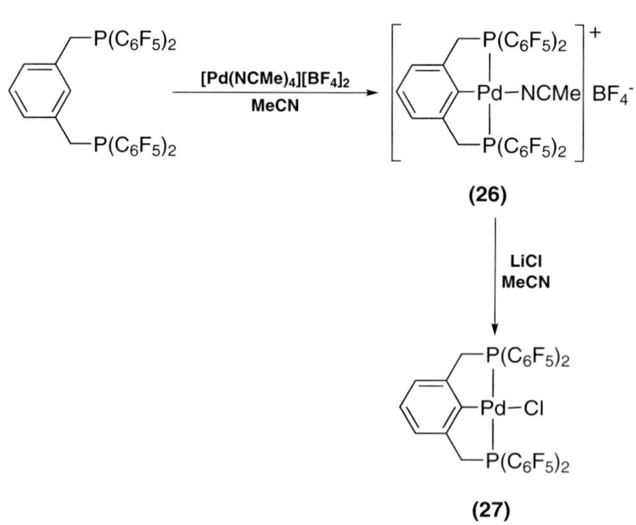

Scheme 1.4

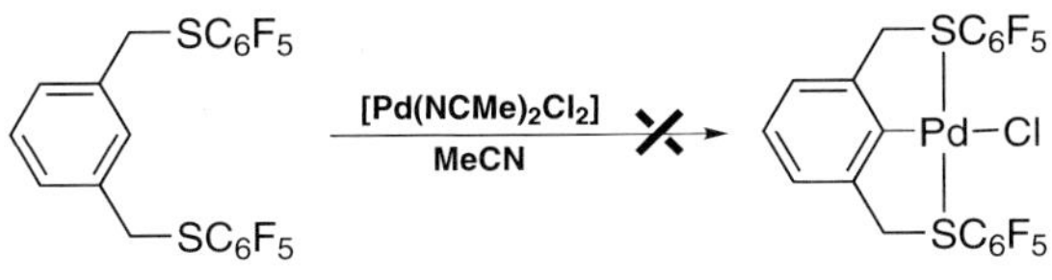

(28)

Scheme 1.5

Nevertheless, a multitude of Y donor atoms have been able to provide an equal number of palladacycles. Hence, palladacycle compounds of the type CY and YCY can be found containing a wide number of functional groups, such as azobenzenes, imines, amines, oximes, phosphines, arsines, thioethers, oxazolines, different heterocycles, including NHC-heterocyclic carbenes, ethers, selenoethers, and so forth. However, despite this rich structural variety, the most common palladacycles are derived from tertiary amines, usually exhibiting five- or six-membered rings. Palladacycles derived from primary and secondary amines are rather rare, since ortho-palladation of primary amines is difficult. In addition, the possibility of further reactions of the acidic protons of the amine with the palladium center or with additional substrates increases the possibility of undesired or side products. Nonetheless, in recent years efficient synthetic methods to attain such compounds have been reported [37], including the efficient cyclometallation of amino-acid derivatives **(29)** [38].

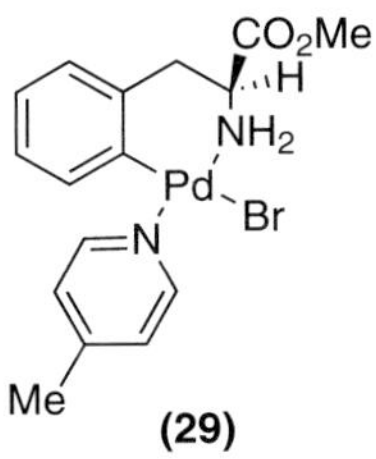

(29)

Additionally, due to their easy synthesis, and modular properties, these compounds have been functionalized to include chiral motifs on their structures. These species have been used in enantioselective transformations and as chiral resolving agents [1r, 1w]. As their achiral counterparts these complexes can be classified according to where the stereogenic center is located in the palladacycle. Thus, there are cyclopalladated compounds that have a stereogenic carbon atom directly σ-bonded to the metal **(30)** [39], those where the stereogenic center is the donor atom (Y), asymmetrically substituted and bound directly to the palladium center; this generally occurs for amine, phosphine, arsine and thioether donor groups **(31)** [13b]. The most common type of chiral functionalized palladacycles, though, are those where the stereogenic center is not directly bonded to the palladium but located elsewhere in the palladated ligand **(32)** [40]. Finally, some compounds exhibit planar chirality, which is generally conferred by the presence of a ferrocene-like moiety forming part of the palladated ligand **(33)** [41].

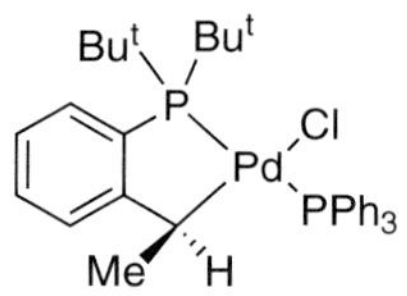

(30)

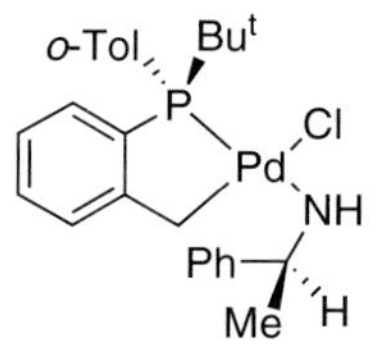

(31)

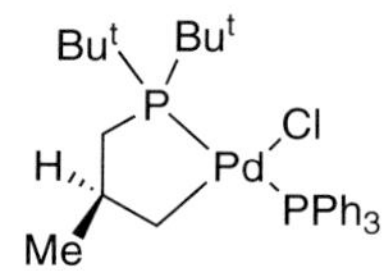

(32)

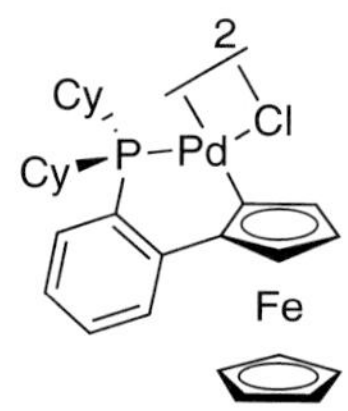

(33)

1.5 Final Remarks

Many palladacycles were first discovered as C—H activation products of a given substrate, and although some specific methods have been designed for the synthesis of other palladacycles not easily available by this method (Chapter 2), the C—H activation process remains the most straightforward method for attaining of these species (Chapters 3 and 4). This is relevant not just because a fairly general and facile method is now available for the synthesis of these compounds but also because in the process of understanding this synthetic method researchers have advanced their knowledge and understanding of the activation of C—H bonds [42]. This is of considerable importance since C—H activation is one of the fundamental steps in alkane dehydrogenation, which has long been considered as one of the holy grails in chemistry [43]. Thus, in recent years researchers have focused on this most interesting fact, attaining recently not dehydrogenative processes with palladium but, as a consequence of the good understanding of the C—H activation process with this metal, C—C couplings without the use of preactivated aromatic carbon fragments [44]. The relevance that palladacycles have acquired in the last decade is reflected in the continuous research and application of these compounds in many different fields, such as medical applications, sensors, optical and electronic devices, catalysis and so forth. This has been manifested in the growing number of publications that include palladacycles (Figure 1.2).

Clearly, the development of the chemistry of palladacycle compounds is both a viable option in the development of new areas of chemistry and a very important

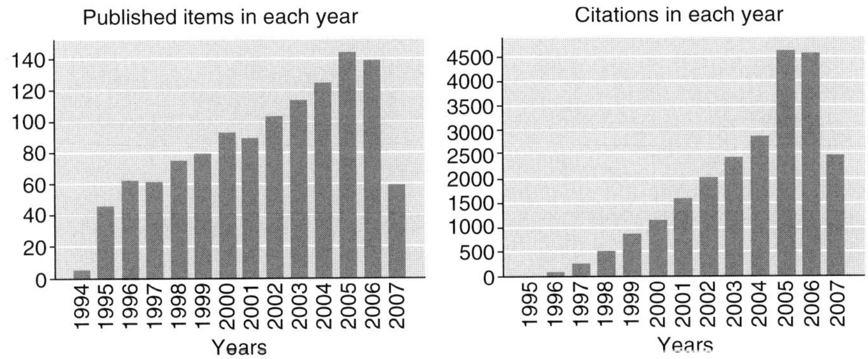

Figure 1.2 Evolution of the number of publications including palladacycles and the steadily growing number of references to these papers in the last 15 years.

tool in the consolidation of present ones. The study of palladacycles, these easy to synthesize, robust and versatile species, represents a very promising and profitable field of research for the future.

Acknowledgments

I gratefully acknowledge the support and enthusiasm of former and current group members and colleagues. The research from our group described in this chapter is supported by CONACYT (J41206-Q; F58692) and DGAPA-UNAM (IN114605; IN227008).

References

1 (a) Parshall, G.W. (1970) *Accounts of Chemical Research*, **3**, 139.
(b) Dehand, J. and Pfeffer, M. (1976) *Coordination Chemistry Reviews*, **18**, 327.
(c) Bruce, M.I. (1977) *Angewandte Chemie (International Edition in English)*, **16**, 73.
(d) Omae, I. (1979) *Coordination Chemistry Reviews*, **28**, 97.
(e) Omae, I. (1979) *Chemical Reviews*, **79**, 287.
(f) Omae, I. (1980) *Coordination Chemistry Reviews*, **32**, 235.
(g) Omae, I. (1982) *Coordination Chemistry Reviews*, **42**, 245.
(h) Constable, E.C. (1984) *Polyhedron*, **3**, 1037.
(i) Ryabov, A.D. (1985) *Synthesis*, 233.
(j) Rothwell, I.P. (1985) *Polyhedron*, **4**, 177.
(k) Newkome, G.R., Puckett, W.E., Gupta, V.K. and Kiefer, G.E. (1986) *Chemical Reviews*, **86**, 451.
(l) Ryabov, A.D. (1990) *Chemical Reviews*, **90**, 403.
(m) Pfeffer, M. (1992) *Pure and Applied Chemistry*, **64**, 335.
(n) Steenwinkel, P., Gossage, R.A. and van Koten, G. (1998) *Chemistry–A European Journal*, **4**, 759.
(o) Albrecht, M. and van Koten, G. (2001) *Angewandte Chemie, International Edition*, **40**, 3750.
(p) Dupont, J., Pfeffer, M. and Spencer, J. (2001) *European Journal of Inorganic Chemistry*, 1917.
(q) Bedford, R.B. (2003) *Chemical Communications*, 1787.

(r) van der Boom, M.E. and Milstein, D. (2003) *Chemical Reviews*, **103**, 1759.
(s) Singleton, J.T. (2003) *Tetrahedron*, **59**, 1837.
(t) Omae, I. (2004) *Coordination Chemistry Reviews*, **248**, 995.
(u) Beletskaya, I.P. and Cheprakov, A.V. (2004) *Journal of Organometallic Chemistry*, **689**, 4055.
(v) Dunina, V.V. and Gorunova, O.N. (2004) *Russian Chemical Reviews*, **73**, 309.
(w) Dupont, J., Consorti, C.S. and Spencer, J. (2005) *Chemical Reviews*, **105**, 2527.
(x) Dunina, V.V. and Gorunova, O.N. (2005) *Russian Chemical Reviews*, **74**, 871.
(y) Szabo, K.J. (2006) *Synlett*, 811.

2 (a) Dyker, G., Körning, J., Nerenz, F., *et al.* (1996) *Pure and Applied Chemistry*, **68**, 323.
(b) Dyker, G. (1997) *Chemische Berichte*, **130**, 1567.
(c) Catellani, M. (2003) *Synlett*, 298.

3 (a) Herrmann, W.A., Bohm, V.P.W. and Reisinger, C.P. (1999) *Journal of Organometallic Chemistry*, **576**, 23.
(b) Herrmann, W.A., Preysing, D.V., Öfele, K. and Schneider, S.K. (2003) *Journal of Organometallic Chemistry*, **687**, 229.
(c) Bedford, R.B., Cazin, C.S.J. and Holder, D. (2004) *Coordination Chemistry Reviews*, **248**, 2283.
(d) Bellina, F., Carpita, A. and Rossi, R. (2004) *Synthesis*, **15**, 2419.
(e) Alacid, E., Alonso, D.A., Botella, *et al.* (2006) *The Chemical Record*, **6**, 117.

4 (a) Jautze, S., Seiler, P. and Peters, R. (2007) *Angewandte Chemie (International Edition in English)*, **46**, 1260.
(b) Weiss, M.E., Fischer, D.F., Xin, Z.Q., *et al.* (2006) *Angewandte Chemie (International Edition in English)*, **45**, 5694.

5 (a) Espinet, P., Etxebarría, J., Marcos, M., *et al.* (1989) *Angewandte Chemie (International Edition in English)*, **28**, 1065.
(b) Baena, M.J., Espinet, P., Ros, M.B. and Serrano, J.L. (1991) *Angewandte Chemie (International Edition in English)*, **30**, 711.
(c) Hegmann, T., Kain, J., Diele, S., *et al.* (2001) *Angewandte Chemie (International Edition in English)*, **40**, 887.
(d) Gimenez, R., Lydon, D.P. and Serrano, J.L. (2002) *Current Opinion in Solid State and Materials Science*, **6**, 527.
(e) Hegmann, T., Kain, J., Diele, S., *et al.* (2003) *Journal of Materials Chemistry*, **13**, 991.
(f) Arias, J., Bardaji, M. and Espinet, P. (2006) *Journal of Organometallic Chemistry*, **691**, 4990 and references therein.
(g) Yenilmez, H.Y., Okur, A.I. and Gül, A. (2007) *Journal of Organometallic Chemistry*, **692**, 940.
(h) Ghedini, M., Aiello, I., Crispini, A., *et al.* (2006) *Coordination Chemistry Reviews*, **250**, 1373 and references therein.

6 (a) Wakatsuki, Y., Yamasaki, H., Grutsch, P.A., *et al.* (1985) *Journal of the American Chemical Society*, **107**, 8153.
(b) Schwartz, R., Gliemann, G., Jolliet, P. and von Zlewsky, A. (1989) *Inorganic Chemistry*, **28**, 742.
(c) Neve, F., Ghedini, M. and Crispini, A. (1996) *Chemical Communications*, 2463.
(d) Ghedini, M., Pucci, D., Colageno, G. and Barigelletti, F. (1997) *Chemical Physics Letters*, **267**, 341.
(e) Neve, F., Crispini, A. and Campagna, S. (1997) *Inorganic Chemistry*, **36**, 6150.
(f) Wu, Q., Hook, A. and Wang, S. (2000) *Angewandte Chemie (International Edition in English)*, **39**, 3933.
(g) Song, D., Wu, Q., Hook, A., *et al.* (2001) *Organometallics*, **20**, 4683.
(h) Aiello, I., Ghedini, M. and La Deda, M. (2002) *Journal of Luminiscense*, **96**, 249.
(i) Neve, F., Crispini, A., Di Pietro, C. and Campagna, S. (2002) *Organometallics*, **21**, 3511.
(j) La Deda, M., Ghedini, M., Aiello, I., *et al.* (2005) *Journal of Organometallic Chemistry*, **690**, 857.

7 (a) Pucci, D., Albertini, V., Bloise, R., *et al.* (2006) *Journal of Inorganic Biochemistry*, **100**, 1575.
(b) Pucci, D., Bloise, R., Bellusci, A., *et al.* (2007) *Journal of Inorganic Biochemistry*, **101**, 1013.

8 Kleinman, J.P. and Dubeck, M. (1963) *Journal of the American Chemical Society*, **85**, 1544.

9 Cope, A.C. and Siekman, R.W. (1965) *Journal of the American Chemical Society*, **87**, 3272.

10 Cope, A.C. and Friedrich, E.C. (1968) *Journal of the American Chemical Society*, **90**, 909.

11 Herrmann, W.A., Brossmer, C., Öfele, K., *et al.* (1995) *Angewandte Chemie (International Edition in English)*, **34**, 1844.

12 Corbet, J.P. and Mignani, G. (2006) *Chemical Reviews*, **106**, 2651.

13 (a) Hiraki, K., Fuchita, Y. and Uchiyame, T. (1983) *Inorganica Chimica Acta*, **69**, 187.
(b) Cheney, A.J. and Shaw, B.L. (1972) *Journal of the Chemical Society Dalton Transactions*, 860.

14 Schwarz, J., Herdtweck, E. and Herrmann, W.A. (2000) *Organometallics*, **19**, 3154.

15 Braunstein, P., Dehand, J. and Pfeffer, M. (1974) *Inorganic and Nuclear Chemistry Letters*, **10**, 581.

16 Bedford, R.B., Hazelwood, S.L., Limmert, M.E., *et al.* (2003) *Chemistry–A European Journal*, **9**, 3216.

17 Abicht, H-P., Issleib, K. and Anorg, Z. (1983) *Allgemeine Chemie*, **500**, 31.

18 (a) Shaw, B.L. and Truelock, M. (1975) *Journal of Organometallic Chemistry*, **102**, 517.
(b) Jones, W.D. and Feher, F.J. (1989) *Accounts of Chemical Research*, **22**, 91.

19 Fedorov, B.S., Golovina, N.I., Strukov, G.V., *et al.* (1997) *Russian Chemical Bulletin, International Edition*, **46**, 1626.

20 Falvello, L.R., Forniés, J., Martín, A., *et al.* (1997) *Inorganic Chemistry*, **36**, 6166.

21 Dupont, J., Basso, N.R., Meneghetti, M.R., *et al.* (1997) *Organometallics*, **16**, 2386.

22 Carbayo, A., Cuevas, J.V. and García-Herbosa, G. (2002) *Journal of Organometallic Chemistry*, **658**, 15.

23 McPherson, H.M. and Wardell, J.L. (1983) *Inorganica Chimica Acta*, **75**, 37.

24 Clark, H.C., Goel, A.B. and Goel, S. (1979) *Inorganic Chemistry*, **18**, 2803.

25 César, V., Bellemin-Laponnaz, S. and Gade, L.H. (2002) *Organometallics*, **21**, 5204.

26 Clot, O., Wolf, M.O. and Patrick, B.O. (2000) *Journal of the American Chemical Society*, **122**, 10456.

27 Maassarani, F., Pfeffer, M. and Le Borgne, G. (1987) *Organometallics*, **6**, 2029.

28 Dupont, J., Pfeffer, M., Rotteveel, M.A., *et al.* (1989) *Organometallics*, **8**, 1116.

29 Benito, M., López, C., Morvan, X., *et al.* (2000) *Journal of the Chemical Society Dalton Transactions*, 4470.

30 Morales-Morales, D. and Jensen, C.M. (eds) (2007) *The Chemistry of Pincer Compounds*, Elsevier, Amsterdam, The Netherlands.

31 Morales-Morales, D., Grause, C., Kasaoka, K., *et al.* (2000) *Inorganica Chimica Acta*, **300–2**, 958.

32 Wang, Z., Eberhard, M.R., Jensen, C.M., *et al.* (2003) *Journal of Organometallic Chemistry*, **681**, 189.

33 Naghipour, A., Sabounchei, S.J., Morales-Morales, D., *et al.* (2007) *Polyhedron*, **26**, 1445.

34 (a) Romeo, R., Arena, G. and Scolaro, L.M. (1992) *Inorganic Chemistry*, **31**, 4879.
(b) Rahman, M.M., Liu, H-Y, Eriks, K., *et al.* (1989) *Organometallics*, **8**, 1.

35 Chase, P.A., Gagliardo, M., Lutz, M., *et al.* (2005) *Organometallics*, **24**, 2016.

36 (a) Arroyo, M., Cervantes, R., Gómez-Benítez, V., *et al.* (2003) *Synthesis*, 1565.
(b) Cervantes, R., Castillejos, S., Loeb, S.J., *et al.* (2006) *European Journal of Inorganic Chemistry*, 1076.

37 (a) Vicente, J., Saura-Llamas, I. and Jones, P.G. (1993) *Journal of the Chemical Society Dalton Transactions*, 3619.
(b) Vicente, J., Saura-Llamas, I., Palin, M.G. and Jones, P.G. (1995) *Journal of the Chemical Society Dalton Transactions*, 2535.
(c) Fuchita, Y., Tsuchiya, H. and Miyafuji, A. (1995) *Inorganica Chimica Acta*, **233**, 91.
(d) Kurzeev, S.A., Kazankov, G.M. and Ryabov, A.D. (2002) *Inorganica Chimica Acta*, **340**, 192.
(e) Vicente, J., Saura-Llamas, I., Palin, M.G., *et al.* (1997) *Organometallics*, **16**, 826.
(f) Albert, J., Cadena, J.M. and Granell, J. (1997) *Tetrahedron Asymmetry*, **8**, 991.
(g) Vicente, J., Saura-Llamas, I., Cuadrado, J. and Ramírez de Arellano, M.C. (2003) *Organometallics*, **22**, 5513.
(h) Vicente, J., Saura-Llamas, I. and Bautista, D. (2005) *Organometallics*, **24**, 6001.
(i) Vicente, J. and Saura-Llamas, I. (2007) *Comments on Inorganic Chemistry*, **28**, 39.

38 Vicente, J., Saura-Llamas, I., García-López, J-A. and Calmuschi-Cula, B. (2007) *Organometallics*, **26**, 2768.
39 Hill, D.F., Mann, B.E. and Shaw, B.L. (1973) *Journal of the Chemical Society Dalton Transactions*, 270.
40 Bottomley, A.R.H., Crocker, C. and Shaw, B.L. (1983) *Journal of Organometallic Chemistry*, **250**, 617.
41 Roca, F.X., Motevalli, M. and Richards, C.J. (2005) *Journal of the American Chemical Society*, **127**, 2388.
42 Alberico, D., Scott, M.E. and Lautens, M. (2007) *Chemical Reviews*, **107**, 174.
43 Goldberg, K.I. and Goldman, A.S. (eds) (2004) *Activation and Functionalization of C-H Bonds; ACS Symposium Series 885*, American Chemical Society, Washington, DC.
44 Stuart, D.R. and Fagnou, K. (2007) *Science*, **316**, 1172.

2
C—H Bond Activation

Martin Albrecht

2.1
General Remarks

Heteroatom-assisted C—H bond activation with palladium to give palladacycles is a reaction of great relevance, both for methodological reasons and due to the application potential of this reaction. Methodologically, direct C—H bond activation of alkenes and arenes is a highly attractive strategy for the insertion of functionality into hydrocarbons. Hence, a thorough understanding of the intimate steps of metal-mediated C—H bond activation is crucial. In particular, the activity of the metal center can be tuned by variation of the nature of the assisting heteroatom. Very low reactivity allows the detailed reaction trajectory to be elucidated, perhaps even enabling the stabilization of crucial intermediates, thus identifying key factors that govern successful metal insertion. Very high reactivity is desirable in a more applied context, since palladation of the C—H bond is a key step in many catalytic reactions such as C—C bond forming and cross-coupling reactions. Hence, a reliable tailoring of the activity of the palladium center is highly desirable, both in laboratory syntheses as well as industrial production processes.

A versatile methodology to control the activity of the metal center and to accomplish C—H bond activation relies on the ability of intramolecular heteroatom lone-pairs to bind (reversibly) to the metal center. This facilitates metallation and, simultaneously, it directs the regioselectivity of this reaction. This process, conceptually related to Directed ortho-metallation (DoM) [1], produces a palladacycle, provided the palladium–heteroatom Pd—E bond is thermodynamically stable (Scheme 2.1).

In such palladacycles, the metal–carbon bond is significantly shielded through chelation as compared to unsupported Pd—C bonds. This increases the stability of the organopalladium product, allowing comprehensive analysis of the properties and reactivity of this important class of compounds. Therefore, unsurprisingly, cyclopalladation is one of the oldest topics in organometallic chemistry. The first reports on cyclopalladation via C—H bond activation appeared in the late 1960s at a time when X-ray diffraction, and likewise NMR spectroscopy, was rarely used to

Palladacycles: Synthesis, Characterization and Applications. Edited by Jairton Dupont and Michel Pfeffer

ISBN: 978-3-527-31781-3

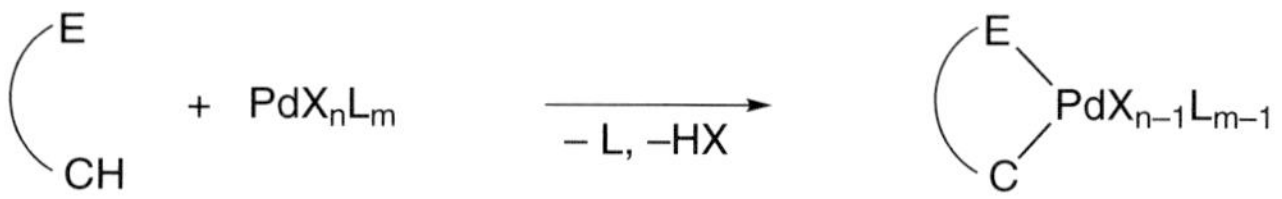

Scheme 2.1

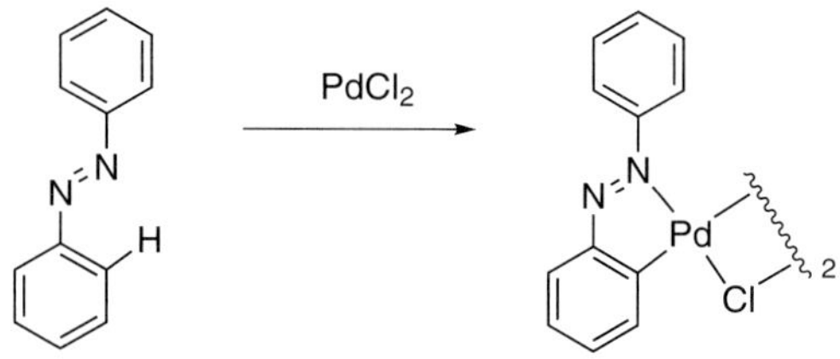

1 **2**

Scheme 2.2

characterize organometallic compounds [2]. Successful cyclopalladation was demonstrated by reacting $PdCl_2$ or Li_2PdCl_4 with diazobenzene **(1)**, thus affording palladacycle **2** (Scheme 2.2). Similar reactivity has been observed with dimethylbenzylamine (dmba).

The vast majority of cyclometallating ligands serve as monoanionic *E,C*-bidentate 4e donors or as a pincer-type monoanionic *E,C,E*-tridentate 6e donors. The coordinating donor group E may be of great variety. Most common are nitrogen-, phosphorus- and sulfur-containing groups such as amines, imines, phosphines, phosphinites, phosphites and thioethers. Palladacycles containing oxygen, selenium, arsenic or carbon donors are also known. The overall charge of the ligand can be modulated. Thus, palladacycles consisting of formally neutral 4e donors, such as bidentate *N,C*-aminocarbenes or *C,C*-dicarbene ligands, have been prepared via C–H bond activation. The wide scope of this reaction with respect to donor groups E emphasizes the potential of the cyclopalladation reaction in synthesis. Moreover, the possibility of adjusting the metal properties via rational and efficient ligand tuning provides access to a very rich chemistry of palladium, particularly in catalysis. Both steric modulations, for example by tailoring the accessibility of the metal center, and electronic modifications to improve the catalytic activity may be introduced without significant alteration of the global Pd(*E,C*) framework.

Considering the above-mentioned aspects, it is not surprising that cyclopalladation has attracted and continues to attract enormous interest in organic and organometallic chemistry [3]. Given the high (and still growing) popularity of palladacycle chemistry, a comprehensive overview of cyclopalladation reactions via C–H bond activation would clearly go beyond the scope of this chapter and, presumably even more relevant, it would be out-dated very rapidly. Therefore, this chapter illustrates the fundamental aspects of cyclopalladation via C–H bond activation. A more comprehensive treatment of the topic can be found in several useful reviews and monographs–specifically accounts summarizing the early developments in cyclopalladation [4], reviews on nitrogen-containing palladacycles [5] and on the synthesis of phosphapalladacycles [6].

Cyclopalladation via C—H bond activation may be considered as a template process that is typically strongly reliant on the intramolecular availability of coordinating heteroatoms. Preliminary bonding of the heteroatom to the palladium center arranges the metal center and the C—H bond in a confined structural motif. Such a heteroatom-assisted preorganization of the reactive components is particularly pronounced with pincer-type, potentially *E,C,E*-tridentate, ligands. In these systems, intramolecular bidentate heteroatom coordination is possible before Pd—C bond formation (see below). Cyclometallation is strongly preferred if five-membered palladacycles are formed, though different ring sizes are also known. This geometry allows for the most ideal accommodation of the 90° bond angle of square planar palladium(II) and the 109–120° angles for the mostly sp^3 and sp^2 hybridized ligand atoms in the metallacycle. This preference for five-membered palladacycles allows one to predict quite safely the C—H bond in a given ligand that is likely to be activated. Owing to this strongly directing effect of the heteroatom, cyclopalladation provides a rational method for the selective activation of an unactivated C—H bond.

2.2 Activation of Aryl C—H Bonds

The details of the cyclopalladation reaction involving aromatic C—H bond activation have been studied particularly well and an outstanding review on the mechanistic features of this reaction has appeared [7]. Very early investigations showed that the reaction rates correlate well with the electron-donating ability of the substituents on the arene. This close analogy to aromatic electrophilic substitution prompted the formulation of a related mechanism for cyclopalladation [8]. Indeed, in many cases a reaction trajectory reminiscent of aromatic electrophilic substitution seems to offer a satisfying mechanistic rational. An alternative and conceptually different mechanism consists of an oxidative addition–reductive elimination sequence. While differentiation of these two pathways is experimentally far from trivial (e.g. in terms of intermediate characterization), the typically electrophilic character of palladium rather supports a substitution sequence. Hence, a reaction profile has been postulated (Scheme 2.3), including initial heteroatom coordination to the metal center followed by the formation of a pi complex, which subsequently rearranges into an arenium intermediate (sigma complex), and finally undergoes proton abstraction to give the cyclopalladated product. Neither a sigma nor a pi intermediate has been isolated thus far. Analogous arenium complexes with platinum – considered to be less electrophilic and kinetically often more inert than palladium – have been prepared and fully characterized [9]. Within certain limitations, such platinum complexes may represent a useful model for the analogous palladium sigma complexes in a substitution reaction.

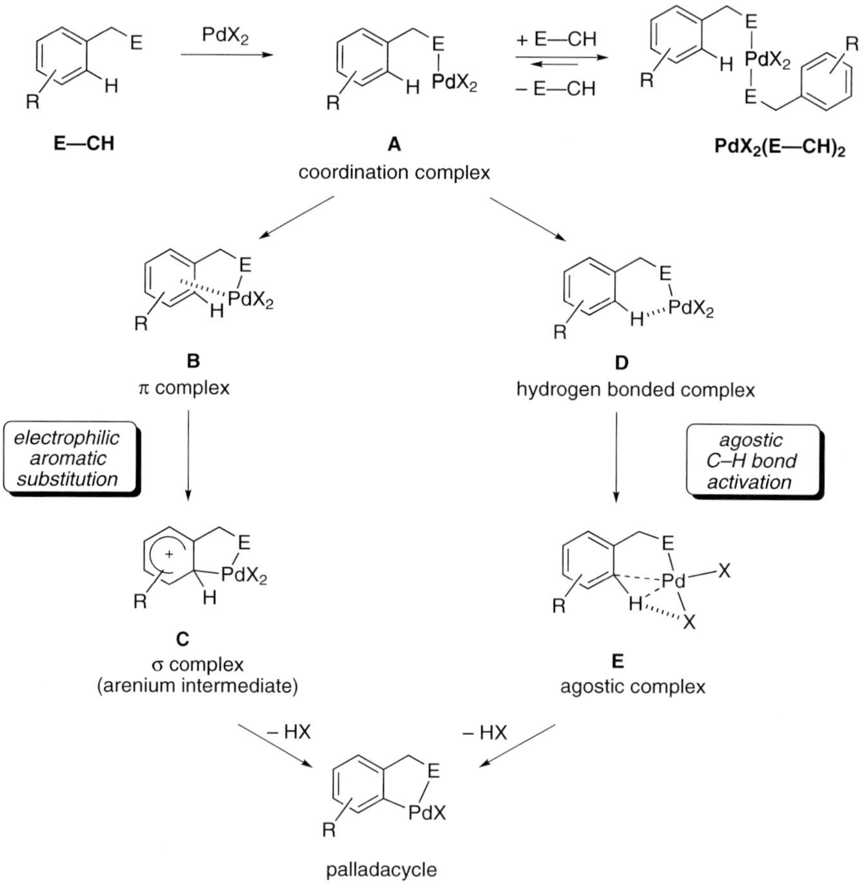

Scheme 2.3

Recent theoretical calculations on the cyclopalladation of dmba with $Pd(OAc)_2$ point to a reaction profile including an agostic interaction as a key structural feature (Scheme 2.3, X = OAc) [10]. The six-membered transition state **D**, including a hydrogen–palladium interaction, has been found to initiate the C—H activation process. Displacement of one oxygen donor of the κ^2-bound acetate by the C—H bond appeared to be rate-determining ($\Delta E = 13\,\text{kcal mol}^{-1}$) and leads to an agostic intermediate **(E)**. Stabilization of this intermediate has been postulated to occur via AcO···H—C_{aryl} H-bonding involving the ortho-hydrogen and the displaced oxygen donor of the acetate. Further reaction to the palladacycle featuring a Pd—C sigma bond was calculated to proceed with virtually no activation energy ($0.1\,\text{kcal mol}^{-1}$). Similar to the assumed electrophilic substitution pathway, the acetate is thought to play a dual role in such an agostic process, acting as ligand for palladium and, simultaneously, as intramolecular base for deprotonation. Furthermore, acetate and related anions can stabilize the supposed intermediate through the formation of a highly ordered six-membered intermediate (Figure

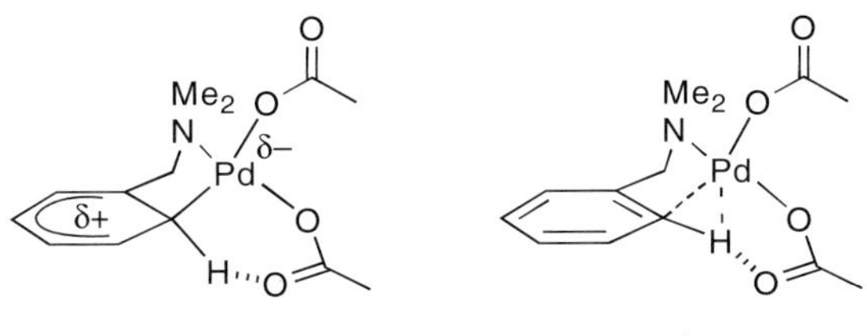

C'   E'

Figure 2.1 Proposed arenium (**C′**, left) and agostic intermediate (**E′**, right) in the cyclopalladation of dimethylbenzylamine with $Pd(OAc)_2$.

2.1). A subsequent proton transfer from the arene to the acetate is obviously very facile. These particular characteristics of acetate (and to a lesser extent also of carbonate) are presumably the main reason why $Pd(OAc)_2$ is often a suitable precursor for cyclopalladation.

Alternative pathways including either a four-membered transition state or an oxidative addition sequence were predicted to be less probable. As a result of an unfavored arrangement of the acetate ligands around the palladium center, the activation barriers for these processes have been calculated to be twice and trice as high, respectively.

Similar agostic complexes have been postulated as intermediates in related cyclometallations with iridium and have been structurally characterized in a pincer rhodium complex [11]. The calculated geometry parameters for agostic intermediate **E′** are very similar to the experimental data and emphasize the distinct differences between the six-membered agostic intermediate and an analogous arenium intermediate (**C′**, Figure 2.1). In the agostic complex, the Pd···H contact is short and the C—H bond distance elongated, while in the arenium system the Pd···H distance and the Pd—C—H bond angle are expected to be comparatively large [9]. Moreover, the calculated atomic charges show alterations at the activated C—H bond only, while in an arenium intermediate charges are expected to change on the entire aromatic system. Given the shallow minimum for **E′** on the energy surface, experimental evidence for such an agostic palladium complex may be difficult to obtain.

Remarkably, many key factors for successful cyclopalladation via an agostic pathway are identical to those of an electrophilic substitution process. These include (i) the coordinating properties of the heteroatom-containing donor group E, (ii) the electron deficiency on the metal center in the precursor salt and (iii) the electron density of the C—H bond (and thus at the aromatic carbon).

2.2.1
Donor Group Coordination

The donor group E is pivotal for determining the regioselectivity of cyclopalladation and also for initiating the C—H bond activation process. Substitution of a

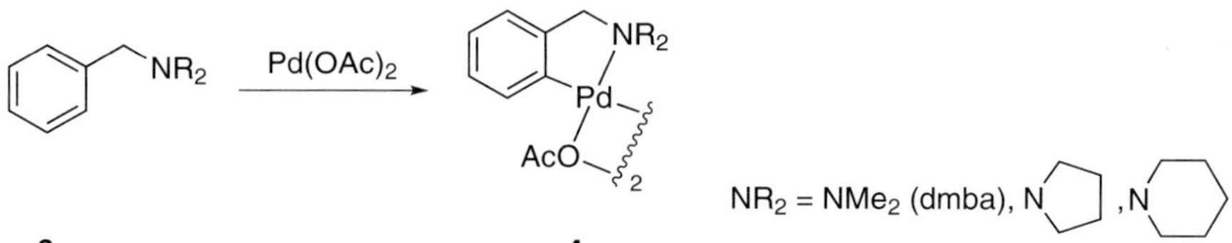

3 4

Scheme 2.4

weakly bound ligand in the metal precursor and formation of a coordination complex is a multifaceted event. More than one heteroatom and hence more than one ligand may coordinate to the metal center. This will form a stable coordination complex of type $[PdX_2(E{-}CH)_2]$ where X is often a halide or a monodentate bound acetate (Scheme 2.3). Subsequent further reaction is generally assumed to occur only upon dissociation of one donor site and formation of a coordinatively unsaturated 14e species $[PdX_2(E{-}CH)]$ **(A)**. An important parameter for cyclometallation is therefore the strength of the Pd–E bond. Strong bonding promotes the formation of the coordination complex $[PdX_2(E{-}CH)_2]$, though it will be detrimental to ligand dissociation to afford the reactive unsaturated species **A**. In contrast, too weak a coordination disfavors ligand substitution in the metal precursor. This balance is typically adjusted by careful choice of the heteroatom and of the surrounding steric bulk. For example, hard amines as in dmba (**3**, R = Me) or softer imines as in diazobenzene have found wide use in cyclometallation since their bonding to soft palladium is not too strong (Scheme 2.4). In amine coordination, the steric shielding of the nitrogen lone pair by the substituents is a crucial parameter. In analogy to the Thorpe–Ingold effect [12], dialkyl substitution of the amine promotes cyclometallation, though metal coordination is typically observed only for small substituents such as in NMe_2. Larger groups coordinate palladium only when their rotational degree of freedom is restricted, for example in cyclic amines such as pyrrolidine or piperidine. In contrast, primary benzylamines are less easily cyclometallated, unless steric bulk is incorporated in the benzylic position. Imines, having an sp^2-hybridized nitrogen, are less sensitive to such steric effects.

Cyclopalladation of phosphine and phosphite analogs via C–H bond activation was reported in the mid-1970s [13]. Benzylphosphines have been noted to undergo internal metallation only with difficulty as compared with their amine analogs. The phosphine–palladium bond is significantly stronger than the Pd–N bond, thus stabilizing the coordination complex. High temperatures, often paired with long reaction times, were required for successful cyclopalladation. The presence of two bulky *t*Bu-substituents at the phosphine facilitates metallation considerably. Apparently, the increase in Pd–P bond strength due to the high basicity of the phosphine is compensated by the steric impact of the bulky *t*Bu-groups. These findings prompted the coinage of the *gem*-di-*tert*-butyl effect in the cyclometallation of phosphines [14]. A similar Pd–phosphine bond weakening has been noted when the donor group has other sterically demanding substituents on the phosphorus and with phosphite donors. In both cases, the pyramidal geometry around phosphorus is significantly distorted at the expense of the Pd—P bond strength.

2.2.2 Metal Precursor

A very useful precursor for metallation is $Pd(OAc)_2$, actually a $[Pd(OAc)_2]_3$ trimer that splits easily into monomeric $[Pd(OAc)_2L_2]$ in the presence of coordinating groups [15]. Monomers are also present in solvents such as benzene at high temperatures. If required, monomeric $Pd(OAc)_2$ may be prepared by reacting the trimer with an excess of NaOAc; however, the reaction is typically slow. Acetic acid is often used as a solvent for cyclopalladation with $Pd(OAc)_2$ since the acetate is a stronger base in this solvent and hence binds better to the palladium(II) center. This precludes reductive elimination of Pd^0 that may engage in (undesired) oxidative addition reactions. The important mechanistic benefits of acetate-containing precursors have been discussed above. A further advantage of using $Pd(OAc)_2$ for cyclopalladation consists in the fact that the HOAc produced as a side product of the C—H bond activation is only a weak acid. Alternative palladium precursors for cyclopalladation include $[PdX_4]^{2-}$, $PdX_2(NCR)_2$ (X = halide, typically Cl) and the highly electrophilic $[Pd(NCR)_4]^{2+}$. Rigidly cis-chelating precursors such as $PdCl_2(cod)$ (cod = 1,5-cyclooctadiene) or precursors consisting of strongly bound ligands, as in $PdCl_2(PPh_3)_2$, have found less wide application for C—H bond mediated cyclopalladation.

2.2.3 Electron Density at the Arene C—H Bond

Generally, electron-releasing substituents at the arene facilitate cyclometallation. Intramolecular competition experiments (e.g. with differently substituted aryl rings in **1**, Scheme 2.2) revealed that cyclopalladation takes place preferably at the arene bearing the more electron-releasing substituents. In the agostic model, this may be rationalized by the increased electron density in the C—H bond, which favors coordination to the electrophilic palladium center. Similarly, the enhanced density at the carbon atom will reduce the activation barrier for an electrophilic aromatic substitution. The observed Hammett correlation for cyclopalladation is not very pronounced, which may be an indication for an agostic rather than an electrophilic substitution pathway.

2.3 Pincer Complexes: A Special Case

Pincer ligands are characterized by a potentially tridentate ECE coordination motif [16]. The presence of two chelating cis-positioned heteroatoms at the palladium center shields the sensitive Pd—C bond. This increases the stability of the cyclopalladated products considerably. However, the presence of two coordination sites complicates the cyclometallation process, in particular the formation of the coordination complex **A** (Scheme 2.3) that precedes Pd–C bond making. Polymeric material may form due to a bridging rather than a chelating coordination mode of the two heteroatoms. Since the cleavage of such polymeric structures requires additional energy, cyclopalladation is typically performed at higher temperatures and longer reaction times than with E,C-bidentate ligand precursors. Heteroatom-directed formation of the C–H activated complex may occur either at the ortho,ortho or at the ortho,para position. The course of this reaction sequence depends strongly on the strength of the Pd–E bond.

Diphosphine ligand precursors such as **5** readily give the monometallic Pd–PCP complex **6** under different reaction conditions and with various metal precursors such as $PdCl_2(NCR_2)$ or $Pd(OTf)_2$ (Scheme 2.5) [17]. Owing to the high trans effect of the phosphines and the strength of the Pd–P bond, chelated structures such as **H** are favored over di- and polymeric products such as **F** or **G**. Preorganization of the ligand in a chelated arrangement **H** is highly beneficial for the regioselectivity of metallation. Transient loss of either a phosphine or a halide X from **H** creates a coordinatively unsaturated species that activates the ortho,ortho-located C–H bond exclusively to give **6**.

Unlike for bidentate ligands, the gem di-*tert*-butyl effect is disadvantageous for the C–H bond activation of tridentate PCP pincer ligands. High temperatures and long reaction times are required for the cyclopalladation of **5b**, while with **5a** palladacycle formation takes place at room temperature within a few hours. Presumably, due to the steric demand of the *t*Bu substituents, the coordination complex

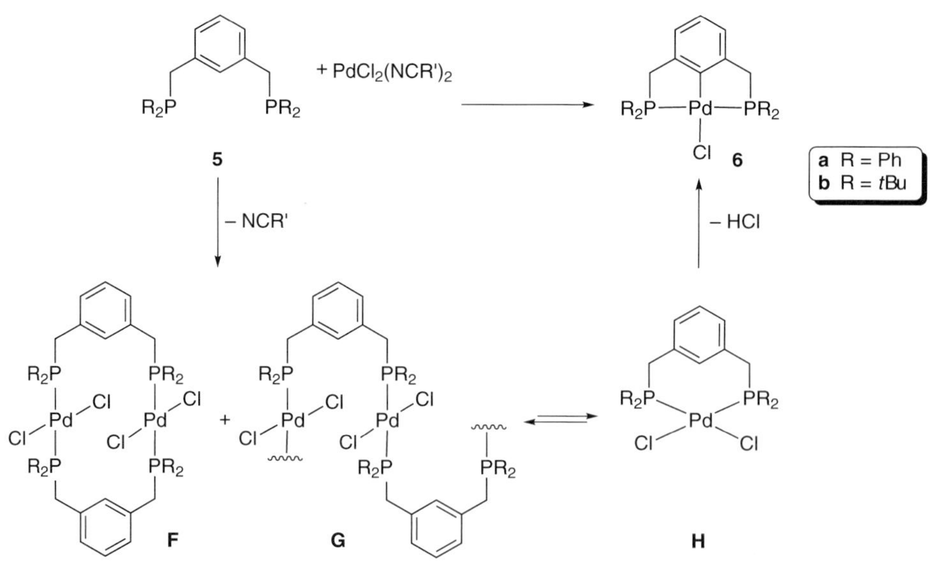

Scheme 2.5

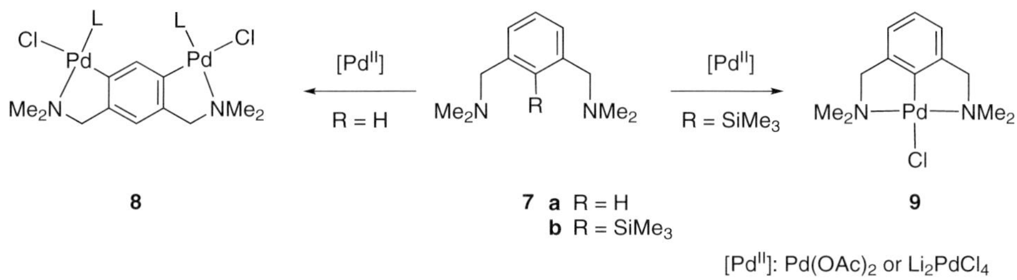

Scheme 2.6

equilibrium is shifted towards di- and polymeric species (**F** and **G**) where mutual trans coordination of the bulky and highly basic phosphines is possible. Structures like **H** are less favored with bulky phosphines. Yet, once formed, such a cis arrangement is assumed to greatly facilitate cyclometallation. For example, not only the C—H bond was observed to be activated but also even much stronger C—O bonds [17b].

In contrast to phosphines, the hard amines in the H-NCN ligand precursor **7a** are only weakly coordinating to palladium due to a hard–soft acid–base mismatch. As a consequence, the chelating structure analogous to **H** is disfavored and steric interactions become predominant rather than the templating effect due to the two coordinating heteroatoms. Thus, congestion arising from cis coordination of the diamines and the spectator ligands at palladium precludes activation of the ortho,ortho position. Instead, cleavage of the kinetically favored C—H bond at the ortho,para position has been observed, which affords the dimetallic species **8** (Scheme 2.6) [18]. The selectivity of cyclopalladation is moved towards the ortho,ortho-metallated product **9** upon modifying the reaction conditions (CH_2Cl_2 instead of MeOH as solvent, $Pd(OAc)_2$ rather than Li_2PdCl_4 as metal precursor). However, formation of the dimetallic complex **8** remains competitive. Selective formation of mononuclear palladium complex **9** has been induced by incorporating a leaving group in the ortho,ortho position that is superior to H^+. Thus, **9** has been synthesized by using the silyl-functionalized ligand precursor **7b** and by performing the reaction in a solvent that promotes hyperconjugation at silicon (such as MeOH). These reactivity patterns agree with the aromatic electrophilic substitution mechanism, while an agostic process may be less probable owing to the steric bulk at the silicon center.

2.4 Transcyclometallation

Transcyclometallation, that is the exchange of cyclometallated ligands on a metal center, is a particular case of C—H bond activation. The process involves both C—Pd bond making and Pd—C bond breaking. Two different processes have been developed, which are distinguished by a dissociative and an associative reaction coordinate, respectively.

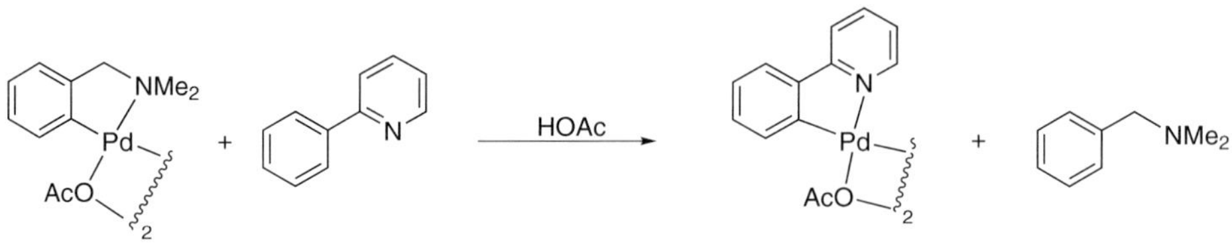

4 10 11 3

Scheme 2.7

Ligand exchange according to Scheme 2.7 has been performed in an acidic medium such as acetic acid [19]. First applied to the exchange of bidentate *C,N*-cyclometallated ligands, the reaction follows a fully dissociative pathway. Protonation of a transiently de-coordinated amine donor group disfavors re-coordination of the heteroatom and concomitant formation of the original metallacycle. Monodentate *C*-bound ligands are more susceptible to acid-mediated Pd—C bond cleavage. Kinetic investigations indeed support such a process involving dissociation of the originally bound ligand and formation of an inorganic Pd salt, followed by cyclometallation with the second ligand according to the classical pathway stipulated above. Equilibria have been observed for Scheme 2.7 that correlate with the basicity of the ligand heteroatom. For example, the C,N_{amine}-chelate in **4** can be exchanged by softer and less basic imines as in phenylpyridine **10** to give **11** and dmba **(3)**. Moreover, cyclopalladation experiments using polydeuterated AcOH-d_4 as solvent provided evidence for deuterium incorporation in both ortho-positions of dmba [20]. This H/D exchange suggests that, most probably, cyclometallation is an equilibrium process in acetic acid and should not be considered to be irreversible. Principally, isotope exchange may also occur in an intermediate consisting of an activated C—H bond and which is relatively long-lived, such as **C′** or **E′** (Figure 2.1).

Interestingly, transcyclometallation has been applied to exchange ligands that are bound via an sp^2-carbon with chelates that are coordinated via an sp^3-carbon. In addition, acid-catalyzed ligand exchange via dissociative Pd—C bond cleavage of **4** has been successfully used for the preparation of As- and Se-containing metallacycles [21]. Further extension of the scope of this reaction may be expected.

A different reaction trajectory has been identified when tridentate coordinating pincer-type ligands are used for transcyclometallation in acid-free media [22]. In this reaction, metal–carbon bond breaking occurs *after* metallation of the incoming ligand, thus following a mechanism that is associative in arene coordination. Experiments using Pt-NCN complex **12** and H-PCP ligand precursor **5a** unraveled some mechanistic details of this process (Scheme 2.8). Again, a key factor is the coordination ability of the heteroatom. Amines are readily displaces by softer phosphines, affording macrocyclic intermediate **13**, which consists of two bridging, κ^2-*P,P′* coordinated H—PCP ligands. Notably, the NCN halide coordination environment favors a mutual trans arrangement of the two phosphine ligands

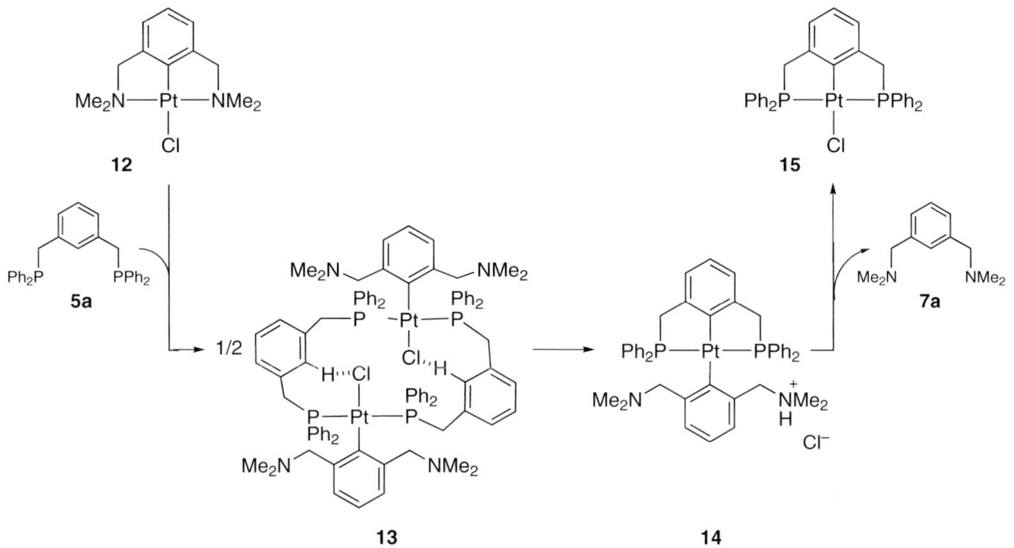

Scheme 2.8

while a cis orientation is preferred with inorganic palladium precursors such as $PdCl_2$(cod) (e.g. **H**, Scheme 2.5). A remarkable and crucial feature of **13** appears to be the formation of intramolecular hydrogen bonds between the metal-bound halide and the aromatic C—H bond. This Pt—Cl···H—C bonding motif activates the C—H bond and concomitantly preorganizes the reactive sites for cyclometallation, as the metal center is confined close to the carbon. Dissociation of one phosphine donor has been proposed to create a coordinatively unsaturated metal center, which–owing to the templating hydrogen bond–is trapped by metal–carbon bond formation, thus affording the cyclometallated product **14**. Intramolecular protonation of the NCN pincer ligand at the carbon center may occur directly or via a metal-bound hydrogen intermediate. This cleaves the metal–carbon bond and yields the Pt—PCP complex **15** along with the neutral H—NCN ligand **7a**, thus formally completing the transcyclometallation.

As a consequence of the highly structured intermediates in this transcyclometallation and the rigid trans orientation of the donor atoms of the incoming ligand, it is possible to also cyclometallate substrates that are difficult to react under standard conditions. For example, the multisite ligand **16** with its phosphine-congested periphery tends to stabilize coordination complexes when treated with inorganic metal precursors [23]. However, with cyclometallated precursors similar to **12**, transcyclometallation yields the polymetallacycle **17** as the predominant product (Scheme 2.9). This selectivity may be attributed to the rigid trans orientation of the heteroatom donors of **16** during the ligand exchange, while in standard cyclometallation processes of such ligands the cis coordinated products are assumed to be too stable to undergo C—H bond activation due to the locally high concentration of heteroatom donor sites.

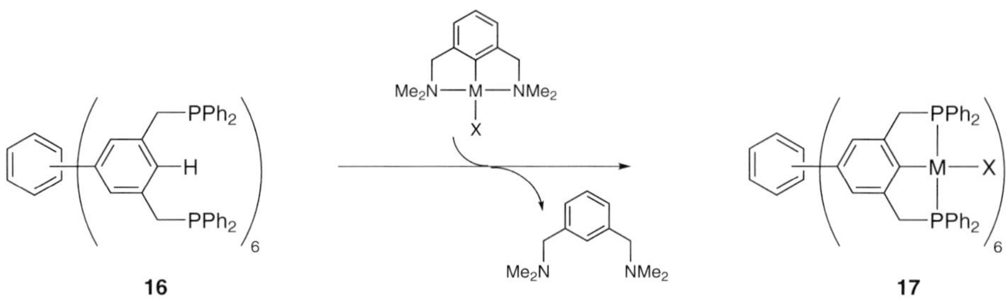

16 17

Scheme 2.9

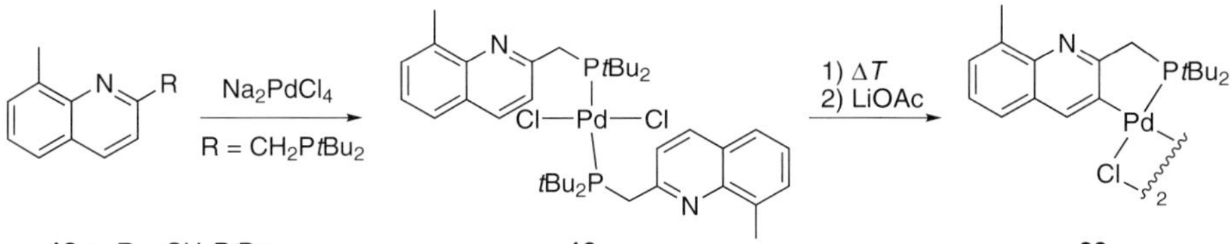

18 a R = CH$_2$P*t*Bu$_2$
b R = CH=NMe
c R = CH$_3$

19 **20**

Scheme 2.10

2.5
Activation of Heterocyclic C–H Bonds, Formation of Pd–Carbene Bonds

A different mechanism of C(sp^2)–H bond activation applies for the activation of C–H bonds in electron-deficient aromatic systems such as in pyridines and related heterocycles. With these electron-poor ligand precursors, neither agostic C–H bonding nor electrophilic substitution seems very probable. Typically, such precursors have been cyclometallated by electron-rich palladium precursors, such as the palladate $[PdCl_4]^{2-}$. The metallation of phosphine-substituted methylquinoline **18a** is illustrative (Scheme 2.10) [24]. In the presence of Na_2PdCl_4, the coordination complex **19** is obtained at room temperature. Increasing the temperature induces cyclometallation and affords, after base-mediated abstraction of the proton, the C(3)-metallated dimer **20**. With the corresponding imine **18b** as ligand precursor, the reaction is less selective and affords in about equal ratios the pyridine-metallated product analogous to **20** and a palladacycle originating from competitive activation of the methyl C–H bond, that is C(8)-metallation (see below).

An extreme case is the cyclopalladation of imidazolium salts such as **21** to yield chelated N-heterocyclic carbene palladium complexes (Scheme 2.11). Pd(dba)$_2$ has been used as precursor for the initial oxidative addition. This process may be directed by pyridine coordination and is in this respect similarly heteroatom-assisted as the cyclometallation discussed previously. The presumed electron-rich Pd(II)-hydride intermediate subsequently engages in a second C–H bond activa-

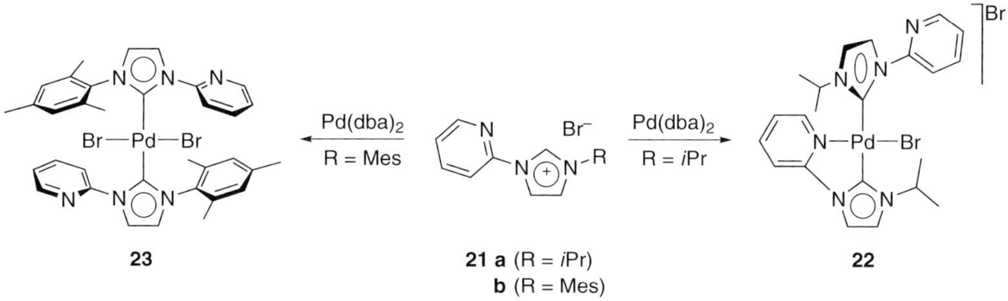

Scheme 2.11

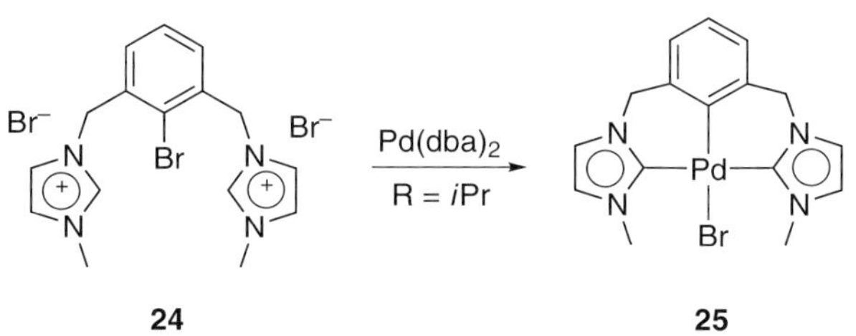

Scheme 2.12

tion to provide, in the case of **21a** (R = *i*Pr), the monocyclometallated complex **22**. It is not clear whether the ligand undergoing the first C—H bond activation is cyclometallated or whether the Pd—N bond is cleaved after formation of the first Pd—C bond. Metallacycle formation is suppressed in the presence of sterically demanding mesityl (Mes) substituents on the imidazolium salt. Palladation of ligand **21b** gives complex **23**, which consists of only monodentate coordinating carbenes.

Similar *in situ* Pd(0) oxidative addition has been applied to initiate cyclometallation of ligand **24**. Double C—H bond activation yields the *C,C,C*-tridentate pincer-type complex **25** (Scheme 2.12). Here, oxidative addition fulfills a similar role as heteroatom coordination in providing the initial interaction that directs the metal center to the C—H bond to be activated.

Intriguingly, cyclopalladation of imidazolium salts also occurs in the absence of directing heteroatoms or preformed Pd—C bonds. The *C,C*-chelated palladacycle **27** is obtained in high yields upon reaction of $Pd(OAc)_2$ with diimidazolium salt **26** (Scheme 2.13). Palladium precoordination by an imidazolium nitrogen lone pair seems energetically highly unfavored, since this would disrupt the aromaticity of the cationic heterocycle. The reaction proceeds with various diimidazolium salts and tolerates large electronic and steric variations in the wing-tip groups R, as well as different linkers connecting the heterocycles (n = 1–3). A likely mechanism that may operate also for the palladation of other azolium salts involves the initial formation of a palladate precursor, $[PdX_2(OAc)_2]^{2-}$. Such a palladate complex may interact with the cationic imidazolium moiety, perhaps by forming anion–π

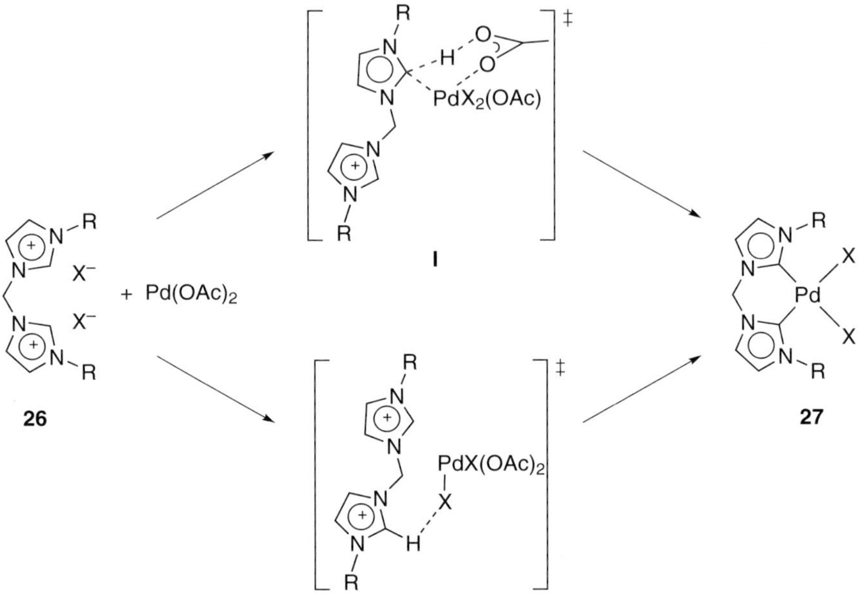

Scheme 2.13

interactions [25]. Subsequent metallation at the most reactive C(2) carbon and formation of one equivalent of acetic acid may occur via a six-membered transition state involving μ^2-κ^2-*O,O*-coordination of the acetate ion to both the palladium center and the most acidic proton attached to C(2) (**I**, Scheme 2.13). Such a transition state is geometrically related to those discussed above (*cf.* Figure 2.1) and is also reminiscent of a Meisenheimer salt. These features suggest a nucleophilic rather than an electrophilic substitution pathway. A nucleophilic process is further supported by the observation that related diimidazolium salts with very weakly coordinating BF_4^- anions fail to be palladated under analogous reaction conditions. Apparently, the corresponding palladate is not produced with BF_4^- anions and, hence, initiation of the cyclopalladation is suppressed. An alternative metallation mechanism may involve hydrogen bonding between the acidic imidazolium proton and a metal-bound halide X (**J**, Scheme 2.13). Subsequent proton dissociation, perhaps as HOAc rather than HX, and coordination of the carbene to the palladium center also results in the formation of **27**.

Less acidic imidazolium C—H bonds may also be activated to give palladacycles. For example, alkylation of the C(2) positions in **26** efficiently protects this site for metallation and directs the cyclopalladation to the imidazolium C(4/5) position [26]. This reactivity is in agreement with a concerted substitution process. A stepwise mechanism involving first proton dissociation and formation of a free carbene is less likely, since C(4)-carbenes are highly unstable species. Alternative mechanisms such as oxidative addition and subsequent reductive elimination are conceivable, though until now little support has been put forward for such pathways.

2.6 Activation of sp³ C—H Bonds

2.6.1 Activation of Benzylic C—H Bonds

Owing to the different nature of C(sp^2)—H and C(sp^3)—H bonds, cyclometallation of sp^3-hybridized carbons is likely to follow a process different from the electrophilic substitution described for arenes. The formation of palladacycles containing a benzylic carbon bound to the metal is a special case of C(sp^3)—H activation due to the enhanced acidity of the benzylic proton, and has been studied extensively. The cyclopalladation of substituted 8-methylquinolines **18** uncovered some key details on the specific reaction trajectory (Scheme 2.14) [27]. For example, coordinating groups such as imines attached to the 2-position **(18b)** favor the formation of bidentate Pd coordination complex **K**, in which the Pd square plane and the quinoline moiety are nearly coplanar. This induces steric congestion due to the close positioning of the acetate ligand cis to the pyridine and the quinoline methyl group. As a consequence, the acetate dissociates easily to give a coordinatively unsaturated palladium center. Furthermore, the C(8) carbon and the palladium center are ideally arranged for a substitution reaction, thus producing the palladacycle **28**. Both agostic and electrophilic mechanisms have been suggested for this reaction, involving anion-assisted abstraction of the benzylic proton and simultaneous coordination of the carbon to the palladium center [5]. No evidence has been obtained for putative M···H—C interactions that may induce an alternative oxidative addition sequence.

Metallation appears to occur only when the arene and the palladium coordination planes coincide. Thus, 2,8-dimethylquinoline **(18c)** does not undergo cyclometallation. Upon coordinating to the pyridine, the sterically demanding and non-coordinating methyl group prompts the palladium square plane to adopt a

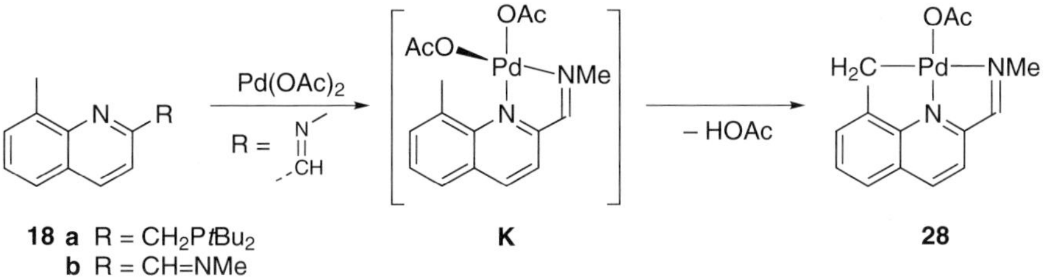

18 a R = CH_2PtBu_2
b R = CH=NMe
c R = CH_3

Scheme 2.14

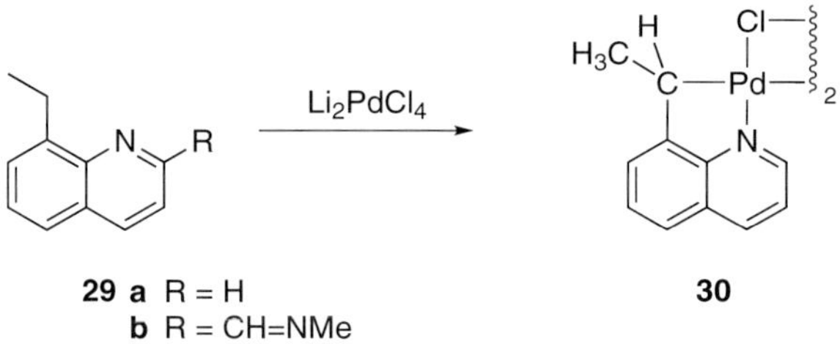

Scheme 2.15

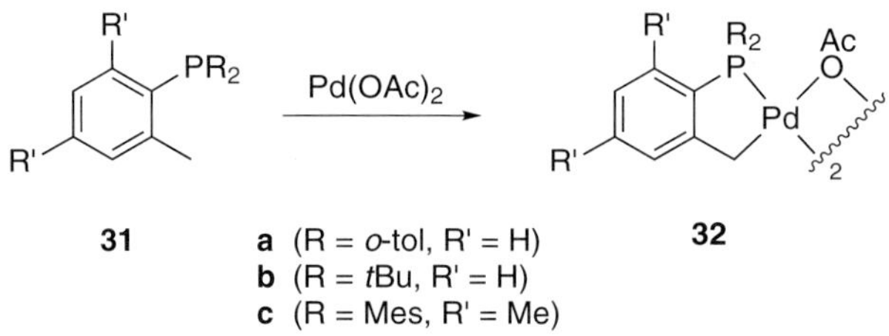

Scheme 2.16

perpendicular orientation with respect to the arene. This geometry does not favor ligand dissociation nor substitution at the sp^3 carbon. Presumably due to similar arguments, the incorporation of flexible donor groups such as R = CH_2PtBu_2 provides coordination products and requires harsher conditions for cyclopalladation.

Cyclopalladation is not restricted to methylquinoline **18**. For example, ethyl-substituted quinoline **29a** undergoes an analogous reaction to give complex **30** (Scheme 2.15) [28]. Here, cyclopalladation creates a new center of chirality at C(8). In the presence of a donor group at C(2) like in **29b**, however, only pyridine C(3)-metallation is observed (*cf.* **20** in Scheme 2.10). Apparently, the increased steric requirements of the ethyl substituents prevent bidentate pyridine-imine coordination of the palladium center, thus prohibiting cyclopalladation at the sp^3-hybridized carbon.

Benzylic C—H bond activation is also well known for the preparation of phosphapalladacycles. This is particularly relevant for P(*o*-tol)$_3$, **31a**, as this ligand has found wide application in catalysis. In the presence of $Pd(OAc)_2$ the ortho-methyl group is metallated very rapidly to give the cyclopalladated and thermally very robust complex **32a** (Scheme 2.16) [29]. Similar C—H bond activation is observed also with Na_2PdCl_4 and **31b** and **31c**. Cyclopalladation is predominantly a consequence of the high basicity of the phosphine. In addition, the steric impact of the bulky substituents at phosphorus exerts a strong repulsion on any exogenous cis-coordinated ligand [phosphine cone angle θ = 194° for P(*o*-tol)$_3$ and 212° for P(Mes)$_3$]. Palladacycle formation provides a pathway for preserving a square-planar geometry around the palladium center rather than a linear P—Pd—P arrangement, which is highly unfavored for palladium(II). In addition, cyclopalladation results

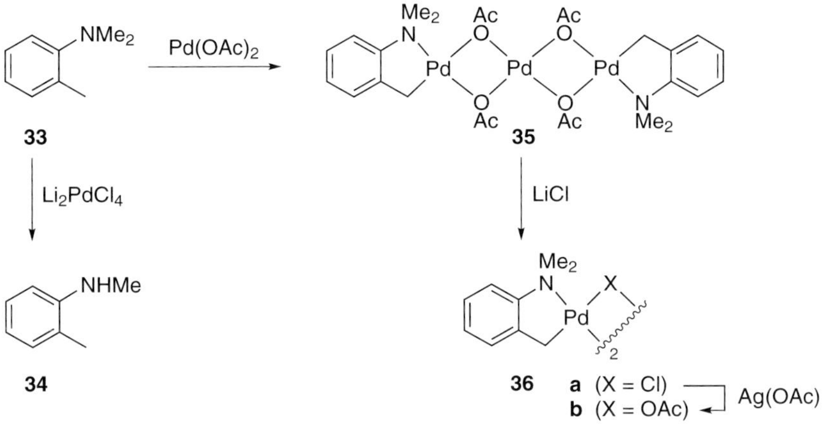

Scheme 2.17

in a loss of entropy and allows the palladium center to accommodate further ligands, thus giving a coordinatively saturated 16e complex such as **32**.

The reactivity pattern with phosphine donors is in striking contrast to that observed with related ligand precursors containing nitrogen donors [30]. With Li_2PdCl_4, the *N,N*-dimethylaniline ligand precursor **33** undergoes an unprecedented de-methylation reaction to give the *N*-methylaniline **34** (Scheme 2.17). With $Pd(OAc)_2$, however, cyclopalladation takes place. The formed five-membered palladacycle appeared to be an unusual trimer **(35)** rather than the typically observed bimetallic species. The intercalated $Pd(OAc)_2$ in **35** can be removed by classical manipulation of ancillary ligands in palladacycle chemistry. Thus, treatment with a chloride salt gives the chloro-dimer **36a** and subsequent halide abstraction mediated by Ag(OAc) affords the dimetallic complex **36b**, which is analogous to **32**.

2.6.2
Activation of Aliphatic C—H Bonds

Cyclopalladation via activation of C_{alkyl}—H bonds has been achieved by using strong phosphine donors. Only few examples are known that involve imine and amine coordination. Generally, phosphine ligands are required that are very basic and which disfavor cis-coordination of other ligands. For example, P*t*Bu_3 (ligand cone angle $\theta = 182°$) undergoes C—H bond activation with various precursors to give the four-membered palladacycle **37** in good yields (Scheme 2.18) [31]. This cyclometallation occurs presumably in most catalytic systems that make use of the very bulky and basic $PtBu_3$ ligand in palladium-catalyzed reactions [31]. In solution, the non-cyclometallated P*t*Bu_3 ligand in complex **37** dissociates relatively easily, thus resulting in an equilibrium between **37** and the catalytically inactive dimer **38**. Hence, for application in catalysis, an excess of P*t*Bu_3 will be highly beneficial in shifting the equilibrium to the former species. The driving forces for the

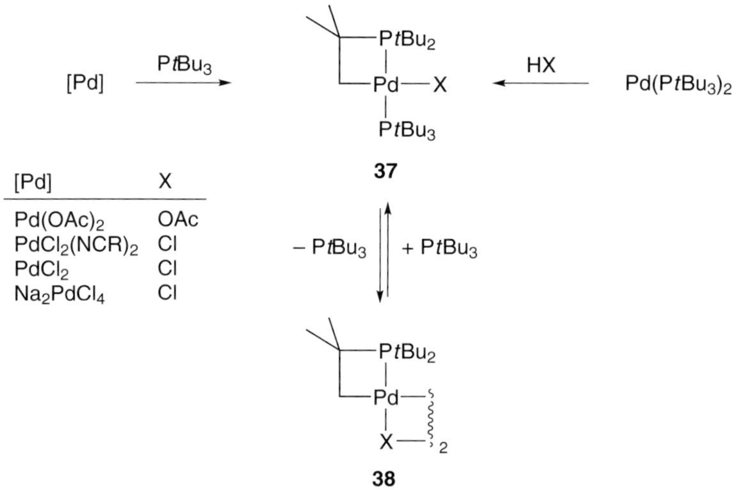

Scheme 2.18

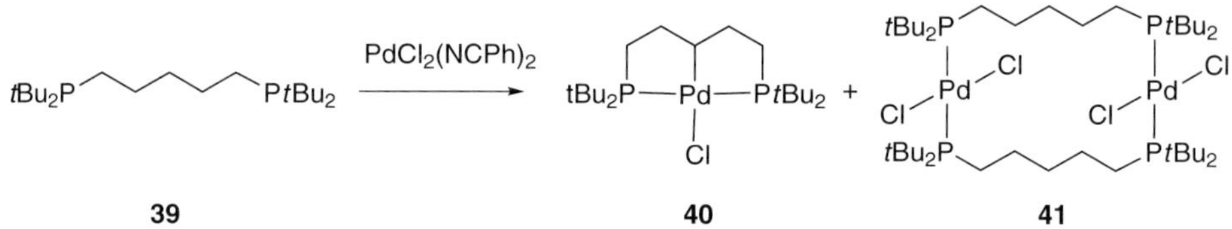

Scheme 2.19

cyclopalladation of $PtBu_3$ and $P(o\text{-tol})_3$ are probably closely related. A major parameter is the accessibility of four-coordinate Pd^{II} centers as a consequence of the restricted ligand rotation upon cyclopalladation (see above). The generation of a four-membered palladacycle in **37** imposes considerable ring strain. Prevention of such strain has been demonstrated by cyclometallation of $P(tBu)_2$ (*neo*-pentyl) with $PtCl_2(NCPh)_2$ [32]. With this ligand, exclusive activation of the neopentyl C—H bond is observed, thus yielding a five-membered platinacycle analogous to **38**.

Aliphatic C—H bond activation is also observed during the cyclopalladation of diphosphinopentane **39**. Reaction of this ligand with $PdCl_2(NCPh)_2$ affords chelate complex **40**, albeit in low yields (Scheme 2.19) [33]. The major product **41** features a dinuclear structure with bridging phosphine coordination. Notably, complex **40** is a rare example of a stable palladium-alkyl species containing β hydrogens that do not undergo β-H elimination. The tridentate coordination mode appeared to be crucial for successful cyclopalladation. Related ligand precursors with only one phosphine donor site such as $P(tBu)_2(nPr)$ only give coordination complexes and do not yield any cyclopalladated products originating from C_{alkyl}—H bond activation.

The cyclopalladation of amine and imine precursors via C_{alkyl}—H bond activation is very rare. A reactivity pattern related to **39** has been observed for a propyl-linked 2,2′-dipyridine system in the presence of $Pd(OAc)_2$ [34]. While no structural data for the cyclopalladated complex have been reported thus far, spectroscopic analyses indicate the activation of an aliphatic C—H bond to give a NCN tridentate coordinating bis-cyclopalladated complex similar to **40**.

2.7 Conclusions and Perspectives

Clearly, heteroatom-assisted C—H bond activation is a very versatile and convenient approach for the synthesis of palladacycles. In recent decades this cyclopalladation reaction has become increasingly popular. As a consequence, important insights are available into the details of this reaction. Nevertheless, some of the intimate steps of the C—H bond activation are still elusive. In particular, experimental differentiation between an electrophilic aromatic substitution and an agostic C—H activation process remains challenging. This, combined with the broad scope of cyclopalladation and the catalytic utility of palladacycles in organic syntheses, represents a major driving force for attracting continued interest in this important organometallic reaction. Without doubt, the future will see significant progress in cyclopalladation chemistry on various levels, including the development of new synthetic cyclopalladation processes, further mechanistic elucidation of known reaction schemes and, certainly, also exciting applications in catalysis and in materials science.

Acknowledgment

The author sincerely thanks Michel Pfeffer (University of Strasbourg) and Gerard van Koten (Utrecht University) for fruitful discussions and the Alfred Werner Foundation for an Assistant Professor Fellowship. The Swiss National Science Foundation and ERA-net chemistry are acknowledged for financial support of our work in this area.

References

1 Snieckus, V. (1990) *Chemical Reviews*, **90**, 879.

2 (a) Cope, A.C. and Siekman, R.W. (1965) *Journal of the American Chemical Society*, **87**, 3272.
(b) Takahashi, H. and Tsuji, J. (1967) *Journal of Organometallic Chemistry*, **10**, 511.
(c) Cope, A.C. and Friedrich, E.C. (1968) *Journal of the American Chemical Society*, **90**, 909.

3 Tsuji, J. (2004) *Palladium Reagents and Catalysts*, John Wiley & Sons, Ltd, Chichester.

4 (a) Dehand, J. and Pfeffer, M. (1976) *Coordination Chemistry Reviews*, **18**, 327.

(b) Omae, I. (1979) *Chemical Reviews*, **79**, 287.

5 (a) Newkome, G.R., Puckett, W.E., Gupta, W.K. and Kiefer, G.E. (1986) *Chemical Reviews*, **86**, 451.
(b) Omae, I. (1986) *Organometallic Intramolecular-Coordination Compounds*, Elsevier, Amsterdam.
(c) Evans, D.W., Baker, G.R. and Newkome, G.R. (1989) *Coordination Chemistry Reviews*, **93**, 155.

6 Dunina, V.V. and Gorunova, O.N. (2004) *Russian Chemical Reviews*, **73**, 309.

7 Ryabov, A.D. (1990) *Chemical Reviews*, **90**, 403.

8 (a) Parshall, G.W. (1970) *Accounts of Chemical Research*, **3**, 139.
(b) Canty, A.J. and van Koten, G. (1995) *Accounts of Chemical Research*, **28**, 406.

9 Albrecht, M., Spek, A.L. and van Koten, G. (2001) *Journal of the American Chemical Society*, **123**, 7233.

10 Davies, D.L., Donald, S.M.A. and Macgregor, S.A. (2005) *Journal of the American Chemical Society*, **127**, 13754.

11 (a) Crabtree, R.H., Holt, E.M., Lavin, M. and Morehouse, S.M. (1985) *Inorganic Chemistry*, **24**, 1986.
(b) Vigalok, A., Uzan, O., Shimon, L.J.W., *et al.* (1998) *Journal of the American Chemical Society*, **120**, 12539.

12 Eliel, E.L., Wilen, S.H. and Doyle, M.P. (2001) *Basic Organic Stereochemistry*, Wiley-VCH Verlag GmbH, Weinheim.

13 (a) Shaw, B.L. and Truelock, M.M. (1975) *Journal of Organometallic Chemistry*, **102**, 517.
(b) Tune, D.J. and Werner, H. (1975) *Helvetica Chimica Acta*, **58**, 2240.

14 Shaw, B.L. (1975) *Journal of the American Chemical Society*, **97**, 3856.

15 Skapski, A.C. and Smart, M.L. (1970) *Journal of the Chemical Society. Chemical Communications*, 658.

16 Albrecht, M. and van Koten, G. (2001) *Angewandte Chemie, International Edition*, **40**, 3750.

17 (a) Moulton, C.J. and Shaw, B.L. (1976) *Journal of the Chemical Society – Dalton Transactions*, 1020.
(b) Rimml, H. and Venanzi, L.M. (1987) *Phosphorus Sulfur*, **30**, 297.
(c) van der Boom, M.E., Lou, S.-Y., Ben-David, Y., *et al.* (1998) *Journal of the American Chemical Society*, **120**, 6531.

18 (a) Trofimenko, S. (1973) *Inorganic Chemistry*, **12**, 1215.
(b) Steenwinkel, P., Gossage, R.A., Maunula, T., *et al.* (1998) *Chemistry – A European Journal*, **4**, 763.

19 (a) Ryabov, A.D. (1987) *Inorganic Chemistry*, **26**, 1252.
(b) Ryabov, A.D. and Yatsimirsky, A.K. (1984) *Inorganic Chemistry*, **23**, 789.

20 Ng, J.K.-P., Chen, S., Li, Y., *et al.* (2007) *Inorganic Chemistry*, **46**, 5100.

21 (a) Yao, Q., Kinney, E.P. and Zheng, C. (2004) *Organic Letters*, **6**, 2997.
(b) Ng, J.K.-P., Tan, G.-K., Vittal, J.J. and Leung, P.-H. (2003) *Inorganic Chemistry*, **42**, 7674.

22 Albrecht, M., Dani, P., Spek, A.L. and van Koten, G. (2000) *Journal of the American Chemical Society*, **122**, 11822.

23 (a) Dijkstra, H.P., Meijer, M.D., Patel, J., *et al.* (2001) *Organometallics*, **20**, 3159.
(b) Dijkstra, H.P., Albrecht, M., Medici, S., *et al.* (2002) *Advanced Synthesis and Catalysis*, **344**, 1135.

24 Deeming, A.J., Rothwell, I.P., Hursthouse, M.B. and Abdul Malik, K.M. (1980) *Journal of the Chemical Society – Dalton Transactions*, 1974.

25 Gamez, P., Mooibroek, T.J., Teat, S.J. and Reedjik, J. (2007) *Accounts of Chemical Research*, **40**, 435.

26 Heckenroth, M., Kluser, E., Neels, A. and Albrecht, M. (2007) *Angewandte Chemie, International Edition*, **46**, 6293.

27 Deeming, A.J. and Rothwell, I.P. (1981) *Journal of Organometallic Chemistry*, **205**, 117.

28 Sokolov, V.I., Sorokina, T.A., Troitskaya, L.L., *et al.* (1972) *Journal of Organometallic Chemistry*, **36**, 389.

29 (a) Cheney, A.J. and Shaw, B.L. (1972) *Journal of the Chemical Society – Dalton Transactions*, 860.
(b) Herrmann, W.A., Brossmer, C., Öfele, K., *et al.* (1995) *Angewandte Chemie (International Edition in English)*, **34**, 1844.

30 (a) Mutet, C. and Pfeffer, M. (1979) *Journal of Organometallic Chemistry*, **171**, C34.
(b) Pfeffer, M., Wehmann, E. and van Knoten, G. (1985) *Journal of Organometallic Chemistry*, **282**, 127.

31 (a) Goel, R.G. and Montemayor, R. (1977) *Inorganic Chemistry*, **16**, 2183. (b) Clark, H., Goel, A.B. and Goel, S. (1979) *Inorganic Chemistry*, **18**, 2803. (c) Geissler, H., Gross, P. and Guckes, B. (1998) Ger. Offen. DE 19647584 (*Chemical Abstracts* **129**, 28078).

32 Mason, R., Textor, M., Al-Salem, N. and Shaw, B.L. (1976) *Journal of the Chemical Society, Chemical Communications*, 292.

33 Al-Salem, N.A., Empsall, H.D., Markham, R., *et al.* (1972) *Journal of the Chemical Society–Dalton Transactions*, 1972.

34 Hiraki, K., Fuchita, Y. and Matsumoto, Y. (1984) *Chemistry Letters*, 1947.

3
Oxidative Addition and Transmetallation

Esteban P. Urriolabeitia

3.1
Introduction

The importance of the metal-mediated C—H bond activation process as a mandatory step in the functionalization of organic molecules is unquestionable, due to the wide range of applications available and also to their atom economy and optimal use of the energetic resources [1]. Simple complexes or salts of Pd(II) are very efficient and versatile materials to promote C—H bond activations on different precursors and, as stated in previous chapters, when the activation is assisted by an ortho functionality the process is termed cyclopalladation or orthopalladation [2]. Such orthometallated compounds are, among many applications, very valuable synthetic intermediates [3].

However, in some instances C—H bond activation can not be achieved, for whatever reason, and the corresponding palladacycle can not be synthesized. Although the C—H activation pathway represents the most elegant form to functionalize a given substrate, there are – fortunately – synthetic alternatives. The present chapter provides the practical chemist with a series of synthetic tools with which to prepare target compounds, especially when direct C—H bond activation is not accessible or when it directs the metallation to a different position than that desired.

Two main groups of reactions are described here, namely oxidative addition and transmetallation. Aspects such as the nature of the precursors, the mechanism of the reaction, the type of resulting products and the range of applicability are covered. Figure 3.1 shows the most important transformations presented in this chapter.

3.2
Oxidative Addition

The oxidative addition of a molecule X–Y to a metal center M can be considered, formally speaking, as the insertion of the metal into the covalent X–Y bond

Palladacycles: Synthesis, Characterization and Applications. Edited by Jairton Dupont and Michel Pfeffer

ISBN: 978-3-527-31781-3

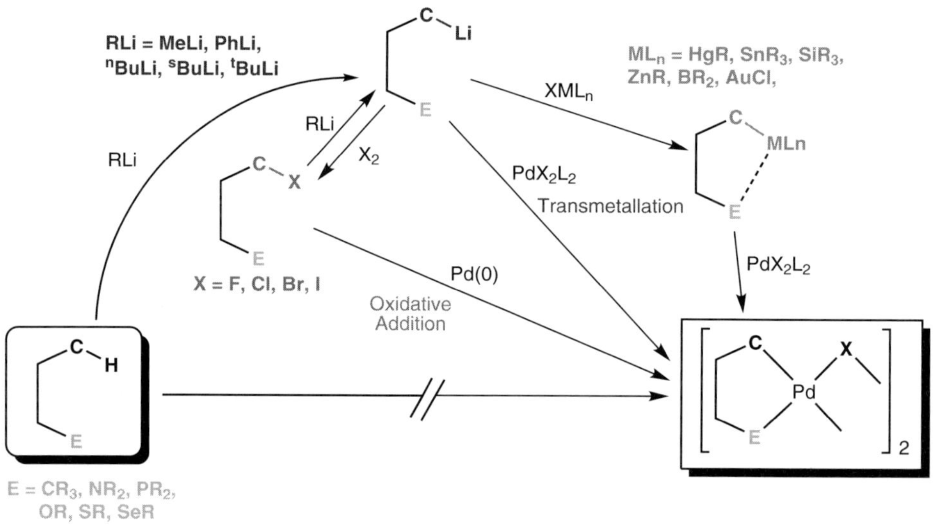

Figure 3.1 Important transformations presented in this chapter.

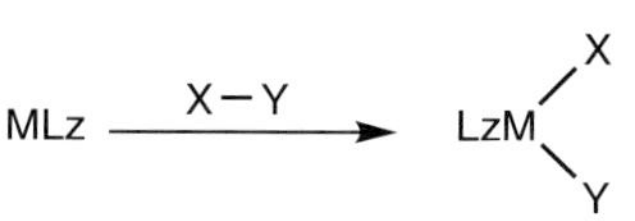

Scheme 3.1

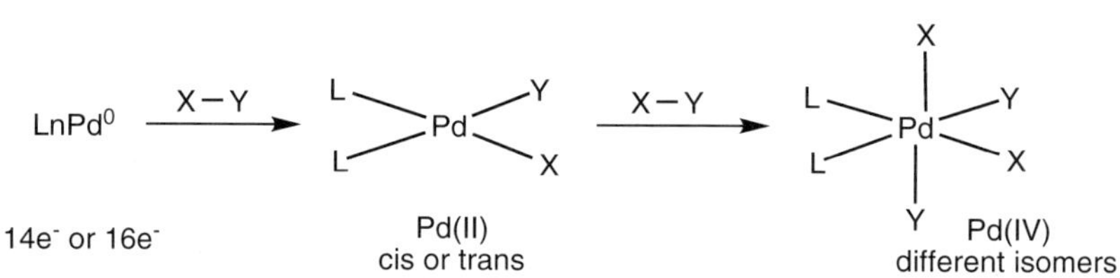

Scheme 3.2

(Scheme 3.1). This reaction implies a two-electron process, as a result of which two new bonds, M–X and M–Y, usually in cis positions, are formed. Therefore, the formal oxidation state and the coordination number of the metal have each been increased by two units [4].

The chemistry of Pd is dominated by the (0), (+II) and, to a lesser extent, the (+IV) oxidation states [5], and the most common oxidative process involve two of these three states. Thus, the oxidative addition on Pd(0) precursors gives square-planar Pd(II) complexes, and further oxidative addition to the latter gives octahedral Pd(IV) derivatives (Scheme 3.2).

The most usual Pd(0) sources are $Pd(PPh_3)_4$ [6a], $Pd(dba)_2$ [6b, c] or $Pd_2(dba)_3$.S (S = solvent, dba = *trans,trans*–dibenzylideneacetone) [7], due to their easy synthesis, thermal stability and their availability on a multigram scale. The phosphane complex must be stored under an inert atmosphere, but the dba derivatives can even be stored without special precaution. In general, the oxidative addition takes place on unsaturated Pd(0) species [4]. Thus, the 18 e^- complex $Pd(PPh_3)_4$ dissociates two PPh_3 ligands in solution, forming coordinatively unsaturated Pd(0) species. In a similar way, $Pd(dba)_2$ corresponds actually to the stoichiometry $Pd_2(dba)_3$.dba, and the Pd atoms in this complex are also unsaturated. The non-bonded dba molecule can be replaced by a solvent molecule by simple recrystallization in the appropriate solvent. For instance, $Pd_2(dba)_3.CHCl_3$ is obtained by recrystallization in $CHCl_3$ [7].

Concerning the organic electrophilic substrates that can undergo oxidative addition, the aryl halides [8], sulfonates, tosylates [9], acetates or carbonates could be considered very efficient precursors, although the halides are probably the most popular, due to their higher stability, good synthetic accessibility and better reactivity. The reactivity order of the aforementioned aryl substrates is ArCl < ArOTs ≪ ArOTf ≈ ArBr < ArI [9]. Thus, at first glance, the iodide-containing substrates are the best choice to attempt an oxidative addition. Vinyl halides, as well as allyl, acyl or alkyl compounds also react cleanly in oxidative addition processes. However, although M(0) complexes (M = Ni, Pd, Pt) activate C—F bonds [10], and even though some complexes of Pt(II) oxidatively add C—F precursors to give orthoplatinated derivatives of Pt(IV) [11], substrates containing C—F bonds are not the best option to obtain palladacycles through oxidative addition.

Several mechanisms account for the two-electron oxidative addition of electrophilic substrates to Pd(0) complexes – concerted, S_N2, free radical or ionic, among others – and many studies have been devoted to ascertain in each case the particular mechanism. The oxidative addition through a concerted pathway always results in a cis arrangement of the new formed bonds, while in a S_N2 mechanism the metal behaves as a nucleophile and the oxidized product could be cis or trans, as also happens in the radical mechanism [12]. Knowledge of the true mechanism of the oxidative addition reaction is crucial, since it is the first step in many important C—C bond forming catalytic cycles, such as Heck, Suzuki or Stille reactions. However, it is difficult to generalize and, in some cases, different mechanisms can be operating on very similar substrates [8, 13]. For instance, the exact nature of the active species involved in the oxidative addition when this step is within a catalytic cycle has generated great controversy [14a]. The "textbook" cycle implies a neutral Pd(0) species, such as PdL_2, which evolves through four-coordinate intermediates, while the "alternative" cycle implies an anionic species such as $[Pd(X)L_2]^-$ (X = halide, acetate, etc.), which seems to evolve through pentacoordinate Pd(II) intermediates [14]. Another delicate point is the stereochemistry of the oxidative addition, which is very important when dealing with chiral X–Csp^3 activations [15]. Practically speaking, the oxidative addition process is a very powerful synthetic tool for the preparation of cyclopalladated complexes, regardless of the organic group (aryl or alkyl), the donor atom [N, P, S, O, C (NHC), etc.] or the nature of the C—X bond to be activated. The reaction is regioselective, and only the functionalized position with the C–X bond will be activated. As has been remarked [3, 16], the main drawback of this process could be the availability of the halide precursors. Two main methods will be described here: direct oxidative addition and modification of a preformed palladacycle or Pd–aryl unit.

The first cyclopalladated complexes obtained through oxidative addition were reported on 1983 [17a]. Complexes of type **1** and **2** were prepared by reaction of the corresponding bromo derivatives with $Pd(dba)_2$ (Scheme 3.3). For **1** this method was proposed since, when direct C–H activation was attempted, methanolysis of the imine ligand occurs. For complexes **2** this method gives a general procedure to obtain cyclopalladated derivatives from tertiary, secondary and primary benzyl amines [17b], which were difficult to obtain through C–H bond activation in the latter cases according with the Cope's rules [17c].

The C,N-cyclopalladated derivatives are the best represented group of complexes, and different types of ligands have been reported. Examples of chelate **(3)** [18] and tridentate **(4, 5)** ligands are shown in Figure 3.2 [19]. Examples in which the ligand acts as a bridging ligand **(6)** have also been described [20].

The C,N-cyclopalladated ligands usually behave as monoanionic ligands although examples in which the ligand can be further deprotonated **(8)** have been reported (Scheme 3.4) [21]. The metallation of *N*-benzylidenebenzylamines has been studied in detail. Several substrates with ortho-halide substituents in the benzyl and/or

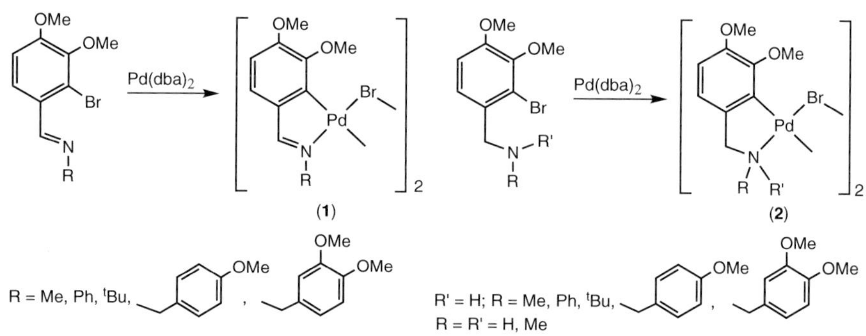

Scheme 3.3

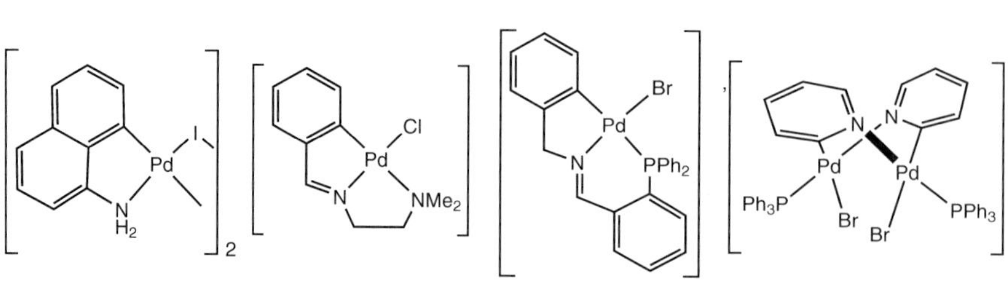

(3) (4) (5) (6)

Figure 3.2 Some examples of C,N-cyclopalladated derivatives.

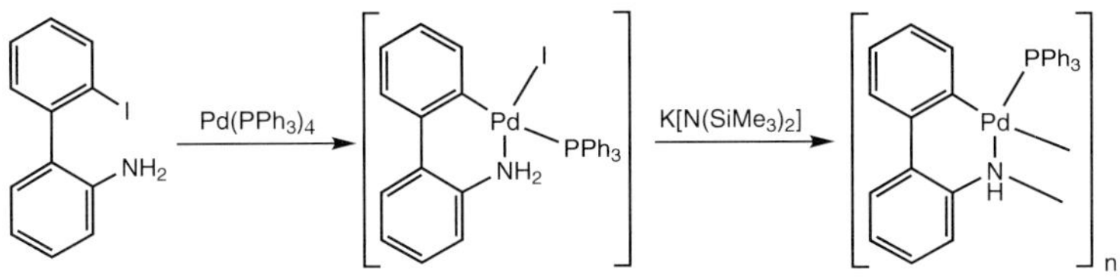

(7) (8)

Scheme 3.4

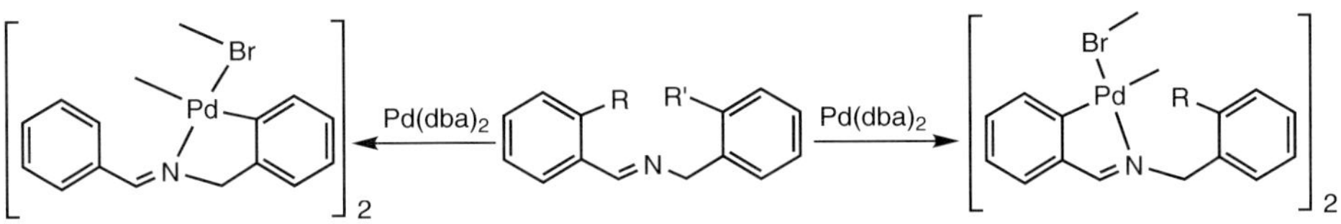

R = H; R' = Br (10) (9) R = Br; R' = H, Br

Scheme 3.5

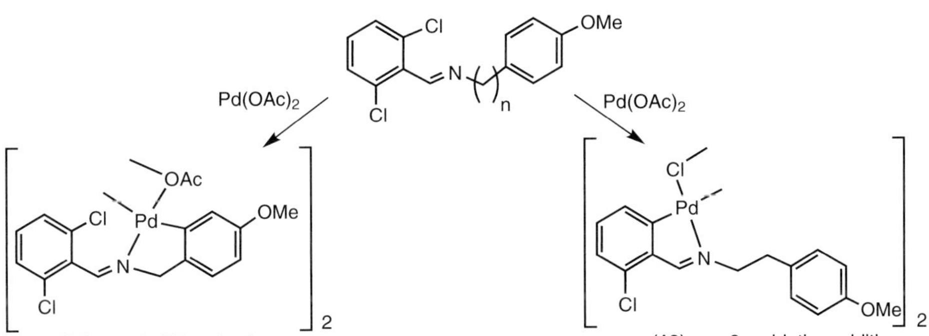

(11); n = 1; CH activation (12); n = 2; oxidative addition

Scheme 3.6

the benzylidene rings have been reacted with Pd(0) [22]. When both rings have ortho-Br substituents, competitive cyclopalladation gives with preference endo-complexes **(9)**, while exo isomers **(10)** have been obtained only in the absence of C—Br bonds at the benzylidene ring (Scheme 3.5) [22a].

Further work shows the complexity of the orientation of the cyclopalladation. When the 2,6-dichlorobenzylidene derivative shown in Scheme 3.6 ($n = 1$) was reacted with $Pd(OAc)_2$, the exo complex **(11)** was obtained, as a result of a standard C—H bond activation. However, for the derivative with $n = 2$, complex **(12)** was obtained; its synthesis is explained through and oxidative addition of the C—Cl bonds of the imine to the Pd(0) formed *in situ* from the reduction of the Pd salt with the imine [22b].

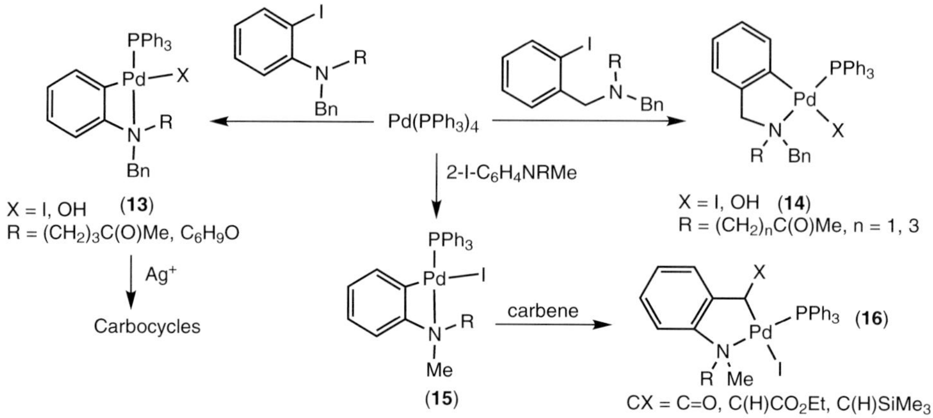

Scheme 3.7

Four-membered azapalladacycles **(13)** were first isolated by Solé *et al.*, as intermediates of the catalytic intramolecular carbocyclization of aryl halides and ketones (Scheme 3.7) [23a]. The formation of **13** seems to occur via bis(phosphane) derivatives, while their stabilization resides in the steric bulk of the N-substituents. A subtle change of substituents on the N atom does not change the formation of the palladacycles – synthesis of very stable **(14)** – but does alter the pathway of the carbocyclization.

Azapalladacycles **15** have been obtained from *N,N*-dialkyl-2-iodoanilines and $Pd_2(dba)_3/PPh_3$ or $Pd(PPh_3)_4$. As expected, the four-membered ring reacts easily to relieve steric stress and affords five-membered rings in various ways, such as aryl–aryl exchange or carbene insertion in the Pd—C bond **(16)** [23b]. The synthesis of **16** is a good example of modification of a preformed palladacycle.

The oxidative addition of haloaryl derivatives with different ortho functional groups has been studied by Vicente [24]. In most cases, the palladacycle is synthesized in two steps (Scheme 3.8). In the first step, oxidative addition affords an aryl derivative, which further reacts with (i) an halide scavenger (Tl^+, Ag^+), allowing the bonding of the functional group or (ii) unsaturated molecules, which insert into the Pd—C bond (Chapter 5) and add new functionalities [24a].

The studied functional groups were formyl, acetyl, cyano, alkenyl [24b] and ureas [24c], among others. Interesting transformations of the functional groups have been reported in the case of formyl and acetyl-containing ligands (Scheme 3.9). Complexes **21** containing the formyl group (R = H) react with semi-stabilized P-ylides in a Wittig process that affords alkenyl derivatives **(22)**, while those containing the acetyl group are deprotonated by the ylide, resulting in the formation of 3-palladaindanones **(23)**.

Unexpected transformations have been observed during the orthopalladation of aryldithioacetals [25]. Oxidative addition of the dithioacetal shown in Scheme 3.10 with $Pd(dba)_2$ gives two products, as a function of the starting conditions. One

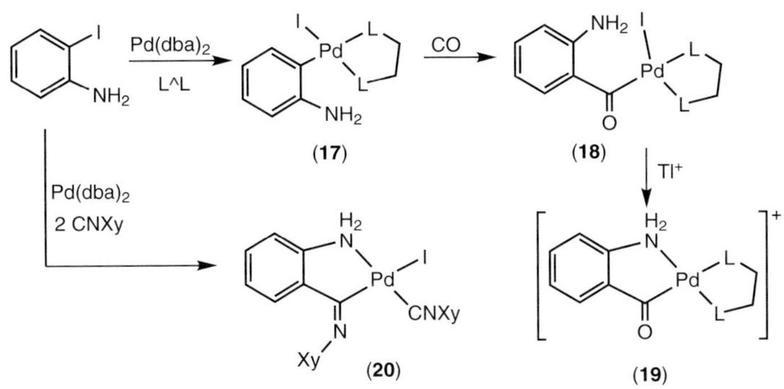

Scheme 3.8

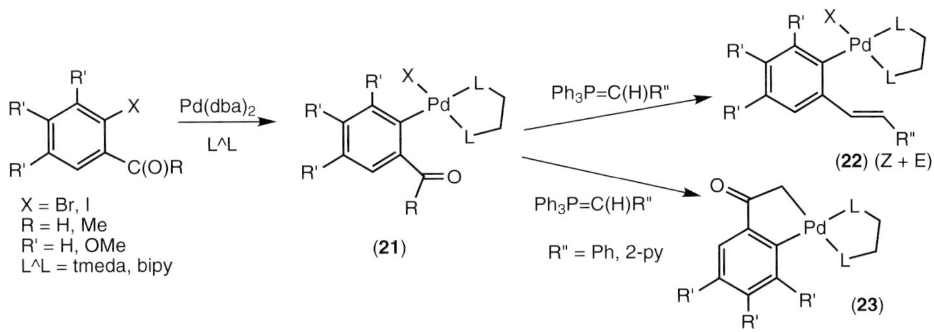

Scheme 3.9

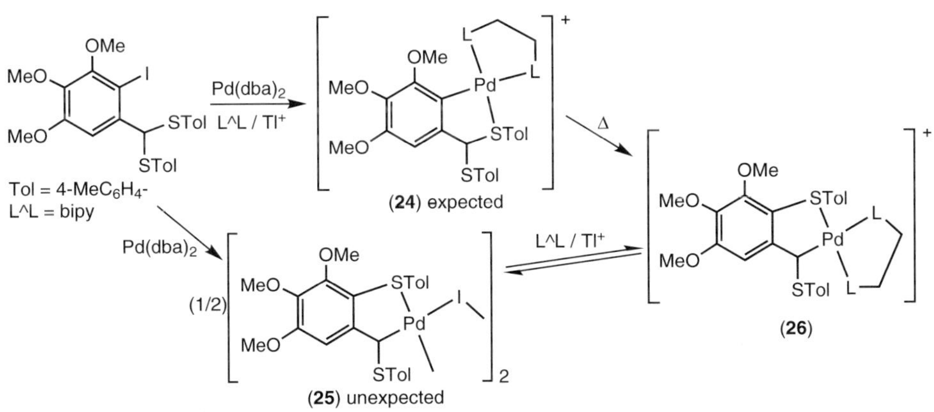

Scheme 3.10

product **(24)** is the "expected" derivative, obtained when the oxidative addition is performed in presence of a chelating ligand (bipy) and a halide scavenger; the "unexpected" product **25** is obtained in the presence of only the Pd source [25a].

The ligand rearrangement in **25** or **26** implies the cleavage of one C_{alkyl}—S and one Pd—C_{aryl} bond, and the formation of one C_{aryl}—S and one Pd—C_{alkyl} bond. The balance seems to be favorable to the rearranged product, since **24** can be transformed into **26** but not conversely, meaning that **24** is the kinetic isomer and **26** the thermodynamic one. Classical sulfur palladacycles **(27)** are obtained from reaction of *o*-iodobenzylthioethers with $Pd(dba)_2$ (Scheme 3.11) [25b].

Cyclopalladated complexes with C,P— and C,O— donor atoms have also been obtained through oxidative addition of aryl halides. Lahuerta *et al.* have reported the synthesis of four-membered palladacycles derived from $Ph_2P(C_6H_4Br\text{-}2)$ [26a, b] (Scheme 3.12a). The anion $[Ph_2PC_6H_4]^-$ acts, in addition to the stable four-membered C,P-chelating mode, as a C,P-bridging ligand and as a P-monodentate ligand [26a, b].

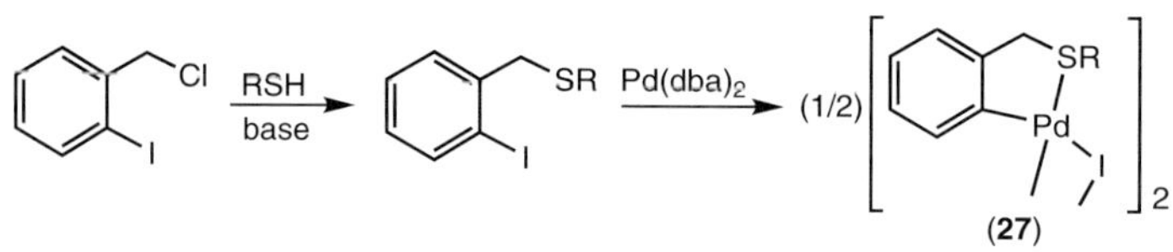

Scheme 3.11

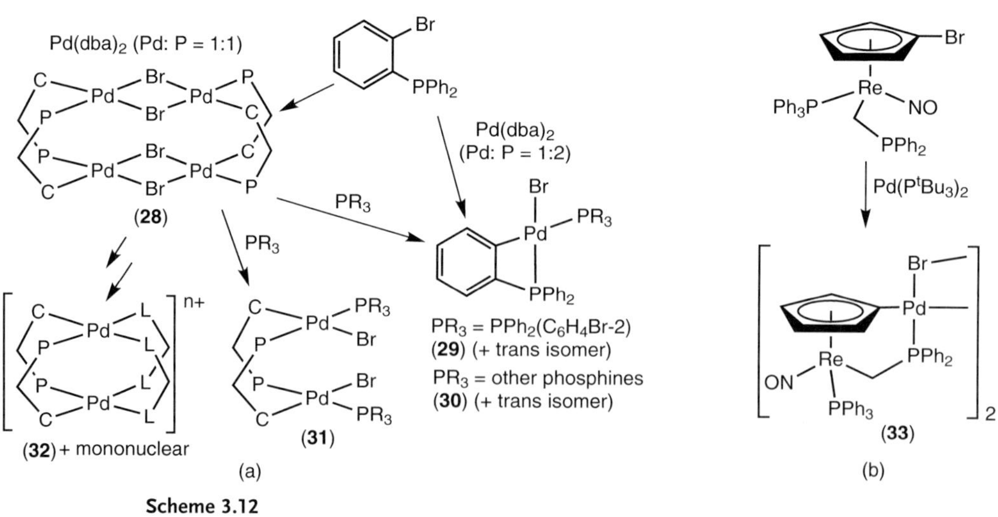

Scheme 3.12

The remarkable stability of the four-membered ring in complexes of type **29** and **30** is related to the presence of bulky substituents on the P bonded atom and steric strain around the metal center. These facts are similar to those described in C,N-complexes **13** and **15**. Classical five-membered C,P-palladacycles have also been obtained by oxidative addition [26c], and interesting chiral derivatives **(33)** (Scheme 3.12b) with chirality located at a Re center have been developed by Gladysz [26d].

The synthesis of C,O-palladacycles is related to research in catalytic C—O coupling processes [27]. Despite the assumed thermodynamic instability of the Pd—O bond due to soft-metal and hard-ligand characters, oxapalladacycles are stable and can be easily isolated from either neutral (alcohol, ether, etc.) or anionic (alkoxide, acid) situations. Complex **34** is obtained from the reaction of 2-(2-bromophenyl)ethanol with $Pd(dba)_2$ in presence of tmeda (Scheme 3.13), and is the precursor of **35** [27a].

Deprotonation of **35** with bases gives **36**, in which the oxapalladacycle acts as a dianionic [27a]. Complex **35** (but not **36**) is involved in C—O couplings, showing the lability of the Pd—O bond in **35** and its high stability in **36**. A very similar reaction scheme has been applied to the synthesis of alkoxide derivatives **37** and **38** (Scheme 3.14a) [27b, c]. Notably, in these two cases the aryl–alcohol chelate intermediate (similar to **35**) has not been isolated. The bridging system in **38** can be cleaved by chelating ligands, affording new monomeric oxapalladacycles. In contrast, aryl derivative **39**, obtained from direct oxidative addition of 2-iodobenzoic acid, reacts with Cs_2CO_3 and $AgBF_4$ to give the tetramer **40** [27c].

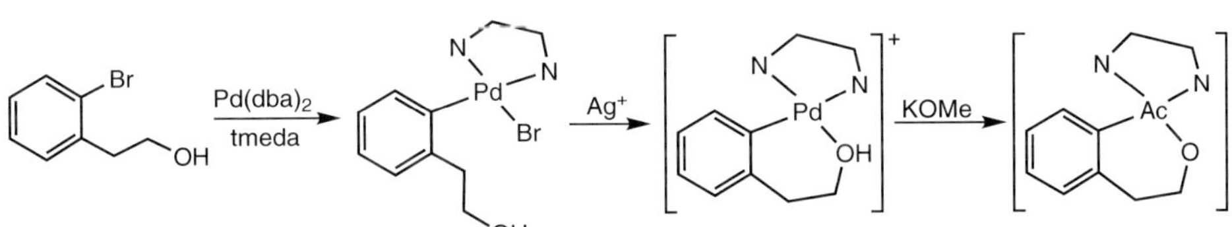

(34) (35) (36)

Scheme 3.13

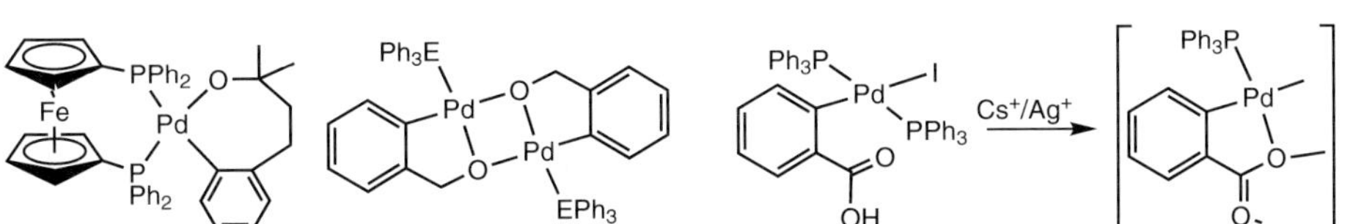

(37) (38) E = P, As (39) (40)

(a) (b)

Scheme 3.14

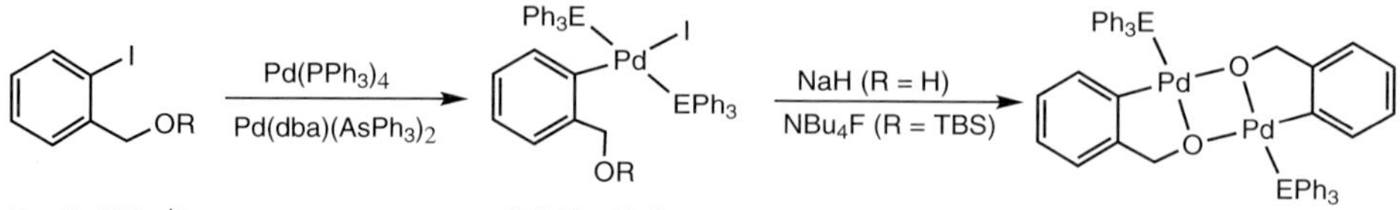

Scheme 3.15

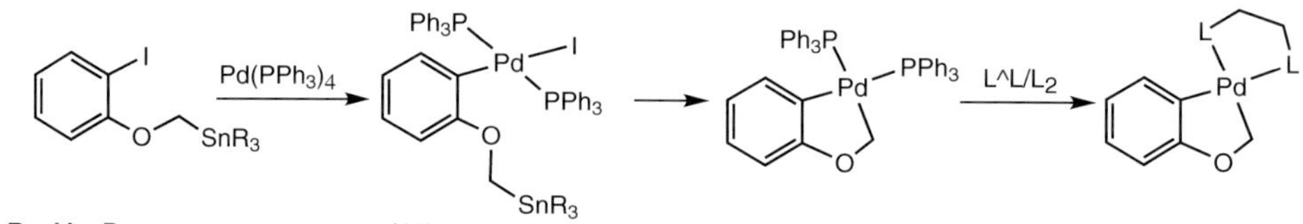

Scheme 3.16

The synthesis of **38** (Scheme 3.15) deserves some comments. The first step is the direct oxidative addition of the iodobenzyl derivative to give the intermediate **41**, which is then deprotonated by a strong base to give the dimer **38**. The authors report that the deprotonation of the benzyl alcohol fragment is hardly reproducible and that, as function of the commercial source of NaH, the yield varies between 0 and 63%. For this reason, the silyl ether derivative was prepared, and treatment of the TBS derivative with a source of F^-, such as *n*-NBu_4F, gives complexes of type **38** cleanly and in very good yield [27b].

C,C-Palladacycles are relevant species in Pd(II) chemistry, usually related with C,C–coupling processes. In most cases, they are proposed reaction intermediates [28], for instance in the extensive chemistry developed by Dyker. However, it is also possible to isolate them, in other cases, as more or less stable solids [29, 30]. Several methods based on oxidative addition allow the synthesis of such palladacycles.

Echavarren *et al.* have reported the synthesis of C,C-palladacycles of type **43** through an oxidative addition process that affords intermediates **42** (usually not isolated) followed by a intramolecular Stille reaction, interrupted at the reductive elimination C,C-coupling [29]. Complexes **43** are stable towards reductive elimination due to the high energy of the resulting 2*H*-benzoxete. The PPh_3 ligands in **43** can be replaced by isocyanides or other classical chelating ligands to give **44** (Scheme 3.16).

The method outlined in Scheme 3.16 also works in the synthesis of azapalladacycles **45** [29]. Sterically crowded oxapalladacycle **46** rearranges to **47** in a process probably promoted by water (Scheme 3.17). This rearrangement is more or less related to that reported for **24** and **26**.

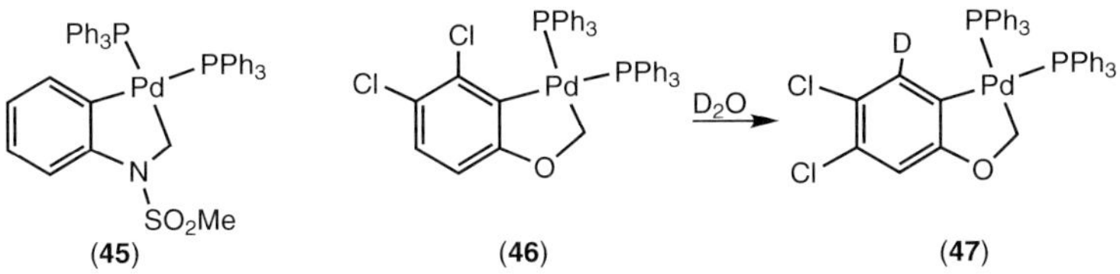

Scheme 3.17

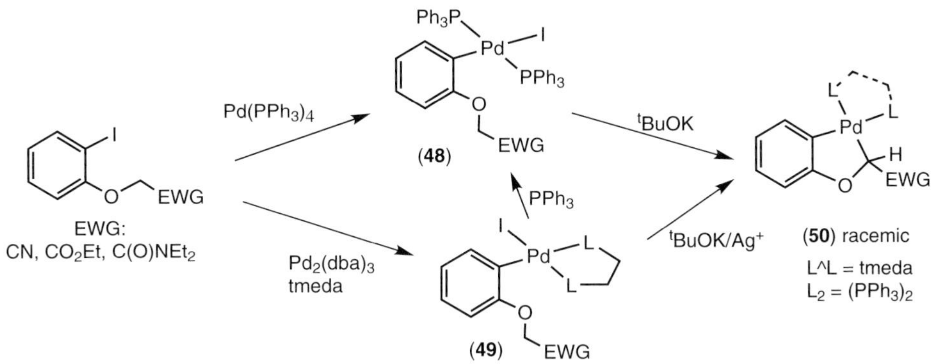

Scheme 3.18

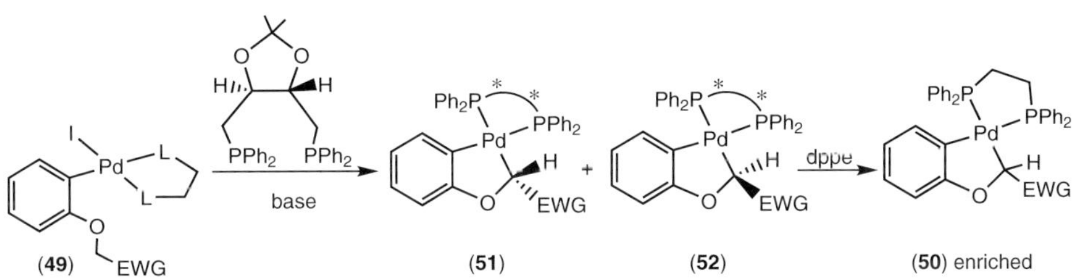

Scheme 3.19

Using a similar approach, Malinakova *et al.* have prepared chiral oxapalladacycles [15, 30]. The starting substrates are 2-iodoarylethers, in which one of the substituents is a $CH_2C(O)R$ (R = OEt, NEt_2) or CH_2CN group, that is, an activated – and easily deprotonable – methylene bonded to an electron-withdrawing group (EWG), whose presence seem to be necessary to achieve cyclization [30a]. The first reaction step is the oxidative addition to give complexes **48** or **49**, which under base treatment cleanly afford oxapalladacycles **50** (Scheme 3.18).

When the cyclization step is carried out in a chiral environment, the carbopalladation reaction could occur with stereoselective induction [30b]. The chiral medium is provided by bis(phosphanes) such as *S,S*-DIOP or (2*S*,4*S*)-bis(dipheny lphosphino)pentane (dppp), while the careful choice of the base allows one to direct the reaction with preference to a given diastereoisomer, **51** or **52** (Scheme 3.19).

Under these conditions, the reported d.e. reaches 80%. Substitution of the chiral sources by achiral versions (dppe) affords oxapalladacycles **(50)** enantiomerically enriched and allows the recovery of the chiral source. The e.e. in enriched **50** is comparable to the d.e. in mixtures of **51–52** and remains constant, showing that enriched **50** does not racemize.

Some other synthetic methods afford C,C–palladacycles, but their approach is slightly different. The first step is a classical oxidative addition of aryl halides, while the second step is the reaction of the aryl-Pd complex with alkenes (norbornene). Under these conditions, the reaction seems to occur as presented in Scheme 3.20 [31]. The process starts with the oxidative addition of phenyl iodide to a source of Pd(0), giving complex **53**. Subsequent treatment of this complex with norbornene affords **54**, after alkene insertion into the Pd—Ph bond, which contains a norbornyl ligand and a η^2-bonded phenyl fragment. Complex **54** is unable to undergo β-hydrogen elimination, and so intramolecular C—H bond activation occurs in the presence of base to afford palladacycle **55**.

From the preceding paragraphs it can be concluded that palladacycles are almost exclusively in the Pd(+II) oxidation state [32]. Renewed interest in Pd(IV) chemistry has emerged with new applications and the involvement of Pd(IV) species in organic synthesis has been reported recently [33]. Concerning palladacycles, few contributions have appeared, but two general methods of synthesis can be discerned: (i) Pd(II) complexes that already contain a palladacycle, and which are oxidized by addition of classical X_2 or XR reagents (Cl_2, Br_2, I_2, MeI, etc.) [34]; (ii) direct oxidative addition of XR (R = orthometallated ligand) over electron-rich Pd(II) compounds [35]. Very interesting examples of the first method are shown in Schemes 3.21 and 3.22 [31, 34], the latter showing two consecutive oxidative additions, although of different nature.

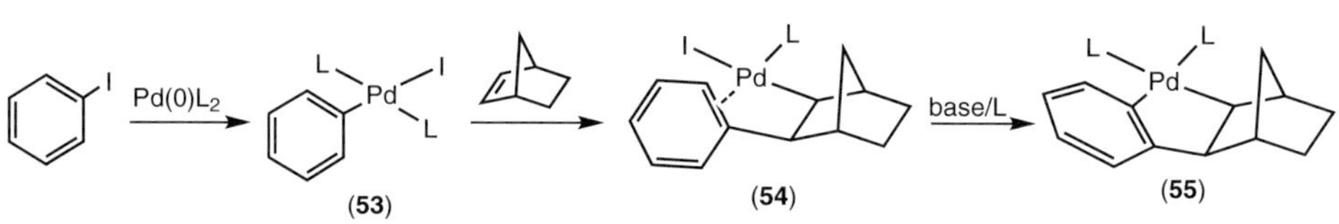

Scheme 3.20

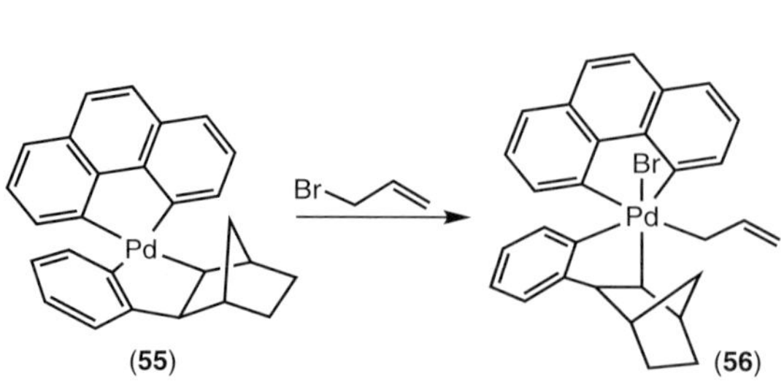

Scheme 3.21

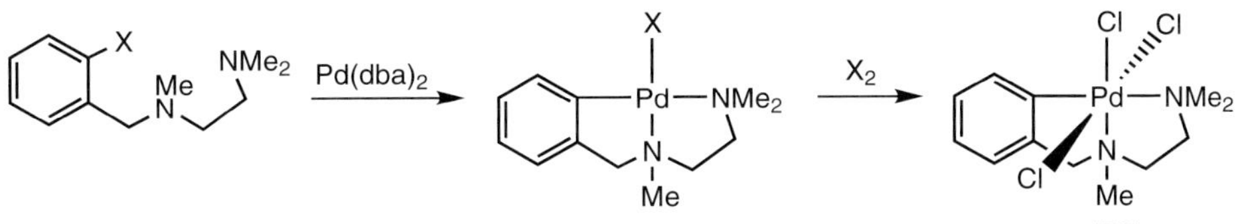

Scheme 3.22

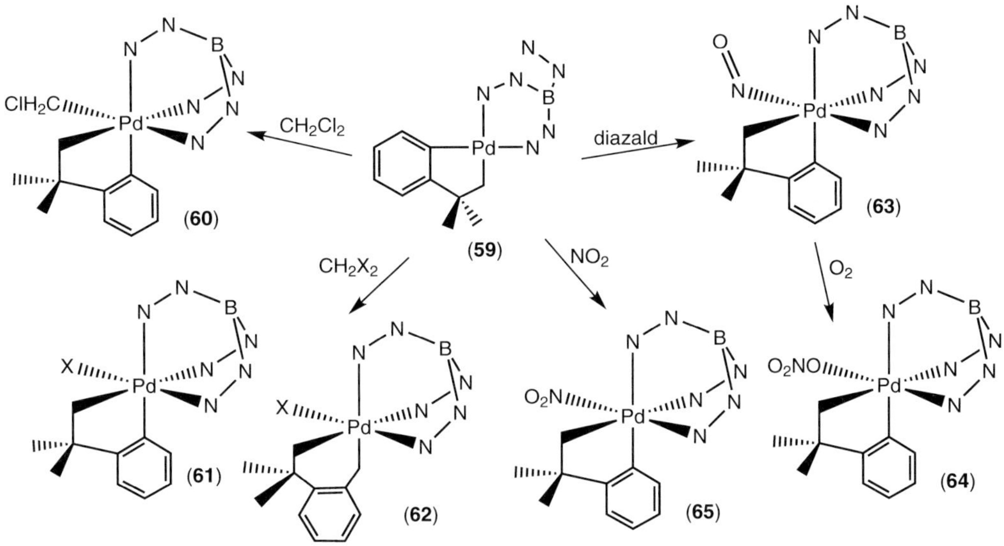

Scheme 3.23

Complexes of type **58** are stable enough to be characterized in solution. The strong electron-donating ability of the pincer C,N,N ligand seems to be the responsible for this stability. This type of pincer ligand has been studied extensively and is discussed below.

Very interesting chemistry has been developed around C,C–palladacycles of Pd(IV) by Cámpora *et al.* (Scheme 3.23) [34b, c]. Stable species have been prepared from anionic Pd(II) derivatives containing Tp ligands [Tp = hydrotris(pyrazolyl)borate].

Only one example of method (ii) has been reported to date (Scheme 3.24). Oxidative addition of 8-bromomethylquinoline to neutral **66** or anionic complexes **67** [Bp = dihydrobis(pyrazolyl)borate] gives the corresponding Pd(IV) derivatives **68** and **69**, respectively, which are the first isolable intramolecular coordination Pd(IV) compounds [35]. In all reported cases, the three C atoms are in a *fac* arrangement. The Bp derivatives are quite stable. However, for the bis(methyl)bipyridine complexes, reductive elimination (giving ethane and ethylquinoline) is the common pathway of decomposition in solution at room temperature.

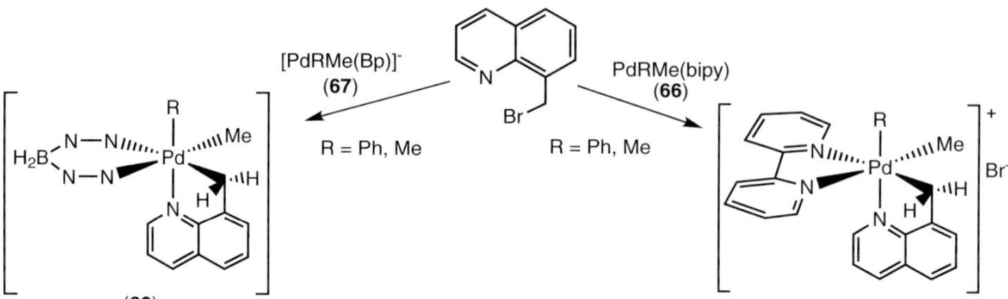

Scheme 3.24

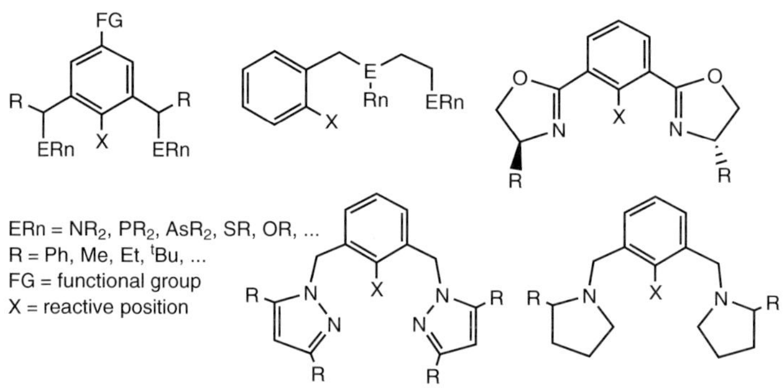

Figure 3.3 Representative examples of pincer molecular skeletons.

Although research on Pd(IV) palladacycles is an area still scarcely represented, recent contributions on cross-coupling and oxidation reactions in which Pd(II)/Pd(IV) catalytic cycles are involved seem to be the prelude to new and interesting developments. These impressive recent advances prove that, despite the wide area yet covered, the chemistry of palladacycles is a continuously growing research field that is far from exhausted.

The best example of the latter statement is the chemistry of pincer complexes based on palladacycles. The synthesis of pincer complexes is worth a separate entry on this chapter, not due to the use of specific preparative methods, but for the wide scope and importance of the applications developed with them, which is covered in appropriate reviews [36], and, of course, in subsequent chapters of this book. Figure 3.3 shows some representative examples of pincer molecular skeletons.

The success of the pincer ligands is due to several factors. The first is the stability of the resulting products, since the pincer arrangement provides very stable chelate structures, able to resist, unchanged, harsh catalytic conditions. Probably, this is one of their most attractive facets. At the same time, the metallated C_{aryl} atom provides a very reactive trans coordination site, taking advantage of the high kinetic trans effect of this atom. In addition, a plethora of substituents can be attached to the central aryl core, aliphatic, carbocyclic or heterocyclic, with different donor

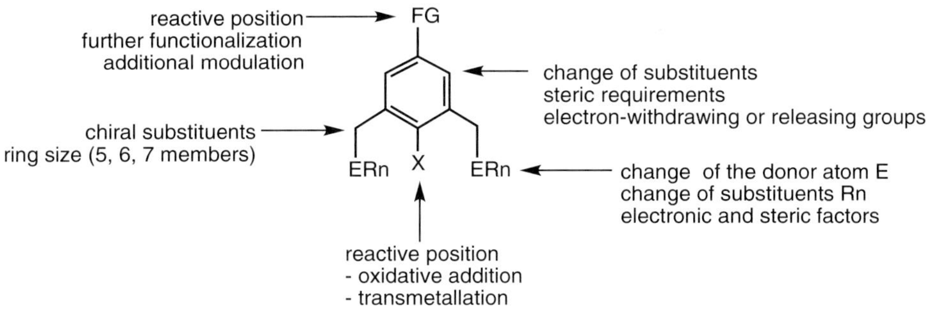

Figure 3.4 Component parts of pincer ligands.

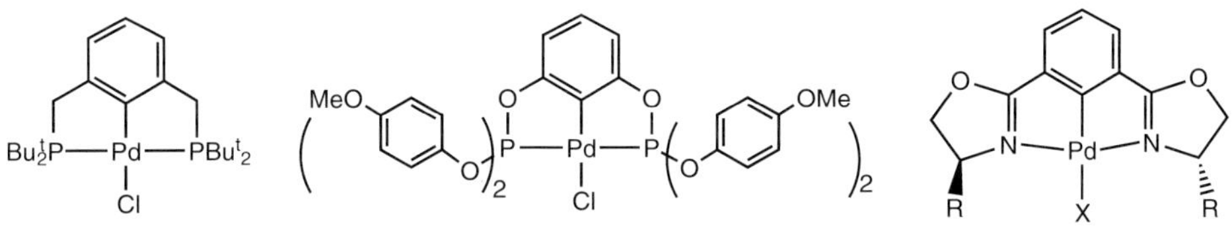

Figure 3.5 Examples of pincer ligands.

atoms (N, P, As, O, S, etc.). The aryl ring, the spacer between the aryl group and the donor atom, the environment of the donor atom and its own nature are thus susceptible to modulation (Figure 3.4) [36b]. These facts enable a tailored synthesis of desired ligands and the corresponding palladacycles, with a more or less complete control of their electronic and steric properties.

The first pincer derivative (**70**, Figure 3.5), reported by Shaw in 1976, was obtained by direct C–H bond activation on a PCP ligand [37]. Most syntheses of PCP and SCS ligands published since then have been performed by this method–with **71** as an exception [38]. The strength of the Pd–S and, mainly, Pd–P bonds seems to be one of the reasons for this behavior. However, the oxidative addition strategy is often employed when reactive and/or sensitive functional groups that are present in the original ligand must also be present in the target molecule. This is because the reaction conditions under which direct C–H bond activation is performed are usually more drastic than those required for the oxidative addition and, moreover, because under these conditions C–H bond activation is less selective and two or more positions can be activated. The method is especially applied on the synthesis of NCN-pincers, for instance with phebox **(72)** [39].

Using the same synthetic procedure as that published for the non-functional system **73** [40a], macrocyclic complexes **(74)** [40b] or sophisticated aryl *p*-substituted compounds **(75, 76)** can be prepared (Figure 3.6) [40c, d].

The reactivity of bifunctional substrates has been developed extensively by van Koten *et al.* through the use of $IC_6H_2(CH_2NMe_2)_2$-3,5-Br-4 [41]. The two C–X bonds in the precursor can be activated, to give **77** or **78**, but the reaction conditions are different. The reaction with $Pd_2(dba)_3$ only activates the C–Br bond at very low

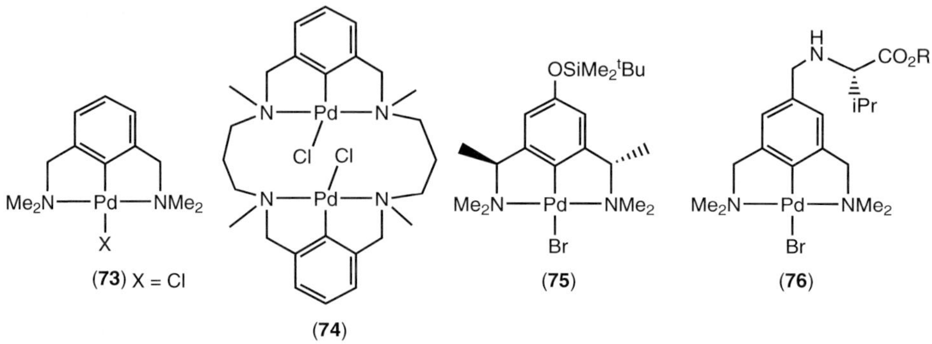

Figure 3.6 Further examples of pincer ligands.

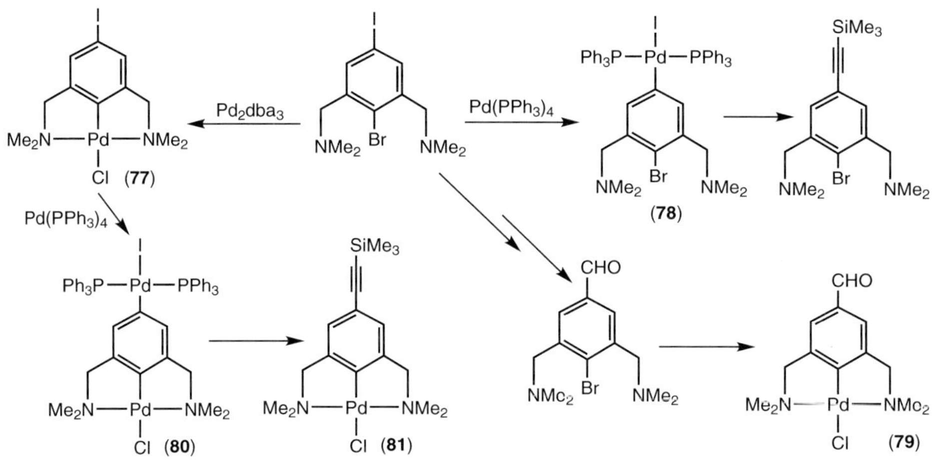

Scheme 3.25

temperature (–80 °C), while reaction with $Pd(PPh_3)_4$ requires a higher temperature (50 °C) to activate the C—I bond (Scheme 3.25). In each case, the other C—X bond remains intact, this allowing several organic **(79)** and organometallic reactions, including the synthesis of binuclear complexes **(80)**.

Two approaches are envisaged in the chemistry represented in Scheme 3.25 to obtain a given functionalized complex. The first includes the metallation of a functionalized substrate, for instance the synthesis of **79** starting from an aldehyde-containing compound. The second involves the functionalization of an already metallated substrate, for instance the synthesis of the silyl derivative **81**. Using the first approach a great variety of substrates can be metallated [41]. The second approach has been scarcely used [42] but it has recently found some applicability for the synthesis of NCN pincer complexes (Scheme 3.26). The 2,6-diformyltriflate precursor is easily metallated, giving **82** in very high yield. After the metallation of the precursor, its functionalization is possible through the two

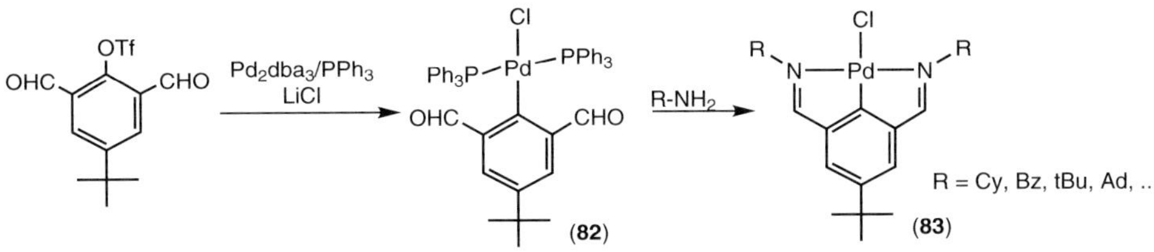

(82) (83)

Scheme 3.26

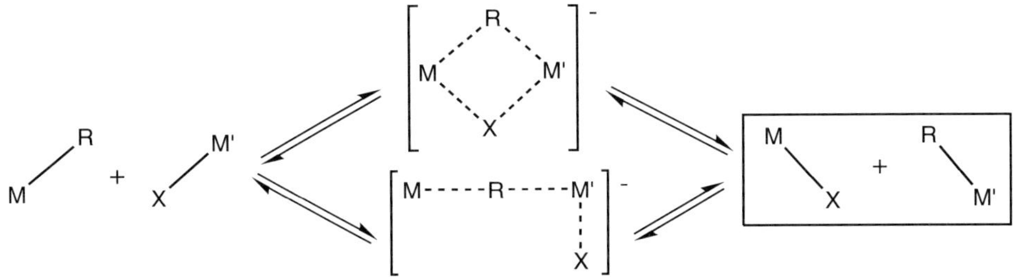

Figure 3.7 Equilibria involved in transmetallation.

formyl groups. Compound **82** reacts with bulky amines RNH_2 or heterocycles, giving NCN pincer-type complexes **(83)** [42].

3.3 Transmetallation

Transmetallation reaction is another fundamental step in both stoichiometric and catalytic processes, as a general and versatile method to create new metal–carbon bonds [43]. The set of equilibria shown in Figure 3.7 contains a very general representation of this process. Looking at the reagents and products, transmetallation involves the transfer of a given hydrocarbon ligand R (alkyl, aryl, acyl, etc.) from one metal M to another different metal M′. At the same time, a ligand X (usually an halide) is transferred from M′ to M. The reaction as a whole is an equilibrium; to obtain a maximum shift of the equilibrium to the right, M should have a lower electronegativity than M′. Usually, M is an alkaline metal (Li or Na), an alkaline earth metal (Mg), a representative element (B, Si, Sn), or even a transition metal (Zn, Hg, Cu, Au, etc.), while M′ is often a transition metal, in our case palladium. The formation of a very insoluble salt of the metal M could be an additional strategy to achieve a good conversion.

Concerning the substrates involved in a transmetallation reaction, we must consider the palladium precursor and the transmetallating reagent. The vast majority of reported processes start from (and finish with) Pd(II) complexes, since this reaction does not alter the oxidation state of the metal. There are no strict prerequisites for the Pd(II) starting materials, although some considerations must be taken into account. The starting Pd complex should contain a Pd–X bond (X = halide), since this would be the reactive position in which the incoming R ligand will be bonded. In addition to the Pd–X bond, the rest of the ligands exert a considerable influence over the course of the reaction, both on electronic and steric grounds. These factors are closely related with the mechanism of the transmetallation process, the latter being subject of notable controversy.

The work developed by Farina, Amatore and Jutand [44a, b] suggests that pre-dissociation of a given ligand is necessary to obtain good conversions on the transmetallation step, since "coordinative insaturation is really the key to high reactivity" [44a]. On the other hand, Espinet *et al.* [13, 44c] have shown that the pre-dissociation is not a requisite, since "the very nature of the transmetallation is that of a ligand substitution on a Pd(II) complex." But even in this case, Figure 3.7 displays an additional uncertainty, such as the nature of the transition state, which could be open or cyclic. Recent studies [45] have shown that a cyclic concerted transition state is more likely but, probably, each reaction (and even each pair of reactants) must be analyzed in detail, since the huge number of variables (nature of groups X and R involved, nature of metals M and M′ implied, solvent, etc.) makes generalization difficult.

Practically speaking, simple halo complexes such as $Alk_2[PdX_4]$ or PdX_2L_2 (X = Cl, Br, I; Alk = Li, Na, K), where the L group is a more or less labile ligand (NCMe, NCPh, py, SC_4H_8, SMe_2; L_2 = 1,5-COD), are very good candidates to undergo a transmetallation reaction in a clean, efficient way. In contrast, the proper choice of the transmetallating reagent M–R is not an easy task, since we have several possibilities. If we center the discussion on the synthesis of palladacycles, the most usual M elements in Figure 3.7 are Li, Mg, Hg, Si, Sn, Zn and, more rarely, other transition metals such as Cu, Ag or Au. The different nature of the metal and, obviously, of the metal–carbon bond implied, results in very different chemical properties, such as stability, reactivity and tolerance to the presence of potential additional functional groups on the same substrate.

Organolithium and magnesium compounds are extremely reactive species, which must be handled under a protected atmosphere. Lithium compounds usually have an oligomeric structure, and they can be synthesized by one of three main methods: (a) from an electrophilic organic halide and Li metal; (b) by Li–halogen exchange, obtained by treatment with other organolithium reagents such as BuLi, PhLi or MeLi; and (c) by lithiation through H abstraction [46, 47a–c] (Scheme 3.27).

Direct metallation can have synthetic difficulties. Using method (c), naphthyl [47d] and ferrocenyl [47e, f] species have been obtained. Metallation through H abstraction seems to be the most practical method, since it is not necessary to functionalize the starting material with a C–halogen group. However, under these conditions, the metallation is not selective in most cases and the lithiation is produced at several places at the same time, giving rise to the formation of isomers. This is evident in the reactivity of the α-substituted naphthyl group, which can be metallated at the 2 and 8 positions (Figure 3.8a), or in reactions performed with ferrocenyl derivatives, which afford derivatives with planar chirality (Figure 3.8b).

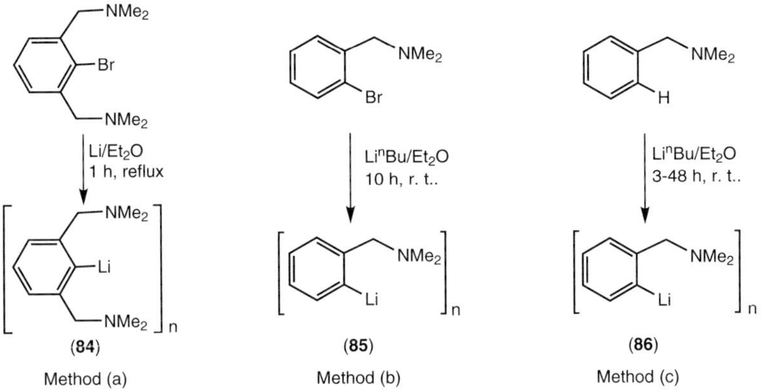

Scheme 3.27

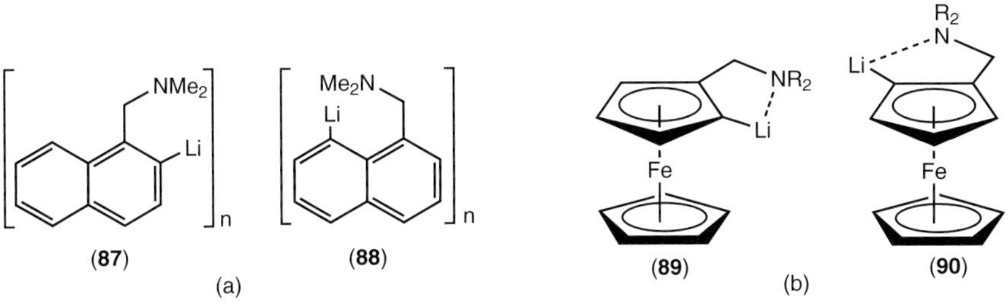

Figure 3.8 Metallation of (a) naphthyl and (b) ferrocenyl derivatives.

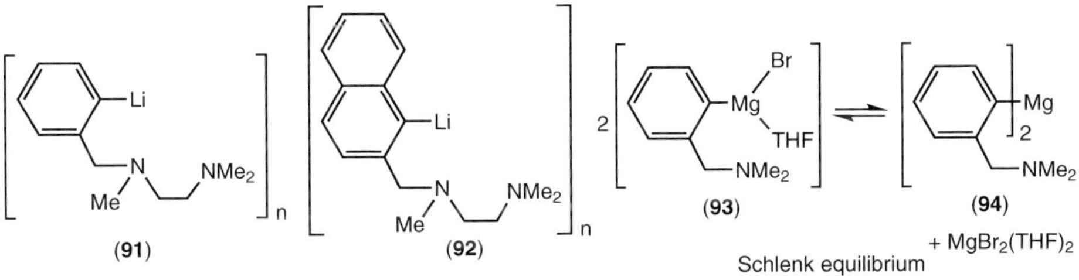

Figure 3.9 Further examples of selective lithiation.

When high selectivity is required, the metallation should be carried out using halide-substituted precursors; such metallation usually proceeds with quantitative yields [3, 47a].

The synthesis of Li-derivatives of pincer ligands provides additional examples of selective lithiation (Figure 3.9) [48]. Grignard reagents are also adequately prepared by reaction of halo derivatives, usually bromo or iodo, with activated Mg metal, a process that can be considered formally as an oxidative addition. Figure 3.9 shows a classical example, involved in the Schlenk equilibrium; notably, lithium derivatives transmetallate organic groups to Mg(II) salts. The chemistry of orthometallated derivatives of groups 2, 12, and 13 has been reviewed exhaustively by Bickelhaupt [49].

The high reactivity of the organolithium and magnesium derivatives could have other drawbacks. Among them, their low tolerance to the presence of other functional groups on the organic substrates [46], and the existence of side reactions (for instance, redox processes) with the metal center (Pd, Pt), which could result in complete decomposition. Both facts can be overcome by substitution of Li by another less electropositive metal (Hg, Zn, Si, Sn, for instance); this method is discussed below in more depth.

Even if halo derivatives of the starting material are available, the selective formation of the lithium derivatives is not always an easy task, since the reaction is strongly dependent of the nature of the halogen atom and the reaction conditions. For instance, treatment of the bis(benzyl)phosphine shown in Scheme 3.28 with 2 equivalents of LiBu gives **95**, the expected substitution product. Further treatment of **95** with LiBu gives **96** through C—H bond activation at the α-benzylic positions [50a].

A subtle change of halide from Cl to Br allows direct lithiation to the α-benzylic position or to the arylic position, as represented in **97** and **98** (Scheme 3.29) [50b].

Further reactivity of the lithiated benzylphosphine ligand in **98** gives bis-substituted Pd(II) complexes **99–102**, with the same or different cyclopalladated ligands. Thus, mixed derivatives with the same (P) or different (N, As) donor atoms can be obtained [50c, d]. The synthesis of complexes **102–105** shows an implicit

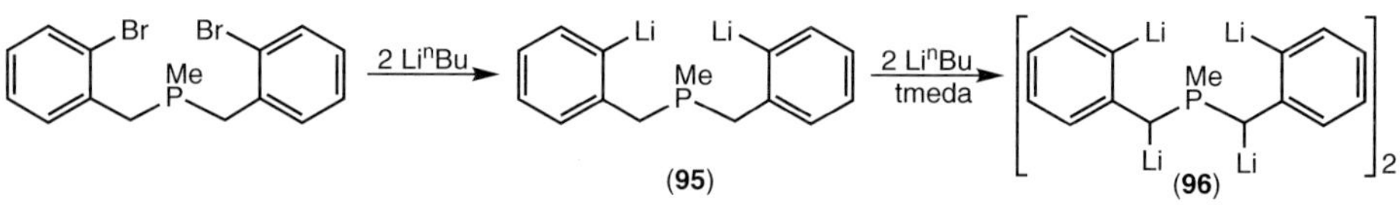

Scheme 3.28

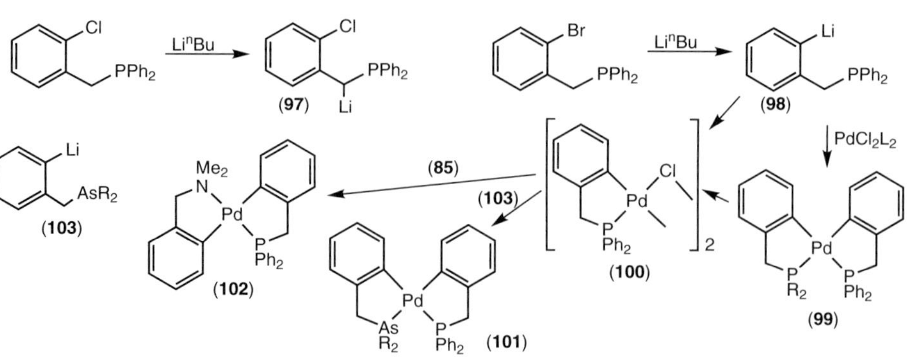

Scheme 3.29

rule in this chemistry. The first palladacycle can be formed by any of the methods reported up to now (C—H bond activation, oxidative addition, transmetallation, etc.) or by other methods. However, the introduction of a second orthopalladated ligand must be carried out using strong nucleophilic reagents, such as the organolithium derivatives. The metallation position can also be changed by addition of ligands. The addition of tmeda (*N,N,N′,N′*-tetramethylethylenediamine) to **98** gives **104** (Scheme 3.30a) [50d]. In **105** and **106**, lithiation at the α-benzyl position occurs by direct metallation (Scheme 3.30b). Other lithium reagents, such as $Ph_2PC_6H_4$-2-Li [51a] **(107)** and $Ph_2AsC_6H_3$-2-Li-4-Me [51b] **(108)**, are easily prepared from the corresponding bromo derivatives and react with Pd(II) to give complexes related to **28–32**.

The synthesis of C,N-palladacycles through transmetallation shares some aspects with that reported for C,P-palladacycles. The first metallacycle is formed quite easily, and can be performed in various ways, but the second one – in neutral complexes – always involves a transmetallation process using lithium derivatives. The doubly metallated complexes **109** and **112** have been synthesized using this strategy (Scheme 3.31) [52a], as have the symmetric and mixed derivatives **113–116** (Figure 3.10) [52b].

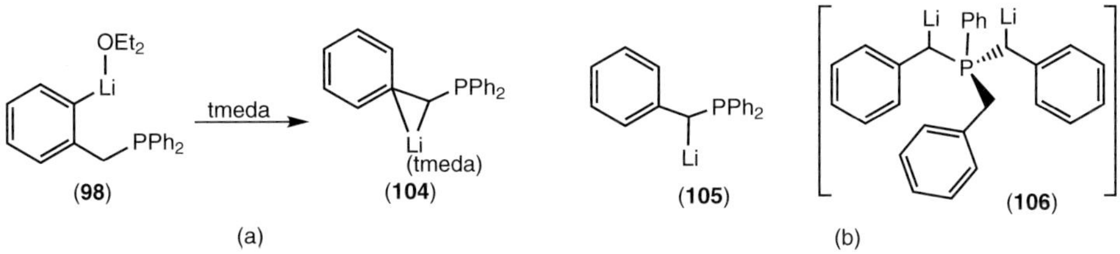

Scheme 3.30

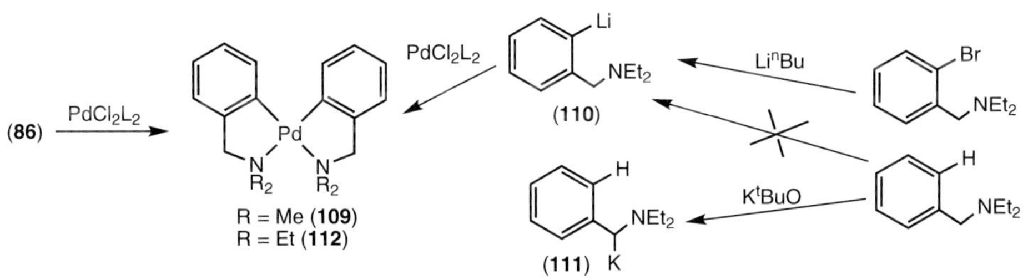

Scheme 3.31

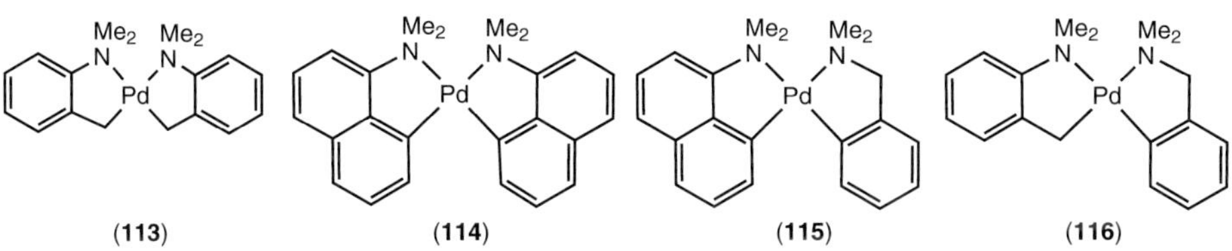

Figure 3.10 Examples of symmetric and mixed doubly metallated complexes.

Van Koten *et al.* have reported [53] the synthesis of naphthyl complexes in the 1- or 3-position under controlled conditions. Direct palladation affords cyclometallation in the 3-position **(120)**, while Br-oriented lithiation allows synthesis of the 1-palladated derivatives **118** (Scheme 3.32).

Other classical C,N-ligands can be cyclopalladated through the prior synthesis of the organolithium derivative and transmetallation to a simple complex of Pd(II). For instance, bisoxazolines such as **72** (one substituent) or **121** (two substituents) [54], ferrocenyl derivatives [55], and, mainly, classical $NC_{aryl}N$ pincer ligands [56] complexes are widely used as catalysts. It has been already stated that interest in complexes containing NCN pincer ligands, and the number of synthetic routes developed to prepare them, is directly related to the potential of their applications. A general survey of structural motifs of NCN pincer complexes **73–81** has been presented in the previous section on oxidative addition, although the transmetallation was their original preparative method [47a, 56a]. Scheme 3.33 shows representative examples of this type of complexes.

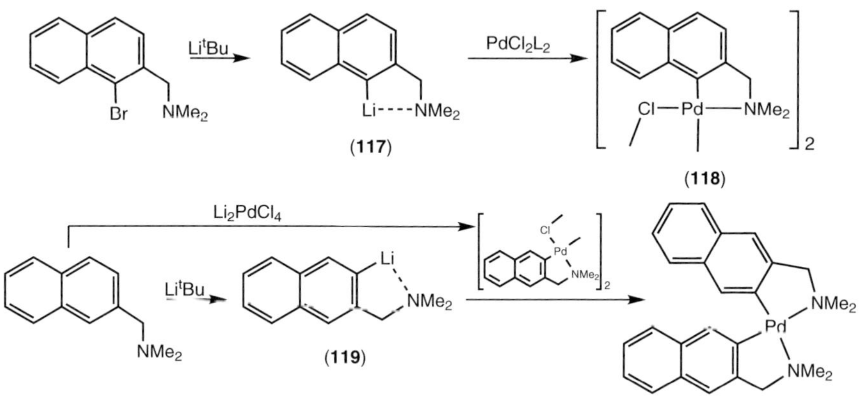

(120)

Scheme 3.32

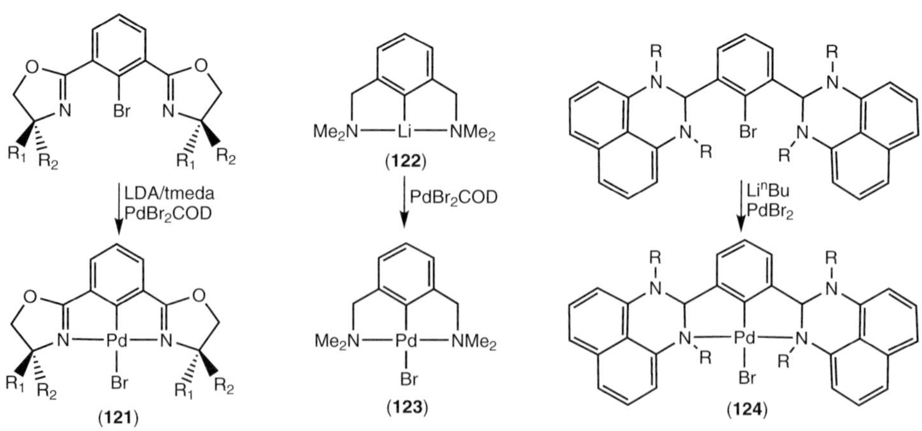

Scheme 3.33

The synthesis of C,C-palladacycles also deserves some comments. It has been already noted that oxa- and azapalladacycles **43–47** can be prepared by oxidative addition followed by intramolecular Stille transmetallation [29]. The transmetallation of doubly lithiated species is also an efficient method of synthesis of four-membered palladacycles **(126)** [57a]; five-membered alkyl cyclopalladated complexes **(129)** are also obtained by transmetallation of the CH_2CMe_2Ph group from the Grignard reagent $ClMgCH_2CMe_2Ph$ **(127)** to $Cl_2Pd(COD)$ followed by C—H activation on the phenyl group (Scheme 3.34) [57b].

Clearly, from the preceding sections, organolithium derivatives are the most widespread transmetallating reagents, due to their high reactivity. This reactivity seems to be related with a highly polarized Li—C bond and a high formal negative charge located at the carbon atom to be transmetallated. For the same reasons they have some drawbacks, such as poor selectivity or undesirable side reactions. The use of other less electropositive elements, which usually form less polarized metal–carbon bonds, gives excellent results since the transfer of the organic group occurs under smooth conditions. The most popular of these metallating reagents are B, Sn, Si and Hg, and almost all are included in a catalytic cycle: Suzuki (B), Stille (Sn) and Hiyama (Si) [13, 58]. Additional advantages of these transmetallating reagents are their high compatibility with a wide scope of functional groups and their stability, which usually allows their isolation. We now develop the most general synthetic methods reported using transmetallating derivatives of Si, Sn and Hg. Notably, as Sn and Hg derivatives are very toxic, environmentally friendly alternatives have been developed (although scarcely) using transition metals such as gold. The pincer derivative **73** can be obtained from aryl–gold(I) complex **130** (Scheme 3.35) [59].

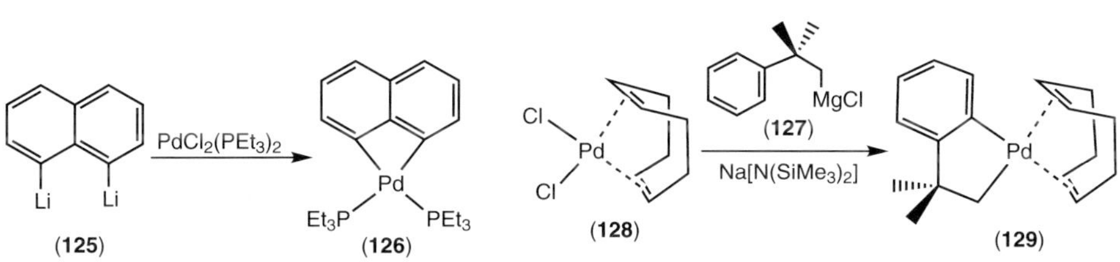

Scheme 3.34

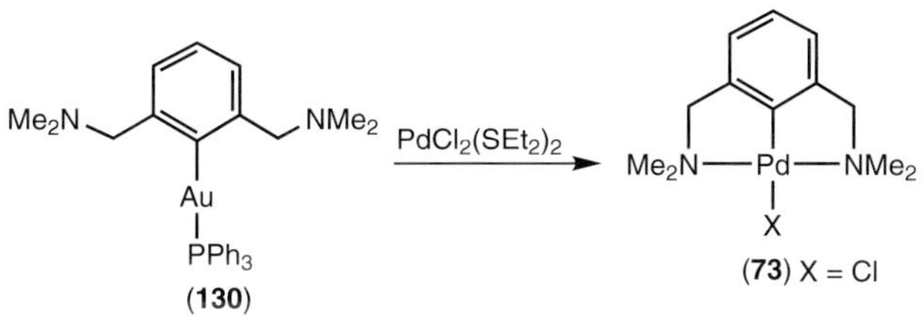

Scheme 3.35

Interest in tin derivatives as useful transmetallating reagents began with the work developed by Stille on Pd-catalyzed C–C coupling using organometallic compounds of tin [60a]. Metallated tin derivatives can be synthesized by different methods. Some of them, such as lithiation, are similar to those reported for palladacycles, while others, such as the use of $R_3Sn{-}SnR_3$ reagents, expand the synthetic possibilities [60c]. Scheme 3.36 shows some representative examples.

Once formed, the transfer of the organic fragment from the Sn(IV) center to the Pd(II) center proceeds very smoothly (CH_2Cl_2, 0 °C), as reported for oxazolines **(72)** [61a] and ketoximes **(136)** (Scheme 3.37) [61b]. These conditions contrast with those reported by Stille for the catalytic C–C coupling, which usually requires harsh conditions.

The group of van Koten has developed extensively the synthesis of palladacycles from Sn or Si derivatives [62]. Chiral derivatives **(137)** [62a], with the Sn atom as a stereogenic center, have been obtained by reaction of $MePhSnBr_2$ with Ar_4Cu_4 or Ar_2AuLi derivatives. Pincer ligands give ionic species **(138)** by displacement of the halide anion [62b]. Naphthyl tin complexes **(139)** have been obtained from

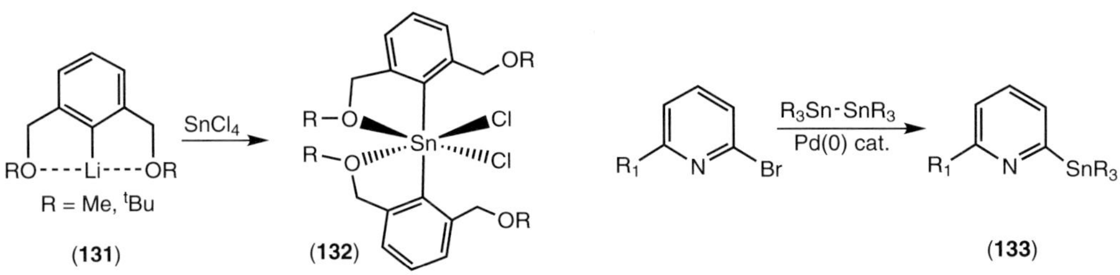

Scheme 3.36

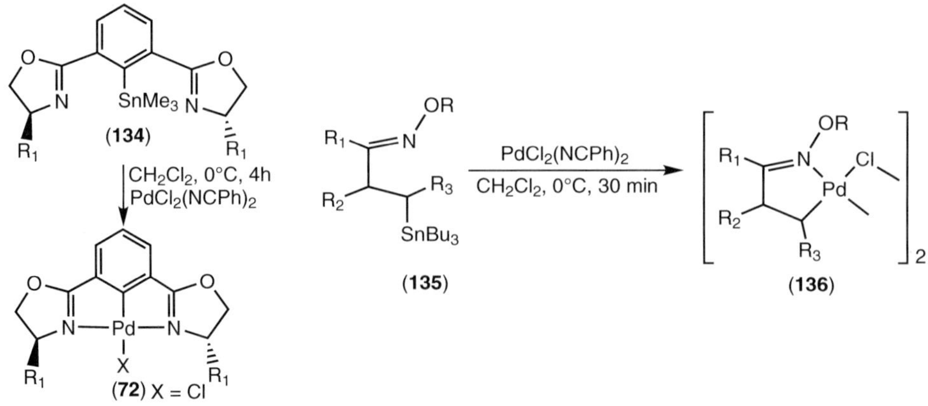

Scheme 3.37

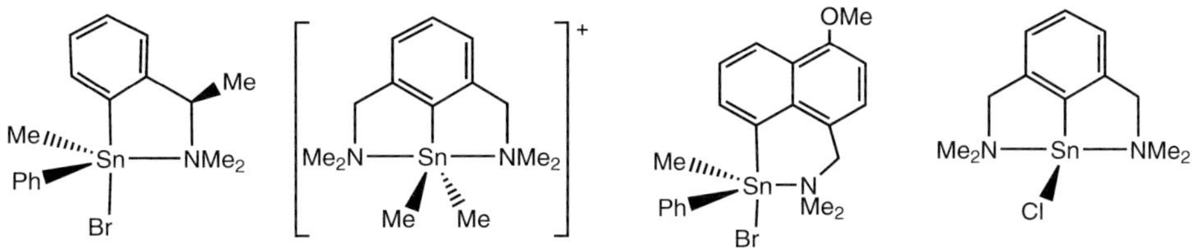

(137) **(138)** **(139)** **(140)**

Figure 3.11 Some Sn derivatives used in the synthesis of palladacycles.

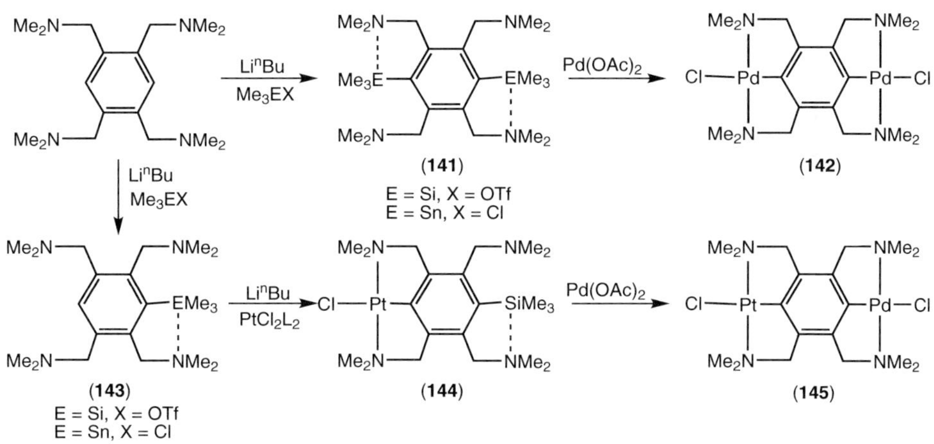

Scheme 3.38

lithium precursors [62c], and the synthesis of Sn(II) complexes **(140)** has also been reported by this group (Figure 3.11) [62d].

The dinuclear complex **142** was prepared from the tetraamine (Scheme 3.38) through the bis-silyl or bis-stannyl derivatives **(141)**. Direct cyclopalladation is not successful in this case. This alternative synthetic pathway, which combines transmetallation (from Li to Si or Sn) and the palladation of the C—Si or C—Sn bonds, proceeds under mild conditions, since all steps are performed at room temperature [63a]. By careful choice of reaction conditions and starting compounds, monosilyl (or stannyl) species **(143)** can be isolated, which are very interesting for the synthesis of mixed Pd—Pt complexes **(145)**. The inertness of the C_{aryl}—Si bond towards the Pt atom is quite remarkable, as it can not undergo electrophilic cycloplatination, in the synthesis of **144**. Subtle effects, such as the presence of intramolecular Si—N coordination, seem to be closely related with this lack of aryl transfer. In contrast, the presence of a silyl group at a given position seems to be determinant in achieving palladation at this position; that is, the silyl group is a powerful directing group [63]. This fact was already known and is quite relevant when a precise metallation is required [64].

A Me_3Si-substituted aryl ligand is selectively palladated at the position of the silyl fragment, using $Li_2[PdCl_4]$ or $Pd(OAc)_2$ as Pd sources and methanol as solvent. This selectivity can be reversed, favoring the activation of C_{aryl}—H bonds instead of the C_{aryl}—Si bonds, through a simple change of solvent and performing the reactions in CH_2Cl_2. In summary, the use of organosilicon derivatives not only provides stable, easy to handle, transmetallating reagents, but also allows the control of the palladation position (C—Si versus C—H activation) as a function of the reaction solvent.

Another class of organometallic complexes widely used as transmetallating reagents are mercury derivatives. They share some properties with those reported for Sn or Si reagents, such as stability and mild transmetallating conditions, although their high toxicity is a very important drawback. Organomercury(II) compounds are known for many classical C,N-ligands, and there are two main synthetic methods to prepare them: (i) transmetallation from a lithium derivative in an inert solvent; (ii) direct mercuriation, by reaction of the free ligand with $Hg(OAc)_2$ in a polar and protic solvent. Compounds **146–148** (Figure 3.12) have been obtained through method (i) [65a] while (ii) is the best choice for **149–151** [65b]. Zinc(II) derivatives similar to **151** are also used as transmetallating reagents [65c].

Transfer of the metallated ligand from the mercury compound to the Pd center is usually performed using $Pd(OAc)_2$ or $PdCl_2L_2$ [65d], although transfers to Pd(0) derivatives have also been reported [65e]. For instance, naphthyl derivative **153** can be obtained after refluxing **152** with $PdCl_2(SEt_2)_2$ in benzene, while a ferrocenyl derivative **(155)** is obtained from **154** by simple stirring at 25 °C (Scheme 3.39).

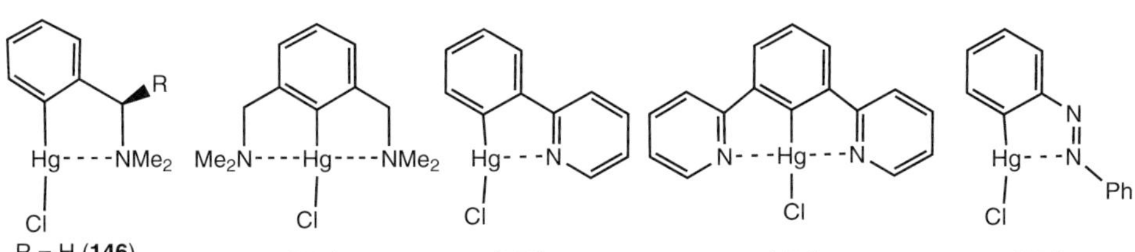

R = H (**146**)
R = Me (**147**)
(**148**) (**149**) (**150**) (**151**)

Figure 3.12 Some examples of organomercury(II) compounds.

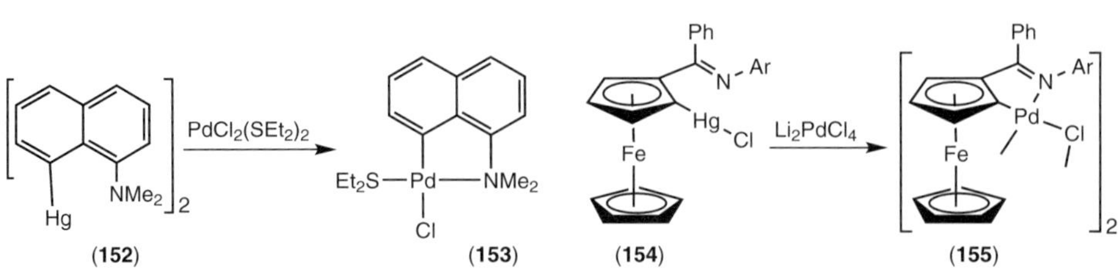

(**152**) (**153**) (**154**) (**155**)

Scheme 3.39

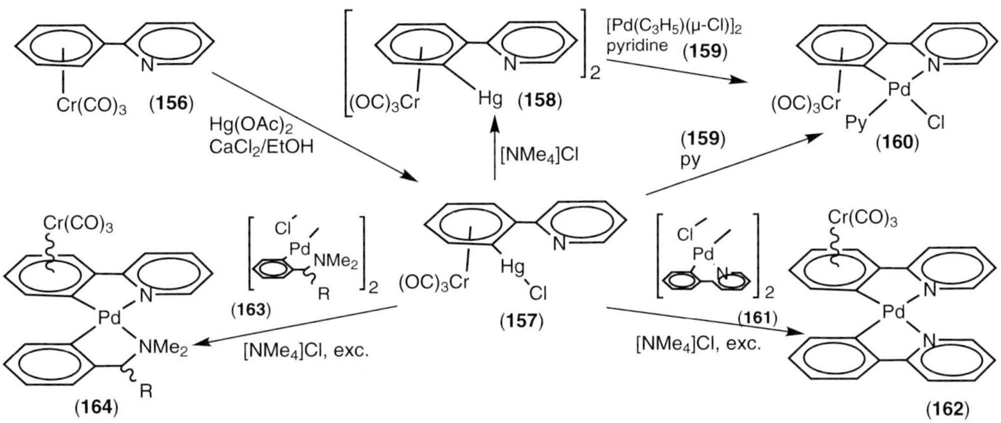

Scheme 3.40

In most cases, transmetallation from Hg to Pd proceeds under very mild conditions, which allows the synthesis of complexes not accessible by other routes (direct C—H activation or oxidative addition). This is the case for the substrates shown in Scheme 3.40, which were developed by Pfeffer, Djukic and coworkers [66]. Direct mercuriation of **156** by reaction with $Hg(OAc)_2$ gives chloromercuriated **157**, which can be conveniently "symmetrized" by reaction with an excess of $[NMe_4]Cl$, giving the bis-aryl derivative **158**. The reaction of **158** with allyl complex **159** gives orthopalladated **160**, which could not be obtained by other methods. The transmetallation occurs under very mild conditions (acetone, −15 °C to room temperature), and can be performed similarly from the mono-aryl **(157)** or bis-aryl **(158)** derivatives, although the bis-aryl seems to be more convenient.

This method has also successfully applied to the synthesis of bis-chelates **162** by reaction of **157** with different chlorine-bridge complexes **(161)**. If enantiomerically pure chiral complexes **(163)** were used, planar chiral derivatives **(164)** can be obtained under the same reaction conditions. Complexes **164**, obtained as mixtures of diastereomers, can be separated chromatographically and show three types of chirality: helical, planar and centered.

Organomercury(II) complexes have been used extensively also by Vicente *et al.* to prepare a wide collection of palladacycles not accessible by other methods. Some of them are shown in Schemes 3.41 and 3.42. For instance, the formyl compound $HC(O)C_6H_2(OMe)_3$-3,4,5 does not react with $Pd(OAc)_2$, but it can be easily mercuriated, giving **165**. It is widely accepted that the mercuriation occurs mainly through an electrophilic substitution S_EAr pathway, balanced with radical mechanisms [66a]. In accord with the S_EAr process, the presence of electron-releasing OMe substituents counterbalances the formyl deactivating effect. Complex **165** reacts with different precursors of Pd(II) to give **166** or its isomerized form **167** [67]. The unexpected formation of **167** is similar to the rearrangement described for C,S-palladacycles **24** and **25** (Scheme 3.10). When the formyl unit was condensed with

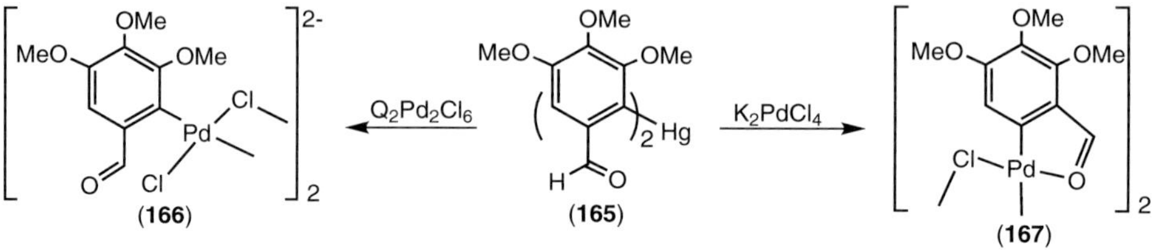

Scheme 3.41

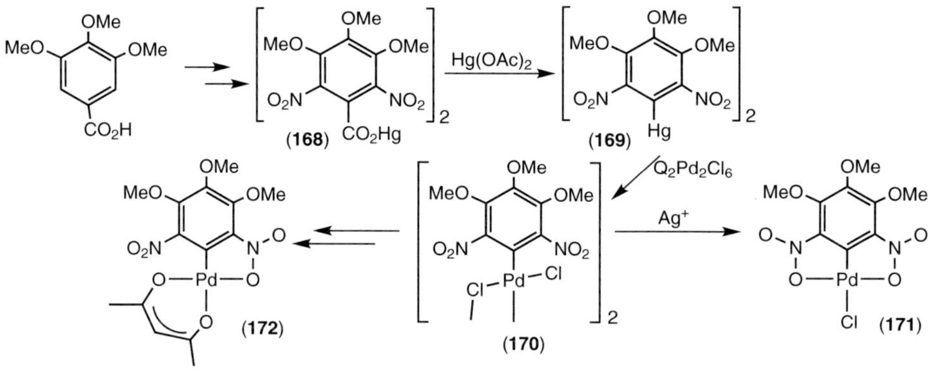

Scheme 3.42

other functional groups (amines, diamines), cyclopalladated complexes were obtained, but without rearrangement. In the same way, acetophenone derivatives have been mercuriated for the first time by direct reaction with $Hg(OAc)_2$, and transmetallated to Pd centers without rearrangement [67b].

Dinitroaryl Pd(II) complexes have been prepared from the mercurial **169**, which, in turn, was synthesized from the acid followed by decarboxylation of benzoate **168**. The dinitroaryl group could act as a σ–aryl ligand, as a C,O-chelate and as a OCO pincer; examples of the three bonding modes have been described **(170–172)** (Scheme 3.42) [68].

The examples described up to now show clearly the tolerance of organomercury compounds to the presence of functional groups on the aryl rings, not only during the synthesis of the Hg compounds but also during the transmetallation step. To the previously reported formyl, keto, imine and nitro units, other functional groups can be added [69]. For instance, aryl rings containing strongly deactivating amides [69a] (Scheme 3.43) can be easily mercuriated **(173)**, allowing the synthesis of **174**, the first carbamoylaryl palladacycle.

The mercuriation of aryl rings, and the subsequent transmetallation, can be performed even if they possess one or two very strongly deactivating substituents (with respect to a S_EAr mechanism), simply by changing the reaction conditions to a more electrophilic scenario. Scheme 3.44 shows a very relevant example [69b].

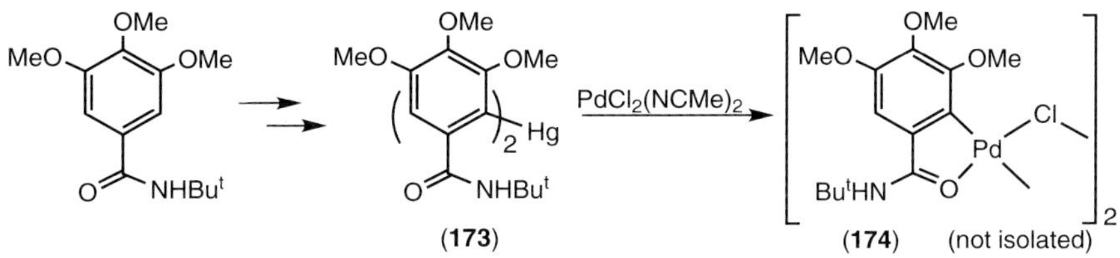

Scheme 3.43

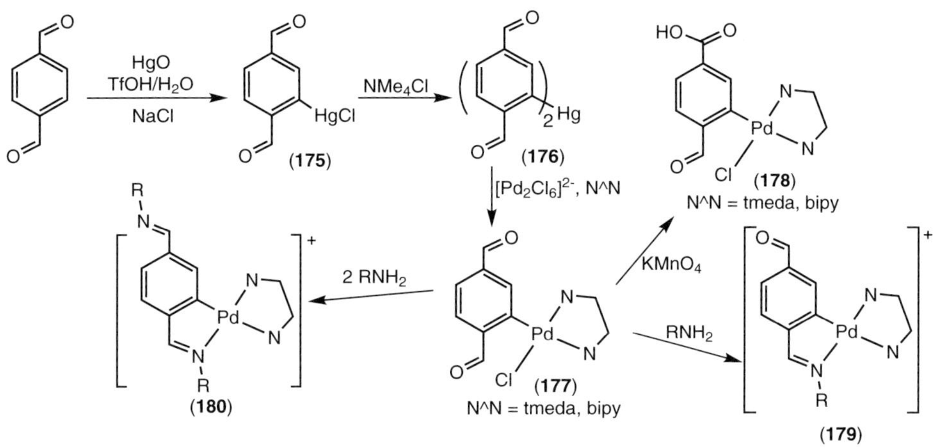

Scheme 3.44

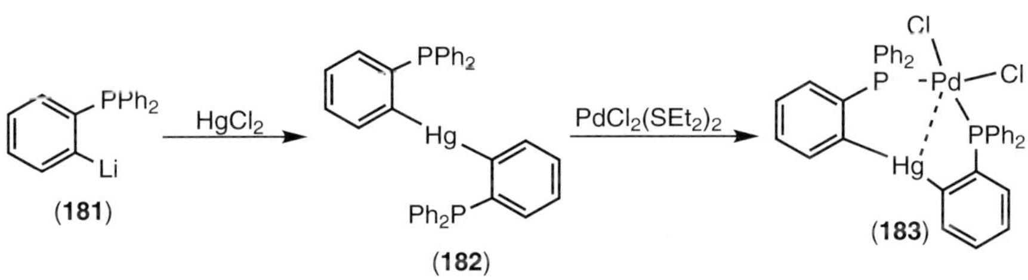

Scheme 3.45

Terephthaldehyde, which contains two formyl substituents in para positions, reacts with HgO in a solution of triflic acid at 90 °C to give mercurial **175** in moderate to low yields. Compound **175** can be symmetrized **(176)** and transmetallated to the Pd(II) center to give stable σ-aryl complexes **(177)**. Complexes **177** show a very interesting reactivity, since the formyl groups can be oxidized to acids **(178)** with $KMnO_4$, and they can be condensed with one or two equivalents of amines to give the corresponding palladacycles **179** and **180**. Bennett *et al.* have reported the synthesis and reactivity of the mercury compound **182**, prepared from **181** (Scheme 3.45) [70]. The reactivity of **182** towards Pd substrates does not progress with complete transfer of the metallated ligand, and so **182** can be viewed as a P,P′-metalloligand.

In summary, the oxidative addition and transmetallation reactions are very useful synthetic procedures. They are versatile methods, with regioselective orientation, and they allow different approaches to the target complexes. The two reactions are complementary to the well-known cyclopalladation reaction, and offer synthetic tools of wide scope with which to prepare the desired compounds.

References

1 Godula, K. and Sames, D. (2006) *Science*, **312**, 67.

2 (a) Mohr, F., Privér, S.H., Bhargava, S.K. and Bennett, M.A. (2006) *Coordination Chemistry Reviews*, **250**, 1851.
(b) Dunina, V.V. and Gorunova, O.N. (2005) *Russian Chemical Reviews*, **74**, 871.
(c) Omae, I. (2004) *Coordination Chemistry Reviews*, **248**, 995.

3 Dupont, J., Consorti, C.S. and Spencer, J. (2005) *Chemical Reviews*, **105**, 2527.

4 Crabtree, R.H. (2005) *The Organometallic Chemistry of the Transition Metals*, 4th edn, John Wiley & Sons, Inc., Hoboken, New Jersey.

5 Maitlis, P., Espinet, P. and Russell, M.J.H. (1982) *Comprehensive Organometallic Chemistry*, Vol. **6** (eds G. Wilkinson, F.G.A. Stone and E.W. Abel), Pergamon Press, Oxford, Chapter 38.1.

6 (a) Coulson, D.R. (1990) *Inorganic Syntheses*, **28**, 107.
(b) Rettig, M.F. and Maitlis, P.M. (1990) *Inorganic Syntheses*, **28**, 110.
(c) Takahashi, Y., Ito, T., Sakai, S. and Ishii, U. (1970) *Journal of the Chemical Society D–Chemical Communications*, 1065.

7 Ukai, T., Kawazura, H., Ishii, Y., *et al.* (1974) *Journal of Organometallic Chemistry*, **65**, 253.

8 Stille, J.K. and Lau, K.S.Y. (1977) *Accounts of Chemical Research*, **10**, 434.

9 (a) Roy, A.M. and Hartwig, J.F. (2004) *Organometallics*, **23**, 194.
(b) Alcázar-Román, L.M. and Hartwig, J.F. (2002) *Organometallics*, **21**, 491.

10 Perutz, R.N. and Braun, T. (2006) *Comprehensive Organometallic Chemistry III*, Vol. **1** (eds R.H. Crabtree and M. Mingos; volume ed. G. Parkin), Elsevier, Oxford, Chap. 1.26.

11 (a) Anderson, C.M., Puddephatt, R.J., Ferguson, G. and Lough, A.J. (1989) *Journal of the Chemical Society D–Chemical Communications*, 1297.
(b) Anderson, C.M., Crespo, M., Ferguson, G., *et al.* (1992) *Organometallics*, **11**, 1177.

12 (a) van Beek, J.A.M., van Koten, G., Wehman-Ooyevaar, I.C.M., *et al.* (1991) *Journal of the Chemical Society–Dalton Transactions*, 883.
(b) Casado, A.L. and Espinet, P. (1998) *Organometallics*, **17**, 954.

13 Espinet, P. and Echavarren, A.M. (2004) *Angewandte Chemie–International Edition*, **43**, 4704.

14 (a) Phan, N.T.S., Sluys, M.V.D. and Jones, C.W. (2006) *Advanced Synthesis Catalysis*, **348**, 609.
(b) Amatore, C., Jutand, A. and Thuilliez, A. (2001) *Organometallics*, **20**, 3241.
(c) Amatore, C., Godin, B., Jutand, A. and Lemaître, F. (2007) *Chemistry–A European Journal*, **13**, 2002.

15 Malinakova, H.C. (2004) *Chemistry–A European Journal*, **10**, 2636.

16 (a) Newkome, G.R., Puckett, W.E., Gupta, V.K. and Kiefer, G.E. (1986) *Chemical Reviews*, **86**, 451.
(b) Dunina, V.V. and Gorunova, O.N. (2004) *Russian Chemical Reviews*, **73**, 309.

17(a) Clark, P.W. and Dyke, S.F. (1983) *Journal of Organometallic Chemistry*, **259**, C17.
(b) Clark, P.W. and Dyke, S.F. (1985) *Journal of Organometallic Chemistry*, **281**, 389.
(c) Cope, A.C. and Friedrich, E.C. (1968) *Journal of the American Chemical Society*, **90**, 909.

18 Pfeffer, M., Sutter-Beydoun, N., De Cian, A. and Fischer, J. (1993) *Journal of Organometallic Chemistry*, **453**, 139.

19 (a) Vila, J.M., Gayoso, M., Pereira, M.T., *et al.* (1997) *Journal of Organometallic Chemistry*, **532**, 171.
(b) Getty, A.D. and Goldberg, K.I. (2001) *Organometallics*, **20**, 2545.

20 Beeby, A., Bettington, S., Fairlamb, I.J.S., *et al.* (2004) *New Journal of Chemistry*, **28**, 600.

21 Driver, M.S. and Hartwig, J.F. (1997) *Journal of the American Chemical Society*, **119**, 8232.

22 (a) Clark, P.W., Dyke, S.F., Smith, G. and Kennard, C.H.L. (1987) *Journal of Organometallic Chemistry*, **330**, 447.
(b) Albert, J. Gómez, M., Granell, J. and Sales, J. (1990) *Organometallics*, **9**, 1405.

23 (a) Solé, D., Vallverdú, L., Solans, X., *et al.* (2003) *Journal of the American Chemical Society*, **125**, 1587.
(b) Solé, D., Vallverdú, L., Solans, X., *et al.* (2004) *Organometallics*, **23**, 1438.

24 (a) Vicente, J., Abad, J.A., Frankland, A.D. and Ramírez de Arellano, M.C. (1999) *Chemistry–A European Journal*, **5**, 3066.
(b) Vicente, J., Abad, J.A., Martínez-Viviente, E., *et al.* (2000) *Organometallics* **19**, 752.
(c) Vicente, J., Abad, J.A., López-Serrano, J., *et al.* (2005) *Organometallics*, **24**, 5044.

25 (a) Vicente, J., Abad, J.A., Hernández-Mata, F.S. and Jones, P.G. (2002) *Journal of the American Chemical Society*, **124**, 3848.
(b) Spencer, J., Pfeffer, M., De Cian, A. and Fischer, J. (1995) *Journal of Organic Chemistry*, **60**, 1005.

26 (a) Aarif, A.M., Estevan, F., García-Bernabé, A., *et al.* (1997) *Inorganic Chemistry*, **36**, 6472.
(b) Estevan, F., García-Bernabé, A., Lahuerta, P., *et al.* (2000) *Inorganic Chemistry*, **39**, 5964.
(c) Gibson, S., Foster, D.F., Eastham, G.E., *et al.* (2001) *Chemical Communications*, 779.
(d) Friedlein, F.K., Kromm, K., Hampel, F. and Gladysz, J.A. (2006) *Chemistry–A European Journal*, **12**, 5267.

27 (a) Alsters, P.L., Boersma, J., Smeets, W.J.J., *et al.* (1993) *Organometallics* **12**, 1639.
(b) Widenhoefer, R.A., Zhong, H.A. and Buchwald, S.L. (1997) *Journal of the American Chemical Society* **119**, 6787.
(c) Fernández-Rivas, C., Cárdenas, D.J., Martín-Matute, B., *et al.* (2001) *Organometallics* **20**, 2998.

28 (a) Dyker, G. (1997) *Chemische Berichte*, **130**, 1567.
(b) Dyker, G. (1999) *Angewandte Chemie–International Edition*, **38**, 1698.

29 (a) Cárdenas, D.J., Mateo, C. and Echavarren, A.M. (1994) *Angewandte Chemie–International Edition in English*, **33**, 2445.
(b) Mateo, C., Cárdenas, D.J., Fernández-Rivas, C. and Echavarren, A.M. (1996) *Chemistry–A European Journal*, **2**, 1596.

30 (a) Portscheller, J.L. and Malinakova, H.C. (2002) *Organic Letters*, **4**, 3679.
(b) Portscheller, J.L., Lilley, S.E. and Malinakova, H.C. (2003) *Organometallics*, **22**, 2961.

31 (a) Catellani, M. (2002) *Pure and Applied Chemistry*, **74**, 63.
(b) Catellani, M. (2003) *Synlett*, 298.
(c) Catellani, M., Motti, E., Faccini, F. and Ferraccioli, R. (2005) *Pure and Applied Chemistry*, **77**, 1243.

32 Canty, A.J. (1993) *Platinum Metals Review*, **37**, 2.

33 (a) Deprez, N.R. and Sanford, M.S. (2007) *Inorganic Chemistry*, **46**, 1924.
(b) Canty, A.J., Rodemann, T., Skelton, B.W. and White, A.H. (2006) *Organometallics*, **25**, 3996.

34 (a) Alsters, P.L., Engel, P.F., Hogerheide, M.P., *et al.* (1993) *Organometallics*, **12**, 1831.
(b) Cámpora, J., Palma, P., del Río, D., *et al.* (2003) *Organometallics*, **22**, 3345.
(c) Cámpora, J., Palma, P., del Río, D., *et al.* (2004) *Chemical Communications*, 1490.

35 Canty, A.J., Hoare, J.L., Patel, J., *et al.* (1999) *Organometallics*, **18**, 2660.

36 (a) Rietveld, M.H.P., Grove, D.M. and van Koten, G. (1997) *New Journal of Chemistry*, **21**, 751.
(b) Albrecht, M. and van Koten, G. (2001) *Angewandte Chemie–International Edition*, **40**, 3751.
(c) van der Boom, M.E. and Milstein, D. (2003) *Chemical Reviews*, **103**, 1759.

37 Moulton, C.J. and Shaw, B.L. (1976) *Journal of the Chemical Society–Dalton Transactions*, 1020.

38 Miyazaki, F., Yamaguchi, K. and Shibasaki, M. (1999) *Tetrahedron Letters*, **40**, 7379.

39 Denmark, S.E., Stavenger, R.A., Faucher, A.M. and Edwards, J.P. (1997) *Journal of Organic Chemistry*, **62**, 3375.

40 (a) Alsters, P.L., Baesjou, P.J., Janssen, M.D., *et al.* (1992) *Organometallics*, **11**, 4124.
(b) Tsubomura, T., Tanihata, T., Yamakawa, T., *et al.* (2001) *Organometallics*, **20**, 3833.
(c) Albrecht, M. Kocks, B.M. Spek, A.L. and van Koten, G. (2001) *Journal of Organometallic Chemistry*, **624**, 271.
(d) Guillena, G., Rodríguez, G. and van Koten, G. (2002) *Tetrahedron Letters*, **43**, 3895.

41 (a) Rodríguez, G., Albrecht, M., Schoenmaker, J., *et al.* (2002) *Journal of the American Chemical Society*, **124**, 5127.
(b) Slagt, M.Q., Rodríguez, G., Grutters, M.M.P., *et al.* (2004) *Chemistry–A European Journal*, **10**, 1331.
(c) Rodríguez, G., Lutz, M., Spek, A.L. and van Koten, G. (2002) *Chemistry–A European Journal*, **8**, 45.

42 (a) Takenaka, K. and Uozumi, Y. (2004) *Advanced Synthesis Catalysis*, **346**, 1693.
(b) Takenaka, K., Minakawa, M. and Uozumi, Y. (2005) *Journal of the American Chemical Society*, **127**, 12273.

43 (a) Osakada, K. (2003) Transmetallation, in *Current Methods in Inorganic Chemistry*, Vol. **3** (eds H. Kurosawa and A. Yamamoto), Elsevier Science, p. 233.
(b) Osakada, K. and Yamamoto, T. (2000) *Coordination Chemistry Reviews*, **198**, 379.

44 (a) Farina, V. (2004) *Advanced Synthesis Catalysis*, **346**, 1553.
(b) Amatore, C., Bashoun, A.A., Jutand, A., *et al.* (2003) *Journal of the American Chemical Society*, **125**, 4212.
(c) Casares, J.A., Espinet, P. and Salas, G. (2002) *Chemistry–A European Journal*, **8**, 4844.

45 (a) Nova, A., Ujaque, G., Maseras, F., *et al.* (2006) *Journal of the American Chemical Society*, **128**, 14571.
(b) Alvarez, R., Nieto Faza, O., Silva López, C. and de Lera, A.R. (2006) *Organic Letters*, **8**, 35.
(c) Sumimoto, M., Iwane, N., Takehama, T. and Sakaki, S. (2004) *Journal of the American Chemical Society* **126**, 10457.

46 Wakefield, B.J. (1974) *The Chemistry of Organolithium Compounds*, Pergamon Press, Oxford.

47 (a) Grove, D.M., van Koten, G., Louwen, J.N., *et al.* (1982) *Journal of the American Chemical Society*, **104**, 6609.
(b) Jones, F.N. and Hauser, C.R. (1962) *Journal of Organic Chemistry*, **27**, 701.
(c) Jastrzebski, J.T.B.H. and van Koten, G. (1988) *Inorganic Syntheses*, **26**, 150.
(d) Gay, R.L. and Hauser, C.R. (1967) *Journal of the American Chemical Society*, **89**, 2297.
(e) Marquarding, D., Klusacek, H., Gokel, G., *et al.* (1970) *Journal of the American Chemical Society*, **92**, 5389.
(f) Sammakia, T. and Latham, H.A. (1996) *Journal of Organic Chemistry*, **61**, 1629.

48 (a) Rietveld, M.H.P., Wehman-Ooyevaar, I.C.M., Kapteijn, G.M., *et al.* (1994) *Organometallics*, **13**, 3782.
(b) Jansen, M.D., Corsten, M.A., Spek, A.L., *et al.* (1996) *Organometallics*, **15**, 2810.

49 Gruter, G.J.M., van Klink, G.P.M., Akkerman, O.S. and Bickelhaupt, F. (1995) *Chemical Reviews*, **95**, 2405.

50 (a) Winkler, M., Lutz, M. and Müller, G. (1994) *Angewandte Chemie–International Edition in English*, **33**, 2279.
(b) Abicht, H.P. and Issleib, K. (1976) *Zeitschrift fur Anorganische Chemie*, **422**, 237.
(c) Abicht, H.P. and Issleib, K. (1980) *Journal of Organometallic Chemistry*, **185**, 265.
(d) Abicht, H.P. and Issleib, K. (1985) *Journal of Organometallic Chemistry*, **289**, 201.
(e) Müller, G., Abicht, H.P., Waldkircher, M., *et al.* (2001) *Journal of Organometallic Chemistry*, **622**, 121.

51 (a) Bennett, M.A., Bhargava, S.K., Ke, M. and Willis, A.C. (2000) *Journal of the Chemical Society–Dalton Transactions*, 3537.
(b) Bennett, M.A., Bhargava, S.K., Bond, A.M., *et al.* (2004) *Inorganic Chemistry*, **43**, 7752.

52 (a) Longoni, G., Fantucci, P., Chini, P. and Canziani, F. (1972) *Journal of Organometallic Chemistry*, **39**, 413.
(b) Arlen, C., Pfeffer, M., Bars, O. and Grandjean, D. (1983) *Journal of the Chemical Society–Dalton Transactions*, 1535.

53 Valk, J.M., Maasarani, F., van der Sluis, P., *et al.* (1994) *Organometallics*, **13**, 2320.

54 (a) Stark, M.A., Jones, G. and Richards, C.J. (2000) *Organometallics*, **19**, 1282.
(b) Chadwick, S.T., Ramirez, A., Gupta, L. and Collum, D.B. (2007) *Journal of the American Chemical Society*, **129**, 2259.

55 Slocum, D.W., Jennings, C.A., Engelmann, T.R., *et al.* (1971) *Journal of Organic Chemistry*, **36**, 377.

56 (a) van Koten, G., Timmer, K., Noltes, J.G. and Spek, A.L. (1978) *Journal of the Chemical Society D–Chemical Communications*, 250.
(b) Jung, I.G., Son, S.U., Park, K.H., *et al.* (2003) *Organometallics*, **22**, 4715.
(c) see Refs. 40b and 40c.

57 (a) Chanda, N. and Sharp, P.R. (2007) *Organometallics*, **26**, 1635.
(b) Cámpora, J., López, J.A., Palma, P., *et al.* (2001) *Inorganic Chemistry*, **40**, 4116.

58 Nishikata, T., Yamamoto, Y. and Miyaura, N. (2004) *Organometallics*, **23**, 4317.

59 Contel, M., Stol, M., Casado, M.A., van Klink, *et al.* (2002) *Organometallics*, **21**, 4556.

60 (a) Milstein, D. and Stille, J.K. (1979) *Journal of the American Chemical Society*, **101**, 4992.
(b) Varga, R.A., Rotar, A., Schürmann, M. , *et al.* (2006) *European Journal of Inorganic Chemistry*, 1475.
(c) Benaglia, M., Toyota, S., Woods, C.R. and Siegel, J.S. (1997) *Tetrahedron Letters*, **38**, 4737.

61 (a) Motoyama, Y., Kawakami, H., Shimozono, K., *et al.* (2002) *Organometallics*, **21**, 3408.
(b) Nishiyama, H., Matsumoto, M., Matsukura, T., *et al.* (1985) *Organometallics*, **4**, 1911.

62 (a) van Koten, G., Jastrzebski, J.T.B.H., Noltes, J.G., *et al.* (1978) *Journal of the American Chemical Society*, **100**, 5021.
(b) van Koten, G., Jastrzebski, J.T.B.H., Noltes, J.G., *et al.* (1978) *Journal of Organometallic Chemistry*, **148**, 233.
(c) van Koten, G., Jastrzebski, J.T.B.H., Noltes, J.G., *et al.* (1980) *Journal of the Chemical Society–Dalton Transactions*, 1352.
(d) Jastrzebski, J.T.B.H., van der Schaaf, P.A., Boersma, J., *et al.* (1989) *Organometallics*, **8**, 1373.

63 (a) Steenwinkel, P., Jastrzebski, J.T.B.H., Deelman, B.J., *et al.* (1997) *Organometallics*, **16**, 5486.
(b) Steenwinkel, P., Gossage, R.A. and van Koten, G. (1998) *Chemistry–A European Journal*, **4**, 759.

64 Eaborn, C. (1975) *Journal of Organometallic Chemistry*, **100**, 43.

65 (a) Attar, S., Nelson, J.H. and Fischer, J. (1995) *Organometallics*, **14**, 4776.
(b) Soro, B., Stoccoro, S., Minghetti, G., *et al.* (2005) *Organometallics*, **24**, 53.
(c) Valk, J.M., Boersma, J. and van Koten, G. (1996) *Organometallics*, **15**, 4366.
(d) Wehman, E., van Koten, G., Jastrzebski, J.T.B.H., Ossor, H. and Pfeffer, M., (1988) *Journal of the Chemical Society–Dalton Transactions*, 2975.
(e) Sokolov, V.I., Bashilov, V.V., Musaev, A.A. and Reutov, O.A. (1982) *Journal of Organometallic Chemistry*, **225**, 57.

66 (a) Berger, A., de Cian, A., Djukic, J.P., *et al.* (2001) *Organometallics* **20**, 3230.
(b) Djukic, J.P., Berger, A., Duquenne, M., *et al.* (2004) *Organometallics* **23**, 5757.

67 (a) Vicente, J., Abad, J.A., Stiakaki, M.A. and Jones, P.G. (1991) *Journal of the Chemical Society D–Chemical Communications*, 137.
(b) Vicente, J., Abad, J.A., Gil–Rubio, J., *et al.* (1993) *Organometallics*, **12**, 4151.

68 Vicente, J., Arcas, A., Blasco, M.A., *et al.* (1998) *Organometallics*, **17**, 5374 and references given therein.

69 (a) Vicente, J., Abad, J.A., Shaw, K.F., *et al.* (1997) *Organometallics*, **16**, 4557.
(b) Vicente, J., Abad, J.A., Rink, B., *et al.* (1997) *Organometallics*, **16**, 5269.

70 Bennett, M.A., Contel, M., Hockless, D.C.R., *et al.* (2002) *Inorganic Chemistry*, **41**, 844.

4
Synthesis via Other Synthetic Solutions

Mario Roberto Meneghetti

4.1
Introduction

As observed in previous chapters, synthetic organometallic chemists can now reach a vast number of different palladacycle complexes, making use of typical organometallic reactions like C—H activation, oxidative addition and transmetallation reactions. To synthesize palladacycles via these reactions, one needs to have at hand ligand precursors with the same molecular skeleton as the desired chelating ligand. However, it is also possible to obtain palladacycles from other type of reactions, whereby the molecular skeleton of the ligand precursor does not belong to the ligand itself but is, in fact, built *in situ* during palladacycle formation. This chapter discusses some of the most representative alternative methodologies for palladacycle complex formation.

4.2
Synthesis of Palladacycles via Nucleophile-Palladation Reaction of Olefins or Alkynes Bearing Electron-Donor Heteroatoms

An elegant way to synthesize palladacycle complexes is based on nucleophile-palladation reactions of unsaturated organic substrates that bear electron-donor heteroatoms. In general, the reaction proceeds first by coordination of the olefin or alkyne, via the electron donor group and the unsaturated moiety (C=C or C≡C bond), at the electrophilic Pd(II). This step is followed by a regioselective nucleophilic attack at one of the unsaturated carbons, leading, respectively, to the more stable σ-alkyl Pd(II) or σ-vinyl Pd(II) palladacycle complex (Scheme 4.1). In general, this nucleophilic addition at the C=C or C≡C bond, coordinated at the metal center, is not typically drawn by electronic or steric effects but mostly by the formation of the more thermodynamically stable five-membered palladacyclic ring, over its six- or four-membered counterparts [1].

Palladacycles: Synthesis, Characterization and Applications. Edited by Jairton Dupont and Michel Pfeffer

ISBN: 978-3-527-31781-3

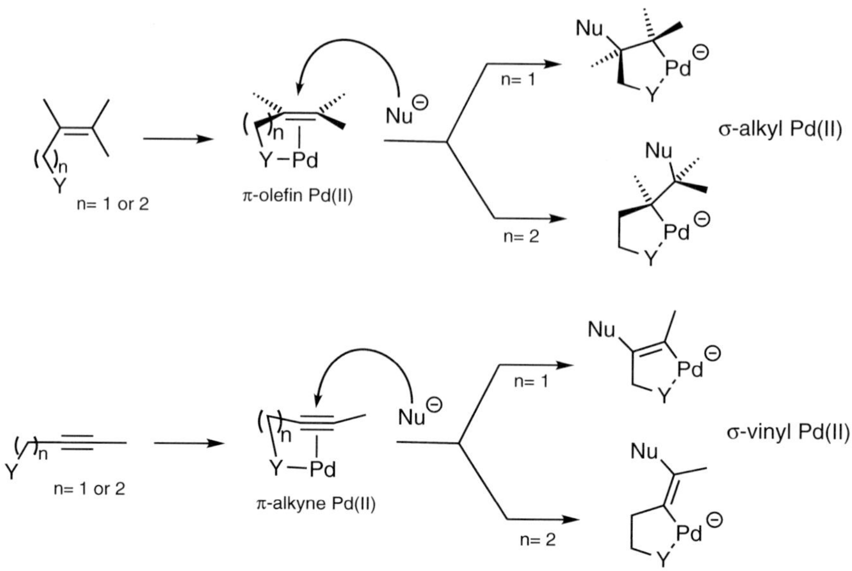

Scheme 4.1

These reactions are normally named according to the nature of the nucleophilic species employed, although actually only few are known, for example, alkoxy-, carbo-, and chloropalladation reactions.

Notably, palladacycles produced by this methodology may be considered as models of intermediate organometallic species of a series of catalytic nucleophile-palladation reactions [2, 3]. In fact, activation of olefins and alkynes by coordination to palladium(II) complexes to undergo nucleophilic addition reactions is among the most studied and used synthetic strategies to catalyze organic transformations [4, 5].

4.2.1
Alkoxypalladation Reaction

The first examples of palladacycles synthesized by nucleophile-palladation were reported by Cope and coworkers at the end of the 1960s. They observed that, in alcoholic media, tertiary allylic amines undergo alkoxypalladation in the presence of Li_2PdCl_4 or $PdCl_2$ to give, in good yields, chloride-bridged forms of σ-alkylpalladacycles complexes (Scheme 4.2). As normally observed in the chemistry of palladacycles, these dimeric complexes are easily cleaved by addition of L-type ligands, which are located trans to the Pd—N bond (Scheme 4.3) [6].

Takahashi and coworkers have prepared analogous σ-alkyl palladacycles complexes of type **3** and **4** from allylic sulfides and sulfoxides derivatives, respectively (Schemes 4.4 and 4.5). However, only palladacycles obtained from allylic sulfides were isolated with high yields [7].

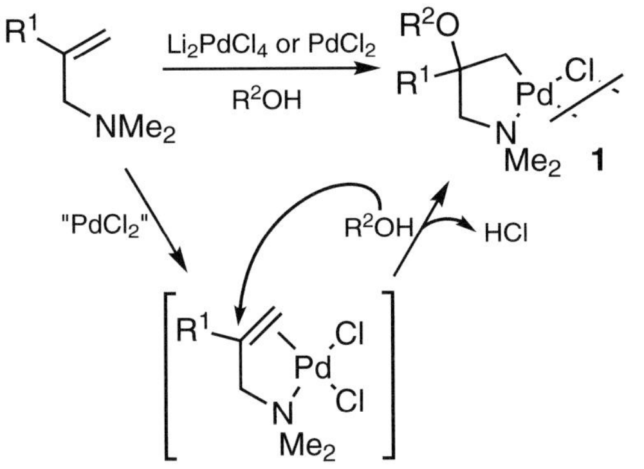

Complex **1**	R^1	R^2
a	H	Me
b	Me	Me
c	Et	Me
d	$HOCH_2CH_2$	Me

Scheme 4.2

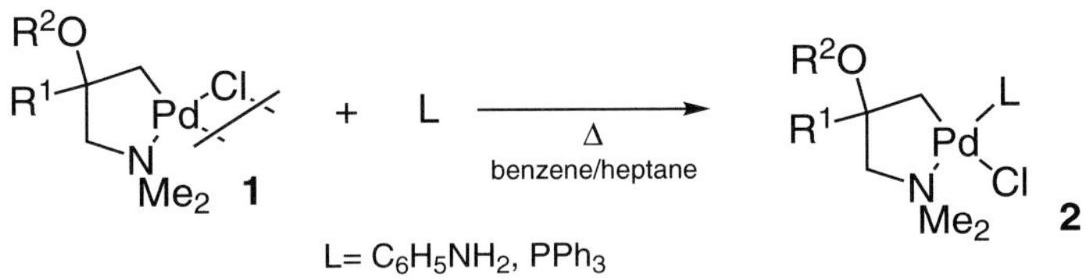

Scheme 4.3

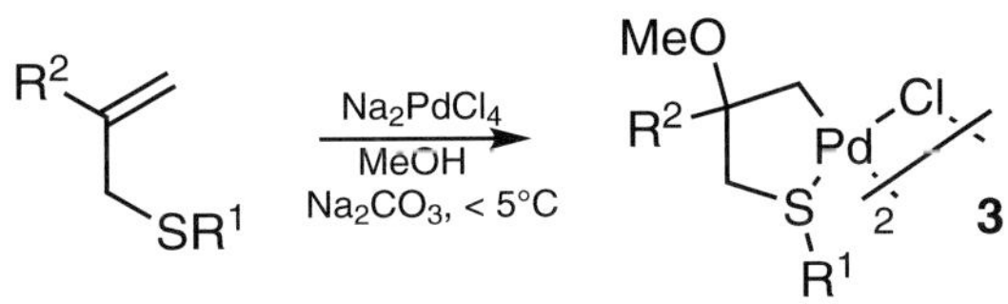

Complex **3**	R^1	R^2
a	*t*-Bu	Me
b	*t*-Bu	H
c	Ph	Me
d	Ph	H
e	Et	Me
f	Et	H

Scheme 4.4

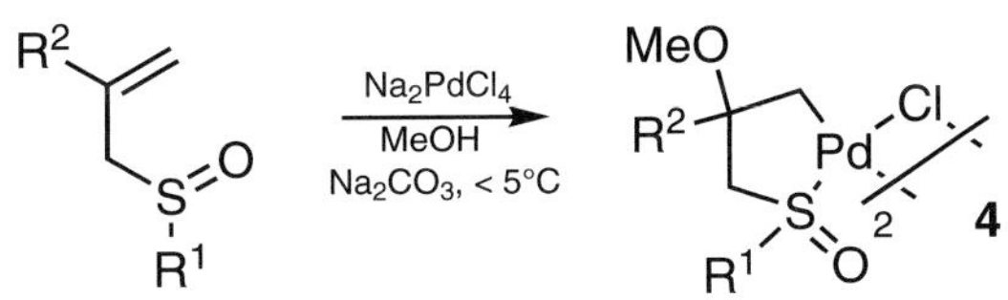

Complex **4**	R^1	R^2
a	Et	H
b	Et	Me
c	Ph	H
d	*t*-Bu	Me

Scheme 4.5

Oxime O-allyl ethers also undergo alkoxypalladation; however, nucleophilic attack occurs at the terminus of the double bond of the allylic fragment (Scheme 4.6) [8]. At about the same time, Holton and coworkers reported the formation of palladacycles by alkoxypalladation of homoallylic amines and sulfides (Scheme 4.7) [9].

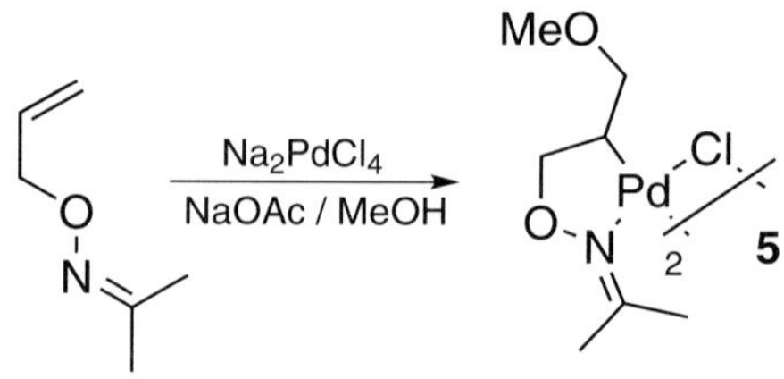

Scheme 4.6

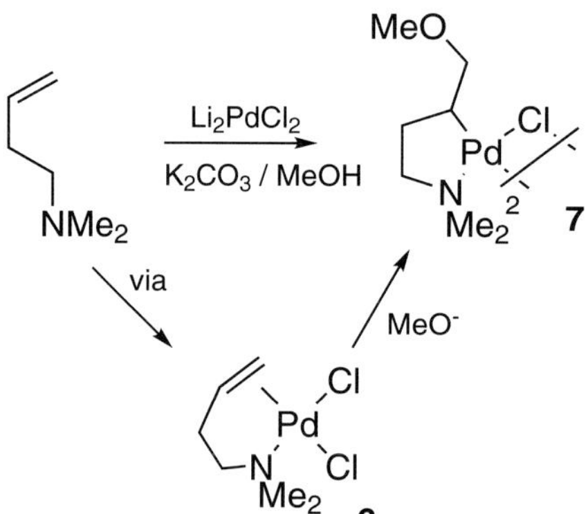

Scheme 4.7

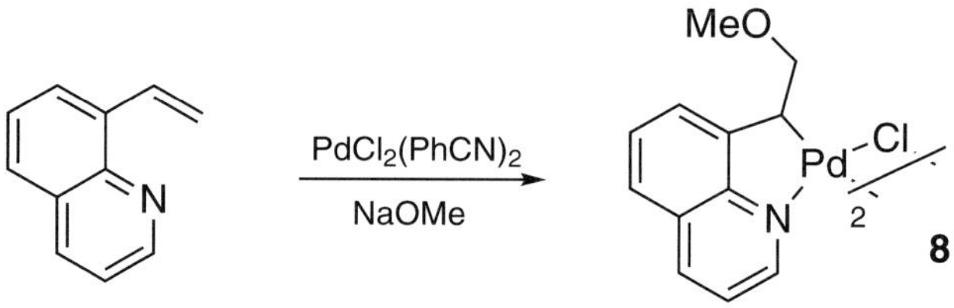

Scheme 4.8

More recently, Dupont and coworkers have extended this type of reaction to 8-vinylquinoline, employing the methoxide anion as nucleophilic agent (Scheme 4.8) [10]. The reaction is very regioselective, leading to compound **8** as the single product.

The same group has also checked that alkoxypalladation reactions can occur with high degree of diastereoselectivity (Scheme 4.9). The structure of the monomeric derivative **10** was determined by X-ray diffraction analysis. The methoxide anion is added to the double bond, leading to the *anti* isomer, in respect of the OMe and Me groups, as the sole product [11].

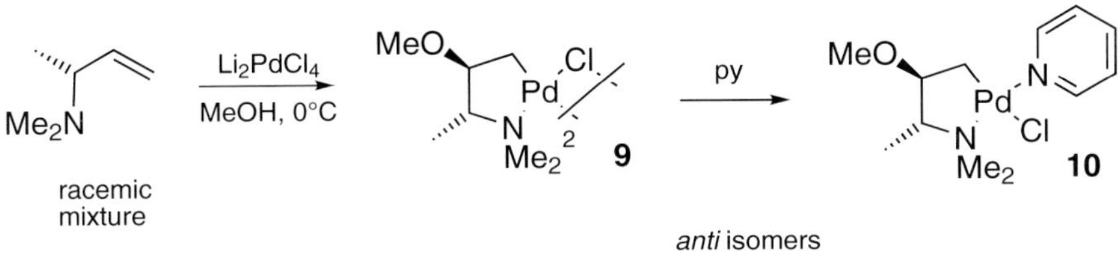

Scheme 4.9

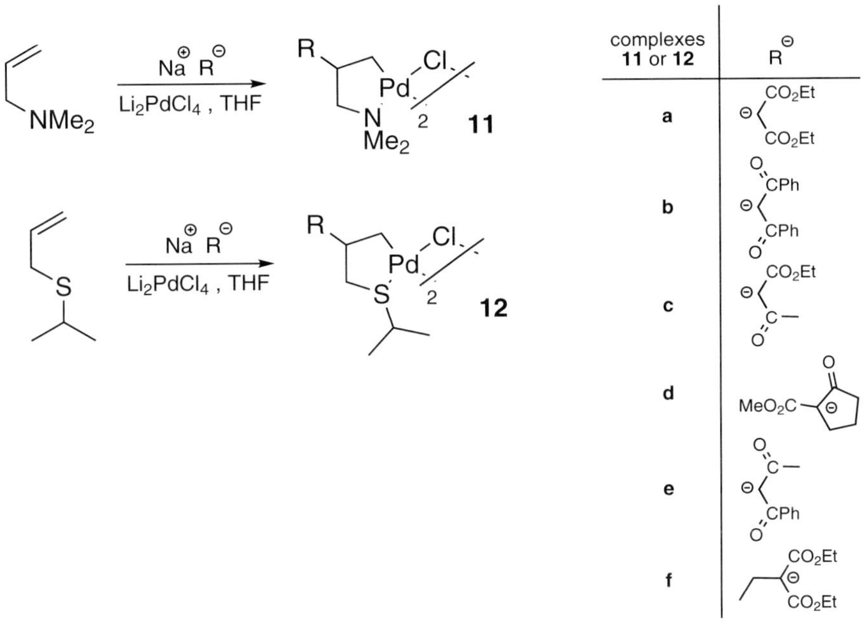

Scheme 4.10

4.2.2
Carbopalladation

Holton's group have also observed that palladacycle complexes can be obtained easily from nucleophilic carbon addition on allylic sulfides and amines, previously activated by coordination to Pd(II) species (Scheme 4.10) [12]. The addition of the carbanion seems to be independent of steric effects of the nucleophile, since palladacycles are produced with high yields even when large nucleophiles are used. However, the reaction is more steric dependent when substituents are present in the allylic substrate. These authors extended the work, including homoallylic amines and sulfides. The nucleophilic attack remains very regioselective, forming five-membered ring palladacycle complexes (Schemes 4.11 and 4.12).

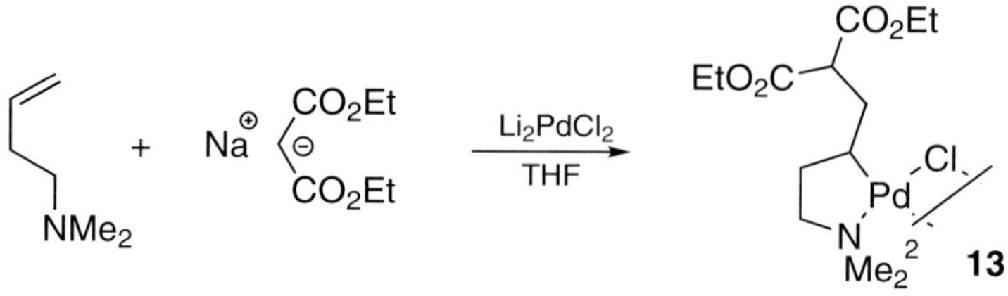

13

Scheme 4.11

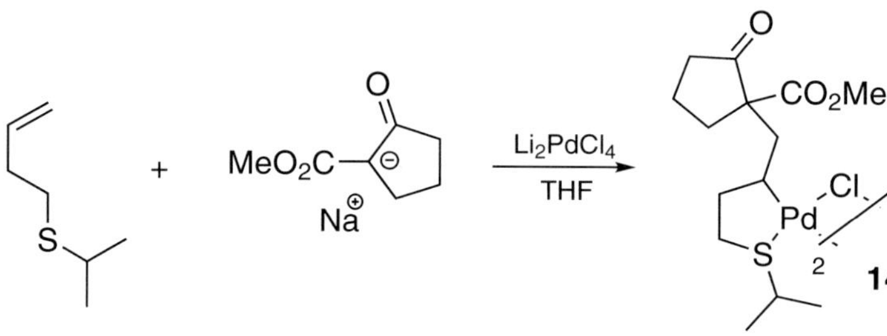

14

Scheme 4.12

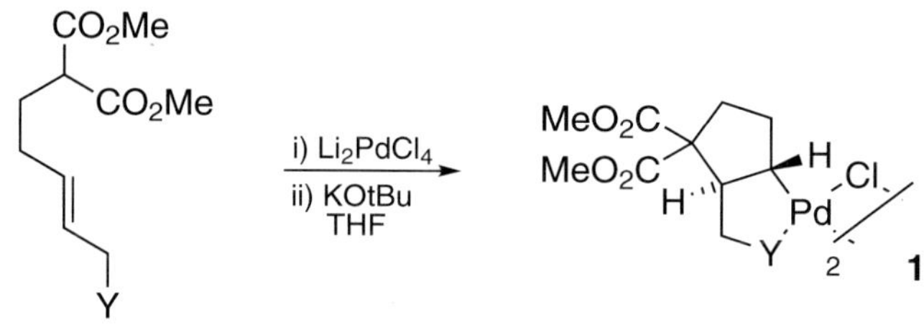

15

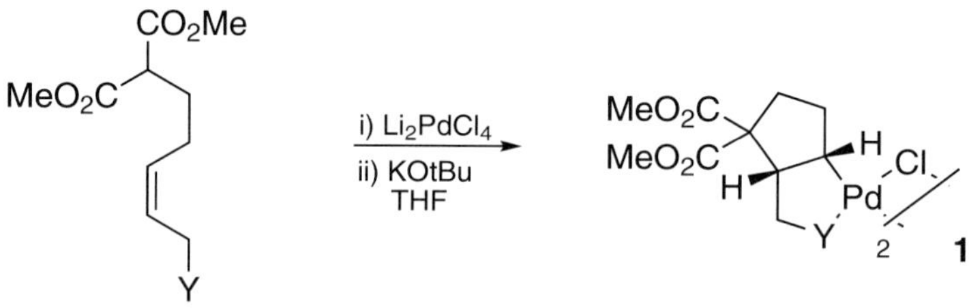

16

Y= SMe, NMe_2

Scheme 4.13

They also verified that palladacycles can be obtained via the intramolecular version of the carbopalladation reaction. Allylic sulfides or amines undergo intramolecular cyclization, affording bicyclic structures whose stereochemistry is governed by the stereochemistry of the allylic double bond (Scheme 4.13) [13].

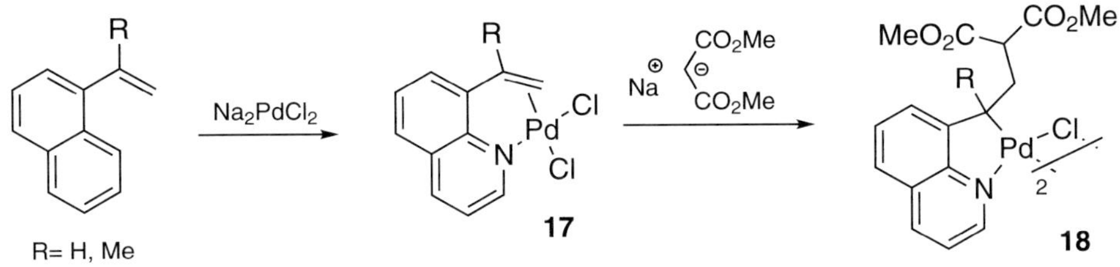

Scheme 4.14

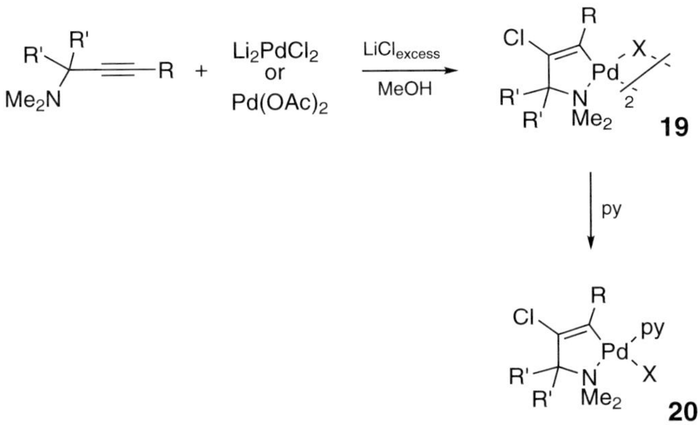

Complex **19**	X	R	R'	Yield (%)
a	Cl	Ph	H	95
b	OAc	Ph	H	73
c	Cl	H	Me	27
d	OAc	H	Me	47

Scheme 4.15

Dupont, again using 8-vinylquinoline derivatives, was able to isolate five-membered ring palladacycle complexes, employing dimethyl malonate as nucleophile (Scheme 4.14) [10]. He also isolated compounds like **17** in the absence of the nucleophilic reagent.

4.2.3
Chloropalladation

The first example of palladacycle synthesis via chloropalladation was reported in 1968 by Yukawa and coworkers [14]. Propargyl amines underwent nucleophilic attack by a chloride anion in the presence of Li_2PdCl_4 or $Pd(OAc)_2$, and excess of LiCl, leading to the respective dimeric palladacycle compounds (**19**), which contain a Pd—C_{vinyl} bond. These dimeric complexes could be split by addition of pyridine, forming stable monomeric structures like **20** (Scheme 4.15).

In this reaction, two possible mechanisms are accepted. In both mechanisms, prior coordination of the electron-donor group and the C≡C bond to palladium(II) species is followed by: (i) insertion of the latter into the Pd—Cl bond (cis-chloropalladation), which takes place preferentially in the presence of low chloride concentrations, or (ii) intramolecular attack of the chloride anion (trans-chloropalladation), which predominantly occurs under high chloride concentrations in the reaction medium [15–17]. Notably, the preference for chloride anion attack instead of other nucleophiles, like alkoxides, must be related to hard–soft acid–base interactions. Alkynes, even coordinated to the metal, are soft enough to allow chloride anion attack.

In the mid-1990s, chloropalladation reactions were studied in more detail as new synthetic strategy for palladacycle complex production were required [17]. Dupont and coworkers investigated the influence of several electronic and steric aspects of different groups linked to the C≡C bond in a series of propargyl amines and thioethers in the synthesis of dimeric palladacycle complexes (**21**). Intermolecular nucleophilic addition of the chloride anion occurs on the carbon that will generate the more thermodynamically stable palladacyclic ring, namely, a five-membered ring complex. The yields in these cases were 70–90%. Monomeric derivatives were easily obtained by addition of pyridine (**22**). Scheme 4.16 displays some representative results [17].

Alkynes containing large R groups lead only to adducts of type **23**, suggesting that bulky groups prevent an effective coordination of the C≡C bond to the palladium center. Dupont *et al.* also verified that propargyl amines appear to react faster than their thioether analogues. This must be related to the relative decrease in electrophilicity of the coordinated C≡C bond of the propargyl moiety that contains the sulfur atom as the electron-donor group. Moreover, when terminal alkynes are employed, the reaction yields are lower [17]. Notably, depending on the R group,

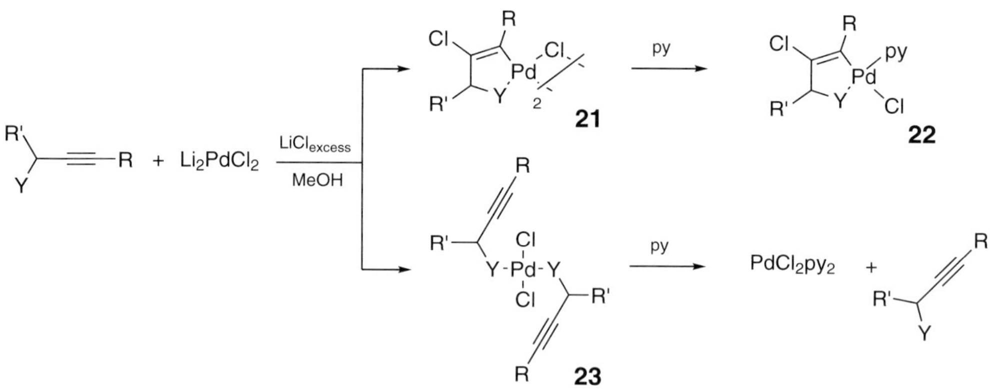

Complex 21	Y	R	R'
a	NMe2	Ph	H
b	NMe2	n-Bu	Me
c	NMe2	H	H
d	SPh	Ph	H
e	S-i-Pr	Me	H
f	SMe	Ph	Me

Complex 23	Y	R	R'
a	NMe2	*t*-Bu	Me
b	S-i-Pr	SiMe3	H
c	SMe	*t*-Bu	Me

Scheme 4.16

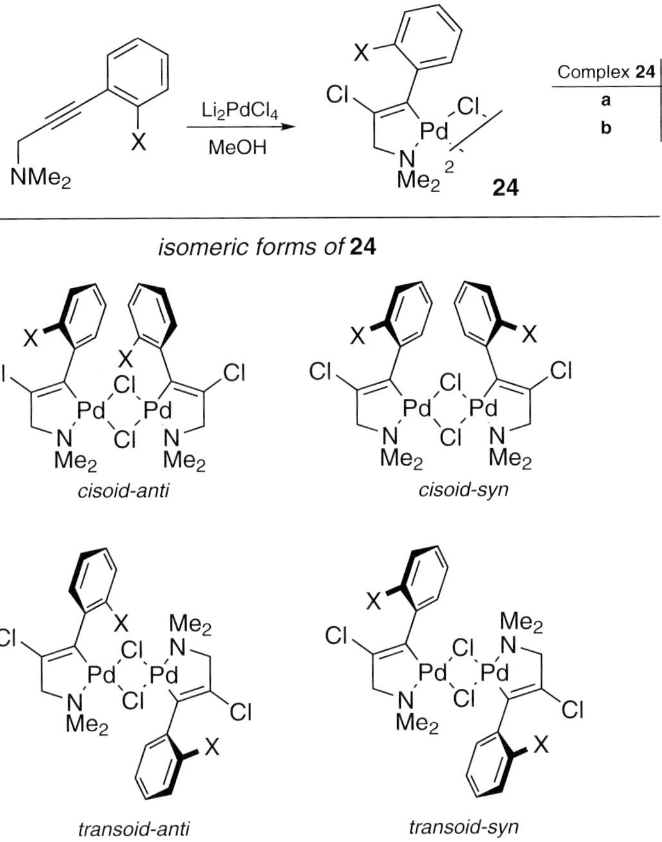

Complex 24	X
a	CF_3
b	OMe

Scheme 4.17

retro-chloropalladation can take place, that is, the reverse reaction of chloropalladation was observed [18].

More recently, Dupont and coworkers have verified that the chloropalladation of 2-phenylsubstituted *N,N*-dimethylpropargylamines can afford atropisomers of the chloro-bridged dimeric palladacycles, either as the *cisoid* or *transoid* forms (Scheme 4.17). These structures present atropism owing to the highly hindered C_{vinyl}—C_{aryl} σ-bond rotation [19].

In the solid state, the molecular structure of type **24** complexes was established by means of X-ray diffraction analysis, and *cisoid-anti* geometries were observed. However, complex **24a** in solution is composed of four isomers (Scheme 4.17). In contrast, only one isomer is observed with complex **24b** (*cisoid* or *transoid*). This must be due to the formation of a "pincer" intermediate by weak coordination of the OMe group (Scheme 4.18). Besides, when **24a** reacts with 2-methylpyridine, monomeric atropisomers are formed (Scheme 4.19).

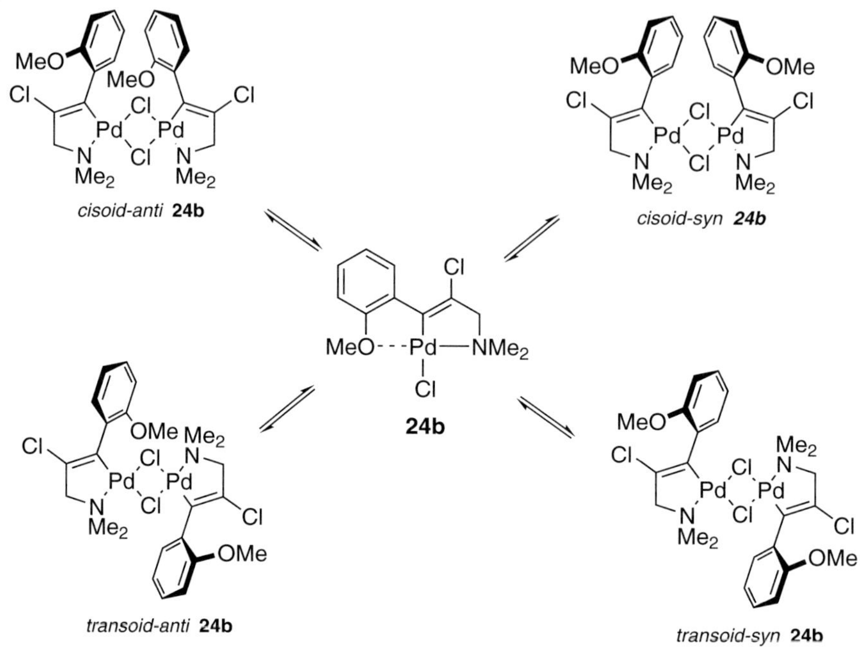

Scheme 4.18

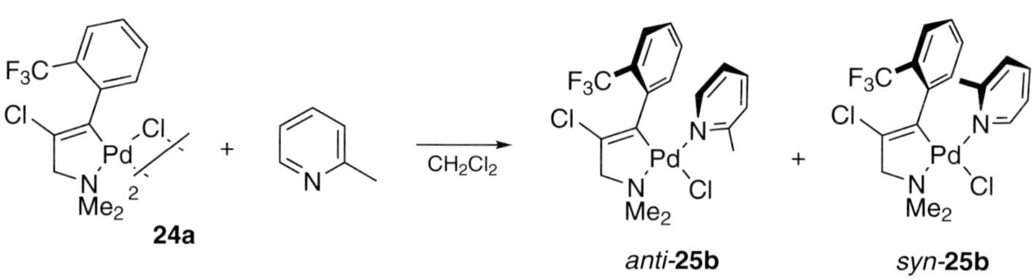

Scheme 4.19

Based on the chloropalladation reaction Dupont's group have also developed an elegant way to synthesize unsymmetrical "pincer" palladacycle complexes using hetero-substituted alkynes as substrates. Various functional donor groups were used to interact with the metallated fragment, such as amines, pyridine, thioethers, phosphines, and phosphinites [19, 20] (Scheme 4.20). With this methodology, "pincer" palladacycles containing 5,5- or 5,6-membered rings can be prepared; the selectivity is under thermodynamic control.

Several examples of catalytic chloropalladation reaction of alkenes, dienes and alkynes have been observed [21, 22]. However, only propargylic and homopropargylic alkynes, bearing electron-donor heteroatoms, are able to generate palladacycle complexes.

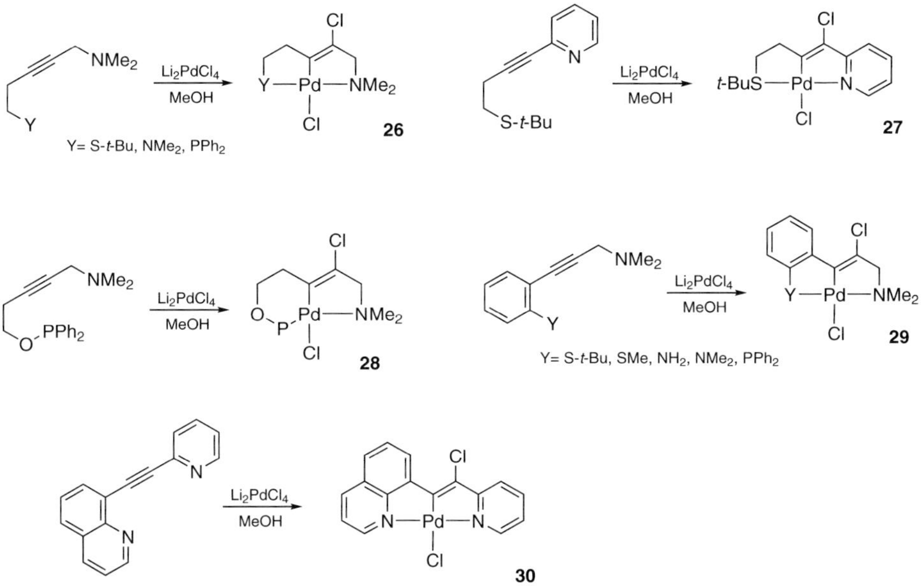

Scheme 4.20

4.3
Carbopalladation Reaction via Insertion of Olefins or Alkynes into the Pd—C σ-Bond of Nonpalladacyclic Species

4.3.1
Insertion of Olefins or Alkynes Bearing Electron-Donor Atoms

There are several examples of palladacycle complexes that undergo insertion reaction of C=C or C≡C bonds into Pd—C σ-bond, leading to other palladacycles with higher membered rings. However it has been also verified that unsaturated organic species, bearing electron-donor atoms, undergo insertion reactions into Pd—C σ-bonds of nonpalladacyclic species, that is, species not stabilized by intramolecular coordination of a heteroatom, leading to the formation of stable palladacycle complexes. Osakada and coworkers have prepared a five-membered ring palladacycle complex by insertion of three acetylene molecules into the $Pd—C_{aryl}$ σ-bond of the complex **31** followed by a cyclization process (Scheme 4.21) [23].

Soon afterwards, Liu *et al.* isolated palladacycle complexes from multiple insertions of alkynes into $Pd—C_{aryl}$ σ-bonds of cationic palladium complexes [24]. Complex **33** undergoes double insertion of alkynes into the $Pd—C_{aryl}$ σ-bond to give six-membered ring complexes like **34** and **35** (Scheme 4.22). However, with excess alkynes, complex **33** gives higher insertion units.

Amatore and coworkers [25] have observed similar results when $Pd—C_{vinyl}$ complexes are in the presence of ethyl propionate (Scheme 4.23).

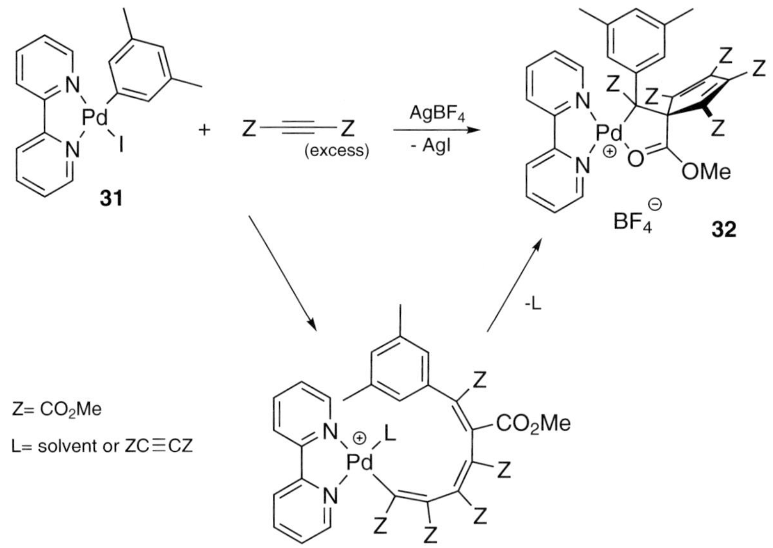

Scheme 4.21

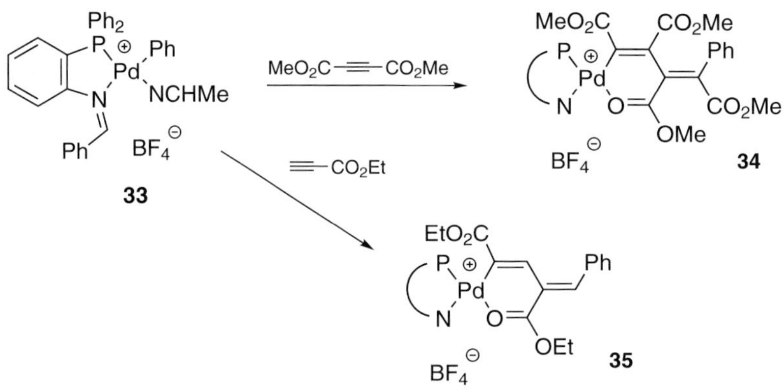

Scheme 4.22

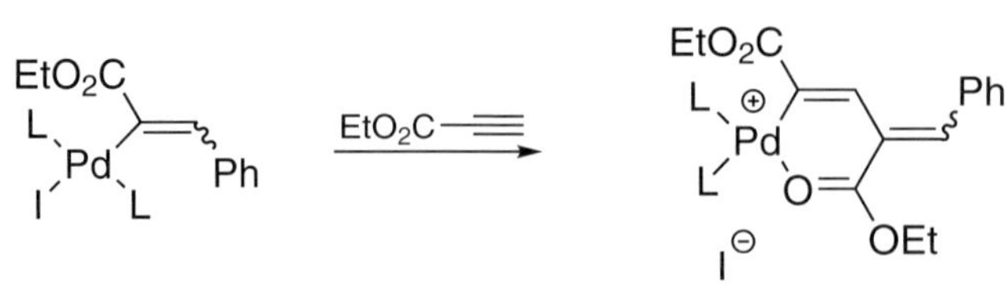

Scheme 4.23

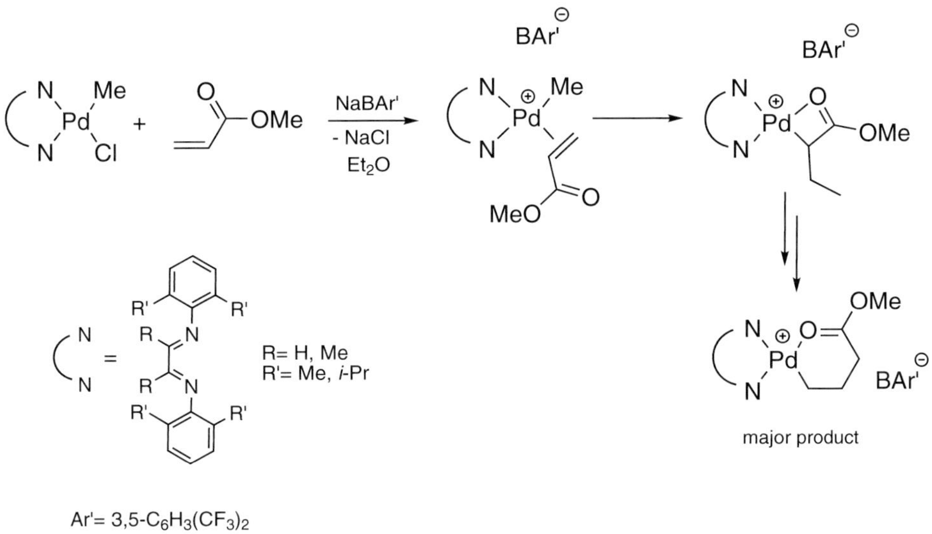

Scheme 4.24

With the current strategy, and taking advantage of the higher stability of organopalladium compounds containing an intramolecular coordination of a donor atom, Brookhart and coworkers have prepared cationic palladacycle complexes with bulky substituted α-diimines ligands. They are air- and temperature-stable catalyst precursors for the copolymerization of ethylene and α-olefins with acrylate, yielding high molecular weight copolymers [26]. Such complexes can be readily prepared from [(N^N)PdMeCl] in the presence of the noncoordinated BAr'_4 counterion and acrylate (Scheme 4.24).

4.3.2
Insertion of Olefins, Allenes or Alkynes into a Pd—C σ-Bond of a Fragment Containing Electron-Donor Atoms

Cationic and neutral Pd—C_{acyl} complexes can generate stable palladacycles by insertion of olefins into the Pd—C_{acyl} σ-bond. Sen [27], Vrieze [28], Boersma [29] and their coworkers have isolated palladacycles, with good yields, from insertion reactions of strained alkenes, like norbornene and norbornadiene, into the Pd—C_{acyl} σ-bond of bipyridine palladium complexes like **36** (Scheme 4.25). Other types of alkenes have also been tested; however, only unstable complexes have been produced, even at low temperatures. Palladacycles resulting from the insertion of strained olefins are more stable since the access of β-hydrogens to the metal center is restricted. In this type of reaction the stability of the insertion products is enhanced by intramolecular coordination of the carbonyl group, and, sometimes, multiple insertions are avoided.

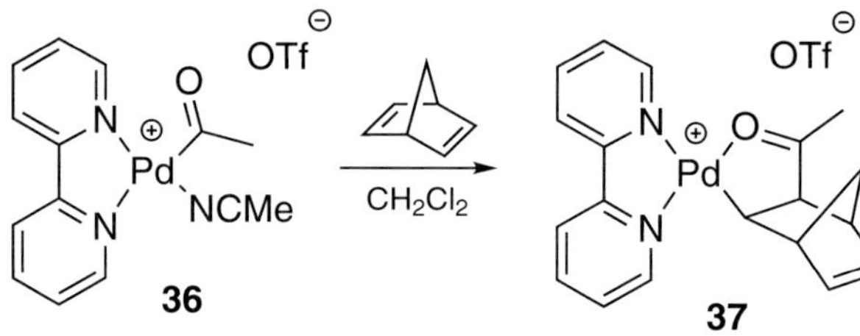

Scheme 4.25

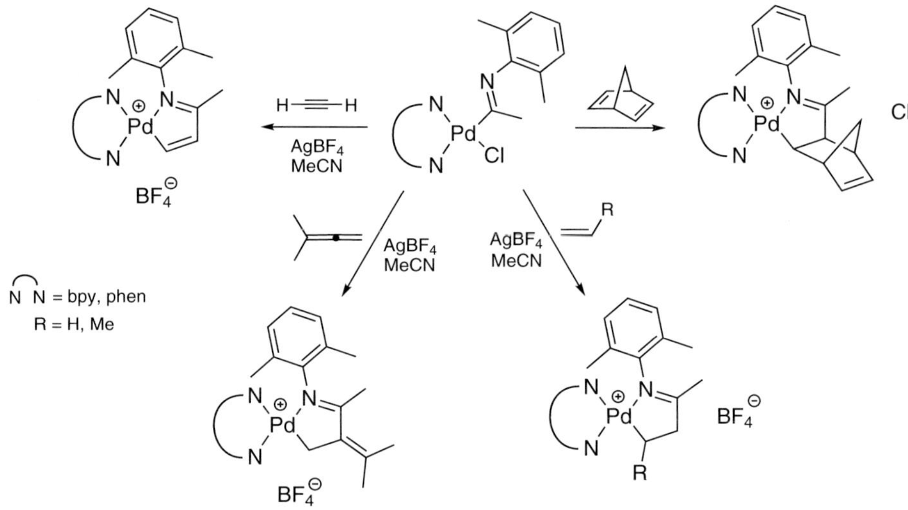

Scheme 4.26

In the same way, five-membered ring palladacycle complexes can be isolated from facile and quantitative insertion of alkenes, allenes and alkynes into the Pd—C σ-bond of neutral and cationic (N^N)Pd(C(=NR)Me)X complexes (Scheme 4.26). Vrieze and coworkers verified that, after the first insertion, no further insertions took place, owing to the strong coordination of the imine nitrogen atom to the palladium center [30, 31].

More recently, Liu and coworkers have isolated five-membered ring palladacycle complexes via single insertion of alkynes into the Pd—C_{acyl} bond of cationic palladium complexes. These palladacycle complexes are very stable and do not undergo further insertion reactions of alkynes or CO (Scheme 4.27) [24].

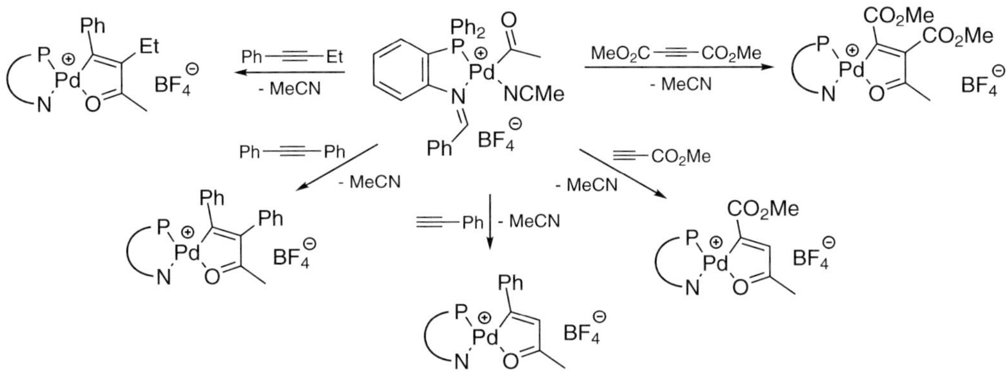

Scheme 4.27

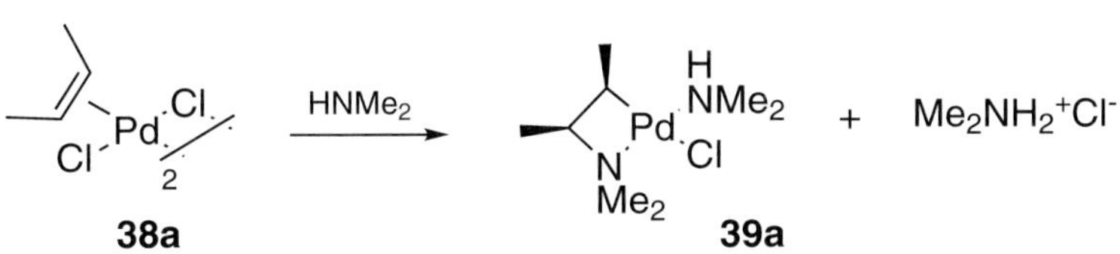

Scheme 4.28

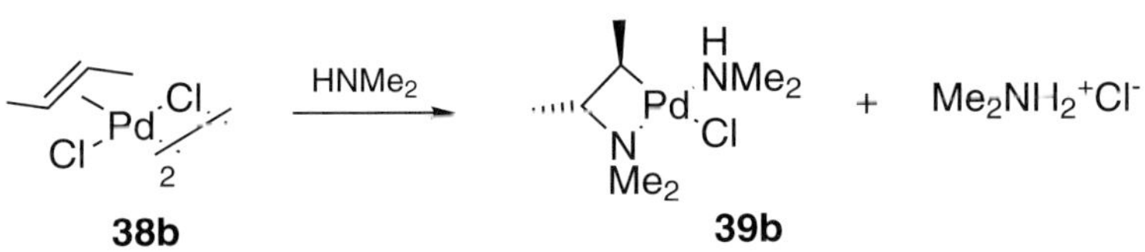

Scheme 4.29

4.4 Nucleophile Palladation of Olefins or Alkynes Not Bearing Heteroatoms

4.4.1 Aminopalladation and Aminoformylpalladation

In general, olefin-palladium(II) complexes not stabilized by chelation do not form stable σ-alkyl complexes upon reaction with nucleophiles but, rather, spontaneously decompose to Pd metal and organic products [2]. However, when the nucleophile is a secondary amine, a stereoselective aminopalladation reaction takes place and may lead to the formation of relatively stable four-membered ring σ-alkyl-palladacycle complexes (Schemes 4.28 and 4.29). These complexes were not isolated but were well characterized by NMR spectroscopy [32, 33].

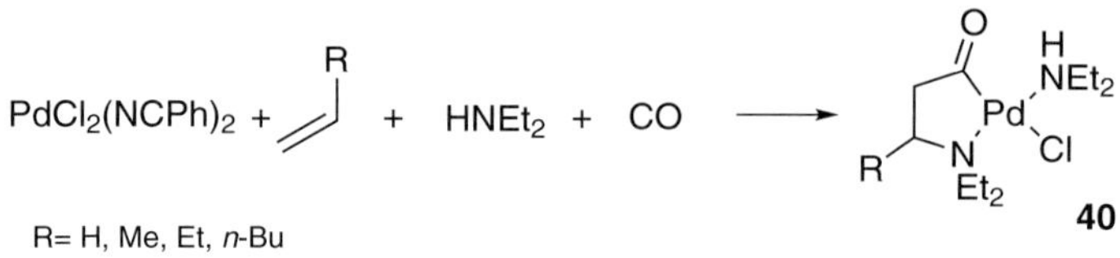

Scheme 4.30

In fact, the product of the aminoformyl-palladation reaction of terminal olefins was observed first. When terminal olefins were in the presence of $PdCl_2(NCPh)_2$ and secondary amines, with the subsequent addition of carbon monoxide, the formation of very stable σ–acylpalladacycle complexes like **40** occurred [34, 35] (Scheme 4.30).

4.5 Conclusion

The present chapter is not just as an overview of synthetic methodologies for palladacycle preparation but is intended to stimulate the development of new synthetic strategies and complexes. In fact, the chemistry of palladacycles, their synthesis and reactivity, remains very rich and far from exhaustion. Numerous palladacycles can be, indeed, attained via different synthetic strategies. Most of them take advantage of organometallic reactions, like C—H activation, oxidative addition and transmetallation, for their synthesis. However, though recent advancements have been achieved, there are other synthetic methods that require further attention. As presented here, palladacycle compounds can also be prepared by employing synthetic strategies whereby the molecular skeleton of the ligand is not the same as the ligand precursor, but is built during the formation of the metallacycle. These synthetic approaches lead to palladacycle compounds containing other functional groups on the organometallic-cyclic moiety, and can undoubtedly be applied for new and different uses.

References

1 Omae, I. (2004) *Coordination Chemistry Reviews*, **248**, 995–1023.
2 Hegedus, L.S. (1999) *Transition Metals in the Synthesis of Complex Organic Molecules*, 2nd edn, University Science Books, Sausalito, CA.
3 Collmann, J., Hegedus, L.S., Norton, J.R. and Finke, R.G. (1987) *Principles and Applications of Organometallics Metal Chemistry*, University Science Books, Mill Valley, CA.
4 Li, J.J. and Gribble, G.W. (2000) Tetrahedron Organic Chemistry Series, Volume **20**, *Palladium in Heterocyclic Chemistry, a Guide for the Synthetic Chemist* (eds J.E. Baldwin, F.R.S. and R.M. Williams), Pergamon Press.
5 Crabtree, R.H. (2005) *The Organometallic Chemistry of the Trasition Metals*, 4th edn, John Wiley & Sons, Inc.
6 Cope, A.C., Kliegman, J.M. and Friedrich, E.C. (1967) *Journal of the American Chemical Society*, **87**, 287.
7 Takahashi, Y., Tokuda, A., Sakai, S. and Ishii, Y. (1972) *Journal of Organometallic Chemistry*, **35**, 415.

8 Constable, A.G., McDonald, W.S., Sawkins, L.C. and Shaw, B.L. (1978) *Journal of the Chemical Society D–Chemical Communications*, 1061.

9 Holton, R.A. and Kjonaas, R.A. (1977) *Journal of Organometallic Chemistry*, **142**, C15.

10 Dupont, J., Halfen, R.A.P., Zinn, F.K. and Pfeffer, M. (1994) *Journal of Organometallic Chemistry*, **484**, C8.

11 Dupont, J., Halfen, R.A.P., Schenato, R., *et al.* (1996) *Polyhedron*, **15**, 3465.

12 Holton, R.A. and Kjonaas, R.A. (1977) *Journal of the American Chemical Society*, **99**, 4177.

13 Holton, R.A. and Zoeller, J.R. (1985) *Journal of the American Chemical Society*, **107**, 2124.

14 Yukawa, T. and Tsutsumi, S. (1968) *Inorganic Chemistry*, **7**, 1458.

15 Bäckvall, J.E., Nilsson, Y.I.M. and Gatti, R.G.P. (1995) *Organometallics*, **14**, 4242.

16 Dupont, J., Basso, N.R. and Meneghetti, M.R. (1996) *Polyhedron*, **15**, 2299.

17 Dupont, J., Basso, N.R., Meneghetti, M.R. and Konrath, R.A. (1997) *Organometallics*, **16**, 2386.

18 Zanini, M.L., Meneghetti, M.R., Ebeling, G., *et al.* (2003) *Polyhedron*, **22**, 1665.

19 Zanini, M.L., Meneghetti, M.R., Ebeling, G., *et al.* (2003) *Inorganica Chimica Acta*, **350**, 527.

20 Ebeling, G., Meneghetti, M.R., Rominger, F. and Dupont, J. (2002) *Organometallics*, **21**, 3221.

21 Lukas, J., van Leewen, P.W.N.M., Volger, H.C. and Kouwenhoven, A.P. (1973) *Journal of Organometallic Chemistry*, **47**, 153.

22 Albelo, G., Wiger, G. and Retting, M.F. (1975) *Journal of the American Chemical Society*, **97**, 4510.

23 Yagyu, T., Osakada, K. and Brookhart, M. (2000) *Organometallics*, **19**, 2125.

24 Reddy, K.R., Surekha, K., Lee, G.-H., *et al.* (2001) *Organometallics*, **20**, 5557.

25 Amatore, C., Bensalem, S., Ghalem, S. and Jutand, A. (2004) *Journal of Organometallic Chemistry*, **689**, 4642.

26 Meeking, S., Johnson, L.K., Wang, L. and Brookhart, M. (1998) *Journal of the American Chemical Society*, **120**, 888.

27 Brumbaugh, J.S., Whittle, R.R., Parvez, M. and Sen, A. (1990) *Organometallics*, **9**, 1735.

28 (a) van Asselt, R., Gielens, E.E.C.G., Rülke, R.E., *et al.* (1994) *Journal of the American Chemical Society*, **116**, 977.
(b) Groen, J.H., Elsevier, C.J. and Vrieze, K. (1996) *Organometallics*, **15**, 3445.

29 Markies, B.A., Kruis, D., Rietveld, M.H.P., *et al.* (1995) *Journal of the American Chemical Society*, **117**, 5263.

30 Delis, J.G.P., Aubel, P.G., Vrieze, K. and van Leeuwen, P.W.N.M. (1997) *Organometallics*, **16**, 4150.

31 Groen, J.H., Delis, J.G.P., van Leeuwen, P.W.N.M. and Vrieze, K. (1997) *Organometallics*, **16**, 68. Notably, in the same period, Vrieze's group observed that acylpalladium diimine complexes react with norbornadiene to afford five-membered ring palladacycle complexes in good yields. However, the palladacycles are very unstable.

32 Akermark, B. and Zetterberg, K. (1984) *Journal of the American Chemical Society*, **106**, 5660.

33 Hegedus, L.S., Akermark, B., Zetterberg, K. and Olsson, L.F. (1984) *Journal of the American Chemical Society*, **106**, 7122.

34 Hegedus, L.S. and Siirala-Hasén, K. (1975) *Journal of the American Chemical Society*, **97**, 1184.

35 Hegedus, L.S., Anderson, O.O., Zetterberg, K., *et al.* (1977) *Inorganic Chemistry*, **16**, 1887.

5
The Pd—C Building Block of Palladacycles: A Cornerstone for Stoichiometric C—C and C—X Bond Assemblage

Jose M. Vila and Mª Teresa Pereira

5.1
Introduction

Among the wide-ranging properties palladacycles display, one of the most salient is their reactivity, which stems from the differing nature of the bonds at palladium, that is, the Pd—C and Pd—Y (Y = donor atom) linkages. The latter not only involves bonding to the metallated ligand but also to bridging or terminal ancillary ligands. Selection of the bond at the metal to be used upon reaction of a palladacycle will depend not only on the nature of the metallated ligand, size of the palladacycle ring and reaction conditions but also to a reasonable degree on the characteristics of the reactant employed. Thus, bond cleavage selectivity is crucial in order to direct the synthesis towards the demanded product.

With organic molecules such as alkenes, alkynes, allenes and acyls, and other small molecules as carbon monoxide and isocyanides, it is usually the Pd—C bond that first undergoes reaction, more often than not in an insertion-type process, before the remaining metal bonds intervene to produce the final species. Despite the overwhelming number of references regarding these processes and the numerous reviews, dedicated totally or in part to this matter, that have appeared [1–4], in the present chapter we nevertheless attempt to give an overview of the most important aspects of this chemistry. To avoid excessive length we will not seek to be comprehensive regarding the variety of products formed after depalladation and we invite the reader to resort to the references for full details of the total number of species formed as depicted herein in the different reactions.

5.2
Reactions with Carbon Monoxide

Treatment of palladacycles with carbon monoxide is a most suitable synthetic method for the production of a myriad of carbo- and heterocyclic organic compounds. The Pd—C bond is highly reactive towards carbon monoxide insertion,

Palladacycles: Synthesis, Characterization and Applications. Edited by Jairton Dupont and Michel Pfeffer

ISBN: 978-3-527-31781-3

and the nature of the final products is mainly governed by the reaction media and by the metallated ligand, which in turn determines the type of donor atom and the size of the palladated ring [5, 6]. After insertion the reaction may proceed further via cleavage of the remaining bonds at palladium and ensuing demetallation, with subsequent formation of the organic cyclic moiety.

The mechanism of CO insertion is yet to be determined beyond doubt, principally owing to the various factors implied in the process. Nevertheless, the mechanism clearly involves an initial step with coordination of CO to the metal trans to the donor atom followed by insertion into the Pd—C bond (Figure 5.1).

For dinuclear palladated azobenzenes, (**1**, Scheme 5.1), Takahashi and Tsuji [5] have proposed coordination of carbon monoxide with halide-bridge splitting, insertion and formation of a palladium-acyl compound. There then follows insertion of the —N=N— bond into the palladium-acyl bond and, finally, hydrogenolysis of the palladium–nitrogen bond (**2**).

A similar mechanism has been proposed by Heck [6] for azobenzene, Schiff base and tertiary benzylamine acetate-bridged palladacycles. For azobenzene (**3**, Scheme 5.2), carbonylation in chlorobenzene proceeds with CO insertion and then addition of the acylpalladium group across the nitrogen–nitrogen double bond. Further reaction with CO and cyclization gives **4**.

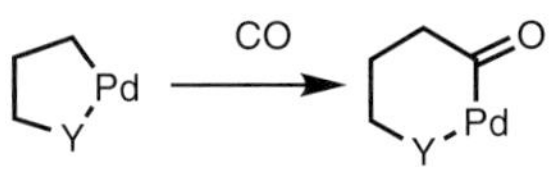

Figure 5.1 Outline of the mechanism of CO insertion into a Pd–C bond.

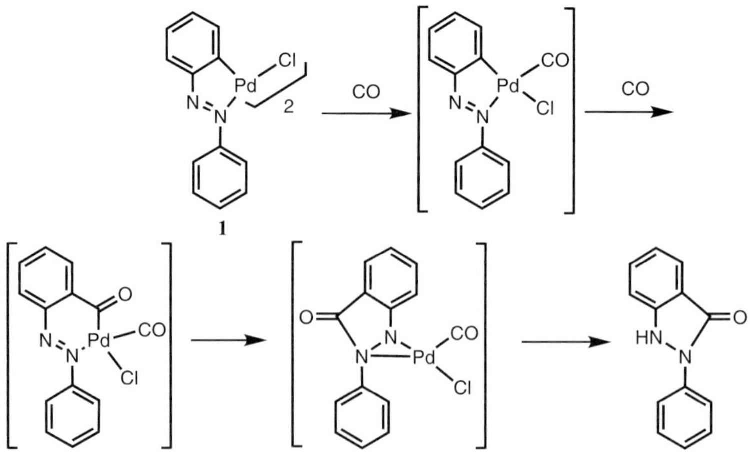

1

2

Scheme 5.1

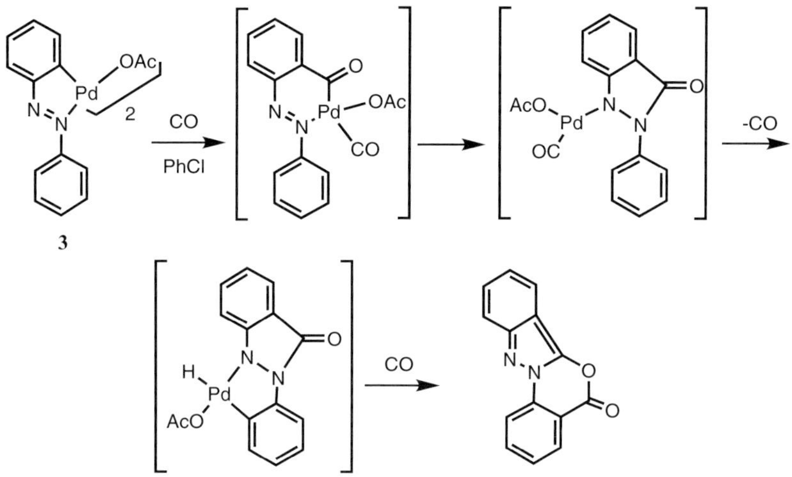

3

4

Scheme 5.2

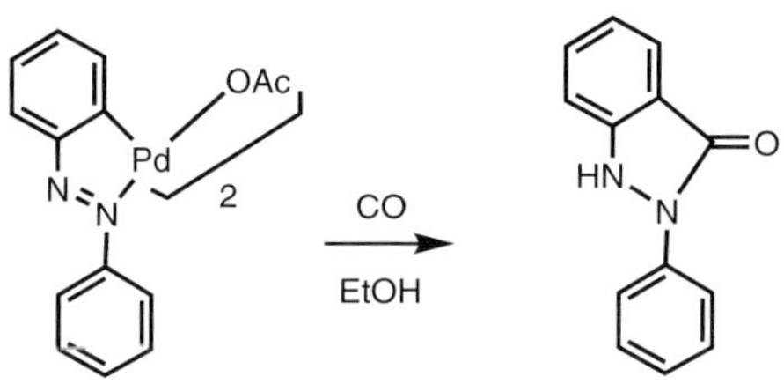

3

2

Scheme 5.3

However, if the reaction is conducted in ethanol (Scheme 5.3), **2** is formed instead of **4**, plus 2-ethoxycarbonylazobenzene; the latter is generated by ethanolysis of an intermediate species.

The tendency of azobenzene to react at both aromatic rings complicates the carbonylation reaction, whereas carbonylations of Schiff base complexes are more straightforward. Thus, through a similar mechanism, the intermediates shown in Scheme 5.4 are produced in xylene. The presence of nucleophiles or a change of the RC=N substituent leads to **8** and **9**; in all instances the reaction progresses by CO insertion, acetate-bridge splitting, and addition of the Pd(AcO)(CO) group to the C=N double bond; in ethanol a non-cyclic ester, **7**, is obtained.

The formation of cyclic compounds, **11**, from benzylamine complexes takes places likewise (Scheme 5.5) [6].

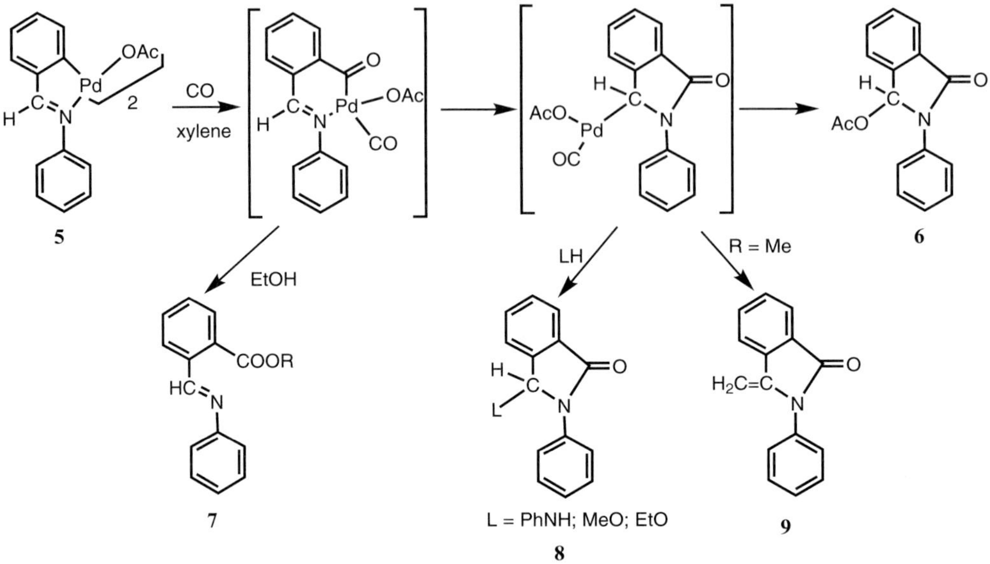

Scheme 5.4

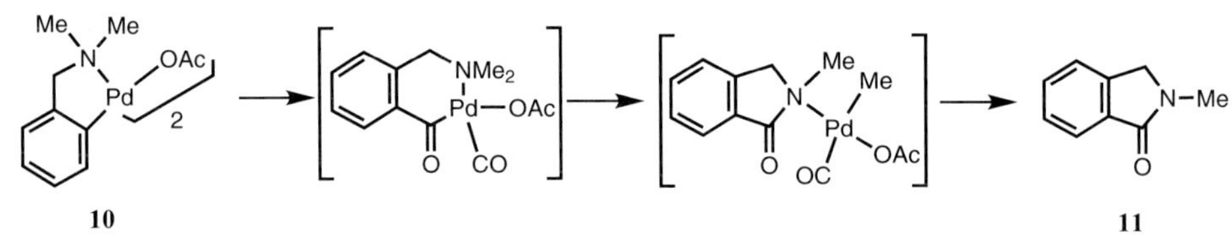

Scheme 5.5

The influence of the donor atom is considered when comparing the reactivity of cyclopalladated benzyl methyl 1-naphthyl sulfide with the corresponding methyl derivative [7]. Thus, treatment of **13** with carbon monoxide yields the insertion product **14**, whereas **12** gives **15**, with a coordinated CO molecule. Heating **14** under mild conditions produces **13**; the insertion–deinsertion of CO into the Pd–C bond may be repeated several times without any observed decomposition of products.

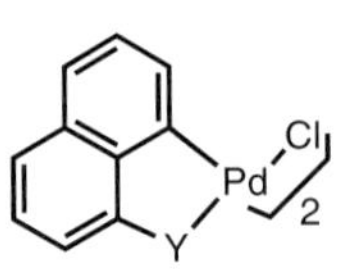

Y = NMe$_2$ **12**
Y = SMe **13**

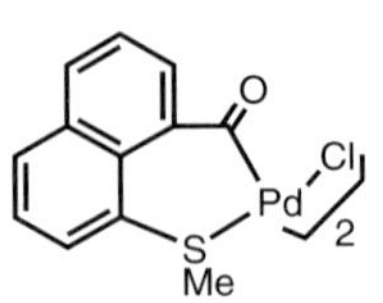

14

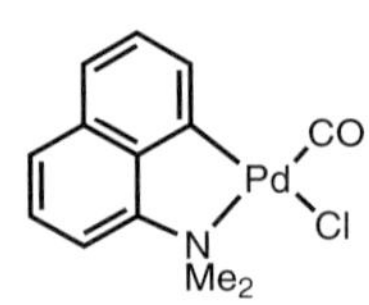

15

Carbonylation of palladated *N*-phenyl-acetophenone hydrazone gives an isoindolinone compound (**18**, Scheme 5.6) similar to **9**. Treatment of **17** with a stoichiometric amount of NaMeO gives **19** after *E*/*Z* isomerization and reductive elimination [8]. Cyclopalladated pyrroles **(20)** [9], amides **(22)** [10], ferrocene derivatives **(24)** [11–14], and other palladacycles bearing nitrogen donors such as Schiff bases **(26)** [15], and indols [16] do not give annulation compounds; instead the respective uncyclized esters are obtained: **21** (Scheme 5.7), **23** (Scheme 5.8), **25** (Scheme 5.9) and **27** (Scheme 5.10). For the ferrocene compounds, owing to the presence of a stereogenic center in the starting material, different diastereoisomers may be produced.

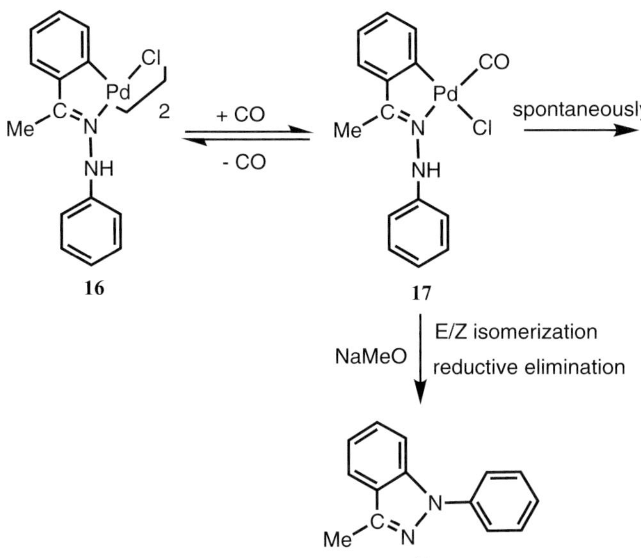

Scheme 5.6

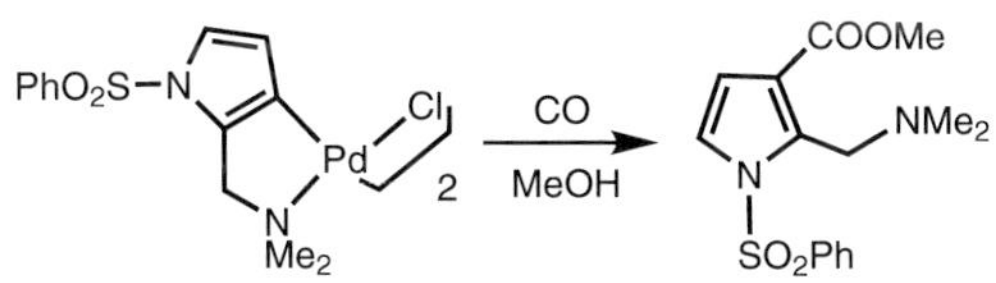

Scheme 5.7

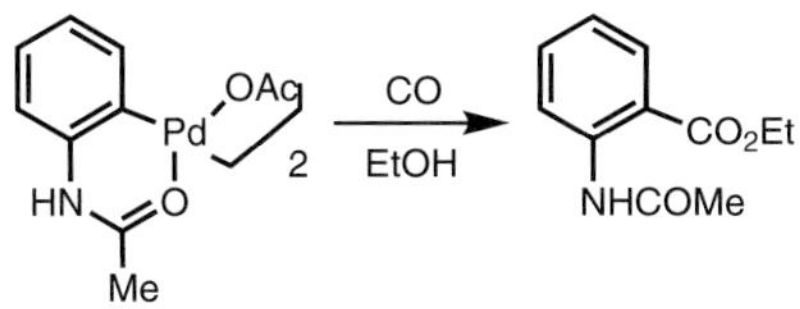

Scheme 5.8

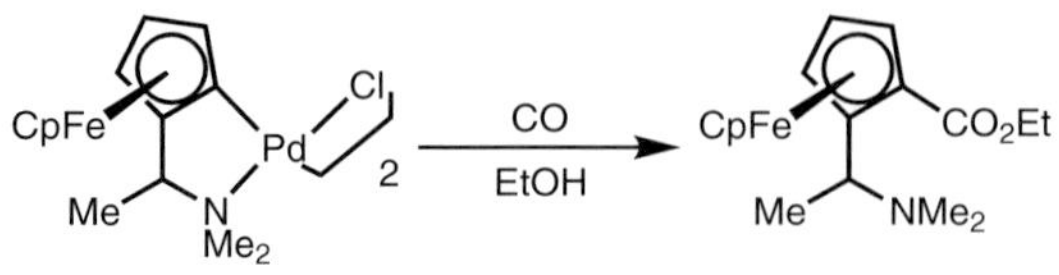

24 25

Scheme 5.9

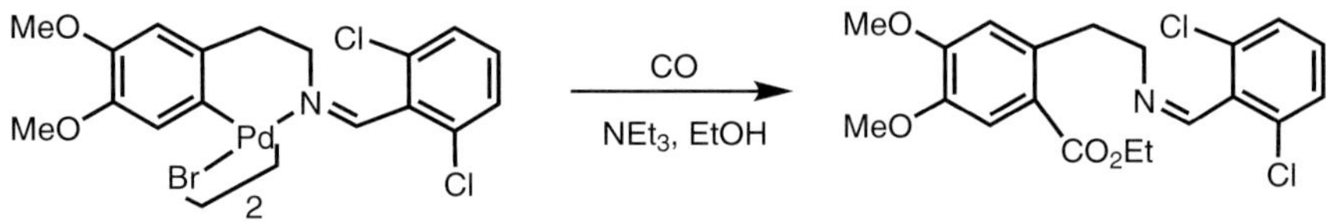

26 27

Scheme 5.10

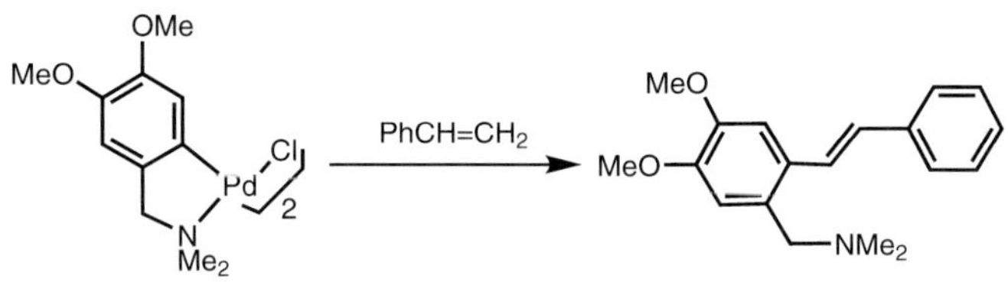

28 29

Scheme 5.11

Although reactions with carbon monoxide usually yield insertion products, at least in an intermediate stage, there are cases when non-inserted compounds are obtained and the reaction does not proceed further, as is the case with derivatives of amines [17], oximes [18, 19] and 8-methylquinoline [20], where the carbonyl ligand is only coordinated to the palladium atom.

5.3
Reactions with Alkenes

Palladacycles usually react with alkenes in 1 : 1 molar ratio to give the corresponding vinylated products, with a two-electron reduction of palladium(II). The only reactive species are terminal olefins, H_2 =CRR′, to afford products where palladium is substituted for the α-alkene carbon; branched alkenes failed to react. The reaction course is governed by the nature of the complexes, rather than by the alkyne itself. Styrene has been used profusely in these reactions with nitrogen [21–23] and sulfur-donor palladacycles [24]; for example, di-μ-chlorobis(*N,N*-dimethyl-3,4-dimethoxybenzylamine-6-*C,N*)dipalladium **(28)** reacts readily with various styrenes, RCH=CH_2, to give the corresponding *o*-aminomethylstilbene derivatives **29** (R=Ph) (Scheme 5.11) [25].

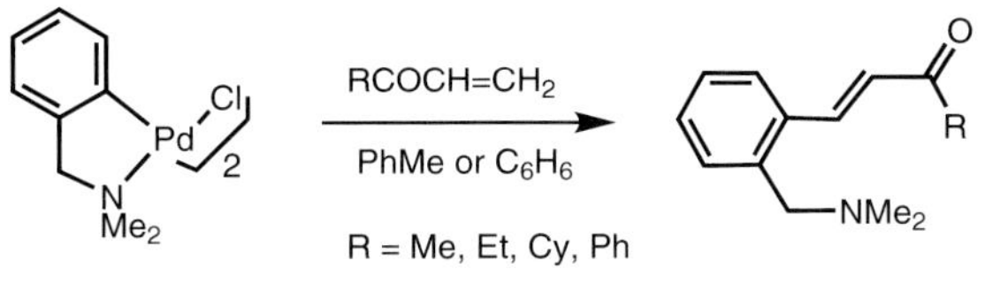

30 **31**

Scheme 5.12

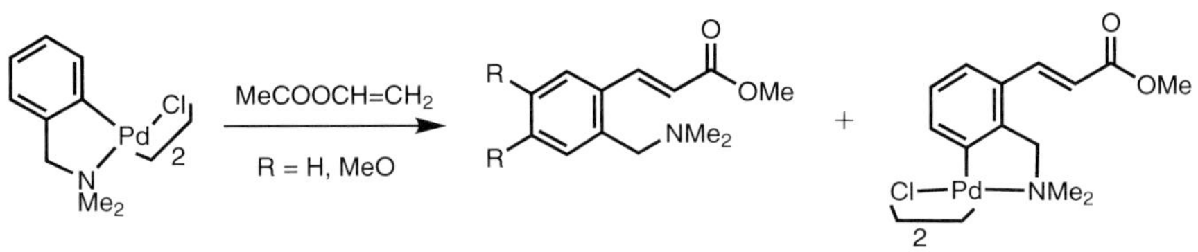

30 **32** **33**

Scheme 5.13

A second-order pathway was found for the reaction of palladated *N,N*-dimethylbenzylamine with styrenes [26]. Many other vinylations are known, which include among others, reactions of oxazoline [27] and ferrocene [28, 29] palladacycles, or the reaction of *N,N*-dimethylbenzylamine palladacycles with ketones (**31**, Scheme 5.12) [30].

Ryabov has reported [4] that the isolation of the vinylated species **32** may proceed in the absence of triethylamine; however, other by-products such as **33** are formed (Scheme 5.13).

Vinylation reactions using five-membered palladacycles are widely used in synthetic organic chemistry and for the formation of heterocyclic compounds by cyclization reactions [23, 31–34]. Ketovinylation of asymmetrically palladated ferrocenes is the key step in the preparation of prostaglandin models with a cyclopentadienyl nucleus [35, 36].

5.4
Reaction with Alkynes

Alkynes have long been known to be very useful building blocks for organic synthesis, since they display a wide reactivity: they may be considered as electrophiles or nucleophiles, depending on the nature of the substituents [3]. In most palladacycles that undergo alkyne insertion, the Y donor is a nitrogen atom because these compounds are much more reactive than the corresponding compounds where Y is another heteroatom such as phosphorus or sulfur. Up to three alkyne molecules may be inserted into the Pd–C bond (Scheme 5.14). The chemistry of alkyne

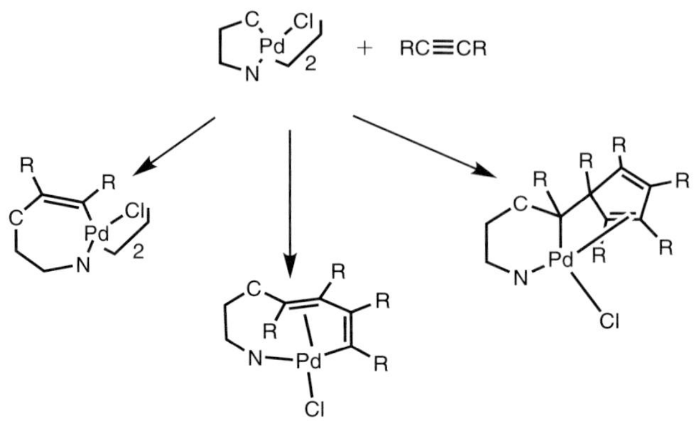

Scheme 5.14

insertion into metal–carbon bonds has been studied to a great extent [37], in particular in the case of palladacycles [38]. Most reactions of palladium cyclopalladated compounds with alkynes give stable complexes because of intramolecular stabilization of the coordinated atom. New metallacyclic units may be synthesized through insertion of one to three alkynes into the Pd–C bond.

Insertion of one alkyne results in cyclopalladated compounds with the palladium atom as part of the new ($n + 2$) membered ring, where n is the number of atoms of the starting metallocyclic unit; rings of up to eight members have been reported. Many organic ligands are prone to undergo alkyne insertion – most have a nitrogen donor atom, although ligands incorporating sulfur and oxygen donors are also known; some examples are *N,N*-dimethylbenzylamine [39], 1-dimethylamino-naphthalene [40], Schiff bases [41, 42], 2-aminobiphenyl [42], azobenzene [43], 2-benzylpyridine [44, 45], methyl-2-biphenylamine [46], *N*-phenyl-2-pyridilamine, phenyl 2-pyridyl ketone [47], benzyl methyl sulfide [48], methyl 2-biphenyl sulfide [46] and ferrocenylimino alcohols [49].

Scheme 5.15 summarizes a likely reaction sequence [3]. The first step possibly involves π coordination of the alkyne to the metal, then the alkyene must bring a carbon atom close to the Pd–C carbon. This is in accordance with the regioselectivity of the process, because the carbon bearing the smaller substituent is always found on the metal, reducing the steric hindrance at palladium. This insertion route is further supported by the fact that changing the Y group may destabilize the alkyne, as occurs when, for example, the NMe_2 unit is exchanged for SMe_2. This is due to the greater trans influence of the latter group, which implies that insertions with sulfur-containing ligands are more difficult. If the reaction is carried out in the presence of a strong coordinating ligand such as a phosphine, no insertion usually takes place. When addressing the problem of heteroannulation with formation of C–C bonds, the reaction between the palladacycle and the alkyne may proceed through a different route, albeit less straightforward, involving nucleophilic addition of the nitrogen atom onto the alkyne, which is activated through coordination to the metal, leading to a zwitterionic organometallic species, from which reductive elimination renders C–C bond formation to yield a heterocycle (Scheme 5.16) [44, 50].

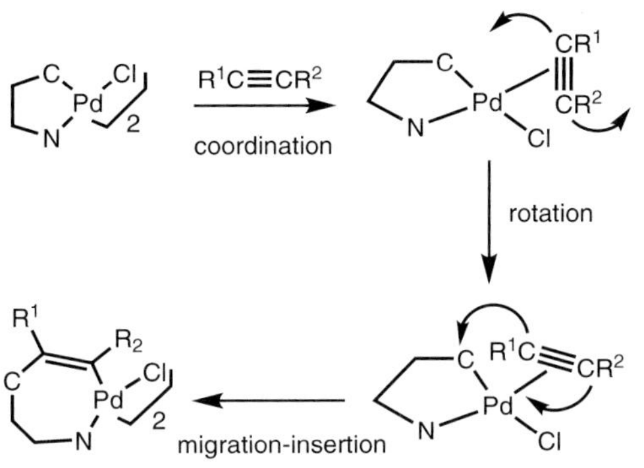

Scheme 5.15

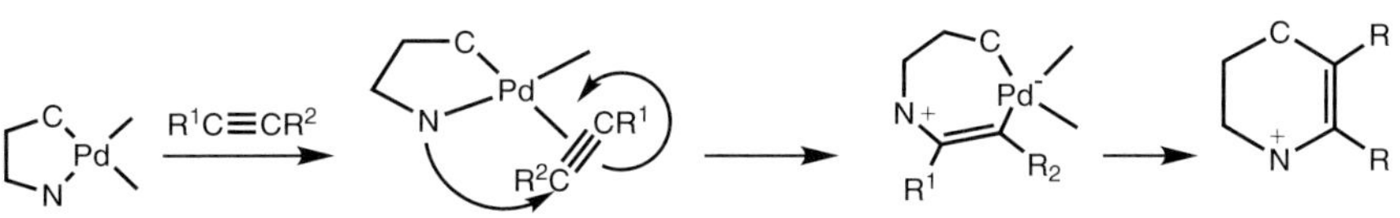

Scheme 5.16

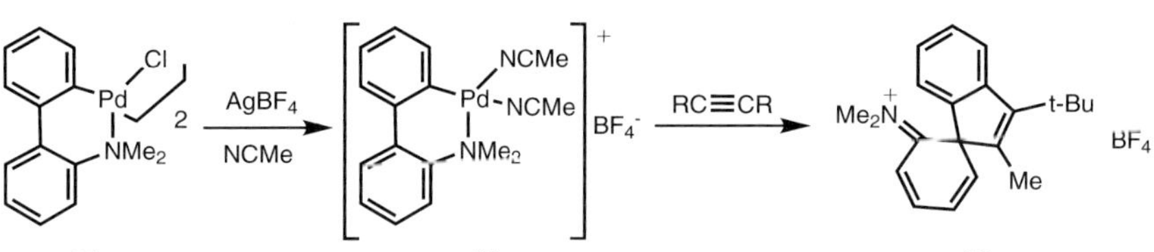

34 **35** **36**

Scheme 5.17

The regioselectivity of the insertion reaction with asymmetrical alkynes is quite high, and although the smallest group should be located on the palladated carbon atom, in some instances the opposite trend has been observed for the insertion of 4,4-dimethyl-2-pentyne to give a spirocyclic compound, with the bulkier group next to the initial metallated carbon atom [39]. The regiochemistry observed may result from the weak chelating ability of the NMe_2 group, which upon decoordination allows a perpendicular approach of the alkyne to the metallated aryl ring, with the flat aryl face presenting little steric hindrance towards the *t*-butyl group [39] (Scheme 5.17).

When steric factors are similar, electronic prerequisites have also been invoked to govern regioselectivity. Thus, for unsymmetrical alkynes such as $PhCC{\equiv}CO_2Et$, it was found, assuming comparable steric requirements for the Ph and CO_2Et groups, that the more electron-withdrawing group (CO_2Et versus Ph) is found adjacent to the previously metallated carbon atom, **37** [51].

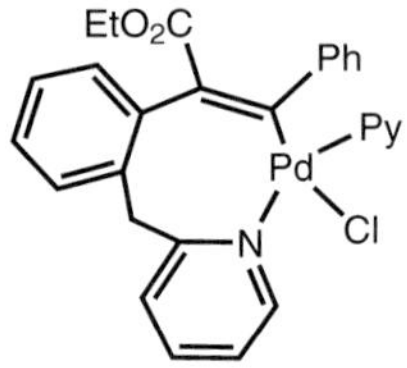

37

The nature of the substituents on the palladium atom in the starting complexes has a strong influence on the insertion reaction and on the stability of the final compounds. For example, changing chlorine-bridging ligands for iodine improves the yield of the final organic product after demetallation [50]. Further activation may be achieved by using cationic starting complexes [52].

The Pd–C bond may be, in some instances, far too reactive to allow only mono-insertion of alkynes and so compounds resulting from bis-insertion are obtained [51]. This is typical of electron-rich acetylenes such as 3-hexyne [53] and diphenylacetylene (Scheme 5.18) [54]. It seems that the Pd–C bond of the mono-insertion products synthesized from this type of alkynes is more reactive than that of the starting material.

These compounds contain nine-membered rings, from which it may be inferred that the initially palladated carbon atom and the palladium atom show a trans + cis arrangement. Consequently, the C=C group next to palladium displays a cis geometry, whereas the other C=C group exhibits a trans disposition. Maitlis [55], in related systems, has proposed a metallocyclic flip that could rationalize the cis/trans isomerization of the latter vinyl group (Scheme 5.19). Although slight steric repulsion between the R groups could be the cause of isomerization, the driving force for the process could be found in the conformation of the nine-membered ring with the four R groups cis to each other. In this cis arrangement the C=C group is not ideally set with respect to the metal atom, precluding good overlap of the olefin π orbitals with the metal d orbitals. Therefore, stabilization is smaller than for the isomer having a trans-cis conformation.

For compounds with six-membered palladacycle rings containing a second heteroatom, the latter may be implicated in coordination to the metal in the final product, after insertion of two alkynes, **40** (Scheme 5.20) [56].

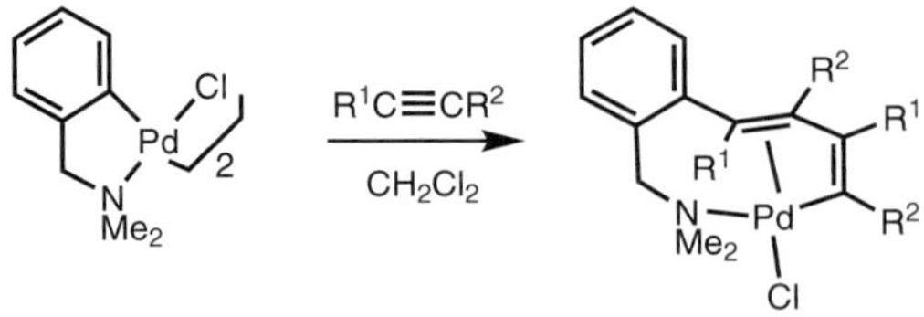

30 **38**

Scheme 5.18

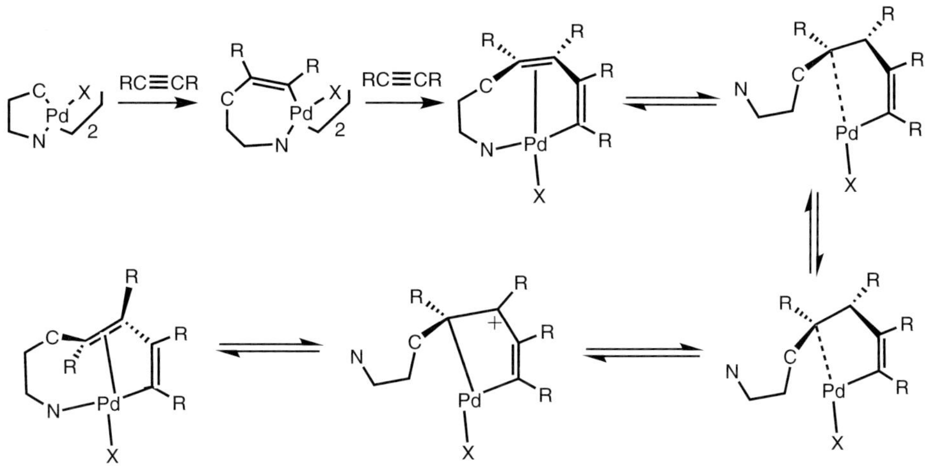

Scheme 5.19

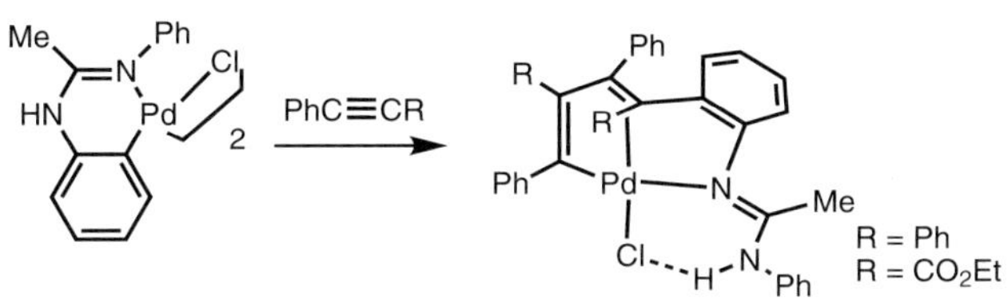

39 40

Scheme 5.20

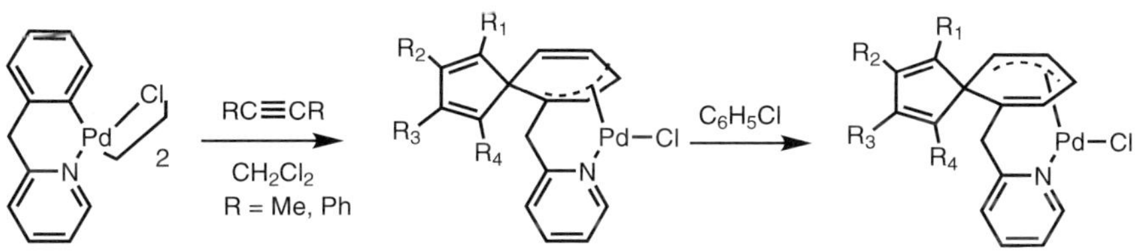

41 42 43

Scheme 5.21

In the absence of such a heteroatom, cyclization of two alkyne molecules with one carbon atom of the benzyl unit is produced. This leads to a spiro junction between a tetraphenylcyclopentadiene and a hexadienyl ring. The palladium atom is linked to this new organic fragment via the nitrogen atom and an η^3-allylic interaction of the benzyl ring through C1–C3 **(43)**. It appears that there is some tendency to avoid formation of ten-membered rings (Scheme 5.21) [51, 56, 57].

Compound **42** (for R_{1-4}=Ph) is not stable at high temperature, isomerizing to **43**. The driving force of this reaction could be assigned to the release of steric hindrance that the 2-picolyl group exerts in **42**, which does not exist in **43**. Notably, the double-alkyne insertion into the Pd–C bond to afford a spirane ligand seems

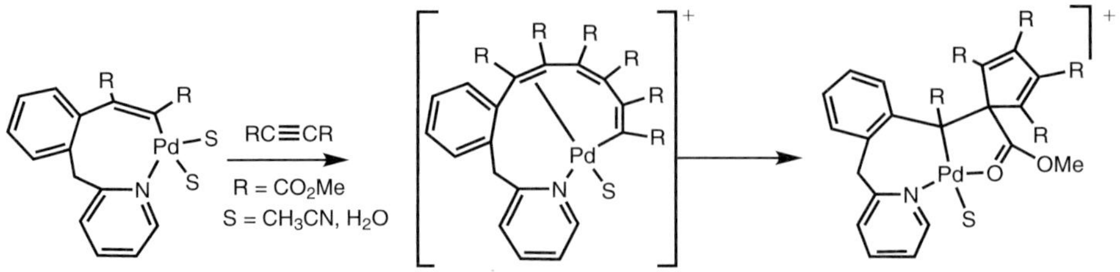

44 45

Scheme 5.22

to be limited to alkynes substituted by methyl or phenyl groups. Related species in which the spiral ring is substituted by at least one CO_2R group have not been synthesized.

Dimethyl acetylenedicarboxylate, and other electron-poor alkynes, react with palladacycles to insert three alkynes into the Pd–C bond. In all cases no compounds with a 12-memebered ring were encountered but, rather, the alkyne moieties cyclized to give a cyclopentadienyl ring. The likely mechanism is via a 12-membered ring compound, which is proposed to exist as an intermediate, that undergoes nucleophilic attack of the carbon σ-bonded to palladium on the ortho carbon of the benzyl ring of the ligand to give **45** [51, 58] (Scheme 5.22).

Nevertheless, a compound with a ten-membered palladacycle ring, generated in a bis(insertion) process, that is contained in a tricyclic[6, 6, 10] system has been reported, **46** [59].

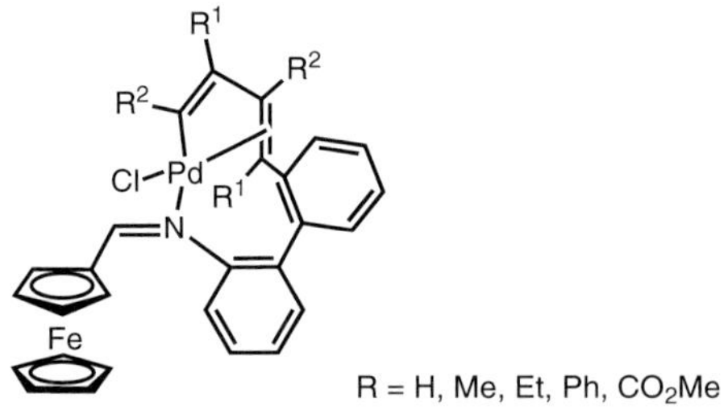

46

Palladacycles derived from dimethylaminoferrocene [60] undergo facile insertion of alkynes into the Pd–C bond; **47** reacts with two equivalents of diphenylacetylene or 3-hexyne in CH_2Cl_2 at room temperature to give **48** (Scheme 5.23).

Treatment of **47** with diphenylacetylene in refluxing chlorobenzene gives palladium-free six- and seven-membered ortho fused rings formed in annulation reactions. Ferrocenylimines containing bidentate [C(sp^2, N] [61], and tridentate [C(sp^2, N, O] [49], [C(sp^2, N, S] [62], also give insertion products; for example **49** reacts

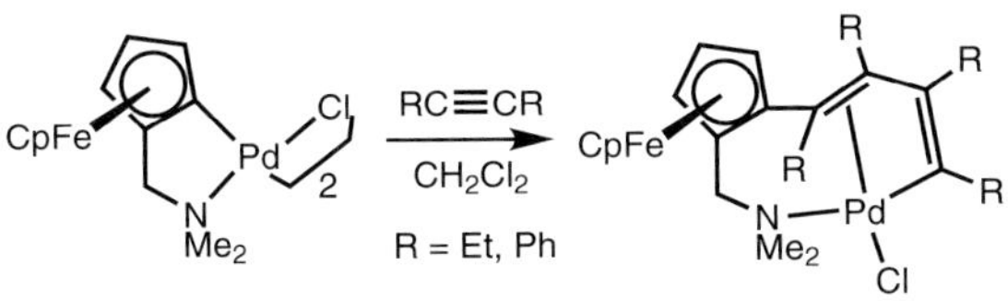

47 48

Scheme 5.23

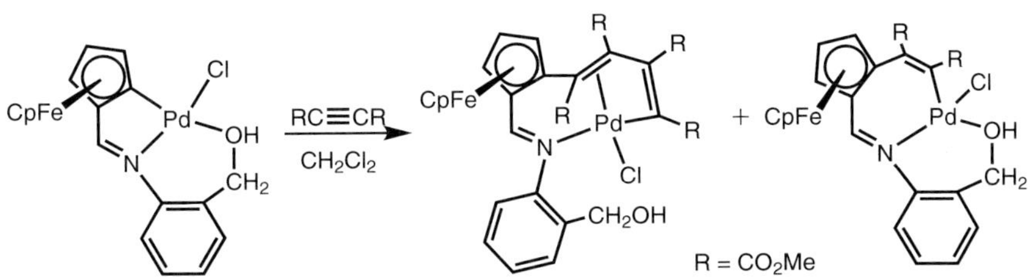

49 50 51

Scheme 5.24

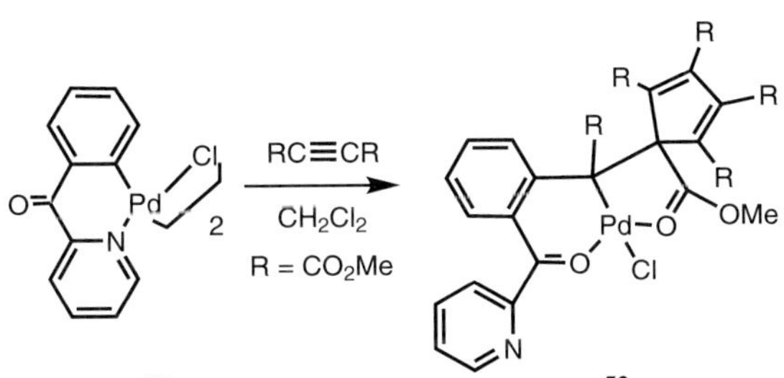

52 53

Scheme 5.25

with $MeO_2CC{\equiv}CCO_2Me$ to give a mixture of mono- and bis-inserted products **51** + **50** (Scheme 5.24).

Comparative studies show that in ferrocenylimines the Pd–C bond is more reactive towards insertion of alkynes than in the Schiff base analogues [62].

Coordination at palladium may be accomplished by bonding to substituents on the cyclopentadienyl unit, as in **53** [47] (Scheme 5.25) and **55** [58] (Scheme 5.26), or to a C=C double bond of the five-membered ring **(56)** [52] (Scheme 5.27).

Syntheses of hetero- and carbocyclic compounds from the aforementioned insertion products has been reviewed by Spencer and Pfeffer [63].

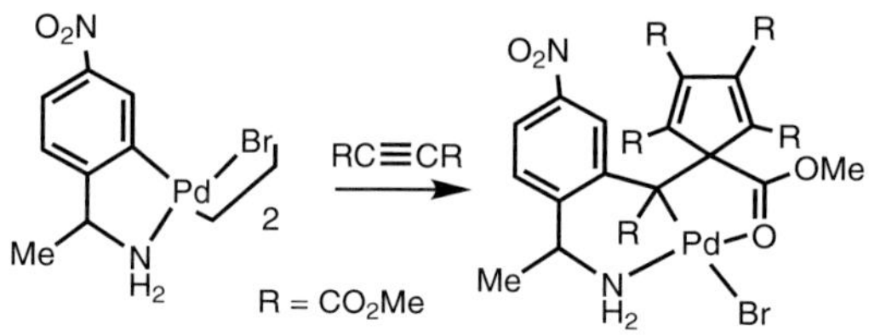

54 55

Scheme 5.26

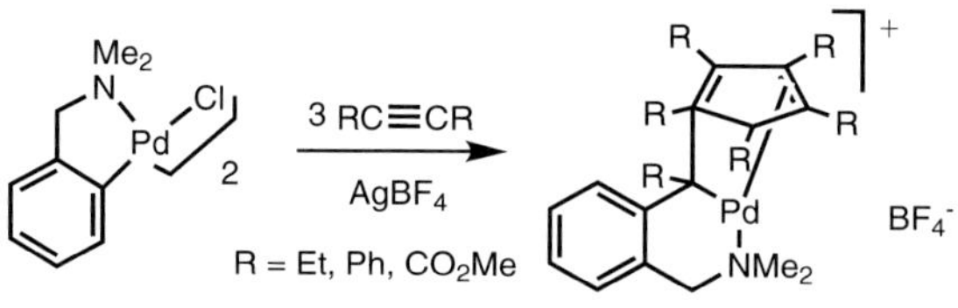

30 56

Scheme 5.27

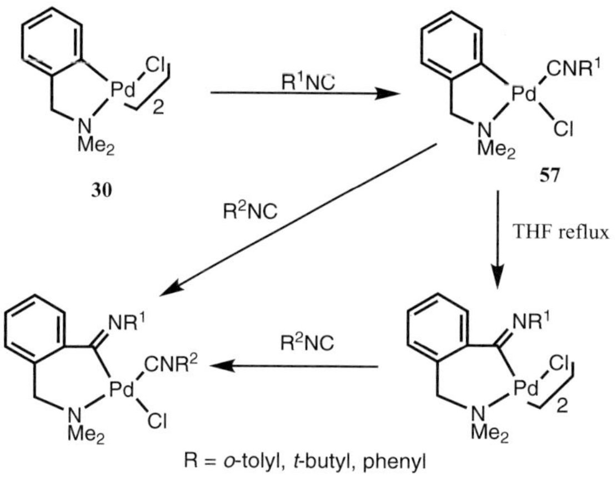

30 57 59 58

Scheme 5.28

5.5
Reaction with Isocyanides

The reaction of palladacycles with isocyanides allows easy formation of C–C bonds. Yamamoto and Yamazaki [64] reported that reactions of azobenzene palladacycles with isocyanides only gave compounds with a coordinated isocyanide group, whereas treatment of *N,N*-dimethylbenzylamine complexes of palladium(II) with isocyanides gave the monocoordinated complexes **57**, which when heated undergo intramolecular insertion of the coordinated isocyanide (R = *o*-tolyl, phenyl) into the carbon–metal bond to give **58**. Also, treatment of **57** with the appropriate isocyanide in a 1 : 1 molar ratio gives **59**, except for Bu^{t}NC, which leads to recovery of the starting material (Scheme 5.28).

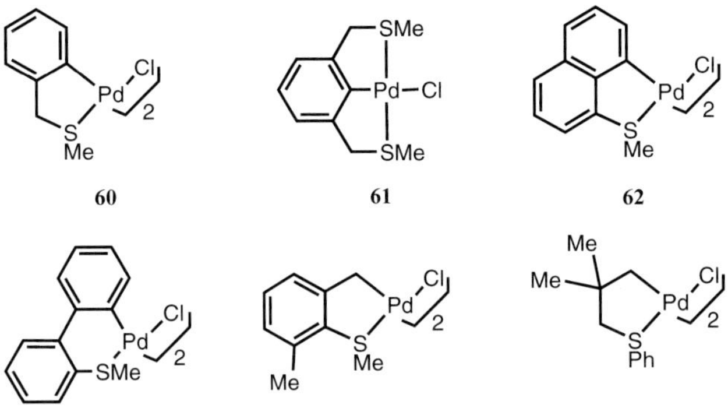

60 61 62 63 64 65

Figure 5.2 Some palladacycles bearing sulfur-donor ligands that have been used in isocyanide insertions.

Diisocyanide products could be obtained from **57** and **58** by further treatment with isocyanides. Decoordination of the NMe_2 group was not observed even in the presence of excess isocyanide. The insertion of isocyanides has also been observed into the Pd—C bond of primary amine palladacycles [65]. As with insertion of carbon monoxide, the insertion of isocyanides into the Pd—C bond occurs much more readily in compounds where an SMe unit has been replaced by an NMe_2 group. Isocyanide insertions have been performed with palladacycles bearing sulfur-donor ligands (Figure 5.2); the most reactive compound being **60**, which is derived from the benzyl methyl thioether ligand [66, 67].

Compound **60** reacts with the corresponding isocyanide to give **67** (Scheme 5.29), which is unstable in solution, rearranging to **69** in 10 h. In several reactions a second equivalent of isocyanide was needed for insertion to take place. Thus, in the reaction of **60** with slightly more than two equivalents of isocyanide the instantaneous formation of **69** is observed. Bridge-splitting reactions of **66** with another equivalent of isocyanide give **68**. Compound **61** is markedly less reactive than **60**. This is obviously due to the effect of the two SMe groups, which prevent migration of the aryl carbon atom on a RNC unit coordinated to palladium. Treatment of **62** with excess benzyl isocyanide gave a dimer compound analogous to **69**. With stoichiometric amounts of RNC (R=Ph, CH_2Ph, *t*-Bu) cleavage of the chloride bridges readily took place, affording monomeric complexes of type **67**. However, when the reaction was carried out in refluxing chlorobenzene, inserted dimeric complexes similar to **66** were obtained. The other sulfur-containing palladacycles show behavior analogous to that of **60**.

These reactions may occur via an associative process, as for the reaction of RNC with $[Pd(X)R'(PPh_3)_2]$ [66]. This process involves the formation of a five-coordinate species from which the σ-bonded aryl group migrates to a cis adjacent isocyanide to afford the insertion product (Scheme 5.30). It is reasonable to assume that decoordination of the sulfur atom is not essential for migration to take place – in agreement with results reported for the related reaction with benzylamine palladacycles. For compound **61** decoordination of one of the sulfur atoms is obviously necessary prior to coordination of the isocyanide to the palladium atom.

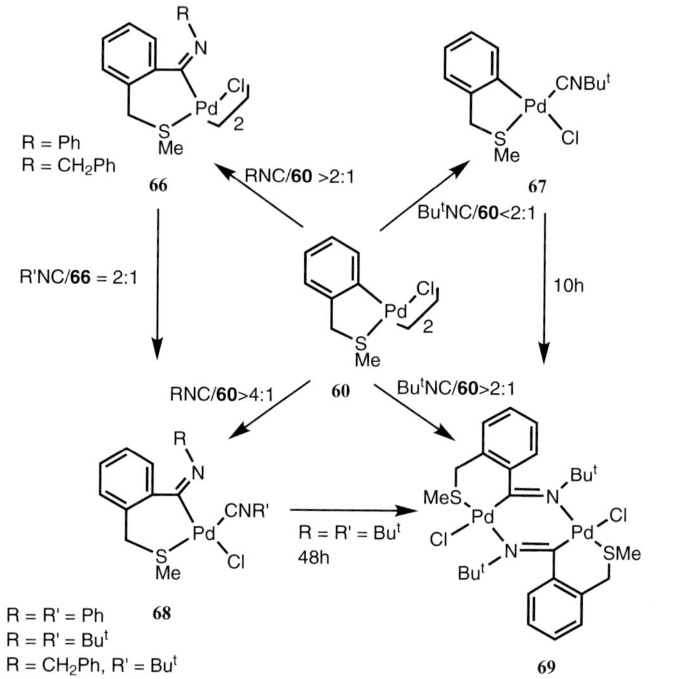

Scheme 5.29

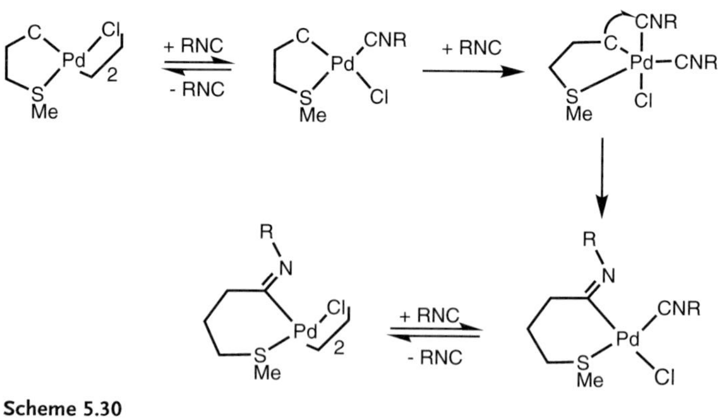

Scheme 5.30

5.6
Reaction with Allenes

The reaction of allenes with palladacycles proceeds through initial formation of a η^3-allylpalladium complex. This intermediate results from the insertion of the allene into the Pd–C bond of the starting cyclopalladated complex [68]. A carbon–carbon bond is formed between the previously metallated carbon and the central electrophilic carbon of the allene molecule. Two types of nucleophilic attack of the

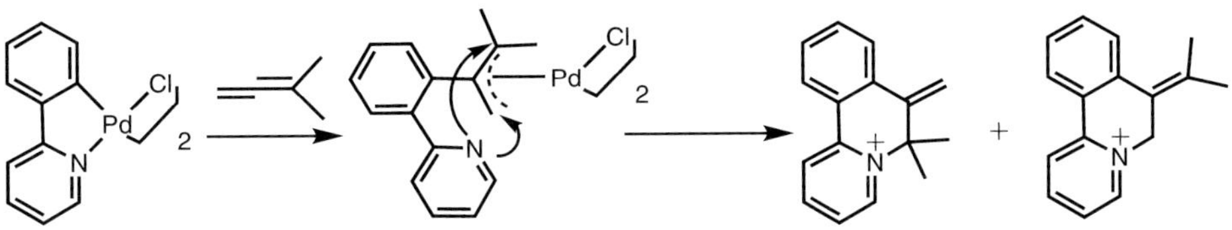

70 71 72 73

Scheme 5.31

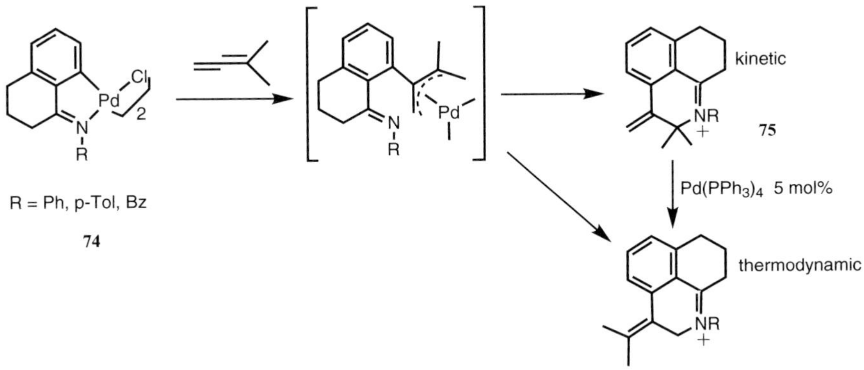

74 75 76

Scheme 5.32

intramolecular nitrogen on the metal-allyl complex are possible, thus giving rise to two regioisomers, **72** and **73** (Scheme 5.31).

The reaction is under kinetic control and attack of the nitrogen atom on the allyl unit is through the more substituted allyl carbon with formation of the less crowded transition state. Thus, **72** is favored as it is the less congested isomer, and thereby the most stable. However, for benzylpyridine and aminopyridine palladacycles, nitrogen attack on the less substituted carbon is preferred and isomer type **73** is formed. A second-order rate constant was found for this process.

Insertion of 1,1-dimethylallene into the Pd—C bond of cyclopalladated α-tetralone ketimines **(74)** affords heterocyclic compounds after nucleophilic attack of the nitrogen-donor atom on the allyl. Ring closure towards **75** is achieved by refluxing in methanol, which shows a high preference for the kinetic product. Refluxing **75** in the presence of a catalytic amount of $Pd(PPh_3)_4$ leads to the thermodynamically more favored **76** [69] (Scheme 5.32).

A mini-library of cationic heterocycles has been prepared by allene insertion into the Pd—C bond of cyclopalladated complexes [70]; the resulting quaternized N-heterocycles were examined in the field of life sciences. Other reactions currently employed that proceed with allene insertion are the synthesis of carbocyclic products [71].

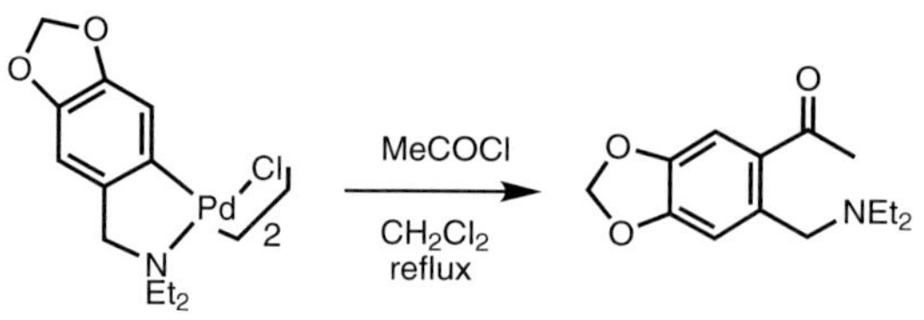

77 78

Scheme 5.33

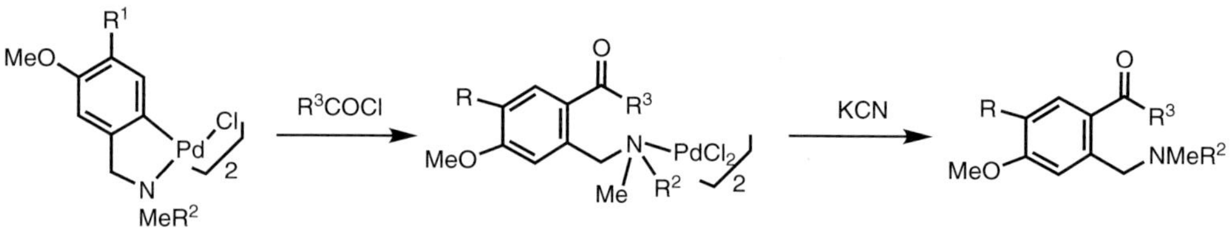

79 80 81

R^1 = H, OMe
R^2 = Me, CH_2CO_2Et

R^3 = CH_2Ph, $CHPh_2$, CH=CHPh, CH_2CH_2Ph, CH(Me)Ph

Scheme 5.34

5.7 Reactions with Acyl Halides

Palladacycles react with acyl halides to give the 2-acyl derivative in good yield (Scheme 5.33) [72].

Benzoyl chloride was found to be less reactive towards palladacycles. Complexes in which the aromatic ring is unactivated react more slowly, and electron-withdrawing substituents on the aromatic ring appear to drastically reduce the rate. Despite the large rate enhancement due to electron-releasing substituents, the regiochemistry of the reaction seems not to change. The introduced acyl group is situated exclusively on the carbon previously bound to palladium. Thus, although electron-releasing groups increase the rate of this reaction they appear to have no effect on the orientation of the entering group [72]. A palladium intermediate **(80)** has been claimed [73], with the final product **(81)** being obtained after treatment with potassium cyanide (Scheme 5.34).

5.8 Reaction with Halogens

Direct halogenation of arenes is an electrophilic reaction that affords almost exclusively para-substituted products; however, the use of palladacycles may drastically change the selectivity in favor of ortho-halogenated compounds, **83** and **84** (Scheme 5.35) [74].

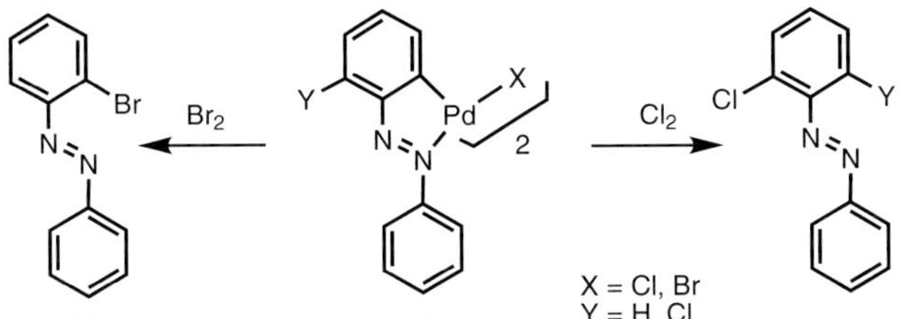

84 82 83

Scheme 5.35

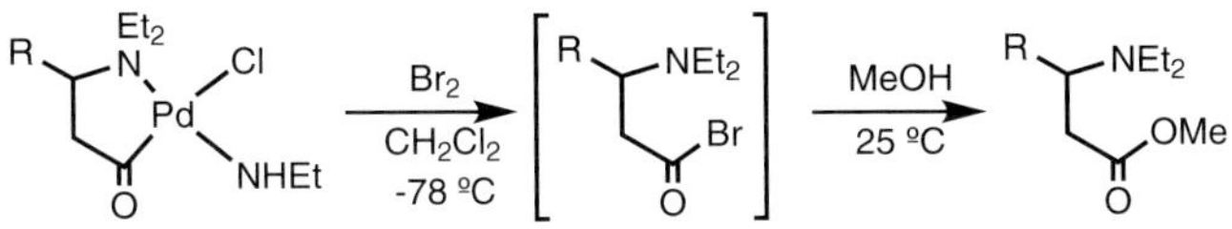

85 86

Scheme 5.36

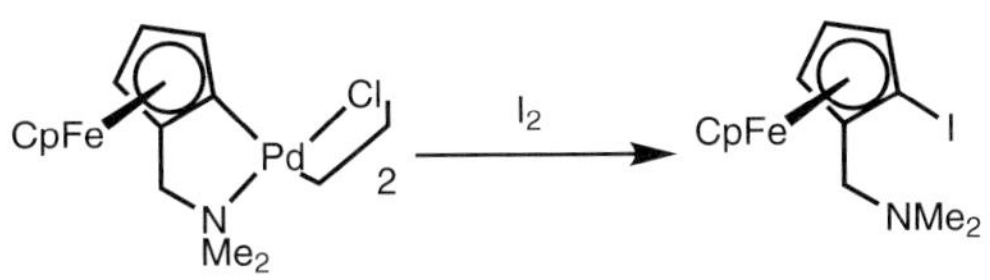

47 87

Scheme 5.37

Halogenation of the platinum analogues proved to be unsuccessful, and the compounds retained the platinum–carbon bond. (β-Aminoacyl)palladium complexes **(85)** were converted efficiently into β-amino acid derivatives **(86)** in quantitative yield by bromination at −78 °C [75] (Scheme 5.36).

Bromination of acetanilide [10] and 2-pyridylferrocene [14] palladacycles gave the corresponding ortho-bromo products. Reaction of cyclopalladated dimethylaminoferrocene with iodine gave the iodo-substituted species **87** [76] (Scheme 5.37).

5.9 Conclusions

After carefully viewing this chapter the reader should have perceived that the reactions of palladacycles that involve cleavage of the Pd—C bond are both large in number and have been profusely investigated; the latter being primarily because they either produce an increase in the links of the metallated chain to give larger rings, thus rendering a plethora of yet unknown palladacycles, or merely because they yield altogether new organic species. In both cases insertion processes have been invoked and the results known so far provide new scope for future discoveries in the field of palladium metallacycles. This is because the overall insertion reaction benefits from the important and interesting properties of the varied palladacycle family, such as their facile synthesis, easy handling and the possibility of modulating their electronic and steric characteristics, properties that may be, totally or in part, extendible to the ensuing compounds, whether they be the inserted products themselves or the resulting organic molecules. The ever growing number of palladacycles and their diversity will broaden the scope of organometallic and organic derivatives prepared from them and thus widen the range for future new applications and/or improvement of already known ones, signifying that this field is far from exhausted. Rather, it moves on to encompass a flourishing field of chemistry of the utmost importance to both inorganic and organic synthetic chemists. Of course there are issues still to be elucidated completely, such as the design of reaction systems with regeneration of the metal in an oxidation state that will allow it to be efficient for a new intramolecular C—H activation, a move further from the stoichiometric processes. Nevertheless, future investigations will be looking at synthesizing many new and otherwise difficult to make inorganic and organic materials from palladacycles, and this is where the properties of the latter shall prove to be paramount; for instance, we may envisage the possibility of making compounds with a functional group by either introducing it on the palladacycle prior to insertion or, conversely, by attaching it subsequently via an appropriate insertion reaction. In this respect, as well as in others already mentioned, the extent of the boundaries to be pursued is limited only by our imagination.

References

1 Dupont, J., Consorti, C.S. and Spencer, J. (2005) *Chemical Reviews*, **105**, 2527.
2 Omae, I. (2004) *Coordination Chemistry Reviews*, **248**, 995.
3 Pfeffer, M. (1990) *Recueil des Travaux Chimiques des Pays-Bas*, **109**, 567.
4 Ryabov, A. (1985) *Synthesis*, 233.
5 Takahashi, H. and Tsuji, J. (1967) *Journal of Organometallic Chemistry*, **10**, 511.
6 Thompson, J.M. and Heck, R.F. (1975) *Journal of Organic Chemistry*, **40**, 2667.
7 Dupont, J., Pfeffer, M., Daran, J.C. and Jeannin, Y. (1987) *Organometallics*, **6**, 899.
8 Carbayo, A., Cuevas, J.V. and García-Herbosa, G. (2002) *Journal of Organometallic Chemistry*, **658**, 15.
9 Cartoon, M.E.K. and Cheeseman, G.W.H. (1982) *Journal of Organometallic Chemistry*, **234**, 123.
10 Horino, H. and Inoue, N. (1981) *Journal of Organic Chemistry*, **46**, 4416.
11 Sokolov, V.L., Troitskaya, L.L. and Reutov, O.A. (1979) *Journal of Organometallic Chemistry*, **182**, 537.
12 Kasahara, A., Izumi, T. and Watabe, H. (1979) *Bulletin of the Chemical Society of Japan*, **52**, 957.
13 Ryabov, A.D., Firsova, Y.N., Goral, V.N., *et al.* (1998) *Chemistry–A European Journal*, **4**, 806.
14 Kasahara, A., Izumi, T. and Maemura, M. (1977) *Bulletin of the Chemical Society of Japan*, **50**, 1878.
15 Tollari, S., Cenini, S., Tunice, C. and Palmisano, G. (1998) *Inorganica Chimica Acta*, **272**, 18.
16 Tollari, S., Demartin, F., Cenini, S., *et al.* (1997) *Journal of Organometallic Chemistry*, **527**, 93.

17 Weinberg, E.L., Hunter, B.K. and Baird, M.C. (1982) *Journal of Organometallic Chemistry*, **240**, 95.

18 Onoue, H., Nakagawa, K. and Moritani, I. (1972) *Journal of Organometallic Chemistry*, **35**, 217.

19 Izumi, T., Katou, T., Kasahara, A. and Hanaya, K. (1978) *Bulletin of the Chemical Society of Japan*, **51**, 3407.

20 Pfeffer, M., Grandjean, D. and Le Borgne, G. (1981) *Inorganic Chemistry*, **20**, 4426.

21 Hiraki, K., Fuchita, Y. and Takakura, S. (1981) *Journal of Organometallic Chemistry*, **210**, 273.

22 Hiraki, K., Fuchita, Y. and Takechi, K. (1981) *Inorganic Chemistry*, **20**, 4316.

23 Girling, I.R. and Widdowson, D.A. (1982) *Tetrahedron Letters*, **23**, 1957.

24 Fuchita, Y., Hiraki, K., Yamaguchi, T. and Maruta, T. (1981) *Journal of The Chemical Society–Dalton Transactions*, 2405.

25 Brisdon, B.J., Nair, P. and Dyke, S.F. (1981) *Tetrahedron*, **37**, 173.

26 Ryabov, A.D., Sakodinskaya, I. and Yatsimirsky, A.K. (1991) *Journal of Organometallic Chemistry*, **406**, 309.

27 Izumi, T., Watabe, H. and Kasahara, A. (1981) *Bulletin of the Chemical Society of Japan*, **54**, 1711.

28 Izumi, T., Endo, K., Saito, O., *et al.* (1978) *Bulletin of the Chemical Society of Japan*, **51**, 663.

29 Pfeffer, M., Sutter, J.P., de Cian, A. and Fischer, J. (1994) *Inorganica Chimica Acta*, **220**, 115.

30 Holton, R.A. (1977) *Tetrahedron Letters*, **24**, 355.

31 Horino, H. and Inoue, N. (1979) *Tetrahedron Letters*, **26**, 2403.

32 Chao, C.H., Hart, D.W., Bau, R. and Heck, R.F. (1979) *Journal of Organometallic Chemistry*, **179**, 301.

33 Girling, I.R. and Widdowson, D.A. (1982) *Tetrahedron Letters*, **23**, 4281.

34 Barr, N., Dyke, S.F. and Quessy, S.N. (1983) *Journal of Organometallic Chemistry*, **253**, 391.

35 Sokolov, V.I. (1983) *Pure and Applied Chemistry*, **55**, 1837.

36 Sokolov, V.I., Troitskaya, L.L. and Khrushchova, N.S. (1983) *Journal of Organometallic Chemistry*, **250**, 439.

37 Maitlis, P.M., Espinet, P. and Russell, M.J.H. (1982), in *Comprehensive Organometallic Chemistry*, Vol. **6** (eds G. Wilkinson, F.G.A. Stone and E.W. Abel), Pergamon Press, Oxford, p. 455.

38 Davies, J.A. (1995) *Comprehensive Organometallic Chemistry*, Vol. **9** (eds G. Wilkinson, F.G.A. Stone and E.W. Abel), Pergamon Press, Oxford, p. 291.

39 Spencer, J., Pfeffer, M., Kyritsakas, N. and Fischer, J. (1995) *Organometallics*, **14**, 2214.

40 Massarani, F., Pfeffer, M. and Le Borgne, G. (1987) *Journal of the Chemical Society D–Chemical Communications*, 565.

41 Wu, G., Geib, S.J., Reihgold, A.L. and Heck, R.F. (1988) *Journal of Organic Chemistry*, **53**, 3238.

42 Albert, J., Granell, J., Luque, A., *et al.* (2006) *Polyhedron*, **25**, 793.

43 Wu, G., Rheingold, A.L. and Heck, R.F. (1987) *Organometallics*, **6**, 2386.

44 Pfeffer, M. (1992) *Pure and Applied Chemistry*, **64**, 335.

45 Massarani, F., Pfeffer, M. and LeBorgne, G. (1990) *Organometallics*, **9**, 3003.

46 Dupont, J., Pfeffer, M., Theurel, L., *et al.* (1990) *New Journal of Chemistry*, **15**, 551.

47 Massarani, F., Pfeffer, M., Spencer, J. and Wehman, E. (1994) *Journal of Organometallic Chemistry*, **466**, 265.

48 Spencer, J., Pfeffer, M., DeCian, A. and Fischer, J. (1995) *Journal of Organic Chemistry*, **60**, 1005.

49 Pérez, S., López, C., Caubet, A., *et al.* (2006) *Organometallics*, **25**, 596.

50 Massarani, F., Pfeffer, M. and LeBorgne, G. (1987) *Organometallics*, **6**, 2029.

51 Massarani, F., Pfeffer, M. and LeBorgne, G. (1987) *Organometallics*, **6**, 2043.

52 Wu, G., Rheingold, A.L., Geib, S.J. and Heck, R.F. (1987) *Organometallics*, **6**, 1941.

53 Tao, W., Silverberg, L.J., Rheingold, A.L. and Heck, R.F. (1989) *Organometallics*, **8**, 2550.

54 Bahsoun, A., Dehand, J., Pfeffer, M., *et al.* (1979) *Journal of The Chemical Society–Dalton Transactions*, 547.

55 Taylor, S.H. and Maitlis, P.M. (1978) *Journal of the American Chemical Society*, **100**, 4700.

56 Dupont, J., Pfeffer, M., Daran, J.C. and Gouteron, J. (1988) *Journal of The Chemical Society–Dalton Transactions*, 2421.

57 Massarani, F., Pfeffer, M. and LeBorgne, G. (1986) *Journal of the Chemical Society D–Chemical Communications*, 489.

58 Vicente, J., Saura-Llamas, I., Palin, M.G. and Jones, P.G. (1995) *Journal of The Chemical Society–Dalton Transactions*, 2535.

59 Benito, M., López, C., Morván, X., *et al.* (2000) *Journal of The Chemical Society–Dalton Transactions*, 4470.

60 Pfeffer, M., Rotteveel, M.A., Sutter, J.P., DeCian, A. and Fischer, J. (1989) *Journal of Organometallic Chemistry*, **371**, C21.

61 López, C., Soláns, X. and Tramuns, D. (1994) *Journal of Organometallic Chemistry*, **471**, 265.

62 López, C., Pérez, S., Soláns, X. and Font-Bardía, M. (2005) *Journal of Organometallic Chemistry*, **690**, 228.

63 Spencer, J. and Pfeffer, M. (1998) *Advances in Metal Organic Chemistry*, **6**, 103.

64 Yamamoto, Y. and Yamazaki, H. (1980) *Inorganica Chimica Acta*, **41**, 229.

65 Vicente, J., Saura-Llamas, I., Grünwald, C. and Alcaraz, C. (2002) *Organometallics*, **21**, 3587.

66 Dupont, J. and Pfeffer, M. (1990) *Journal of The Chemical Society–Dalton Transactions*, 3193.

67 Canty, A.J. (1995) *Comprehensive Organometallic Chemistry*, Vol. **9** (eds G. Wilkinson, F.G.A. Stone and E.W. Abel), Pergamon Press, Oxford, p. 225.

68 Chengebroyen, J., Linke, M., Robitzer, M., *et al.* (2003) *Journal of Organometallic Chemistry*, **687**, 313.

69 Diederen, J.J.H., Frühauf, H.W., Hiemstra, H., Vrieze, K. and Pfeffer, M. (1998) *Tetrahedron Letters*, **39**, 4111.

70 Sirlin, C., Chengebroyen, J., Konrath, R., *et al.* (2004) *European Journal of Inorganic Chemistry*, 1724.

71 Gai, X., Grigg, R., Collard, S. and Muir, J.E. (2000) *Chemical Communications*, 1765.

72 Holton, R.A. and Natalie, K.J. Jr (1981) *Tetrahedron Letters*, **22**, 267.

73 Clark, P.W., Dyke, H.J., Dyke, S.F. and Perry, G. (1983) *Journal of Organometallic Chemistry*, **253**, 399.

74 Hahey, D.R. (1971) *Journal of Organometallic Chemistry*, **27**, 283.

75 Hegedus, L.S., Anderson, O.P., Zetterberg, K., *et al.* (1977) *Inorganic Chemistry*, **16**, 1887.

76 Onishi, M., Hiraki, K. and Iwamoto, A. (1984) *Journal of Organometallic Chemistry*, **262**, C11.

6
C—H Activations via Palladacycles

John Spencer

6.1
Introduction: C—C Bond Formation via Cyclopalladation Reactions

A C—H activation reaction is perhaps the most commonly employed synthetic route to palladacycles [1]. The thus-formed palladacycle products have been shown historically to be effective stoichiometric agents [2] (see also Chapter 5), and more recently to be precatalysts for C—C and C—heteroatom bond formations involving aryl halides, preferably chlorides due to their greater availability, lower cost and lower MW (albeit lower reactivity) compared with their bromo or iodo congeners [3]. Recently, palladacycles have emerged as intermediates or precatalysts in a host of interesting *atom economical* catalytic transformations involving the activation or functionalization of a C—H bond as opposed to the insertion of the metal into a C—X (X=Cl, Br, I, etc.) bond [4]. Hereafter, extremely important, C—H activation will be presented (exemplified in Scheme 6.1), and it will be demonstrated that intramolecular coordination to the metal and the formation of palladacycles as either precatalysts or intermediates can help drive the high selectivity in these synthetically powerful processes.

6.2
Stoichiometric C—H Activation Chemistry

Several earlier reports from Holton's group highlighted the significant synthetic achievements that are possible using a stoichiometric palladacycle template **1**, formed by a C—H activation, to direct a C—C bond formation, as in the biomimetic synthesis of narwedine (Scheme 6.2) [5].

A series of C—H activations of sp^3 C—H bonds combining a transmetallation with a vinylboronic acid moiety and a carbonylation were reported by Sames' group in a synthesis leading to teleocidin BIX4 (Scheme 6.3). Palladacycles of the N,C,O-

Palladacycles: Synthesis, Characterization and Applications. Edited by Jairton Dupont and Michel Pfeffer

ISBN: 978-3-527-31781-3

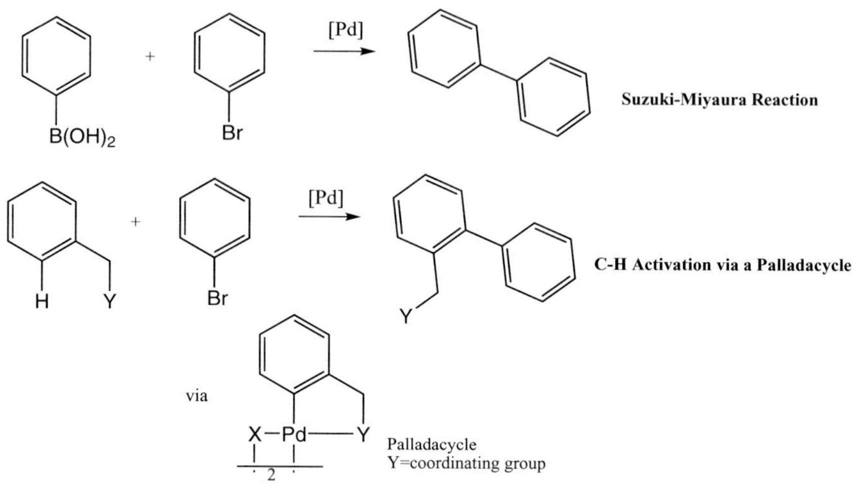

Scheme 6.1 Traditional Suzuki–Miyaura couplings versus the C—H activation pathway.

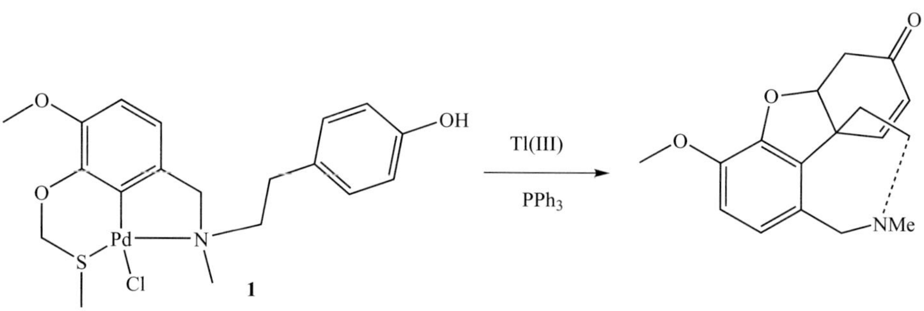

Scheme 6.2 Palladacycle-mediated synthesis of narwedine.

pincer type (**2** and **3**) were used as stoichiometric templates to enable these C—C bond-forming processes [6a].

Stoichiometric vinylations of imine-based palladacycles, reported barely twenty years ago, led to ortho-substituted benzaldehydes following acid hydrolysis [7]. These processes have progressed to catalytic vinylations of haloanilines [8] that may involve palladacyclic intermediates, and to catalytic C—H activations (see Scheme 6.5 below).

The stoichiometric coupling of haloolefins with palladacycles has been described recently. For the reaction of palladacycle **4** with the bromo-olefin depicted in Scheme 6.4, the reactivity of the olefin component was in the order trans-bromo > cis-bromo > cis-iodo [9].

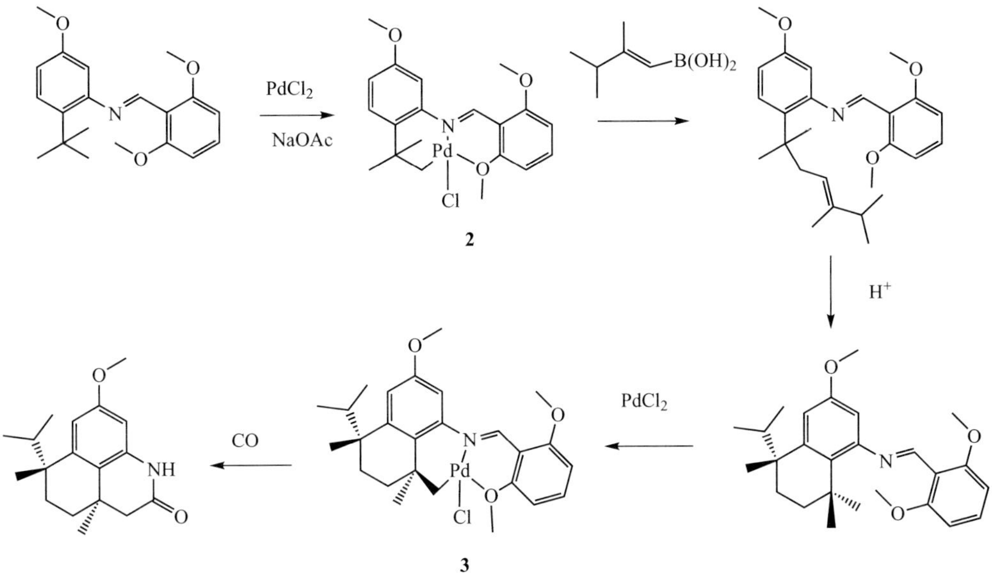

Scheme 6.3 A series of stoichiometric C—H activation reactions via palladacycles.

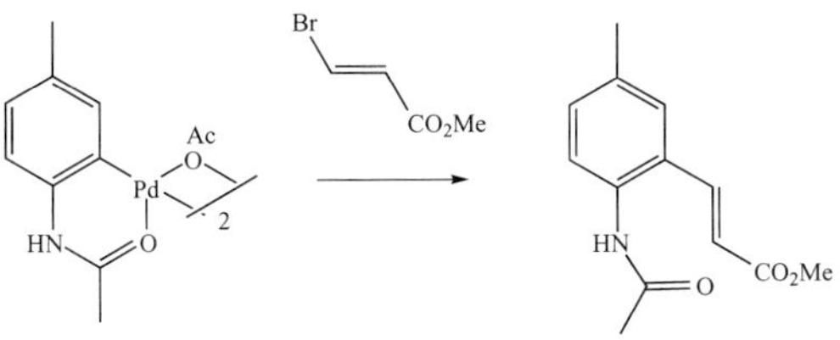

4

Scheme 6.4 Stoichiometric vinylation of a palladacycle with a bromo-olefin.

6.3 Catalytic Chemistry

6.3.1 Vinylations

A catalytic Pd-catalyzed vinylation has been reported that employs benzoquinone as a reoxidant and a vinylsilane. An N,C,S-pincer palladacycle intermediate resulting from C—H activation of the *tert*-butyl group of the starting material was postulated (Scheme 6.5) [6b].

Ambient temperature ortho-alkenylations of anilides catalyzed by palladium have been reported, highlighting the huge strides achieved in catalytic C—H activation chemistry [9a]. Similar Heck-type reactions (Scheme 6.6), involving a C—H activation, have been disclosed, where the regiochemistry was governed by the

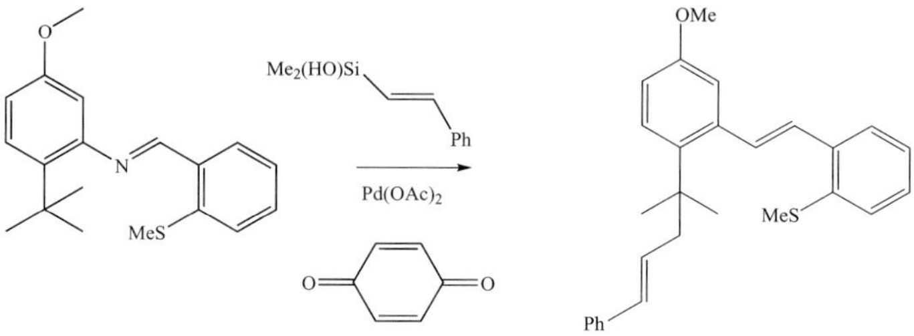

Scheme 6.5 Catalytic C–H activation/vinylation chemistry via a palladacycle intermediate.

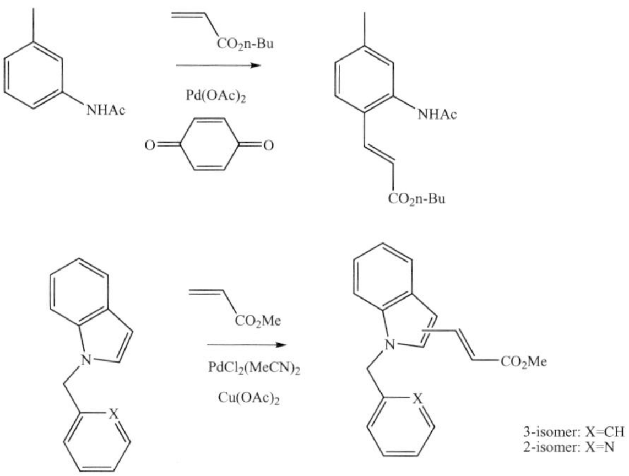

Scheme 6.6 Catalytic palladium-mediated Heck-type reactions involving C–H activation via palladacycles.

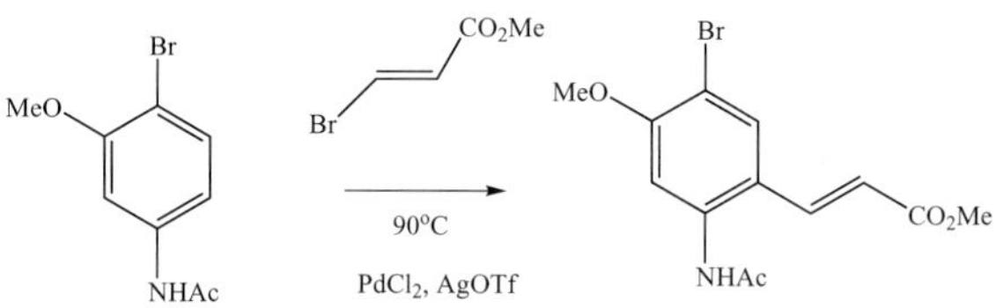

Scheme 6.7 Catalytic vinylations of bromo-olefins.

presence or absence of a directing group (2-isomer versus 3-isomer, respectively) [9b].

A catalytic version of the reaction presented in Scheme 6.4 has been developed. A bromide substituent is tolerated on one of the aryl coupling partners. Similar values for intra- and intermolecular deuterium isotope effects were observed (3.6 versus 3.7), suggesting that C–H bond cleavage was the turnover limiting step. These reactions, however, require a stoichiometric amount of a silver(I) salt (Scheme 6.7) [10].

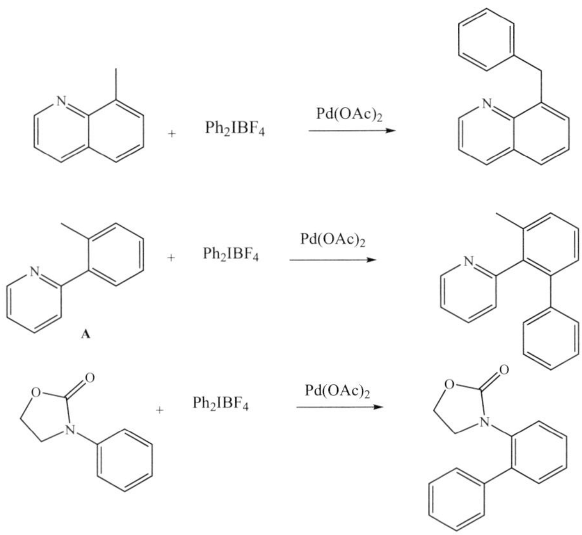

Scheme 6.8 C–H activation/arylations.

6.4 Arylations

The direct coupling of an aryl group with a C–H bond is a very attractive, atom economical process and many of these reactions are highly regioselective and are directed by the coordination of, for example, an amine or amide group to palladium. Both sp^3 and sp^2 C–H activation/arylation processes have been developed employing palladium catalysis and iodine(III) derivatives (Scheme 6.8). The arylation of derivative **A** is an important observation in that an sp^2 C–H group is preferentially activated over and sp^3 C–H. Moreover, in some cases, as we will see later, diarylations can be observed and a methyl group can act as a "blocker." A likely mechanism involves a chelation-assisted C–H activation to form a palladacycle, oxidation of the palladacycle [Pd(II) to Pd(IV)] by the iodine(III) agent and reductive elimination to furnish the arylated product (cf. Scheme 6.16 below) [11].

Benzodiazepines undergo C–H activation to yield ortho-arylated analogues, probably via palladacycles (Scheme 6.9). Examples of the latter were characterized by mass spectrometry and by X-ray crystallography [12].

Amides and carbamates also undergo C–H activation/alkylation or arylation reactions, and in some cases diarylation occurs (Scheme 6.10) [13, 14].

The orthoarylation of benzylamines, by iodobenzene derivatives, is also catalyzed by palladium, in the presence of silver acetate and trifluoroacetic acid (Scheme 6.11). An iodide group on the benzylamine coupling partner is tolerated and, surprisingly, reactions were faster for benzylamines carrying electron-donating substituents, which is the opposite trend to that observed for a process impli-

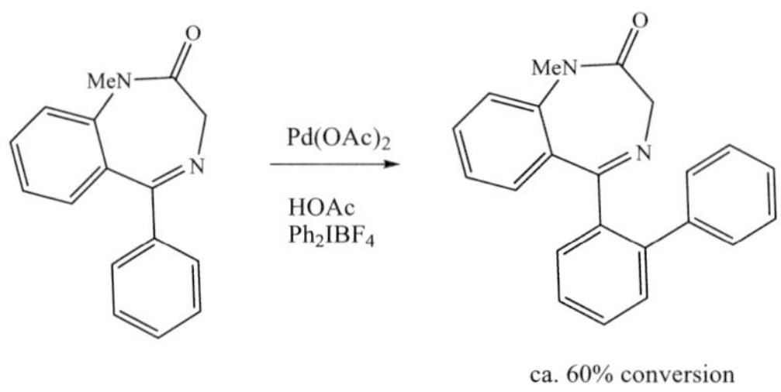

Scheme 6.9 Arylation of a benzodiazepine.

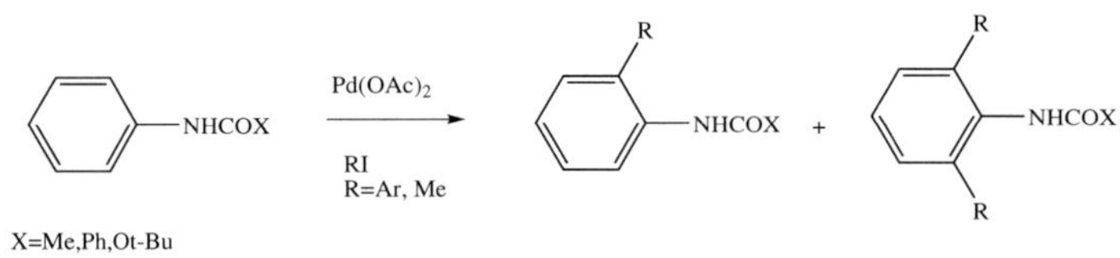

Scheme 6.10 Orthoalkylations and arylations of amides and carbamates.

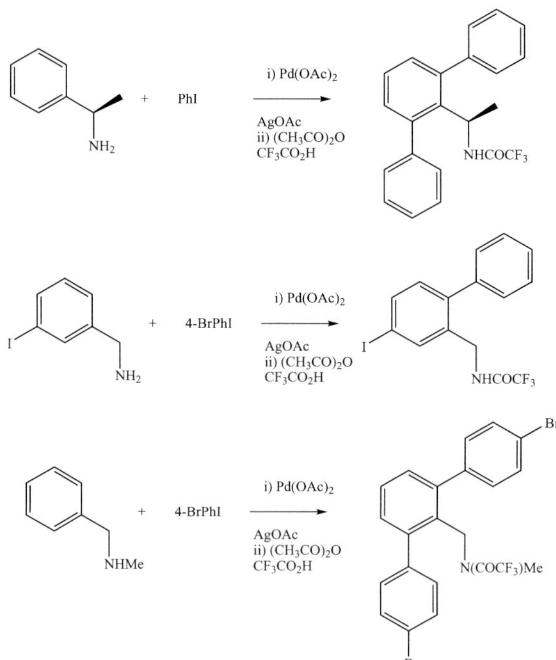

Scheme 6.11 Palladium-mediated arylation of benzylamines.

cating a Pd(0)–Pd(II) cycle. The latter two points suggest a reaction pathway involving a Pd(II)–Pd(IV) cycle [15, 16]. In many cases, diarylation was observed and a bromide substituent could be tolerated on the halide coupling partner.

Various C—H activations of sp^3 C—H bonds that are probably mediated by palladacycles can also be performed; Scheme 6.12 shows representative examples. A mechanistic rationale involves the formation of a palladacycle by C—H activation, oxidation of Pd(II) to Pd(IV) by oxidative addition of ArI. This is followed by reductive elimination of the product and anion metathesis (I to OAc) by the stoichiometric silver salt [16].

The chelation-assisted arylation of sp^3 CH bonds is mediated by palladium catalysts and both mono- and diarylations have been reported (Scheme 6.13) [17].

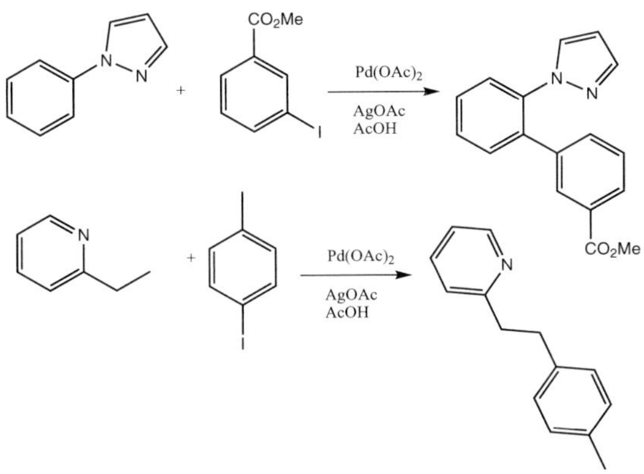

Scheme 6.12 Activation/arylation of sp^2 and sp^3 CH bonds.

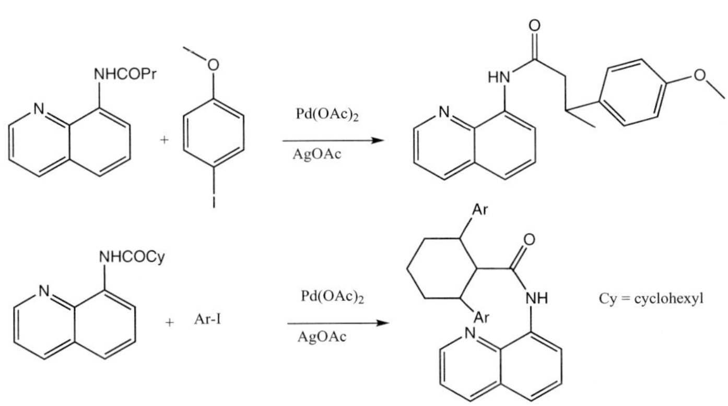

Scheme 6.13 Arylation of sp^3 C—H bonds.

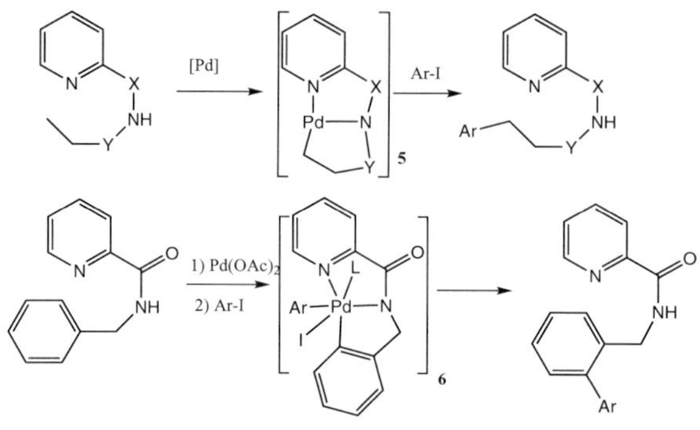

Scheme 6.14 Arylation mechanism involving a pyridine tether.

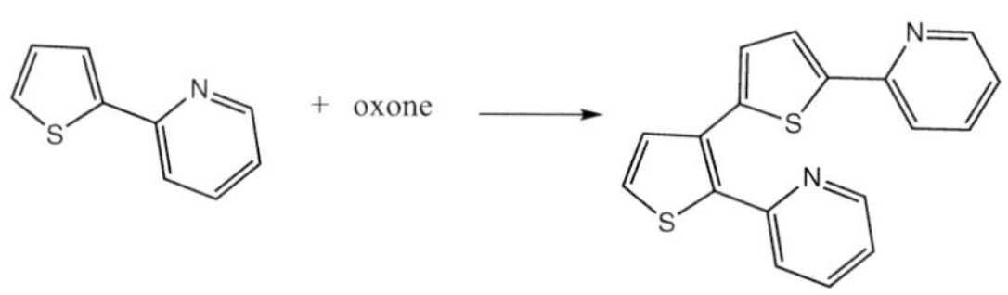

Scheme 6.15 Oxidative couplings leading to biaryls.

High turnovers (TON) of up to 650 were achieved for this reaction and a speculative mechanistic proposal involves palladium amides **5** and **6**, respectively (X, Y = CH_2, CO, NH) (Scheme 6.14). The pyridine auxiliary is removable.

6.5
Direct C—H C—H Coupling Reactions

A further, desirable, step in the quest for efficient coupling reactions is to eliminate the need for an aryl halide coupling partner and to activate two arene C—H bonds. Selective palladium-catalyzed oxidative coupling reactions have been employed for the formation of biaryls via a homocoupling process (Scheme 6.15).

The authors proposed an initial C—H activation process, involving Pd(II), oxidation of the palladacycle **7** to Pd(IV), formation of Pd(IV) palladacycle intermediate **8** and reductive elimination (Scheme 6.16) [18]. Such reactions may be considered to be aryl halide-free variants of Ullmann-type homocoupling processes.

More recent examples involve the cross coupling of aromatics via palladacycles involving the C—H activation of sp^3 and sp^2 C—H bonds (Scheme 6.17). No homocoupling products were observed for these reactions, which, however, require stoichiometric amounts of silver salts and several equivalents of arene coupling partner. Future studies may improve the atom economy of these still highly impressive reactions.

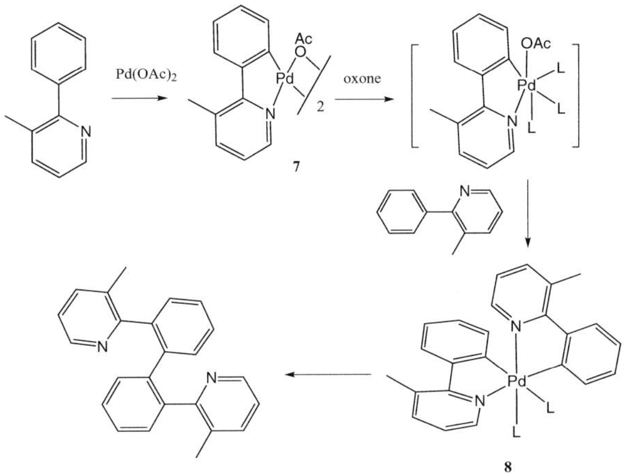

Scheme 6.16 Oxidative coupling mechanism.

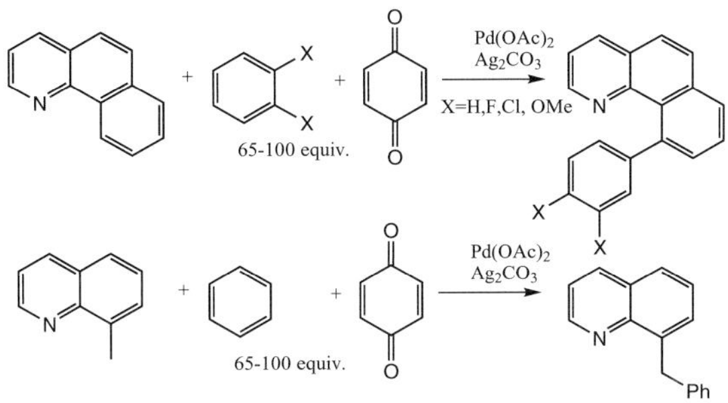

Scheme 6.17 Cross couplings involving C—H bond activations.

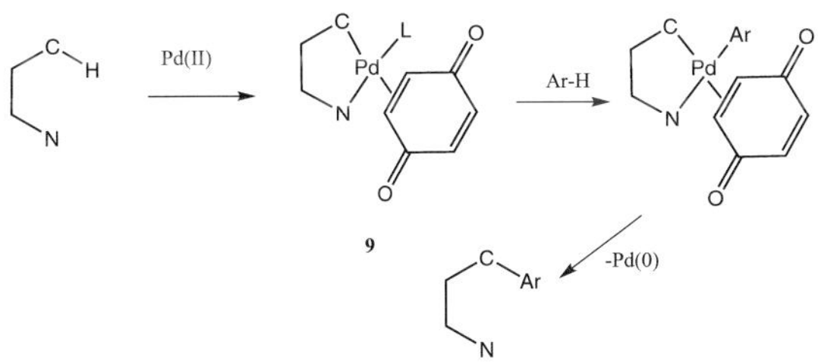

Scheme 6.18 Proposed cross-coupling mechanism.

A putative mechanism involves a directed C—H activation to give the palladacycle **9**, followed by a non-directed C—H activation (Scheme 6.18). Reductive elimination furnishes the ortho-arylated product and Pd(0), which is reoxidized to Pd(II) by the silver salt [19].

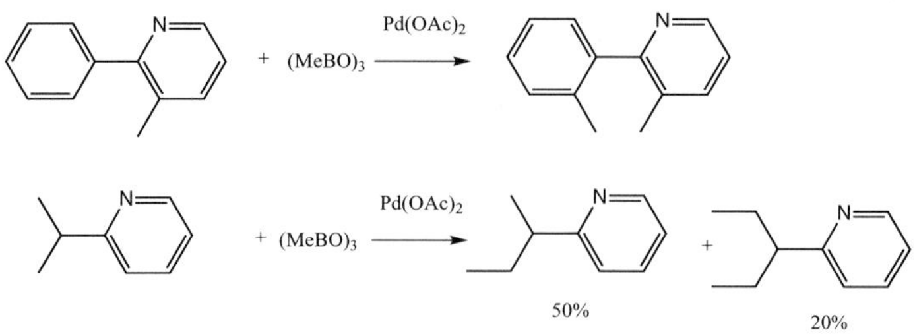

Scheme 6.19 Activation of sp^2 and sp^3 C—H bonds.

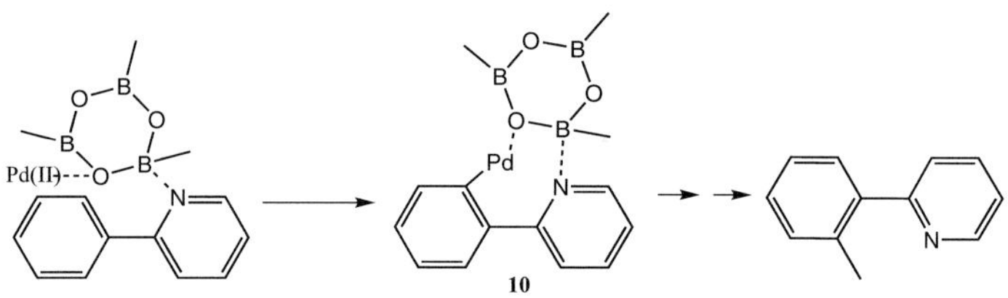

Scheme 6.20 Palladacycle intermediate in the C—H activation/alkylation reaction.

6.6 Alkylations

It is now possible to alkylate aromatic or aliphatic C—H bonds via Pd catalysis. Palladium-catalyzed C—H activation, transmetallation processes have also been reported (Scheme 6.19) using tin reagents [20a] or methylboroxine as transmetallating agent [20b].

A mechanistic rationale suggested the involvement of palladacycle **10** and that C—H cleavage is the rate-limiting step. The authors suggest that an oxygen of the methylboroxine unit can coordinate to Pd(II), a C—H activation (see **10**) and intramolecular transmetallation processes (Scheme 6.20).

6.7 Other Reactions

6.7.1 Carbonylations

Palladacycle intermediate **11** was proposed for a related carbonylation process, and this is supported by the high degree of regioselectivity observed when using a bis-chelating substrate (Scheme 6.21) [21].

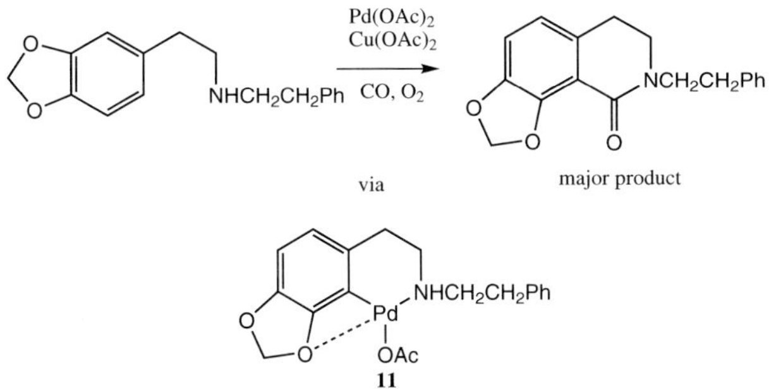

Scheme 6.21 Catalytic C—H activation/carbonylation via a palladacycle intermediate.

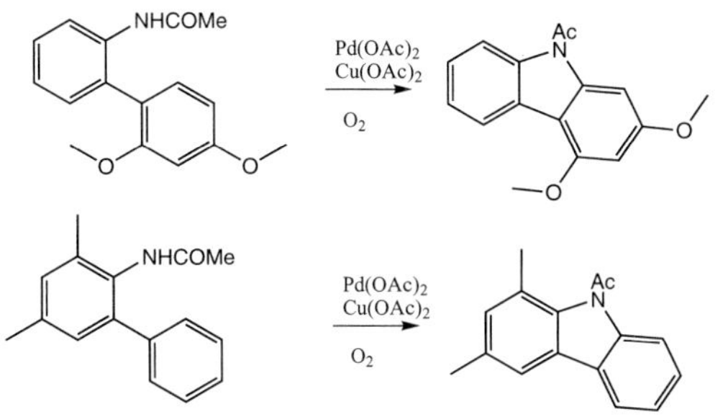

Scheme 6.22 Carbazoles via C—H activation and C–N bond formation.

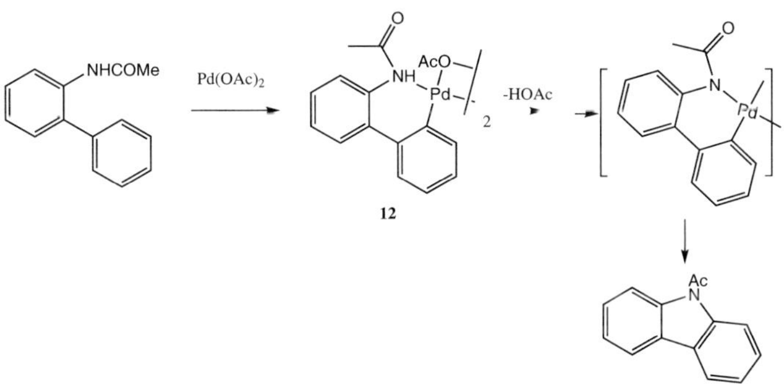

Scheme 6.23 Proposed mechanism for carbazole formation involving a palladacycle.

6.7.2
C—N Bond Formation

The regioselective synthesis of carbazoles involving C—H activation and C—N bond formation has been reported recently (Scheme 6.22) [22].

The proposed mechanism involves a C—H activation (orthometallation) leading to a palladacycle intermediate **12**, which loses acetic acid and undergoes reductive elimination to yield the final carbazole product. A copper(II) salt acts as a reoxidant of the Pd(0) to Pd(II) in a Wacker-type process (Scheme 6.23).

6.8
Conclusion

Synthetically powerful atom economical transformations have been carried out by capitalizing on the unique well-established properties of palladacycles, such as their facile synthesis by chelation-assisted C—H activation, their ease of handling and air stability as well as their use as well-defined intermediates for mechanistic investigations of carbon–carbon coupling processes. Currently, new, somewhat speculative properties, of palladacycles are emerging, such as Pd(IV) chemistry [23] and the surprising revelation that aryl bromides and even iodides are tolerated on one or more coupling partner in novel palladium-mediated coupling reactions, which is uncommon in Pd(0)–Pd(II) chemistry and more likely to involve a Pd(II)–Pd(IV) cycle. This has led to a new era of palladium-catalyzed chemistry involving palladacycles as reagents or intermediates, with exciting prospects for cleaner organic synthesis, applications in fine chemical and pharmaceutical chemistry as well as theoretical, mechanistic and structural organometallic chemistry.

References

1 (a) Parshall, G.W. (1970) *Accounts of Chemical Research*, **3**, 139.
(b) Dehand, J. and Pfeffer, M. (1976) *Coordination Chemistry Reviews*, **18**, 327.
(c) Bruce, M.I. (1977) *Angewandte Chemie – International Edition in English*, **16**, 73.
(d) Omae, I. (1979) *Coordination Chemistry Reviews*, **28**, 97.
(e) Omae, I. (1979) *Chemical Reviews*, **79**, 287.
(f) Omae, I. (1980) *Coordination Chemistry Reviews*, **32**, 235.
(g) Omae, I. (1982) *Journal of the Society of Synthetic Organic Chemistry of Japan*, **40**, 147.
(h) Omae, I. (1982) *Coordination Chemistry Reviews*, **42**, 245.
(i) Constable, E.C. (1984) *Polyhedron*, **3**, 1037.
(j) Rothwell, I.P. (1985) *Polyhedron*, **4**, 177–200.
(k) Newkome, G.R., Puckett, W.E., Gupta, V.K. and Kiefer, G.E. (1986) *Chemical Reviews*, **86**, 451.
(l) Steenwinkel, P., Gossage, R.A. and van Koten, G. (1998) *Chemistry – A European Journal*, **4**, 759.
(m) Herrmann, W.A., Bohm, V.P.W. and Reisinger, C.P. (1999) *Journal of Organometallic Chemistry*, **576**, 23.
(n) Albrecht, M. and van Koten, G. (2001) *Angewandte Chemie – International Edition*, **40**, 3750.
(o) van der Boom, M.E. and Milstein, D. (2003) *Chemical Reviews*, **103**, 1759.
(p) Singleton, J.T. (2003) *Tetrahedron*, **59**, 1837.
(q) Bellina, F., Carpita, A. and Rossi, R. (2004) *Synthesis*, **15**, 2419.
(r) Bedford, R.B., Cazin, C.S.J. and Holder, D. (2004) *Coordination Chemistry Reviews*, **248**, 2283.
(s) Omae, I. (2004) *Coordination Chemistry Reviews*, **248**, 995.
(t) Dunina, V.V. and Gorunova, O.N. (2004) *Russian Chemical Reviews*, **73**, 309.
(u) Dupont, J., Consorti, C.S. and Spencer, J. (2005) *Chemical Reviews*, **105**, 2527.

2 (a) Ryabov, A.D. (1985) *Synthesis*, 233.
(b) Pfeffer, M. (1990) *Recueil des Travaux Chimiques des Pays-Bas*, **109**, 567–76.
(c) Pfeffer, M. (1992) *Pure and Applied Chemistry*, **64**, 335.
(d) Spencer, J. and Pfeffer, M. (1998) *Advances in Metallo-Organic Chemistry*, **6**, 103.

3 (a) Beletskaya, I.P. and Cheprakov, A.V. (2004) *Journal of Organometallic Chemistry*, **689**, 4055.
(b) Farina, V. (2004) *Advanced Synthesis Catalysis*, **346**, 1553.
(c) Dupont, J., Pfeffer, M. and Spencer, J. (2001) *European Journal of Organic Chemistry*, 1917.
(d) Bedford, R.B. (2003) *Chemical Communications*, 1787.

4 (a) Recent reviews: Pfeffer, M. and Spencer, J. (2006) *Comprehensive Organometallic Chemistry III*, Ch 10.2.1 and references cited therein.
(b) Godula, K. and Sames, D. (2006) *Science*, **312**, 67 and references cited therein.
(c) Tosibu, M. and Chatani, N. (2006) *Angewandte Chemie–International Edition*, **45**, 1683 and references cited therein.
(d) Labinger, J.A. and Bercaw, J.E. (2002) *Nature*, **417**, 507 and references cited therein.
(e) Shilov, A.E. and Shul'pin, G.B. (1997) *Chemical Reviews*, **97**, 2879 and references cited therein.
(f) Ritleng, V., Sirlin, C. and Pfeffer, M. (2002) *Chemical Reviews*, **102**, 1731 and references cited therein.
(g) Kakiuchi, F. and Chatani, N. (2003) *Advanced Synthesis Catalysis*, **345**, 1077.

5 Holton, R.A., Sibi, M.P. and Murphy, W.S. (1988) *Journal of the American Chemical Society*, **110**, 314.

6 (a) Dangel, B.D., Godula, K., Youn, S.W., *et al.* (2002) *Journal of the American Chemical Society*, **124**, 11856.
(b) Sezen, B., Franz, R. and Sames, D. (2002) *Journal of the American Chemical Society*, **124**, 13372.

7 Girling, I.R. and Widowson, D.A. (1988) *Journal of the Chemical Society–Perkin Transactions 1*, 1317.

8 Adams, D.R., Duncton, M.A.F., Roffey, J.A.R. and Spencer, J. (2002) *Tetrahedron Letters*, **43**, 7581.

9 (a) Boele, M.D., van Strijdonck, G.P.F., de Vries, A.H.M., *et al.* (2002) *Journal of the American Chemical Society*, **124**, 1586.
(b) Capito, E., Brown, J.M. and Ricci, A. (2005) *Chemical Communications*, 1854.

10 Zaitsev, V.G. and Daugulis, O. (2005) *Journal of the American Chemical Society*, **127**, 4156.

11 Kalyani, D., Deprez, N.R., Desai, L.V. and Sanford, M.S. (2005) *Journal of the American Chemical Society*, **127**, 7330.

12 (a) Spencer, J., Chowdhry, B.Z., Mallet, A.I., *et al.* (2008) Tetrahedron, **64**, 6082.
(b) Spencer, J., Sharratt, D.P., Dupont, J., *et al.* (2005) *Organometallics*, **24**, 5665.

13 (a) Tremont, S.J. and Rahman, H.U. (1984) *Journal of the American Chemical Society*, **106**, 5759.
(b) Kametani, Y., Satoh, T., Miura, M. and Nomura, M. (2000) *Tetrahedron Letters*, **41**, 2655.
(c) Campeau, L.C. and Fagnou, K. (2006) *Chemical Communications*, 1253.

14 Daugulis, O. and Zaitsev, V.G. (2005) *Angewandte Chemie–International Edition*, **44**, 4046.

15 Lazareva, A. and Daugulis, O. (2006) *Organic Letters*, **8**, 5211.

16 (a) Shabashov, D. and Daugulis, O. (2005) *Organic Letters*, **7**, 3657.
(b) Daugulis, O., Zaitsev, V.G., Shabashov, D., *et al.* (2006) *Synlett*, 3382 and references cited therein.

17 (a) Zaitsev, V.G., Shabashov, D. and Daugulis, O. (2005) *Journal of the American Chemical Society*, **127**, 13154.
(b) Bowie, A.L. Jr, Hughes, C.C. and Trauner, D. See also a recent Rhazinilam synthesis, which may involve a palladacyclic intermediate: (2005) *Organic Letters*, **7**, 5207.

18 Hull, K.L., Lanni, E.L. and Sanford, M.S. (2006) *Journal of the American Chemical Society*, **128**, 14047.

19 Hull, K.L. and Sanford, M.S. (2007) *Journal of the American Chemical Society*, **129**, 11904.

20 (a) Chen, X., Li, J.J., Hao, X.S., *et al.* (2006) *Journal of the American Chemical Society*, **128**, 78.
(b) Chen, X., Goodhue, C.E. and Yu, J.Q. (2006) *Journal of the American Chemical Society*, **128**, 12634.

21 Orito, K., Horibata, A., Nakamura, T., *et al.* (2004) *Journal of the American Chemical Society*, **126**, 14342.

22 Tsang, W.C.P., Zheng, N. and Buchwald, S.L. (2005) *Journal of the American Chemical Society*, **127**, 14560.

23 Canty, A.J., Patel, J., Rodemann, T., *et al.* (2004) *Organometallics*, **23**, 3466.

7
Cyclopalladated Compounds as Resolving Agents for Racemic Mixtures of Ligands

Jean-Pierre Djukic

7.1
Introduction

As most palladacycles are by nature Lewis acids, the chemistry that lies behind their use as resolving agents implies a thorough understanding of the kinetics and thermodynamics of ligand coordination. Dimeric halogeno-bridged cyclopalladated complexes are a particular class of transition metal Lewis-acidic compounds. Owing to the lability of the halogeno bridge, reasonable monodentate σ-donating ligands can cleave a Pd—X bond, thus leading to new monomeric species whose stereochemistry around the Pd(II) center is controlled by the prevalence of Pearson's anti-symbiotic effect [1] also named in some cases "transphobia" [2, 3]. Furthermore, depending on the steric and electronic requirements of the incoming ligand, the coordination of the latter to the palladium center may well not be irreversible. The main outcome of the reaction may be the formation of a mixture of equilibrating species, namely dimers, monomers and doubly ligated cationic species. Excess ligand can even effectively displace this equilibrium and lead to the dechelation of the Pd(II) center. With bidentate ligands, the formation of a cationic bischelate is, in principle, favored by entropy: with sterically demanding ligands chelation of the incoming ligand may recede to monodentation. Such cases are dealt with in the part dedicated to monodentate ligand resolution.

These features have historically played an important role in the development of the use of enantiopure cyclopalladated compounds, which can considered a handy class of Lewis-acidic agents for the preparation of scalemic and enantiopure ligands [4]. The ready synthesis of relatively air- and heat-stable scalemic cyclopalladated complexes such as those prepared from enantiopure *N,N*-dimethyl,1-phenylethylamine and *N,N*-dimethyl,1-naphthylamine has offered a means for the so-called *optical resolution* of ligands.

This chapter describes the methods used for the resolution of a large array of ligands: it covers those ligands spanning from main group hetero-element monodentate ligands to bidentate ligands. In these cases the stereogenic centers may

Palladacycles: Synthesis, Characterization and Applications. Edited by Jairton Dupont and Michel Pfeffer

ISBN: 978-3-527-31781-3

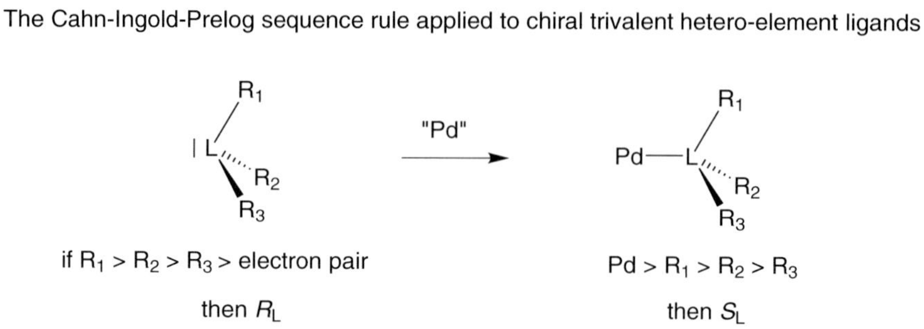

Scheme 7.1

either be at the hetero-element itself (hetero-element-centered chirality) (Scheme 7.1) or remote (carbo-centered and axial chiralities).

7.2
Resolution Methods

Two methods have mainly been used to separate enantiomers from a given racemate *rac*-L consisting of enantiomers L and *ent*-L (Schemes 7.2 and 7.3).

The first method is based on the physical separation of covalent diastereomers [5] produced by the coordination of *rac*-L to an enantiopure palladacycle. This method requires at least three steps:

1. The formation of a mixture of covalent diastereoisomers: this step relies very much on the nature of the incoming ligand, differences being expected between monodentate and bidentate ligands. In some cases chiral recognition is expressed by a markedly unbalanced diastereomeric ratio.
2. The separation of diastereomers by physical means (fractional crystallization, chromatography): this certainly is the most critical step when the physical properties of diastereomers (solubility, polarity) are not known in advance.
3. The release of the enantioenriched/pure ligand: the conditions chosen for release of the enantioenriched ligand rely on a good knowledge of the configurational stability and reactivity of the ligand itself to prevent racemization and overall chemical degradation.

The second method, mostly encountered with monodentate ligands, is based on the resolution of a racemate of which one of the enantiomers, say L, expresses a higher affinity (chiral recognition) towards the host, that is the palladium(II) center, whereas the enantiomer *ent*-L remains mostly unbound. The keys to optimal resolution are (i) pertinent choice of experimental conditions (ligand's concentration, choice of chiral palladium auxiliary) and (ii) selective extraction of the monomeric palladium diastereomer from the solution (generally by precipita-

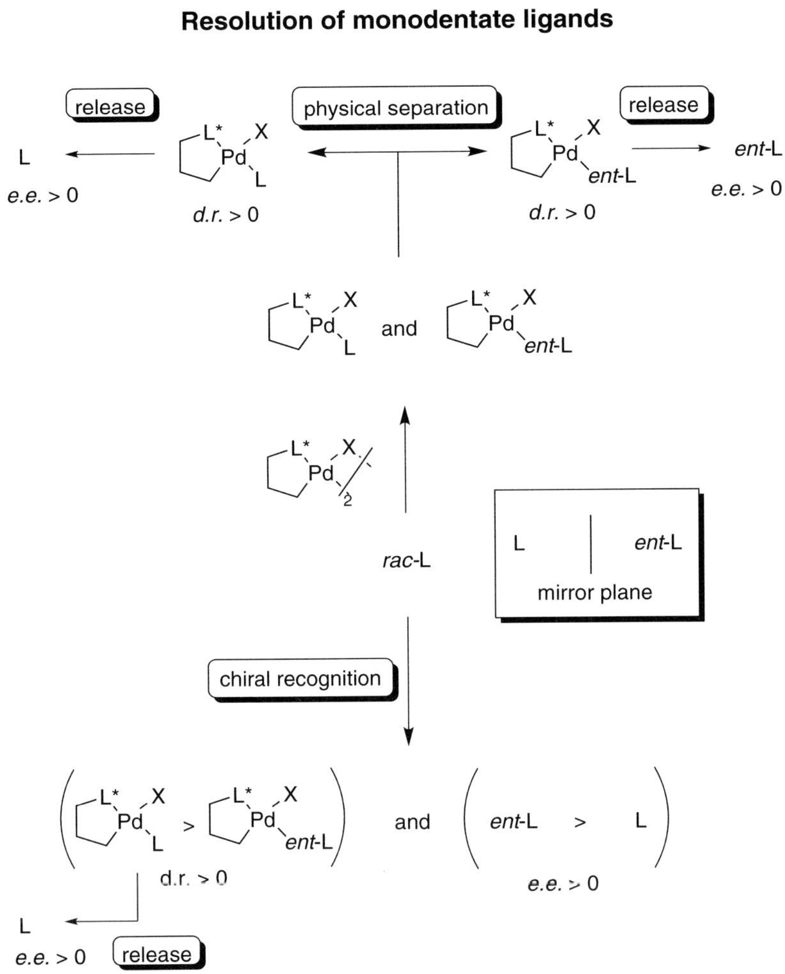

Scheme 7.2

tion) as it is being formed. Of utmost importance is knowledge of the factors that influence the efficiency and selectivity of the resolution.

7.3
Chiral Palladacyclic Auxiliaries

One may distinguish four main classes of palladacyclic chiral auxiliaries, which have been used in various ways for the resolution of mono- and bidentate ligands (Figure 7.1). The first class embraces a wide array of substituted benzylamine derivatives in which chirality is most often centered at the benzylic position. Several cases containing N-centered chirality have also been also developed.

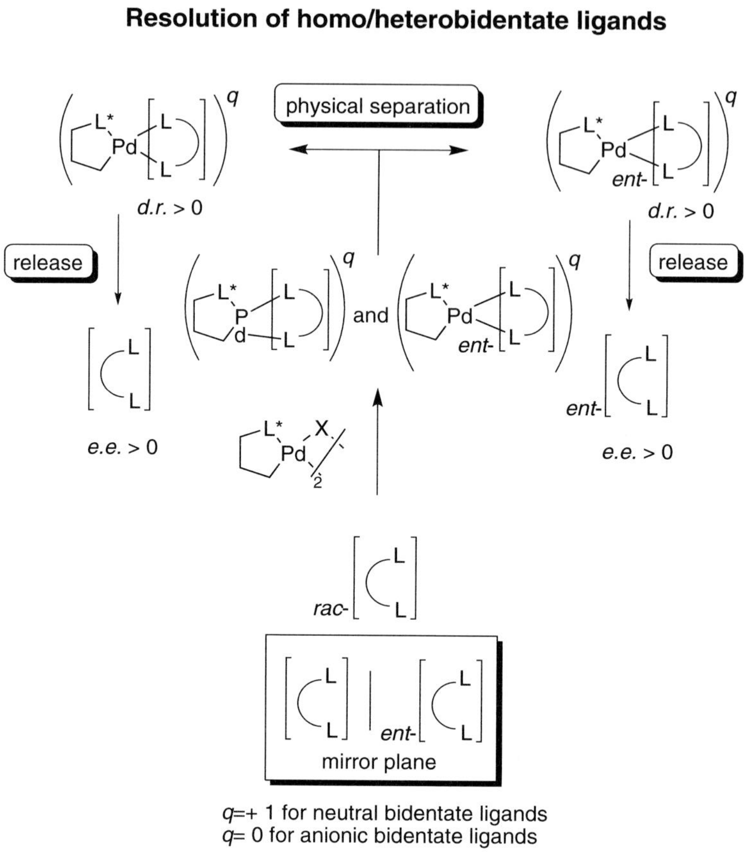

Scheme 7.3

The second class of enantioenriched/pure palladacycles consists of several derivatives of α- and β-naphthylethylamine.

The third and fourth classes of compounds contain only one representative each, that is a planar- and centro-chiral ferrocenyl palladium complex and a phenylpyrrolidine derivative, respectively.

Analysis of literature resources gives an informative view of the most used palladium complexes, which might be useful to experimentalists wishing to devise a starting-from-scratch resolution procedure for racemic ligands that are not mentioned herein. Figure 7.2 graphically represents the occurrences of the compounds mentioned in Figure 7.1 in reports dealing with the resolution of either monodentate or homo- and heterobidentate ligands from 1971 to 2006. Considering a lower limit of "popularity" to be arbitrarily a minimum of five appearances in articles published within this time-frame, (S_C)-**1**, (R_C)-**1**, (S_C)-**7** and (R_C)-**7** (Figure 7.1) lie way above this threshold; the first and last being by far the most "popular" chiral auxiliaries, all classes of racemic ligands considered. This trend is reproduced

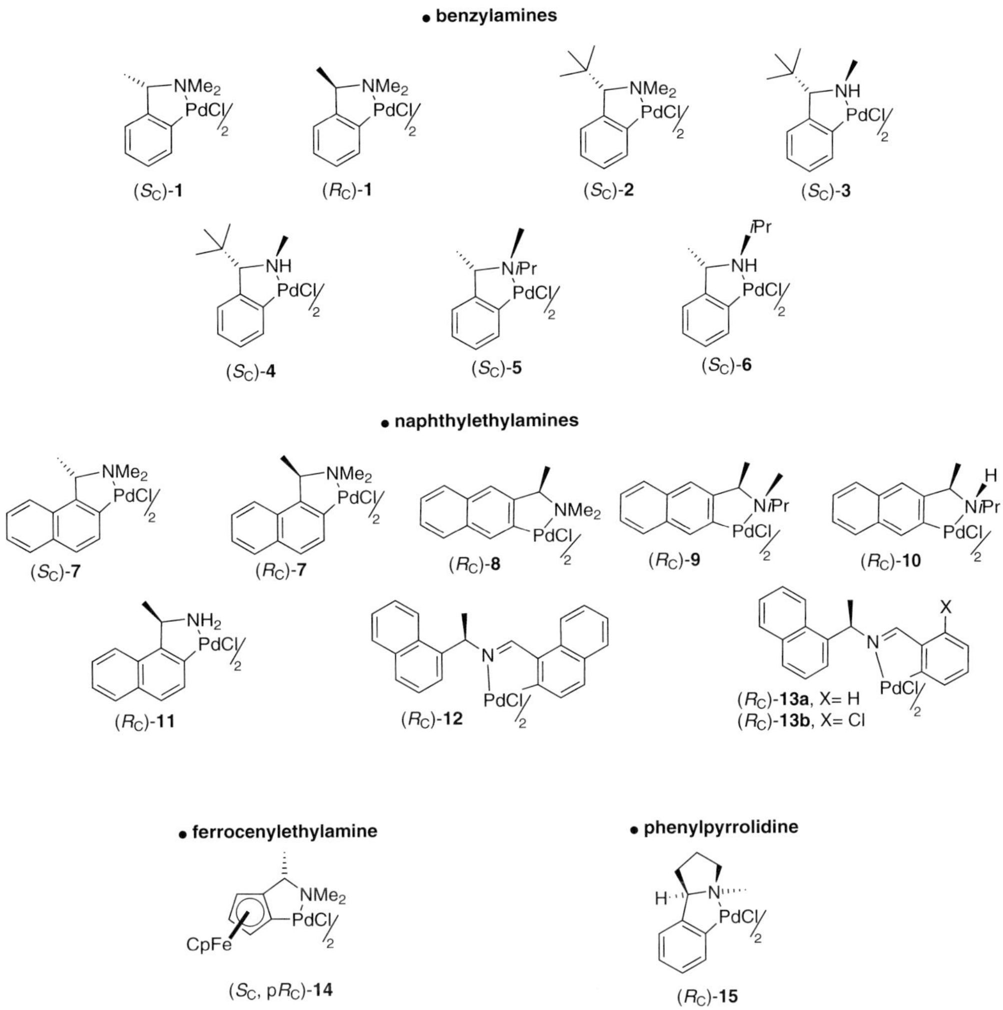

Figure 7.1 The four main classes of palladacyclic chiral auxiliaries.

faithfully even in a more detailed analysis that distinguishes between those appearances connected to the resolution of monodentate ligands (Figure 7.3a) and those connected to the resolution of bidentate ligands (Figure 7.3b). Despite intensive efforts to elaborate new selective and efficient palladium chiral auxiliaries, as shown below, (S_C)-**1** and (R_C)-**7** have received the highest consideration from experimentalists. The frequent use of naphthylethylamine-derived complexes may be explained by their pronounced ability to operate chiral recognition, which is in part due to the "locked asymmetric envelop conformation" of their palladacycles as suggested by Brown *et al.* [6].

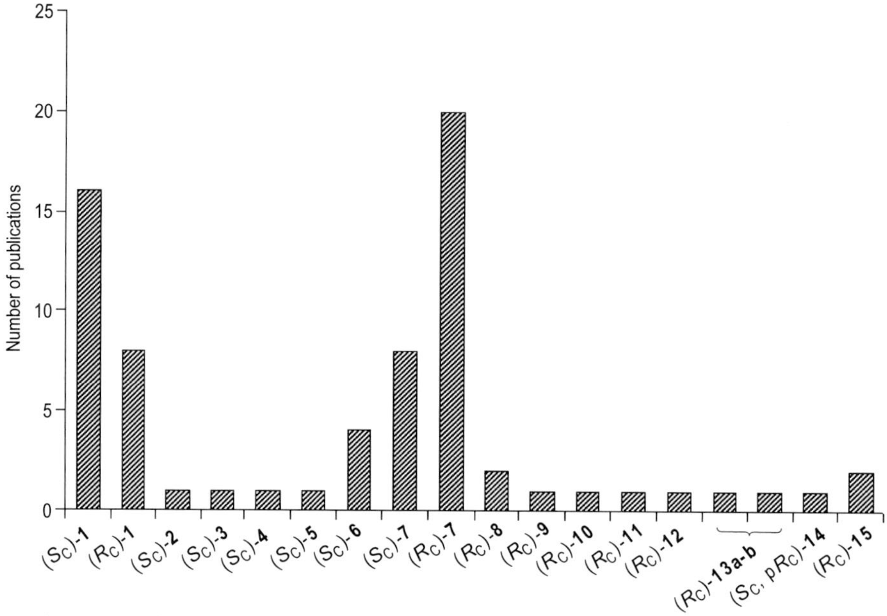

Figure 7.2 Relative occurrence in the literature of the various types of enantiomeric palladium complexes used as resolving agents (Figure 7.1).

7.4 Monodentate Ligands

A typical feature of the reaction of monodentate ligands with chloro-bridged palladacycles is the reversibility of the coordination of the exogenous (incoming) ligand to the palladium center, which may to some extent be accompanied by partial de-chelation of the metal center.

Figure 7.4 displays the structures of racemic organo-phosphorus, arsenic and antimony monodentate ligands that have been resolved optically by means of chiral Pd(II) complexes since 1971.

7.4.1 Resolution of Phosphines and Arsines

The use of cyclopalladated complexes as chiral auxiliaries for the resolution of monodentate ligands was first reported by Otsuka *et al.* in a communication published in 1971, which established the foundations of the methodology used subsequently by many other researchers [7]. The resolutions of racemic phosphines **P1** and **P2** (Figure 7.4) were first described using dimer (S_C)-**1** (Figure 7.1). In a typical experiment the latter was treated with a four-fold excess of racemic ligand and the resulting Pd(II) adduct was separated from unreacted amounts of ligand

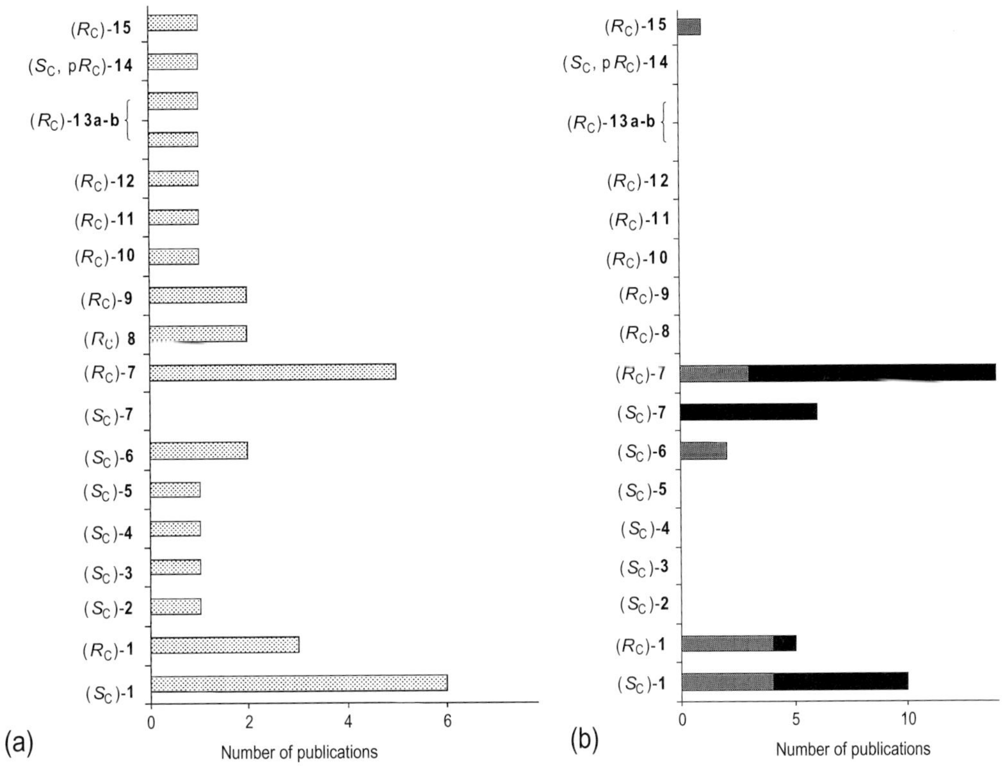

Figure 7.3 Occurrence of palladium resolving agents in articles dealing with the resolution of (a) monodentate and (b) bidentate ligands: dark gray sections correspond to the resolution of homobidentate ligands, black sections to resolutions of heterobidentate ligands.

by precipitation induced by a controlled addition of a non-polar solvent. A first fraction of enantioenriched ligand was recovered from the mother liquor whereas a second batch of enantioenriched ligand with opposite specific rotation was released by treatment of the scalemic mononuclear Pd(II) adduct with dppe [1,2-bis(diphenylphosphino)ethane].

This first report, which did not mention either the enantiomeric purity of the obtained scalemic ligands or their absolute configuration, was followed by a series of replicates of the resolution of monodentate phosphines possessing phosphorus-centered chirality. For instance, complex (R_C)-**8** was found to be more efficient than (S_C)-**1** by Otsuka *et al.* in the resolution of racemic dialkylarylphosphines **P3** and **P4** (Figure 7.4) following the above-mentioned procedure, with *optical purities* for the isolated enantiomers ranging from 63 to 100% [8]. Scalemic phosphine **P5** (Figure 7.4) was reportedly optically enriched to 44% with the help of (R_C)-**8** [8]. Notably, (R_C)-**8** was not found suitable for the resolution of triarylphosphines such

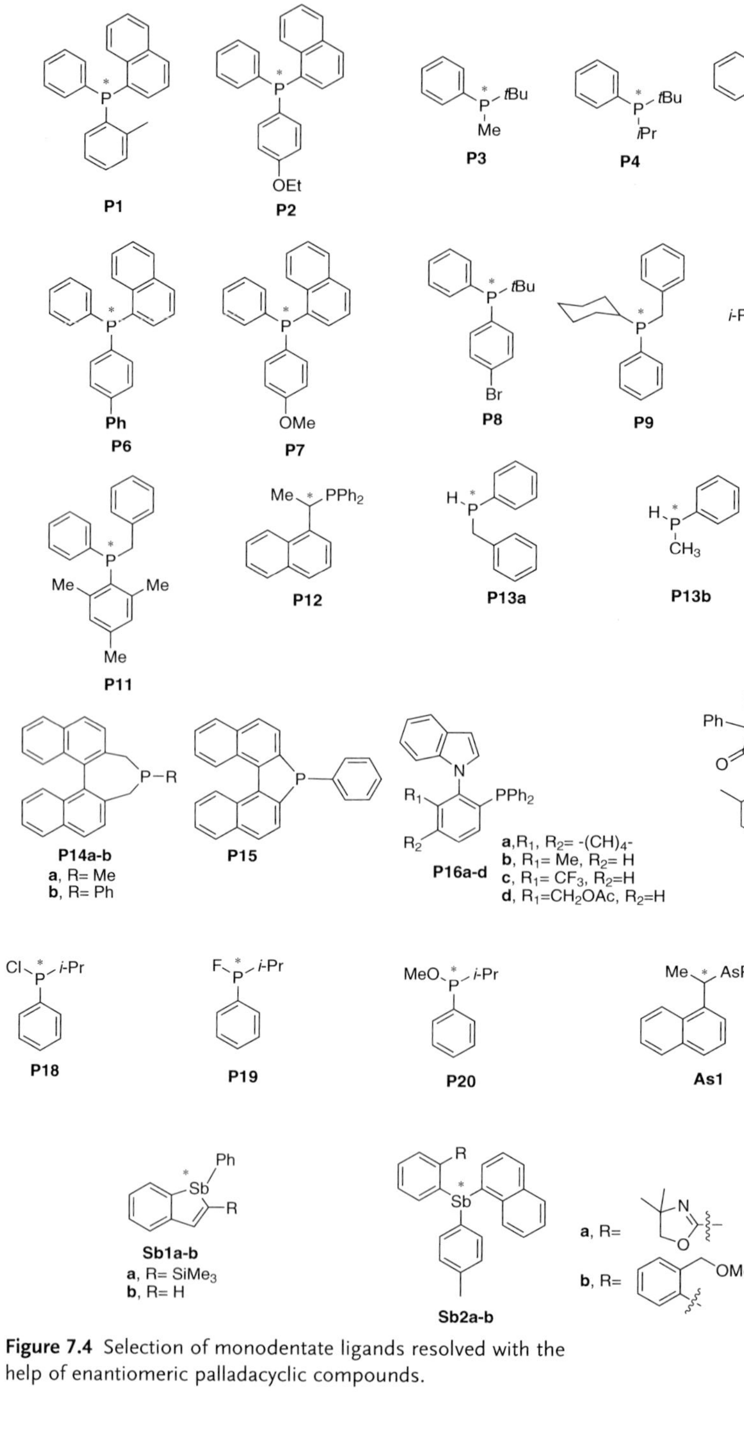

Figure 7.4 Selection of monodentate ligands resolved with the help of enantiomeric palladacyclic compounds.

as **P6** and **P7** (Figure 7.4) as they were assumed to react unselectively with the palladium complex and to yield adducts of similar solubility.

Few studies have investigated the origin of the chiral recognition process that leads to the selective coordination of a given enantiomer.

In an effort to understand the parameters acting upon the efficiency of the resolution of chiral dialkylphosphines, Dunina *et al.* systematically diversified the structure of the cyclopalladated auxiliaries by introducing N-centered chirality and bulky substituents at the benzylic position of benzylamine-based ligands [9]. This approach led to a comparative study of the coordination and chiral recognition abilities of (S_C)-**2–6**, (R_C)-**8–10** and (R_C)-**15** (Figure 7.1) towards racemic phosphine **P3** (Figure 7.4), which established the influence of steric hindrance at the nitrogen center over the enantioselectivity of chiral recognition in solution [9, 10]. The reversible coordination of phosphine **P3** with enantiopure (S_C)-**2–6**, (R_C)-**8–10** and (R_C)-**15** and with their racemic equivalents was subject to a thorough investigation of the thermodynamic aspects of the coordination process (Scheme 7.4).

The most important point of this study was that the degree of chiral discrimination was directly related to the reversibility of a ligand's coordination to the Pd center. To avoid the predominant formation of diastereomeric bis-phosphine palladium complexes under the non-stoichiometric conditions used previously by Otsuka *et al.*, the phosphine-to-Pd ratio was kept close to 1:1, which allowed an accurate determination of the corresponding equilibrium constant K by ^{31}P NMR spectroscopy at low temperature. The value of K provided a direct evaluation of the enantiodiscriminating ability, which was determined for a series of palladium complexes. With increasing steric demand at the stereogenic benzyl position the value of K increased significantly. Furthermore, the introduction of N-centered chirality induced, roughly, a two-fold enhancement of diastereoselectivity as

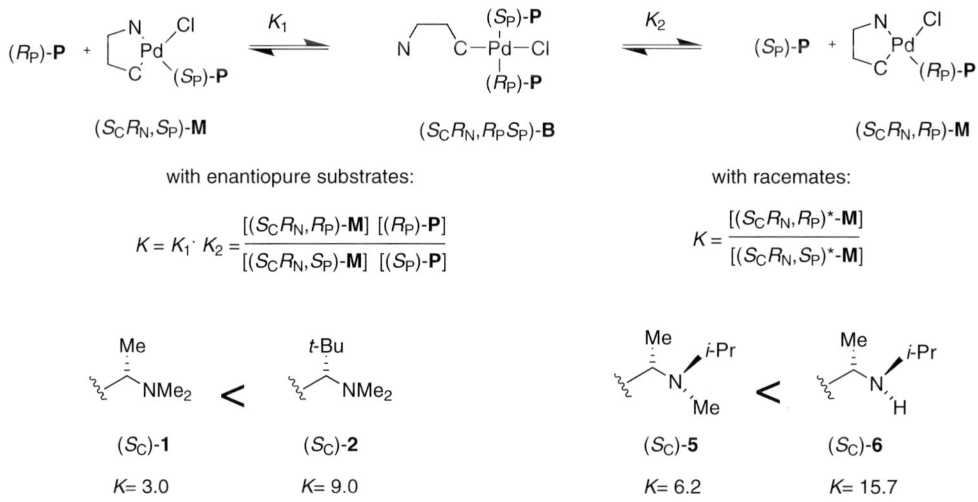

Scheme 7.4

compared to palladacycles possessing exclusively C-centered chirality. Again, not only the increase of steric bulk at one of the substituents of the nitrogen atom resulted in a large enhancement in *K*, but the largest difference of steric volume between the two substituents of the nitrogen atom enhanced the enantiodiscrimination of the ligand's coordination. Among the unexpected consequences of the introduction of N-centered chirality, a reversal of enantioselectivity for the coordination of the phosphine to the palladacyclic chiral auxiliary was also noticed.

These trends were subsequently applied to the resolution of phosphine *rac*-**P3** by using (S_C)-**6** (Figure 7.1) as chiral resolution agent, which enabled the separation of (R_P)-**P3** (e.e. 63%) and (S_P)-**P3** (e.e. 100%) in 84 and 85% yield, respectively [10]. Complex (S_C)-**2** was later used by Dunina *et al.* for the resolution of *rac*-**P8** (Figure 7.4) using chromatography as a means to separate diastereomers [11]. In this case the formation of a nearly 1 : 1 mixture of thermodynamically stable (S_C,R_P) and (S_C,S_P) adducts was followed by their chromatographic separation on silica gel. Scalemic phosphines were released by treatment of the latter with 1,2-diaminoethane and subsequently trapped with [$Pd(PhCN)_2Cl_2$].

7.4.2
Resolution of Air-Sensitive Ligands

The resolution of air-sensitive phosphines was approached in a few cases by using labile adducts made from nickel(II) salts as the source of ligand in metal-to-palladium ligand exchange reactions. For instance, adducts of air-sensitive phosphine *rac*-**P9** (Figure 7.4) with complexes (R_C)-**12** and (R_C)-**13a-b** (Figure 7.1) were formed by ligand-extraction from the coordinatively labile complex [(*rac*-**P9**)$_2$$NiCl_2$]. In this case, the corresponding diastereomeric Pd adducts were separated by conventional chromatography with d.e.s higher than 95% [12]. The nickel adduct of air-sensitive phosphine *rac*-**P10** (Figure 7.4), that is [(*rac*-**P10**)$_2$$NiCl_2$], was similarly treated with (R_C)-**11** (Figure 7.1): the corresponding diastereomeric adducts were separated efficiently by chromatography, with final d.e.s spanning 77–95% [13]. Dichloro-substituted imine complex (R_C)-**13c** was found to be a reasonable chiral shift reagent for the determination of enantiomeric excesses of chiral-at-carbon phosphine ligands by ^{1}H NMR spectroscopy [14].

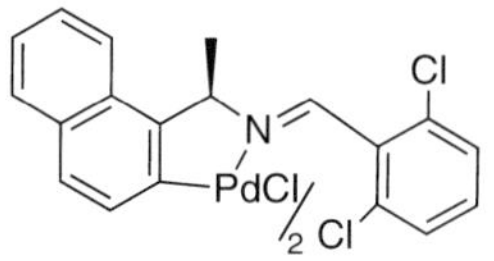

(R_C)-**13c**

Similarly, when the air-sensitivity of monodentate ligands prevents their isolation in a pure form, direct trapping of the scalemic ligand with a labile metal complex may be advocated. This strategy is well exemplified by Granell's resolution of phosphine *rac*-**P11** (Figure 7.4) using either complexes (R_C)-**7** or (R_C)-**11** (Figure 7.1) [15]. With both complexes, diastereomeric adducts were separated efficiently by chromatography, affording scalemic adducts with d.e.s of 60–95%. The release and trapping of the free scalemic phosphine was carried out by treating these adducts sequentially with dppe and either [$(PhCN)_2PdCl_2$] or [$(\eta^3$-2-$MeC_3H_4)Pd(\mu$-$Cl)]_2$ [15].

Another elegant extension of this strategy is that reported by Leung's team on the resolution of phosphine *rac*-**P12** and arsine *rac*-**As1** (Figure 7.4), both of which are air sensitive [16]. Treatment of these monodentate ligands with (R_C)-**7** yielded 1 : 1 mixtures of the corresponding (R_C,R_C) and (R_C,S_C) adducts. The latter were subsequently separated by fractional recrystallization: the (R_C,S_C) diastereomers precipitating out from the mother liquor in 60 and 28% yield in a pure form (Scheme 7.5).

The chiral *N,N*-dimethylnaphthylethylamine ligand was removed in 98% yield by treating both (R_C,S_C) adducts with concentrated hydrochloric acid. The products, that is new palladium(II) coordination complexes of (S_C)-**P12** and (S_C)-**As1**, were then heated in the presence of sodium acetate, which promoted palladation of the naphthyl group and the formation of two new complexes both displaying dextrogyric properties at 589 nm in CH_2Cl_2. A similar treatment applied to the phosphorus-containing adduct of (R_C,R_C) configuration afforded the corresponding levogyric enantiomer of cyclopalladated (R_C)-**14** [16].

The resolution of secondary phosphine ligands of general formula (R)(R′)PH was attempted in a few cases, mostly by Granell and coworkers [17, 18]. Such

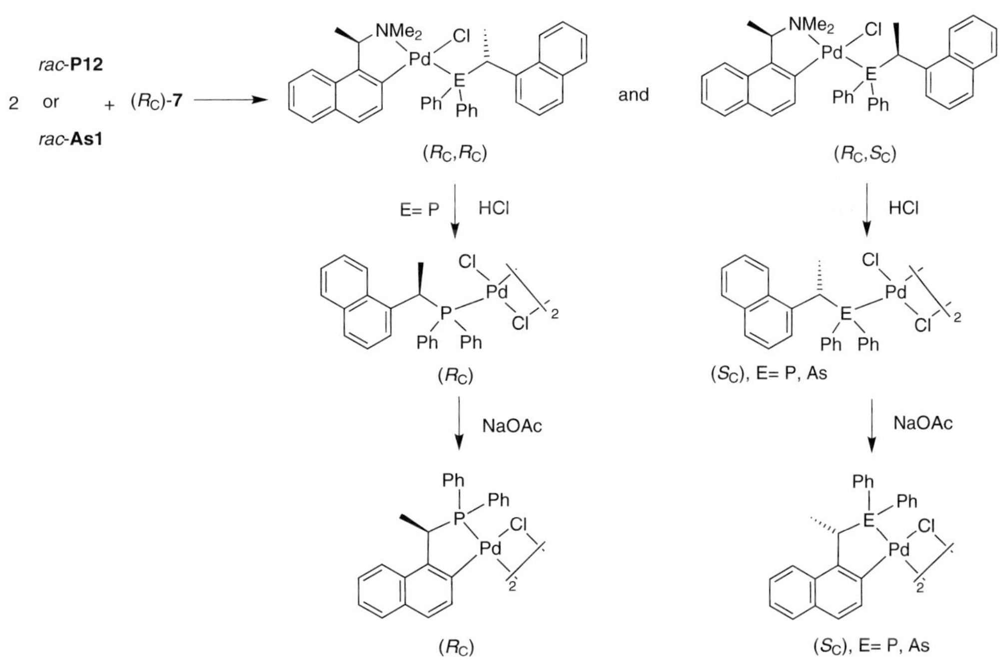

Scheme 7.5

ligands display a marked sensitivity to oxygen and their obtention in enantiomerically pure form is difficult due to their high sensitivity to protic reagents, which promote racemization by the formation of an achiral phosphonium salt (Scheme 7.6). Treatment of racemic **P13a** and **P13b** (Figure 7.4) with amino palladium complexes (R_C)-**16** and (R_C)-**11** (Figure 7.1) allowed the formation of the corresponding diastereomeric adducts, which were separated either by recrystallization or chromatography in diastereomeric excesses higher than 95%. The free enantioenriched phosphines were released by treatment of the latter adducts with dppe. Of the two enantiomeric phosphines obtained by this method, **P13a** displayed a higher propensity to racemize than **P13b**, which was found to retain its configuration for over 20 min.

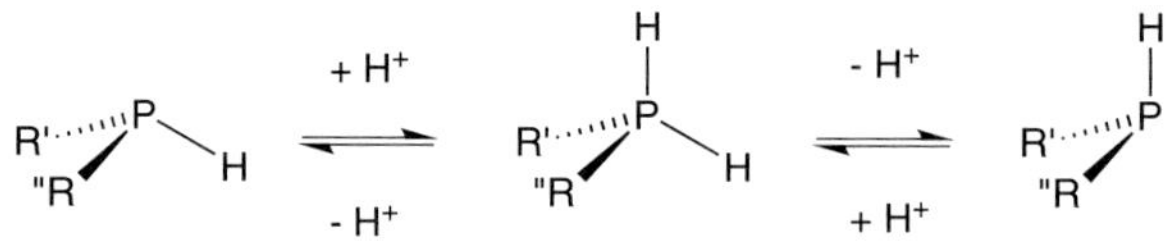

Scheme 7.6

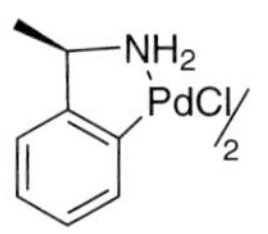

(R_C)-**16**

7.4.3
Resolution of Atropoisomeric Phosphines

Atropoisomeric phosphines can also be resolved, thereby providing easy access to highly valuable ligands, especially in the field of homogenous catalysis. The configurational stability of the axial-chiral phosphine itself is an issue that deserves attention though. Atropoisomeric binaphthalene-core phosphacyclic derivatives *rac*-**P14a,b** (Figure 7.4) containing both axial and phosphorus centered chiralities were obtained using (R_C)-**1** as resolution agent [19, 20]. In this case, a 2 : 1 ratio of racemic phosphine vs. palladacycle afforded the corresponding diastereomeric adducts, which were separated by fractional crystallization. Phosphepine (S_P)-**P14b** (Figure 7.4) was reportedly recovered enantiopure upon treatment of the less soluble adduct with dppe [20]. Atropisomeric dinaphthophosphole *rac*-**P15** (Figure 7.4) was also investigated but, owing to its low barrier to racemization in solution at room temperature, coordination to a non-racemic palladacycle was essentially used to gain information on the configurational stability of this ligand. The groups of both Gladiali [21] and Tani [22] noticed that the reaction of *rac*-**P15** with either (R_C)-**1** or (S_C)-**1** produced a sole adduct of reasonable kinetic stability. Tani established with sound crystallographic evidence the *P* configuration of the ligand in the adduct arising from a reaction with (S_C)-**1**, which was found to be configurationally stable according to variable-temperature NMR experiments [22].

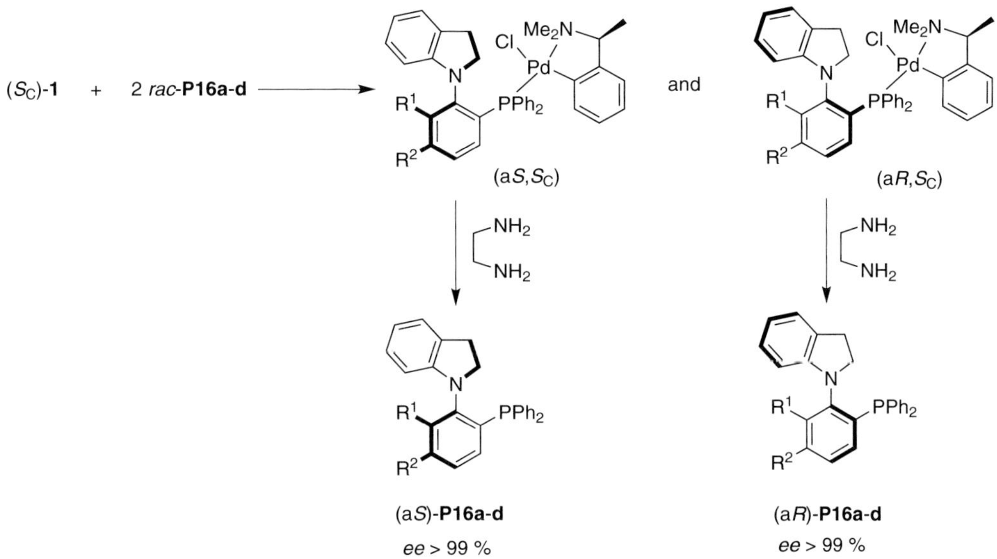

Scheme 7.7

The kinetic stability of adducts of atropoisomeric phosphines *rac*-**P16a–d** (Figure 7.4) was also established more recently by Mino *et al.* (Scheme 7.7) [23]. These adducts were efficiently prepared by coordination of the racemic phosphines to (S_C)-**1** and subsequently separated by conventional chromatography. The (a*S*,*S*) diastereomers are less polar than the (a*R*,*S*). The (a*S*) and (a*R*) phosphines were released by treatment of the adducts with 1,2-ethylenediamine: the enantiomeric excess of the corresponding phosphines was determined by chiral phase HPLC to be higher than 99% in each case. Shimizu *et al.* have described a similar successful resolution of atropoisomeric ligand **P17** (Figure 7.4) using (R_C)-**1** as resolution agent [24]. The resulting adducts were separated by flash chromatography, and subsequently treated with hydrochloric acid and 1,2-diaminoethane to release the corresponding (–)-(a*R*) and (+)-(a*S*) enantiomers, whose absolute configurations were ascertained by structural X-ray diffraction analysis.

7.4.4 Resolution of Halogenophosphines

Resolution of halogenophosphines such as **P18** and **P19** (Figure 7.4) is a challenge owing to the electrophilic character conferred to the phosphorus center by the presence of the electronegative halogeno substituent. Wild *et al.* have elegantly shown that chloro- [25] and fluoro- [26] phosphines were prone to interact with a scalemic palladacycle to give rather chemically stable adducts (Scheme 7.8). In fact, the reaction of a 2 : 1 mixture of **P18** [25] and (R_C)-**7** afforded an unbalanced mixture of two diastereomeric adducts, of which the less soluble major component possessed the (R_C,R_P) absolute configuration according to structural X-ray diffraction

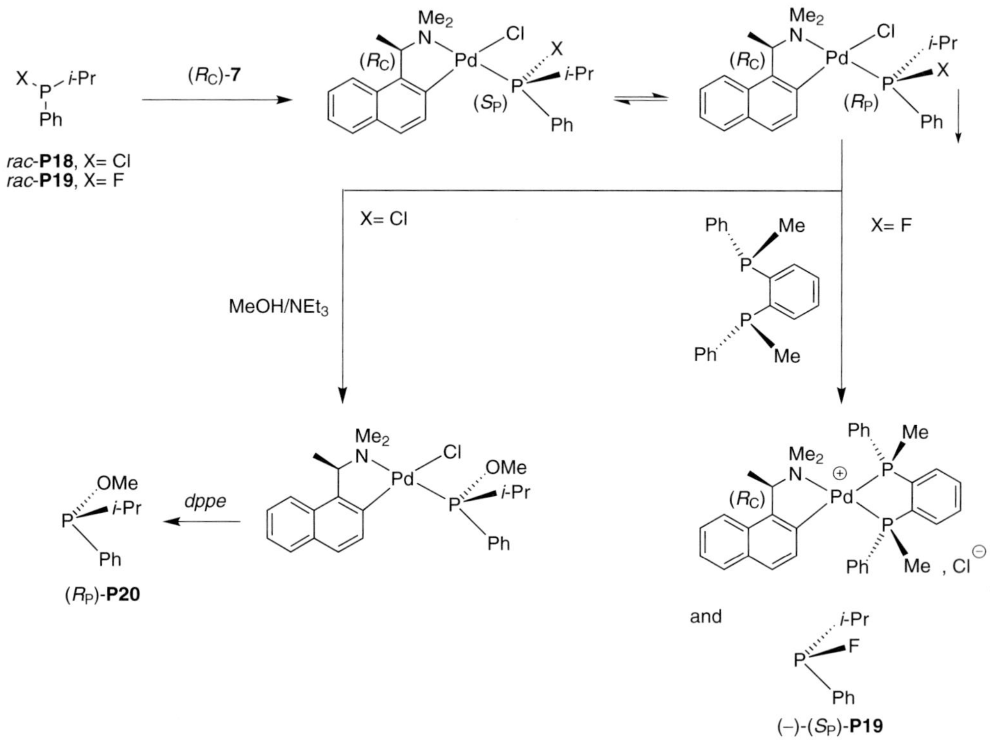

Scheme 7.8

structural analysis. Treatment of this compound with *meso*-1,2-bis(methylphenylphosphino)benzene (mppb) would not release scalemic **P18** but rather its racemate due to swift racemization in solution. To circumvent this, the palladium adduct was treated with methanol under basic conditions, thus allowing the preparation, with 92% e.e., of configurationally stable methoxyphosphine (R_P)-**P20** (Figure 7.4), which was released upon subsequent treatment with mppb. In this case, the substitution of the chloro substituent by a methoxy was established to proceed with inversion of configuration at the phosphorus center [25].

Racemic fluorophosphine **P19** was found to form an equimolar mixture of diastereomeric adducts upon reaction with (R_C)-**7** in dichloromethane [26]. Upon evaporation of the latter and dissolution of the residue in diethyl ether the ratio of adducts changed to 75 : 25, suggesting a solvent-dependent equilibration. Extraction of the major component was made possible as it precipitated in a reportedly enantiopure form upon concentration of the mother liquor. Its (R_C,R_P) absolute configuration was ascertained by structural X-ray diffraction analysis. In this case, scalemic (S_P)-**P19** was liberated from the (R_C,R_P) adduct by treatment with mppb in benzene, without any racemization being noticed within a 30 min period. Full racemization was, however, noticed after 6 h in solution [26].

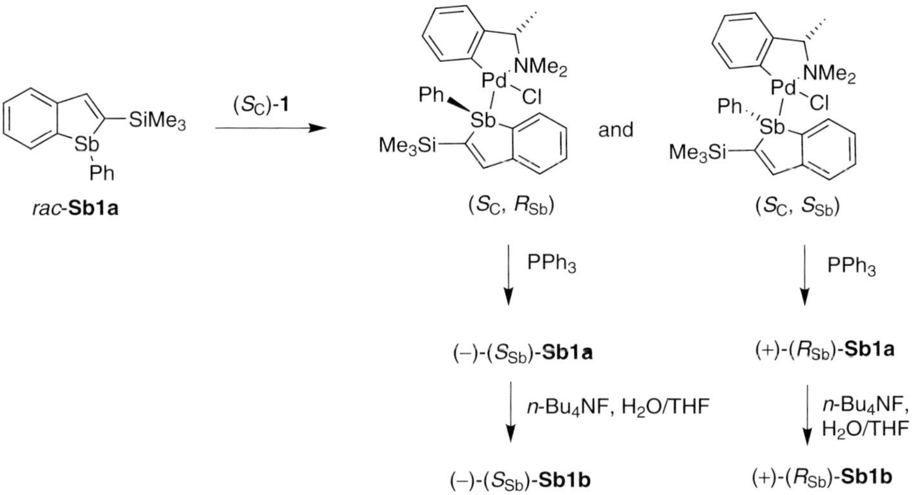

Scheme 7.9

7.4.5 Resolution of Stibines

A racemic organoantimony ligand was first resolved by Kurita and coworkers, starting from the stibindole *rac*-**Sb1a** (Figure 7.4) [27]. The latter was treated with (S_C)-**1** to afford the corresponding 1 : 1 pair of diastereomeric adducts, which were separated by conventional silica-gel chromatography (Scheme 7.9). Each diastereomer was then treated with a slight excess of PPh_3 to release enantioenriched (+)-(R_{Sb})- and (–)-(S_{Sb})-**Sb1a** ($[\alpha]_D$ = ±415 in MeOH), which could be separately treated with a fluoride source to remove the $-SiMe_3$ group and afford (+)-(R_{Sb}) and (–)-(S_{Sb})-**Sb1b** (Figure 7.4). In a separate set of experiments carried out by reaction of an eight-fold excess of *rac*-**Sb1a** with (S_C)-**1** a net preferential coordination of the (R_{Sb}) enantiomer of **Sb1a** to (S_C)-**1** over the (S_{Sb}) enantiomer was noticed.

A similar treatment was applied to stibine **Sb2a,b** (Figure 7.4) using (S_C)-**7** as resolving agent [28]. Upon separation of the corresponding diastereomeric adducts, and their treatment with either dppe or triphenylphosphine, the free enantiopure stibines (–)-(R_{Sb})-**Sb2a,b** and (+)-(S_{Sb})-**Sb2a,b** were recovered with enantiomeric excesses of >89%.

7.4.6 Resolution of Cluttered Chiral Bidentate Ligands

Chiral bidentate ligands presenting marked steric cluttering or relatively weak bonding affinity for palladium may in some cases behave as monodentate ligands if double chelation of the metallic center is impossible or disfavored. For instance, chiral-at-phosphorus BIPNOR (R^*,R^*)-**P21** [29] behaves with $(PhCN)_2PdCl_2$ like a typical bidentate ligand, leading to the formation of a Pd(II) chelate, whereas upon treatment with (R_C)-**1** a mixture of two dinuclear Pd adducts was formed in which each phosphorus atom of the BIPNOR ligand was bonded to a distinct palladium center (Scheme 7.10).

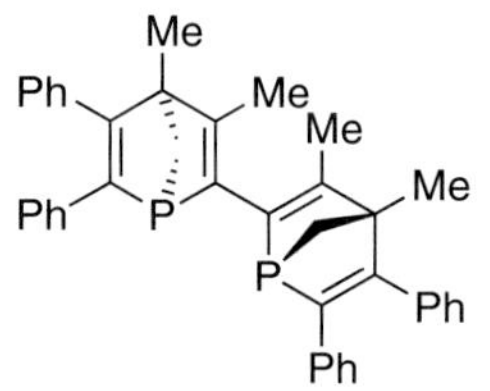

BIPNOR, (R_P*,R_P*)- **P21**

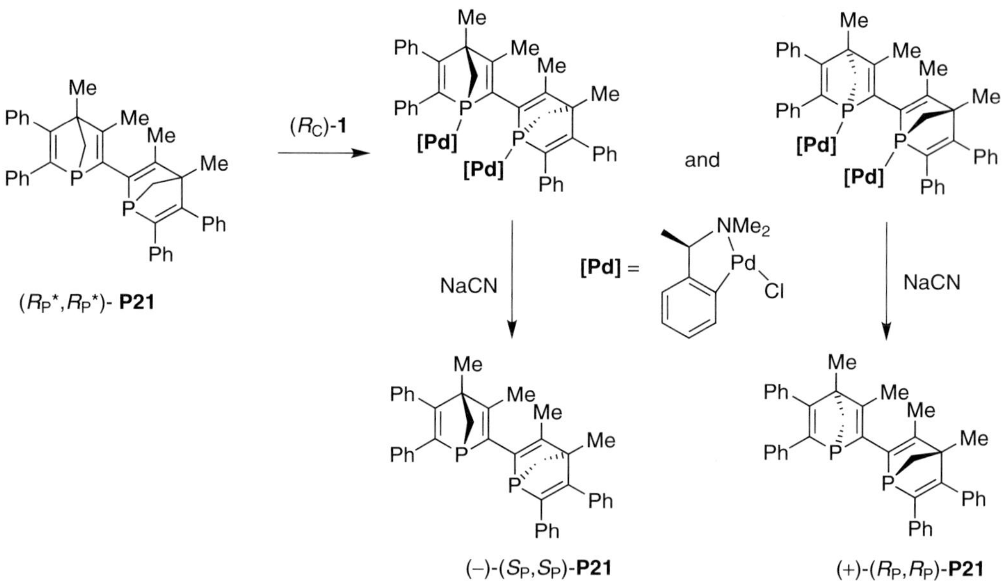

Scheme 7.10

These two diastereomers were separated by chromatography and treated with NaCN to release scalemic homochiral (+)-(R_P,R_P) and (−)-(S_P,S_P) BIPNOR **P21**.

Similarly, axial-chiral biquinoline *rac*-**N1** [30] and spiro-oxime *rac*-**N2** [31] were found to produce a mixture of cationic μ-chloro bridged dipalladium diastereomeric adducts upon reaction with complexes (S_C)-**1** and (R_C)-**1** (Scheme 7.11).

Notably, not all spiro ligands behave as double monodentate ligands, as demonstrated by the case of Sasai's axial- and centro-chiral SPRIX ligand **NN1**, which displays classical bidentate behavior. The latter was successfully resolved with the mediacy of (R_C)-**1** into its enantiopure (−)-(P,R_C,R_C) enantiomer [32].

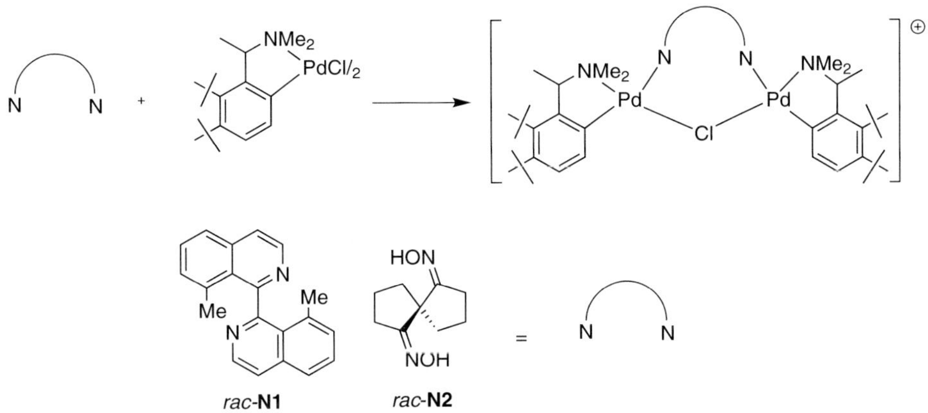

Scheme 7.11

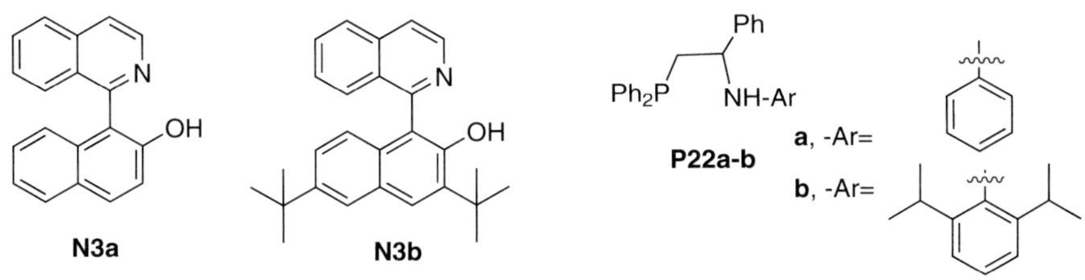

Figure 7.5 Some heterobidentate ligands.

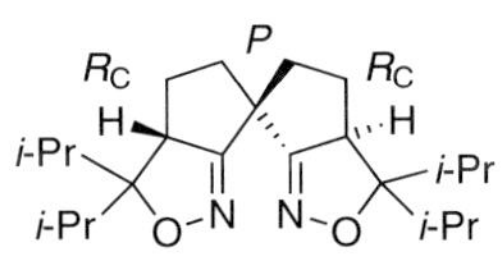

i-Pr-S*PRIX*, (−)-**NN1**

Complex (R_C)-**7** (Figure 7.1) was reportedly more convenient as chiral auxiliary [28] in the resolution of **N1** since it may operate an efficient *chiral recognition*, leading to selective precipitation of the adduct of $(R_C,\mathrm{a}R)$ configuration, with the diastereomer of $(R_C,\mathrm{a}S)$ configuration remaining in the mother liquor. From these two compounds, the highly racemizable scalemic free biquinoline ligands (+)-(a*R*)- and (−)-(a*S*)-**N1** were released by treatment with dppe.

With heterobidentate ligands such as **N3a,b** and **P22a,b** (Figure 7.5) steric cluttering causes the ligand to behave exclusively as a monodentate ligand, binding through nitrogen and phosphorus atoms respectively. Practically, *rac*-**N3a,b** [33] was resolved following Otsuka's historical procedure, using a substoichiometric amount of (R_C)-**7**, which allowed the recovery of substantial amounts of enantio-enriched uncoordinated ligand: the latter was found to display a strong propensity to racemization. With *rac*-**P22a,b** (Figure 7.5) [34], resolution was achieved by successive recrystallization of the Pd adducts made from (S_C)-**17** with diastereomeric excesses ranging from 57 to 98%. The nature of the solvent used in the crystallization was found to impact on the efficiency of the separation: chlorinated solvents favored the co-crystallization of diastereomers due to strong intermolecular hydrogen-bonding interactions, whereas coordinating solvents ensured complete separation by recrystallization. The enantioenriched ligands were released by treatment of diastereomers with dppe.

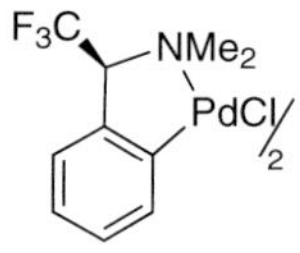

(S_C)-**17**

7.5
Bidentate Ligands

Two classes of bidentate ligands are considered here. The first consists of neutral ligands that can chelate the Pd center of an enantiopure chiral substrate to produce cationic diastereomeric adducts. The second consists of those ligands that are potentially acidic and capable of chelating the Pd(II) center of a palladacyclic enantiopure substrate to lead to neutral diastereomeric bischelates upon deprotonation. Most of the ligands addressed in the literature contain at least one heteroatom that belongs to group 15 of the periodic table. As the formation of bis-chelates is favored thermodynamically in most cases, the release of the enantioenriched ligands is particularly difficult and relies on *ad hoc* procedures that may combine ligand-exchange reactions and the protonolysis of the palladium bound [C,N] chiral chelate unit.

7.5.1
Neutral Ligands

Figure 7.6 displays examples of neutral homo- and heterobidentate ligands[1] that have been resolved successfully using enantiopure chloro-bridged palladacycles. Table 7.1 provides detailed information on the conditions of the resolution and indicates the methods used to release the free ligand in an enantioenriched or pure state.

The resolution of symmetric homobidentate ligands presenting either a C_2 axis of symmetry or axial chirality, such as **NN2,3**, **PP1–3** and **AsAs1**, poses no major difficulty. In most reported cases, the diastereomeric adducts have been separated

1) The term *homobidentate* is used here to designate ligands in which the atoms binding the metal are identical in nature, notwithstanding the structure and the substitution pattern existing at these atoms. Conversely, the term *heterobidentate* is used for ligands in which the atoms binding the metal are different in nature.

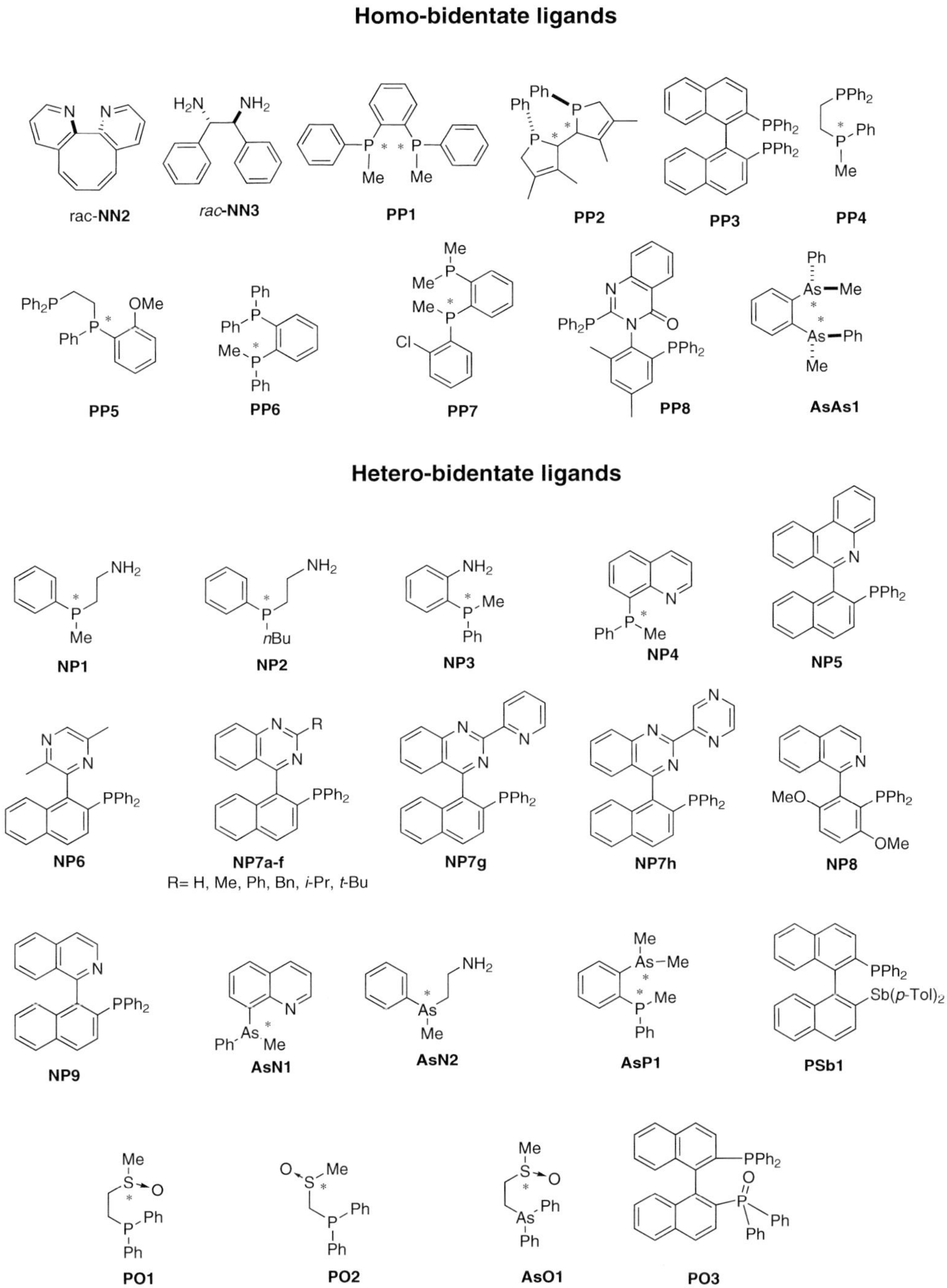

Figure 7.6 Homo- and heterobidentate ligands.

Table 7.1 List of experimental conditions used for the resolution of selected racemic homo- and heterobidentate neutral ligands.

Entry	Ligand	Palladacycle	Pd : ligand	Method for the separation of diastereomers[a]	Relative yields (%), d.e. (%)	Release/trapping method (% yield)	Ligand's configuration (% e.e.)	Ref.
	[N,N]							
1	**NN3**	(S_C)-**6**	1:6	**A**, MeCN	66, 98	HCl_{aq}/CH_2Cl_2 (85)	S_C,S_C (98)	36, 37
	[P,P]							
2	**PP1**	(R_C)-**1**	1:1	**B**, $NH_4[PF_6]$	93, >99	HCl_{aq} & KCN	S_P,S_P (>99)	38
3	**PP2**	(S_C)-**7**	1:1	**B**, $Ag[ClO_4]$	82, >99 88, >99	HCl_{aq} & KCN (90, 94)	S_P,R_C (>99) R_P,S_C (>99)	39
4	**PP3**	(S_C)-**1**	1:1	**A**	72–78, na	$LiAlH_4$ (75–82)	a*R* (>99) a*S* (>99)	40, 41
5	**PP4**	(R_C)-**7**	1:1	**C**, $NH_4[PF_6]$	distribution	H_2SO_4, LiCl & KCN	S_P (>99)	42
6	**PP5**	(R_C)-**7**	1:1	**C**, $Na[BF_4]$	distribution	HCl_{aq} & KCN	S_P (>98)	43
7	**PP6**	(S_C)-**1**	–	**C**, $NH_4[PF_6]$	49, >99 24, >99	$HCl_{acetone}$ & KCN (95–78)	S_P (>99) R_P (>99)	44
	[As,As]							
8	**AsAs1**	(S_C)-**1**	1:1	**B**, $NH_4[PF_6]$	95, >99 91, >99	$HCl_{acetone}$ & KCN (100)	R_{As},R_{As} (>99) S_{As},S_{As} (>99)	45

	[N,P]							
9	**NP1**	(R_C)-**7**	1:1	**C**, $NH_4[PF_6]$	85, >99 72, >99	pbmpp[b] (98–81)	S_P (>99) R_P (>99)	46
10	**NP2**	(S_C)-**1**	1:1	**D**, $Na[PF_6]$	na	NaCN & $[Pd(MeCN)_2Cl_2]$	S_P (na) R_P (na)	47, 48
11	**NP3**	(S_C)-**7**	1:1	**C**, $NH_4[PF_6]$	33, >99 28, >99	H_2SO_{4aq}, LiCl & KCN or pbmpp	R_P (>99) S_P (>99)	49
12	**NP4**	(R_C)-**7**	1:1	**C**, $NH_4[PF_6]$	90, >99 90, < 99	H_2SO_{4aq}, LiCl & KCN (91–92)	S_P (>99) R_P (>99)	50
13	**NP5**	(R_C)-**7**	1:1	**D**, $K[PF_6]$	86, >99 85, >99	dppe[c] (a*R*) dppe[c] & HCl_{aq} (a*S*, 94)	a*R* (>99) a*S* (>99)	51
14	**NP6**	(R_C)-**7**	1:1	**C**, $K[PF_6]$	10, >99	dppe[c]	a*S* (highly racemizable)	52
15	**NP7a-f**	(R_C)-**7**	1:1	**A & C**, $K[PF_6]$	variable, >99	dppe[c] (>90)	a*S* and a*R* (>99)	53
16	**NP7g-h**	(R_C)-**7**	1:1	**A & C**, $K[PF_6]$	variable, >99	dppe[c] (85–95)	a*S* and a*R* (>99)	54
	[As,N]							
17	**AsN1**	(R_C)-**7**	1:1	**C**, $NH_4[PF_6]$	90, >99 91, >99	dae[d] (91–90)	S_{As} (>99) R_{As} (>99)	49
18	**AsN2**	(R_C)-**7**	1:1	**C**, $NH_4[PF_6]$	98, >99 82, >99	dae[d] (92–90)	S_{As} (>99) R_{As} (>99)	45

Table 7.1 *Continued*

Entry	Ligand	Palladacycle	Pd : ligand	Method for the separation of diastereomers[a]	Relative yields (%), d.e. (%)	Release/trapping method (% yield)	Ligand's configuration (% e.e.)	Ref.
	[As,P]							
19	**AsP1**	(S_C)-**1** and (R_C)-**1**	1 : 1	C, $NH_4[PF_6]$	93, >99 63, >99	$HCl_{acetone}$ & KCN	R_{As} (>99) S_{As} (>99)	55
	[P,Sb]							
20	**PSb1**	(S_C)-**7**	1 : 1	E	84, >99 99, >99	PPh_3	a*R* (>99) a*S* (>99)	56
	[P,O]							
21	**PO1**	(S_C)-**7**[e]	1 : 1	A	84, >99 75, 85	dppe[c] (91–89)	S_S (>99) R_S (>99)	57
22	**PO2**	(S_C)-**7**	1 : 1	A	55, >99	dppe[c] (67)	R_S (>99)	58
	[As,O]							
23	**AsO1**	(S_C)-**7**[e]	1 : 1	A	84, >99 70, 90	KCN (91–93)	S_S (>99) R_S (>99)	56

a A, conventional fractional crystallization; **B**, fractional crystallization initiated by anion metathesis; **C**, anion metathesis followed by fractional crystallization; **D**, recrystallization of the less soluble diastereomer and metathesis of the anion of the more soluble diastereomer followed by recrystallization; **E**, chromatography.
b pbmpp: *rac*-1,2-phenylenebis(methylphenylphosphine).
c dppe: 1,2-diphenylphosphinoethane.
d dae: 1,2-diaminoethane.
e The complex was converted into its bis(acetonitrile) cationic perchlorate by chloride abstraction.

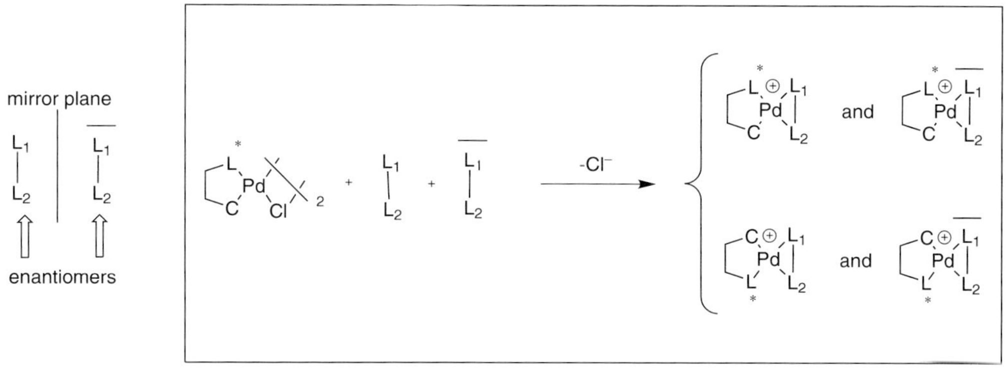

Scheme 7.12

by fractional recrystallization in various solvents, by precipitating the corresponding cationic palladium adducts [35–41, 45] (Table 7.1, entries 1–4 and 8); changes in the nature of the counter-anion were frequently made to optimize the difference in solubility between two related diastereomers, often using weak coordinative anions such as PF_6^-, BF_4^- and ClO_4^-.

In cases where the ligand is not symmetric, such as with heterobidentate (Table 7.1, entries 9–23) and with some homobidentate ligands (**PP4–7, PP8**), complications as to the separation of the related diastereomers may arise (Scheme 7.12). Indeed, in a square-planar coordination system (designated by the *SP*-4 symbol) various orientations are possible for the ligands binding the metal center: stereoisomerism of the *SP*-4 coordination geometry combined with the chiralities of each chelating unit may potentially lead to intricate mixtures of diastereomers, which may preclude further efficient resolution.

Several methods have been evaluated to cope with such problems. Unfortunately only the *sacrificial* fractional recrystallization of diastereomers is possible in cases were the distribution of diastereomeric adducts is homogenous (Table 7.1, entries 5 and 6): evidently, the main drawback may lie in a poor reproducibility as most of the material (consisting of minor or soluble diastereomers) is either lost or difficult to recover in a pure state. Such was the case with ligands **PP4,5** and to a lesser extent with **PP7** [59].

With asymmetric homobidentate ligands, when both the steric volumes and steric requirements of the substituents located at each hetero-element are significantly different, stereochemical discrimination may occur during coordination – the less encumbered and more stable cationic palladium bis-chelate being supposedly favored. A good example is the resolution of atropoisomeric ligand **PP8** (Figure 7.6), which was achieved using (S_C)-**1** [60].

With heterobidentate ligands (Table 7.1, entries 9–23), transphobia or antisymbiotic electronic effects operate preponderantly, which results, generally, in fewer possible *SP*-4 configurations and, consequently, in fewer diastereomeric adducts (Scheme 7.13). With enantiopure benzyl or naphthyl-amine palladium

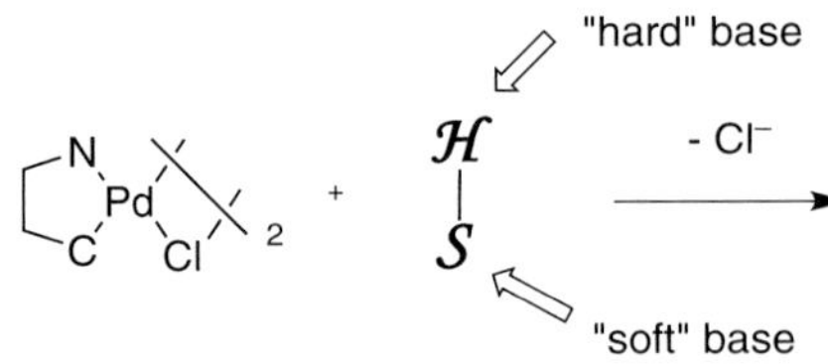

Scheme 7.13

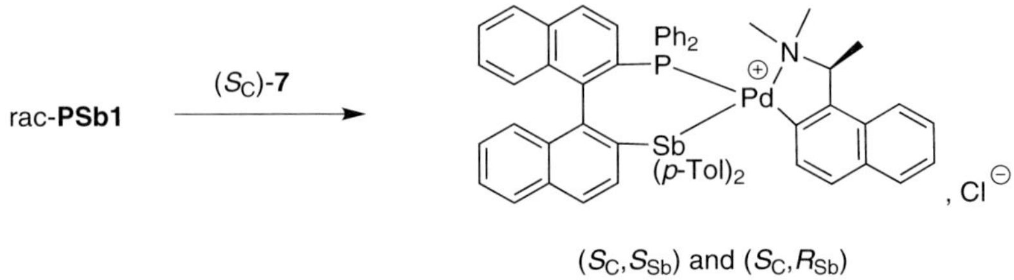

Scheme 7.14

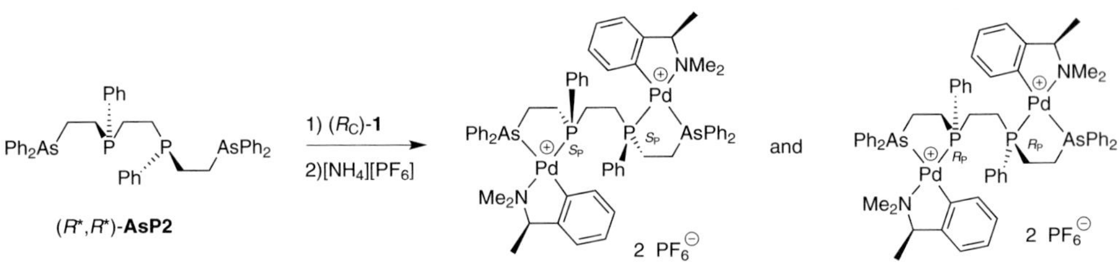

Scheme 7.15

agents for example, [N,P]-type racemic ligands **NP1–8** (Figure 7.6) invariably form two adducts, in which the "hard" nitrogen centers (relative to phosphorus) bound to the Pd center are essentially in a cis relationship (Table 7.1, entries 9–16) [46–54, 61].

A similar trend can be noted also for typical [As,N] (Table 7.1, entries 17 and 18), [As,O] (Table 7.1, entry 23), [As,P] (Table 7.1, entry 19), [P,O] [Table 7.1, entries 21 and 22 and BINAPO **(PO3)** [62]] and [P,Sb] (Table 7.1, entry 20) ligands: according to Pearson's "hard and soft acids bases" principle [63] and anti-symbiotic effect, the harder base of the bidentate ligand, namely the N, O, P, O and P atoms of the latter series of ligands, is positioned cis with respect to the nitrogen atom of the benzyl- or naphthyl-amine unit in the final adduct (Scheme 7.14).

An exception to this predominance of the anti-symbiotic effect is the behavior of the linear [As,P,P,As] tetradentate ligand (*R**,*R**)-**AsP2** (Scheme 7.15) when it is treated with (R_C)-**1** [64]. This ligand behaves like a bis-bidentate ligand that chelates two palladacyclic units to afford a mixture of two dicationic and dinuclear adducts of (R_C,S_P,S_P) and (R_C,R_P,R_P) absolute configurations. In the latter two products the "soft" As center binds the Pd(II) center in a cis fashion with respect to the nitrogen center of the chiral [C,N] ligand.

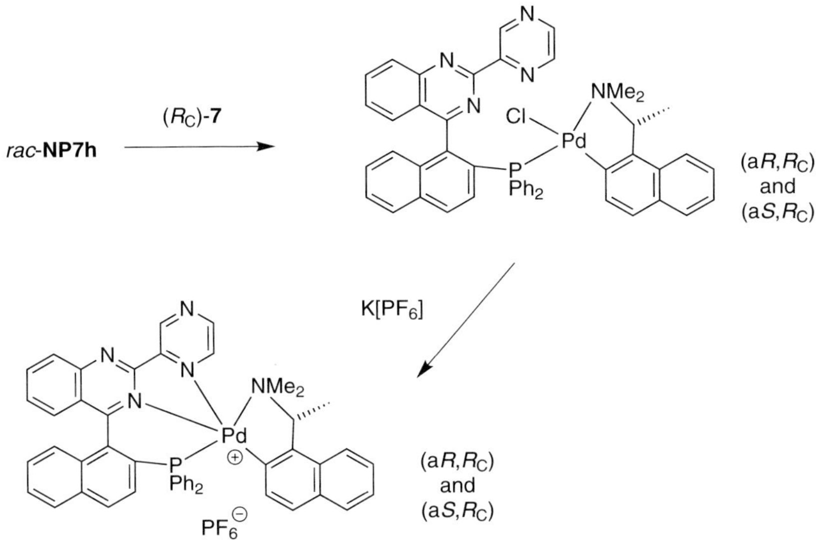

Scheme 7.16

With encumbered ligands such as **NP7a–h** (Figure 7.6), the coordination of the palladium reportedly failed to lead to the expected cationic bis-chelate in the first instance. The reaction, rather, proceeded by monodentate coordination of the incoming ligand to the metal at the phosphorus center: the chelation was essentially triggered by abstraction of the chloro ligand by a treatment with $K[PF_6]$.

Furthermore, analogous ligands **NP7g,h** showed a typical tridentate behavior (Scheme 7.16): the resulting diastereomeric adducts adopt a square-base pyramid coordination geometry around the five coordinate Pd center, the pyridyl and pyrazinyl moieties occupying the apical position.

Similarly, ligand **NP5** (Figure 7.6) reacts with complex (R_C)-**7** to yield, in the first instance, a neutral adduct resulting from the monodentate coordination of (a*R*)-**NP5**, which selectively precipitates out of the reaction solution. The second adduct present in the mother liquor arises from (a*S*)-**NP5** and possesses a similar structure: it is converted into the bis-chelate form upon addition of $K[PF_6]$ and recovered pure by recrystallization.

The particular popularity of complexes (R_C)- and (S_C)-**7** in the resolution of bidentate ligands that is illustrated by Table 7.1 is related to the conclusions drawn by Brown *et al.* on the contrasting behavior of naphthyl versus phenyl ethylamine Pd resolving agents [65–67]. In attempts to resolve the axial chiral ligand *rac*-**NP9** (Figure 7.6), Brown *et al.* noticed that the separation of the cationic adducts by precipitation was nearly impossible with phenylethylamine complex (S_C)-**1** whereas it succeeded fairly well with naphthylethylamine derivative (R_C)-**7**. Notably, the latter complex was even able to selectively coordinate the a*R* enantiomer of **NP9**

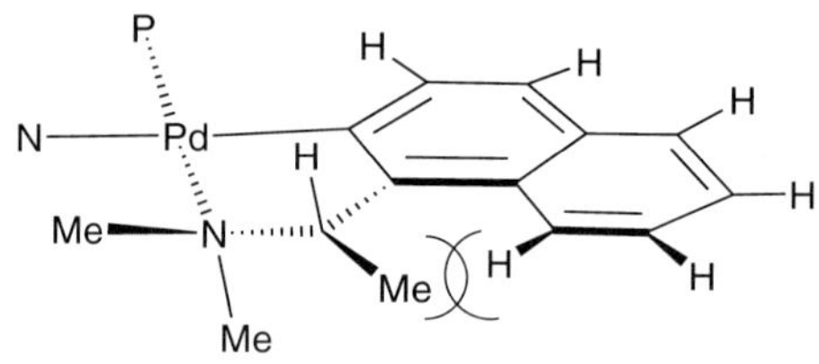

Figure 7.7 Unfavorable steric interaction in the equatorial methyl conformation of a metallacycle.

(Figure 7.6) when reacted with a large excess of racemic ligand, thus indicating a rather pronounced chiral recognition ability.

The hypothesis brought forward to explain this difference in resolving ability was that the ease of separation of diastereomers by precipitation was somewhat related to the conformational flexibility of the *N,N*-dimethylamine aryl fragments in both complexes [6]. A high flexibility, such as observed with phenylethylamine ligands, would favor the crystallization of pseudo-racemates, that is the co-crystallization of the corresponding diastereomeric palladium bischelates. Conversely, low flexibility – or locked conformation – of the metallacycle, such as noticed with naphthylethylamine ligands, would favor the crystallization of pure diastereomers: the reason for a locked conformation resides in the equatorial methyl conformation of the metallacycle, which is possible with phenylethylamine ligands, but is greatly disfavored by steric repulsion between the benzylic methyl group and the naphthyl moiety (Figure 7.7).

Therefore, Brown *et al.* anticipated that enantiomers of **7** would express superior chiral discriminating ability, particularly with bulky incoming ligands, as their coordination would affect even more the distortion of the square plane geometry around the Pd center and greatly destabilize the equatorial methyl conformation of the metallacycle. For an updated procedure for the synthesis and resolution of **NP9** the reader is referred to reference [68].

7.5.2
Anionic Ligands

The reaction of mono-anionic bidentate ligands with μ-chloro-bridged palladacycles produces neutral products, whose *SP*-4 configuration depends on the relative Pearson's hardness of the hetero-elements binding the metal center. In heterobidentate anionic ligands, the negatively charged atom possesses a higher hardness, which generally leads it to bind cis with respect to the nitrogen atom that belongs to the vicinal [C,N] chelating ligand. Consequently, in most cases the chelation reaction yields two neutral diastereomeric adducts, of which the structure can be predicted and controlled.

As several types of anionic chiral heterobidentate ligands have been resolved using enantiopure palladacyclic agents, only a short outline is given here.

The resolution of racemic amino acids, for instance, has attracted some attention as access to a non-natural specimen of *R* configuration was possible. Wild *et al.* [69] resolved *rac*-piperidine-2-carboxylic acid using (S_C)-**7** as resolving agent. In a typical experiment, the latter complex was reacted with the sodium pipecolate. Concentration of the solution allowed the selective precipitation of the (S_C,$S_C$$S_N$) diastereomer in 80% yield. Further workup of the mother liquor afforded the pendant stereoisomer of (S_C,R_C,R_N) configuration in 73% yield. In both cases, absolute configurations were established by structural X-ray diffraction analyses. The corresponding enantiopure (−)-(*S*) and (+)-(*R*) enantiomers of **NO1** were readily recovered by acidic treatment of the bis-chelates.

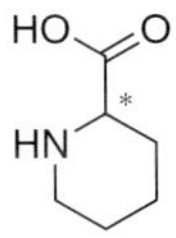

NO1

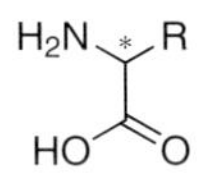

NO2a-c **a**, R= Me
b, R= Bz
c, R= *i*-P

A different approach was investigated by Navarro *et al.* [70], who attempted the resolution of racemic alanine (**NO2a**), phenylalanine (**NO2b**) and valine (**NO2c**) using the acetylacetonato derivative of (S_C)-**1** as resolution agent, the acac ligand playing the role of a base in this case. Unfortunately, all the reactions that were attempted did not show any diastereoselectivity.

Notably, cyclopalladated complexes, such as fluorinated complex (S_C)-**17** in the presence of K_2CO_3 [71], have proven to be particularly efficient as NMR chiral shift resolution agents for the determination of the enantiomeric excess of scalemic amino-acids.

In a comprehensive investigation of the synthesis of optically active macrocycles containing resolved asymmetric trivalent arsenic stereocenters, Wild *et al.* studied the resolution of racemic 2-(mercaptoethyl)methylarsine **AsS1** (Scheme 7.17) [72] and its phosphine parent **PS1** [73]. Coordination of the ligand to the palladium center of (R_C)-**7** was promoted by triethylamine as base. The reaction, carried out with a 1 : 2 ratio of ligand relative to Pd, afforded a mixture of two dinuclear diastereomeric adducts containing a palladium bischelate unit and palladacyclic unit bonded via Pd to the sulfur atom of the vicinal chelate (Scheme 7.17), thus introducing an additional stereogenic center. The reason for not observing more diastereomers was putatively related to the enforced conformational restrictions introduced by the 2-naphthylethylamine chelating unit. Notably, the stereoselectivity of palladium chelation was complete, the harder donor atom of **AsS1** and **PS1**, that is sulfur, being positioned cis to the NMe_2 moiety. The two diastereomeric products derived from **AsS1** were separated by repeated fractional crystallization and sequentially treated with 1,2-diaminoethane and aqueous KCN to remove the ancillary palladacyclic entity and release enantiopure (−)-(R_{As})- and (+)-(S_{As})-**AsS1**.

Isolation of enantiopure (−)-(R_P)-**PS1** required more steps, starting from the dinuclear adduct: the removal of the ancillary palladacycle using 1,2-

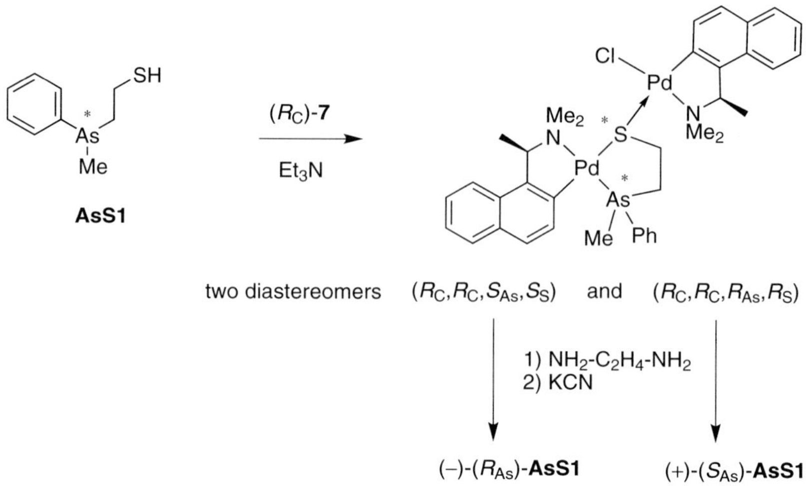

Scheme 7.17

diaminoethane, the benzylation of the sulfido-S atom with $PhCH_2Br$, the release of S-benzylated (R_P)-**PS1** by ligand-exchange with (R*,R*)-1,2-bis(methylphenylph osphino)benzene and, finally, a debenzylation step with Na/NH_3.

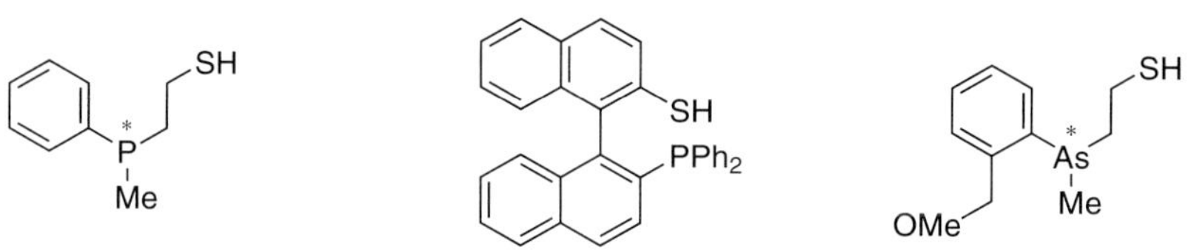

PS1 **PS2** **AsS2**

Enantiopure (+)-(R_{As})-**AsS2** and scalemic (−)-(S_{As})-**AsS2** were obtained following a resolution procedure similar to that used for **AsS1** using (R_C)-**7** [74]. Scalemic (−)-(S_{As})-**AsS2** was eventually brought to enantiopurity in a second round of resolution using (S_C)-**7**. As a further example of resolved anionic [P,S] ligand, one may mention Gladiali's BINAPS atropoisomeric ligand **PS2**, which was resolved with the help of (S_C)-**1** into the corresponding enantiopure (−)-(aS) and (+)-(aR)-**PS2** enantiomers [75].

Enantiopure enantiomers of planar-chiral carborane ligand **PH1** have been obtained for the first time by Brunner *et al.* by resolution of the corresponding racemate using (R_C)-**1** (Scheme 7.18) [76]. Separation of the diastereomeric adducts of what is formally an anionic [P,H] ligand was achieved by fractional crystallization. Notably, the stereospecificity of the chelation of **PH1** places the negatively charged H atom cis to the NMe_2 moiety. Subsequent treatment of each diastereomer with hydrochloric acid and NaCN afforded the enantiomers. The absolute configuration of each enantiomer was then ascertained by structural X-ray diffraction analysis.

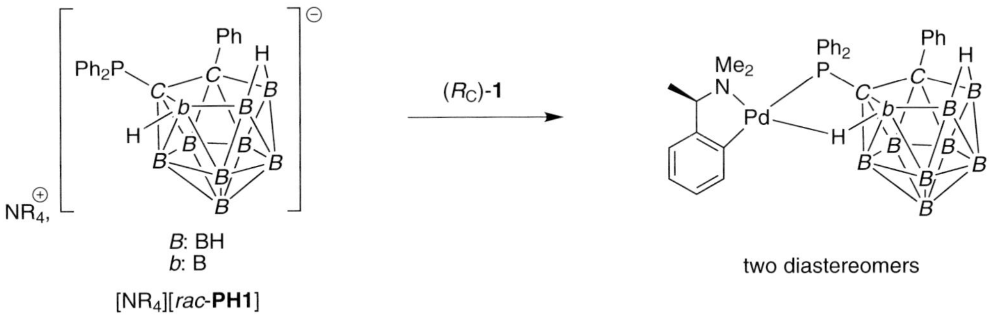

Scheme 7.18

7.6
Conclusion

The use of enantiopure palladacycles as resolution agents can be considered as a convenient and time-saving way to obtain either scalemic or enantiopure ligands for screening purposes in catalysis and for laboratory-scale fine synthesis.

As shown above, the setup of an effective method of resolution relies somewhat on empiricism. There is no universal method, only plausible methodological threads on which an experimentalist may rely when starting from scratch with a new chiral ligand. Resolving agents such as (S_C)- and (R_C)-**7**, the potential of which is well documented, may constitute a good starting point.

All the resolution methods presented here require substantial amounts of scalemic palladium chiral resolving agents, an experimental constraint that may be a matter of concern. Efficient release of scalemic ligands, efficient recycling of Pd(II) and recovery of valuable chiral auxiliary ligands become critical issues when experimentalists consider scaling-up procedures that were originally elaborated with grams or most often with milligrams of substrate. The release of the free scalemic ligand from a diastereomeric palladium salt may require the use of noxious reagents (KCN) and entail tedious subsequent treatments of the residual metal-containing wastes.

References

1 Pearson, R.G. (1973) *Inorganic Chemistry*, **12**, 712–13.

2 Vicente, J., Abad, J.A., Frankland, A.D. and Ramirez de Arellano, M.C. (1999) *Chemistry–A European Journal*, **5**, 3066–75.

3 Vicente, J., Arcas, A., Bautista, D. and Jones, P.G. (1997) *Organometallics*, **16**, 2127–38.

4 Wild, S.B. (1997) *Coordination Chemistry Reviews*, **166**, 291–311.

5 Fogassy, E., Nogradi, M., Kozma, D., *et al.* (2006) *Organic and Biomolecular Chemistry*, **4**, 3011–30.

6 Alcock, N.W., Hulmes, D.I. and Brown, J.M. (1995) *Journal of the Chemical Society D–Chemical Communications*, 395–7.

7 Otsuka, S., Nakamura, A., Kano, T. and Tani, K. (1971) *Journal of the American Chemical Society*, **93**, 4301–3.

8 Tani, K., Brown, L.D., Ahmed, J., *et al.* (1977) *Journal of the American Chemical Society*, **99**, 7876–86.

9 Dunina, V.V., Golovan, E.B., Gulyukina, N.S. and Buyevich, A.V. (1995) *Tetrahedron: Asymmetry*, **6**, 2731–46.

10 Dunina, V.V. and Golovan, E.B. (1995) *Tetrahedron: Asymmetry*, **6**, 2747–54.

11 Dunina, V.V., Kuz'mina, L.G., Rubina, M.Yu., *et al.* (1999) *Tetrahedron: Asymmetry*, **10**, 1483–97.

12 Albert, J., Cadena, J.M., Granell, J.R., *et al.* (2000) *Tetrahedron: Asymmetry*, **11**, 1943–55.

13 Albert, J., Bosque, R., Cadena, J.M., *et al.* (2000) *Tetrahedron: Asymmetry*, **11**, 3335–43.

14 Granell, J.R. and Muller, G. (2001) *Contribution to Science*, **2**, 87–94.

15 Albert, J., Bosque, R., Cadena, J.M., *et al.* (2002) *Chemistry–A European Journal*, **8**, 2279–87.

16 Ng, J.K.P., Tan, G.K., Vittal, J.J. and Leung, P.H. (2003) *Inorganic Chemistry*, **42**, 7674–82.

17 Albert, J., Cadena, J.M., Granell, J., *et al.* (2000) *European Journal of Organic Chemistry*, 1283–6.

18 Albert, J., Granell, J. and Muller, G. (2006) *Journal of Organometallic Chemistry*, **691**, 2101–6.

19 Gladiali, S. and Fabbri, D. (1997) *Chemische Berichte/Recueil*, **130**, 543–54.

20 Gladiali, S., Dore, A., Fabbri, D., *et al.* (1994) *Tetrahedron: Asymmetry*, **5**, 511–14.

21 Gladiali, S., Fabbri, D., Banditelli, G., *et al.* (1994) *Journal of Organometallic Chemistry*, **475**, 307–15.

22 Tani, K., Tashiro, H., Yoshida, M. and Yamagata, T. (1994) *Journal of Organometallic Chemistry*, **469**, 229–36.

23 Mino, T., Tanaka, Y., Hattori, Y., *et al.* (2006) *Journal of Organic Chemistry*, **71**, 7346–53.

24 Chen, Y., Smith, M.D. and Shimizu, K.D. (2001) *Tetrahedron Letters*, **42**, 7185–7.

25 Pabel, M., Willis, A.C. and Wild, S.B. (1995) *Tetrahedron: Asymmetry*, **6**, 2369–74.

26 (a) Pabel, M., Willis, A.C. and Wild, S.B. (1994) *Angewandte Chemie–International Edition in English*, **33**, 1835–7.
(b) Pabel, M., Willis, A.C. and Wild, S.B. (1994) *Angewandte Chemie*, **106**, 1917.

27 Kurita, J., Usuda, F., Yasuike, S., *et al.* (2000) *Chemical Communications*, 191–2.

28 Yasuike, S., Kishi, Y., Kawara, S.I., Yamaguchi, K. and Kurita, J. (2006) *Journal of Organometallic Chemistry*, **691**, 2213–20.

29 Robin, F., Mercier, F., Ricard, L., *et al.* (1997) *Chemistry–A European Journal*, **3**, 1365–9.

30 Chelucci, G., Cabras, M.A., Saba, A. and Secchi, A. (1996) *Tetrahedron: Asymmetry*, **7**, 1027–32.

31 Ebeling, G., Gruber, A.S., Burrow, R.A., *et al.* (2002) *Inorganic Chemistry Communications*, **5**, 552–4.

32 Takzawa, S., Yogo, J., Tsujihara, T., *et al.* (2007) *Journal of Organometallic Chemistry*, **692**, 495–8.

33 Tucker, S.C., Brown, J.M., Oakes, J. and Thornthwaite, D. (2001) *Tetrahedron*, **57**, 2545–54.

34 Camus, J.M., Garcia, P.R., Andrieu, J., *et al.* (2005) *Journal of Organometallic Chemistry*, **690**, 1659–68.

35 Wang, X.C., Cui, Y.X., Mak, T.C.W. and Wong, H.N.C. (1990) *Journal of the Chemical Society D–Chemical Communications*, 167–9.

36 Dunina, V.V., Kuz'mina, L.G., Parfyonov, A.G. and Grishin, Yu. K. (1998) *Tetrahedron: Asymmetry*, **9**, 1917–21.

37 Dunina, V.V., Kuz'mina, L.G., Parfyonov, A.G. and Grishin, Yu. K. (1999) *Russian Chemical Bulletin*, **48**, 183–94.

38 Roberts, N.K. and Wild, S.B. (1979) *Journal of the American Chemical Society*, **101**, 6254–60.

39 He, G., Mok, K.F. and Leung, P.H. (1999) *Organometallics*, **18**, 4027–31.

40 Miyashita, A., Yasuda, A., Takaya, H., *et al.* (1980) *Journal of the American Chemical Society*, **102**, 7932–4.

41 Lopez, C., Bosque, R., Sainz, D., *et al.* (1997) *Organometallics*, **16**, 3261–6.

42 Leitch, J., Salem, G. and Hockless, D.C.R. (1995) *Journal of the Chemical Society–Dalton Transactions*, 649–56.

43 Ramsden, J.A., Brown, J.M., Hursthouse, M.B. and Karalulov, A.I. (1994) *Tetrahedron: Asymmetry*, **5**, 2033–44.

44 Gabbitas, N., Salem, G., Sterns, M. and Willis, A.C. (1993) *Journal of the Chemical Society – Dalton Transactions*, 3271–6.

45 Roberts, N.K. and Wild, S.B. (1979) *Journal of the Chemical Society – Dalton Transactions*, 2015–21.

46 Martin, J.W.L., Palmer, J.A.L. and Wild, S.B. (1984) *Inorganic Chemistry*, **23**, 2664–8.

47 Kashiwabara, K., Kinoshita, I. and Fujita, J. (1978) *Chemistry Letters*, 673–6.

48 Kinoshita, I., Kashiwabara, K. and Fujita, J. (1980) *Bulletin of the Chemical Society of Japan*, **53**, 3715–16.

49 Barclay, C.E., Beeble, G., Doyle, R.J., *et al.* (1995) *Journal of the Chemical Society – Dalton Transactions*, 57–65.

50 Allen, D.G., McLaughlin, G.M., Robertson, G.B., *et al.* (1982) *Inorganic Chemistry*, **21**, 1007–14.

51 Valk, J.M., Claridge, T.D.W., Brown, J.M., *et al.* (1995) *Tetrahedron: Asymmetry*, **6**, 2597–610.

52 McCarthy, M. and Guiry, P.J. (1999) *Tetrahedron*, **55**, 3061–70.

53 Connolly, D.J., Lacey, P.M., McCarthy, M., *et al.* (2004) *Journal of Organic Chemistry*, **69**, 6572–89.

54 Flanagan, S.P., Goddard, R. and Guiry, P.J. (2005) *Tetrahedron*, **61**, 9808–21.

55 Doyle, R.J., Salem, G. and Willis, A.C. (1995) *Journal of the Chemical Society – Dalton Transactions*, 1867–72.

56 Yasuike, S., Kawara, S., Okajima, S., *et al.* (2004) *Tetrahedron Letters*, **45**, 9135–8.

57 Chooi, S.Y.M., Siah, S.Y., Leung, P.H. and Mok, K.F. (1993) *Inorganic Chemistry*, **32**, 4812–18.

58 Leung, P.H., Quek, G.H., Lang, H., *et al.* (1998) *Journal of the Chemical Society – Dalton Transactions*, 1639–43.

59 Chatterjee, S., George, M.D., Salem, G. and Willis, A.C. (2001) *Journal of the Chemical Society – Dalton Transactions*, 1890–6.

60 Dai, X. and Virgil, S. (1999) *Tetrahedron: Asymmetry*, **10**, 25–9.

61 Alcock, N.W., Brown, J.M., Pearson, M. and Woodward, S. (1992) *Tetrahedron: Asymmetry*, **3**, 17–20.

62 Gladiali, S., Pulacchini, S., Fabbri, D., *et al.* (1998) *Tetrahedron: Asymmetry*, **9**, 391–5.

63 Ho, T.L. (1975) *Chemical Reviews*, **75**, 1–20.

64 Cook, V.C., Willis, A.C., Zank, J. and Wild, S.B. (2002) *Inorganic Chemistry*, **41**, 1897–906.

65 Alcock, N.W., Brown, J.M. and Hulmes, D.I. (1993) *Tetrahedron: Asymmetry*, **4**, 743–56.

66 Brown, J.M., Hulmes, D.I. and Layzell, T.P. (1993) *Journal of the Chemical Society D – Chemical Communications*, 1673–4.

67 Brown, J.M., Hulmes, D.I. and Guiry, P.J. (1994) *Tetrahedron*, **50**, 4493–506.

68 Lim, C.W., Tissot, O., Mattison, A., *et al.* (2003) *Organic Process Research & Development*, **7**, 379–84.

69 Hockless, D.C.R., Mayadunne, R.C. and Wild, S.B. (1995) *Tetrahedron: Asymmetry*, **6**, 3031–7.

70 Navarro, R., Garcia, J., Urriolabeitia, E.P., *et al.* (1995) *Journal of Organometallic Chemistry*, **490**, 35–43.

71 Levrat, F. Stoeckli-Evans, H. and Engel, N. (2002) *Tetrahedron: Asymmetry*, **13**, 2335–44.

72 Leung, P.H., McLaughlin, G.M., Martin, J.W.L. and Wild, S.B. (1986) *Inorganic Chemistry*, **25**, 3392–5.

73 Leung, P.H., Willis, A.C. and Wild, S.B. (1992) *Inorganic Chemistry*, **31**, 1406–10.

74 Kerr, P.G., Leung, P.H. and Wild, S.B. (1987) *Journal of the American Chemical Society*, **109**, 4321–8.

75 Gladiali, S., Medici, S., Pirri, G., *et al.* (2001) *Canadian Journal of Chemistry – Revue Canadienne de Chimie*, **79**, 670–8.

76 Brunner, H., Apfelbacher, A. and Zabel, M. (2001) *European Journal of Inorganic Chemistry*, 917–24.

8
Application of Cyclopalladated Compounds as Catalysts for Heck and Sonogashira Reactions

Carmen Nájera and Diego A. Alonso

8.1
Heck Reaction

8.1.1
Introduction

Palladium-catalyzed cross-coupling reactions represent one of the most important processes in organic chemistry. They have been studied extensively since they constitute a powerful method for the formation of C—C and C—heteroatom bonds and their scope continues to increase year on year (Figure 8.1) [1]. Among palladium-catalyzed transformations, the Mizoroki–Heck reaction [2] occupies a special place since its discovery in the early 1970s [3] as an indispensable method to prepare arylated and vinylated olefins (Scheme 8.1). The reaction is broadly defined as a Pd(0)-catalyzed coupling of an aryl or vinyl halide or sulfonate with an alkene under basic conditions. However, as depicted in Scheme 8.1, the synthetic value of the Heck coupling resides in the wide range of functionalized substrates that can be successfully employed as starting materials.

The Mizoroki–Heck reaction is one of the simplest ways to obtain variously substituted olefins, dienes and other unsaturated compounds, many of which are useful polymers, dyes, natural products and biologically active non-natural compounds. Therefore, the reaction has developed significantly, as is evident by the numerous publications appearing in this field since its discovery. Consequently, this useful reaction is one of the most used transformations in modern times (Figure 8.2).

After the discovery of the reaction, the main focus of research was on the scope of the new synthetic tool. However, during the past few years, considerable attention has been devoted to mechanistic investigations and the development of new catalytic systems. Significant advances have been made in this direction, with most of the success owed to the development of new, very active and, at the same time, more stable palladium catalysts [1a, 4]. This chapter provides an overview of the

Palladacycles: Synthesis, Characterization and Applications. Edited by Jairton Dupont and Michel Pfeffer

ISBN: 978-3-527-31781-3

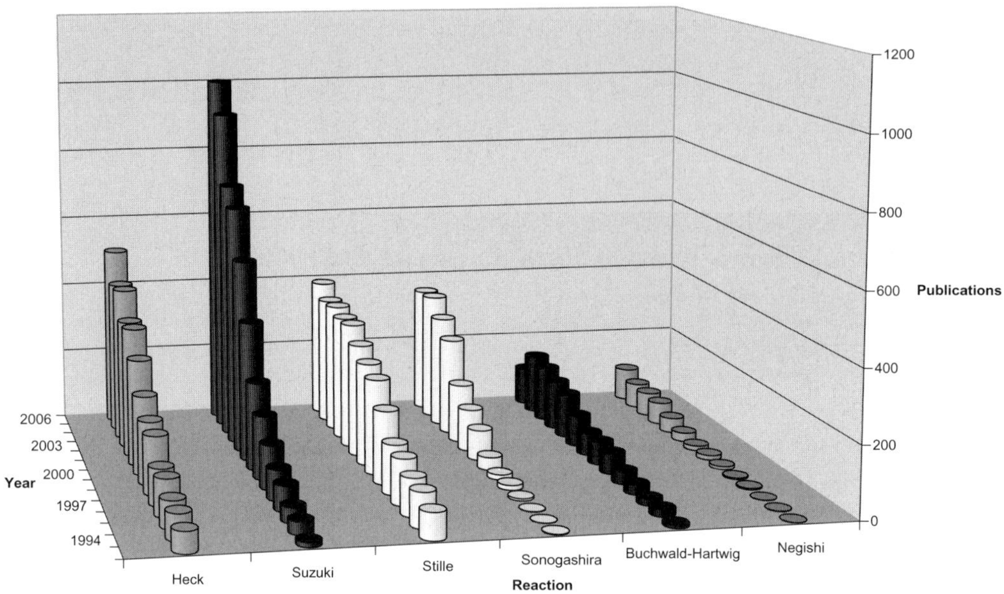

Figure 8.1 Palladium-catalyzed reactions (Source: Scifinder).

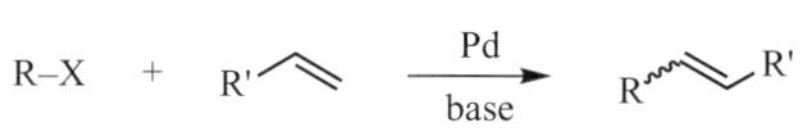

R = aryl, vinyl, benzyl, allyl
X = Cl, Br, I, OTf, OTs, N_2^+
R' = EWG, EDG

Scheme 8.1 Mizoroki–Heck reaction.

development in this area, with particular emphasis on cyclopalladated complexes.

8.1.2
Mechanism

While the Mizoroki–Heck cross-coupling reaction is a very important part of the synthetic chemist's toolbox, being applied to a wide variety of different substrates, the mechanism of the process is less studied, and therefore it has not been proved in all details. Despite intensive research on the Heck reaction, the conventional mechanism of the main catalytic cycle initially suggested by Heck [3b] remains practically unchanged. Thus, the generally accepted mechanism involves an assumed homogeneous palladium catalyst that cycles between the Pd(0) and Pd(II) oxidation states during the course of the catalytic reaction (Scheme 8.2) [5]. Usually, a Pd(II) precatalyst is employed and is assumed to be reduced to coordinatively unsaturated Pd(0) species, which are usually coordinated by weak donor ligands

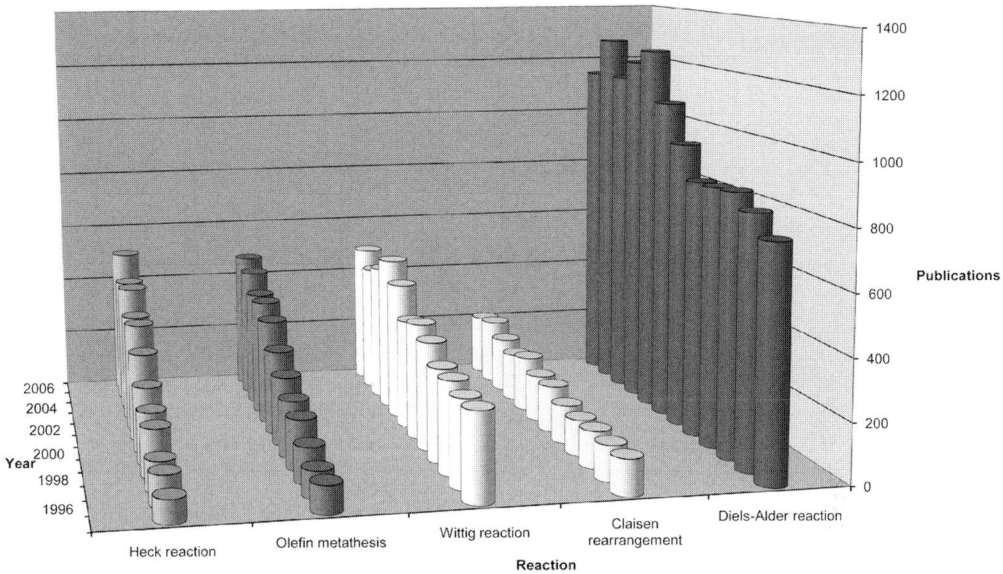

Figure 8.2 Most used synthetic transformations (Source: Scifinder).

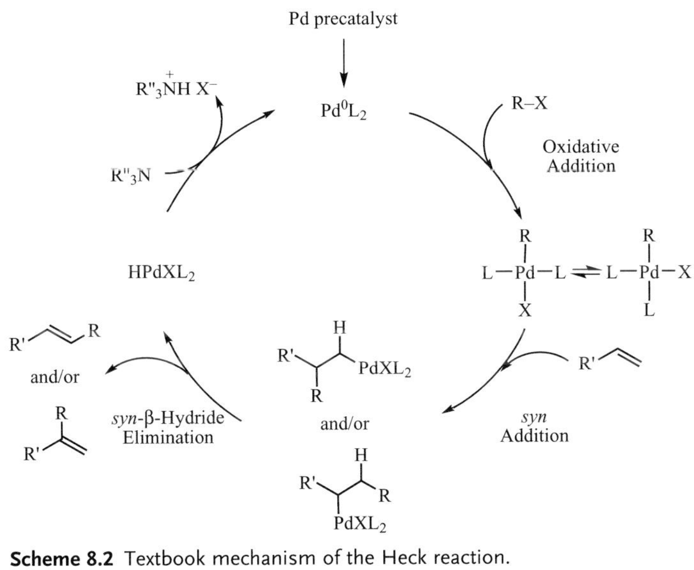

Scheme 8.2 Textbook mechanism of the Heck reaction.

such as tertiary phosphanes and are considered as the real catalytically active species [6]. The reduction of the Pd(II) precursor to Pd(0) is usually performed by the phosphane ligand [7], the amine base [8] and the olefin [9]. As shown in Scheme 8.2, oxidative addition to the halide forms a σ-alkenyl or σ-arylpalladium(II) intermediate [10]. Usually, the rate of reactions of aryl chlorides and most reactions

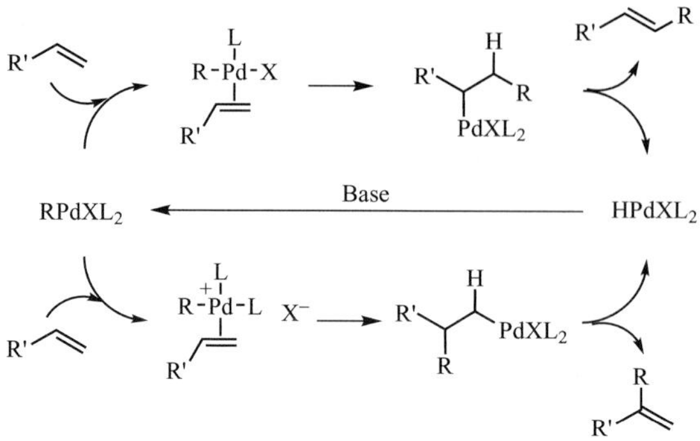

Scheme 8.3 The coordination–insertion process.

of aryl bromides are controlled by the rate of oxidative addition. After η^2-coordination to the olefin, the σ-alkenyl or σ-arylpalladium(II) intermediate suffers *syn*-insertion of the σ-alkenyl or σ-arylpalladium bond into the C=C double bond, generating the new C—C bond. This step controls the regioselectivity of the reaction. Subsequent *syn*-β-hydride elimination yields the alkene product and a hydridopalladium halide ($HPdXL_2$) that, after reductive elimination, regenerates the coordinatively unsaturated $Pd(0)L_2$ with the aid of the base. The reductive elimination step usually controls the yield and scope of the reaction.

With respect to the regioselectivity of the Heck reaction, the results of different studies on the coordination–insertion process of an olefin on a Pd(II) complex support a mechanism based on two reaction pathways involving neutral or cationic species (Scheme 8.3). The mechanism of insertion is suggested to involve the dissociation of either a neutral ligand (neutral pathway) or anionic ligand (X^-, cationic pathway) to allow the coordination of the olefin [11]. Then, the nature of the product obtained from the oxidative addition step greatly influences the rest of the catalytic cycle. For instance, cationic palladium species formed from the oxidative addition of triflates [11b] and diazonium salts [12] behave differently to the neutral species generated from halides [13]. In a neutral pathway, steric effects have a large impact upon regiocontrol and tend to favor β-substitution products. However, when cationic species are involved, electronic effects can dominate the regioselectivity of the process [14]. Thus, the electronic nature of the alkene substrate affects the regioselectivity of the reaction. Electron-rich olefins react faster via the cationic pathway, while electron-poor olefins react faster via the neutral pathway (Scheme 8.3). In the cationic mechanism olefins such as vinyl ethers and allyl alcohols [15] coordinate to the cationic palladium atom, favoring migration to the α-carbon [16]. Acrylates, however, always favor complete β-selectivity [14].

The existence of an anionic catalytic cycle for the Heck reaction has been proposed by Amatore and Jutand (Scheme 8.4) [6]. According to the authors this cycle is highly likely to operate when ligating anions such as halides or acetate are present in the reaction, while the widely accepted textbook catalytic cycle should be considered when non-ligating species such as triflates are the existing anions.

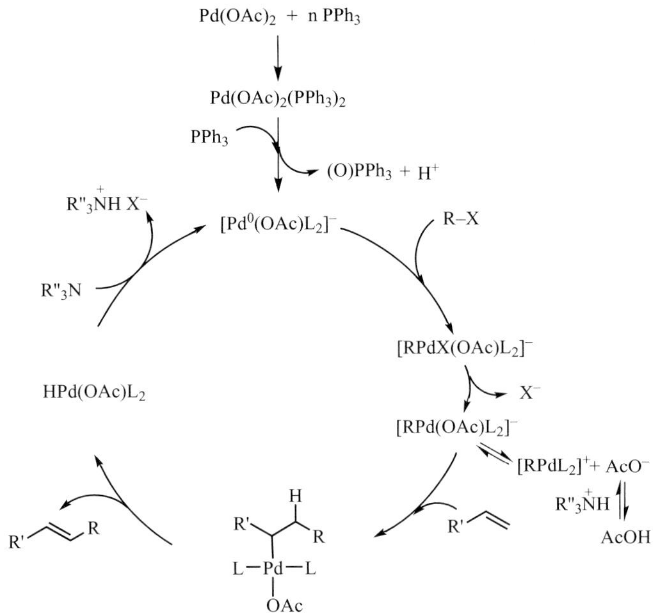

Scheme 8.4 Anionic mechanism of the Heck reaction.

The anionic species actually participate in the oxidative addition and following reaction steps (Scheme 8.4).

An anionic mechanism has been also proposed for the ligand-free palladium-catalyzed Heck reaction with aryl iodides [17]. Different characterization techniques such as electrospray mass spectrometry (ESI-MS) and extended X-ray absorption fine structure (EXAFS) have demonstrated the presence of several anionic monomeric and dimeric palladium species that are the real catalysts and coexist with palladium nanoparticles rapidly formed at the onset of the reaction. In the case of aryl bromides the oxidative addition is the rate-determining step and most of the palladium is in the zero state, forming soluble palladium nanoparticles [18] whose agglomeration to palladium black is avoided by working under low catalyst concentrations [19] or by addition of certain additives such as ammonium and phosphonium salts [20] or nitrogen ligands [21] that stabilize the colloids.

The irruption of palladacycles [22] as very efficient and stable catalyst precursors at high temperatures for the Heck and various cross-coupling reactions brought some exciting debate about the possible involvement of oxidation states +II and +IV in the catalytic cycle of the Heck reaction, particularly when a reducing agent could not be initially clearly identified in the process [23]. The original work by Herrmann, Beller and coworkers described cyclopalladated complexes **1** and **2** (Figure 8.3) as very active catalysts in cross-coupling reactions [22a, 23a]. The fact that in those reactions no apparent decomposition of the palladacycle was observed and that the catalyst was recovered unchanged at the end of the reaction led to the authors to mention a not detailed mechanistic option that would involve a Pd(II)/

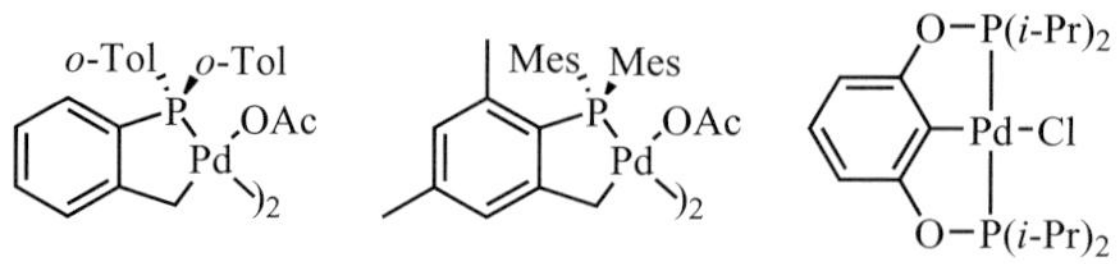

1 **2** **3**

Figure 8.3 Herrmann's phosphane-derived palladacycles.

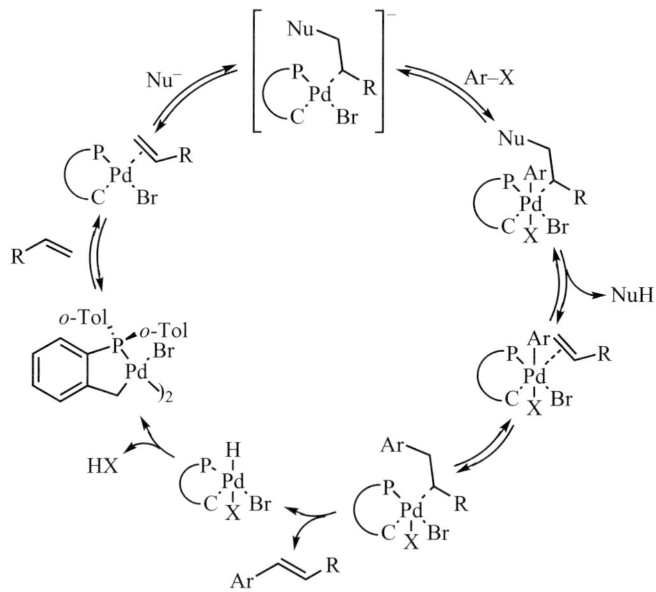

Scheme 8.5 Pd(II)/Pd(IV) catalytic cycle for the Heck reaction with phosphane palladacycles.

Pd(IV) catalytic cycle. Nevertheless, the authors indicated that unknown reduction pathways to Pd(0) could not be ruled out, particularly when they noticed that the new palladacycles reacted with bromoarenes in the Heck reaction only when the olefin was already present [23a]. Scheme 8.5 ($Nu^- = AcO^-, Br^-, \ldots$) depicts a plausible Pd(II)/Pd(IV) catalytic cycle proposed by Shaw based on experimental evidence described in the literature at that point. The concept of a new catalytic cycle circulated through the scientific literature rapidly and a non-experimentally supported Pd(II)/Pd(IV) cycle was again suggested, for palladium pincer-type ligands [24] such as the phosphinite PCP complex **3** [23e] (Figure 8.3), since strong multidentate ligands were believed not to be capable of full or partial deligation even under harsh reaction conditions.

So far, there is no direct experimental evidence in favor of a Pd(II)/Pd(IV) mechanism, while there is plenty of evidence in opposition. Early on, the hypothesis was questioned when Hartwig and Louie provided two different pathways whereby phosphane palladacycle **1** was easily transformed into phosphane-ligated

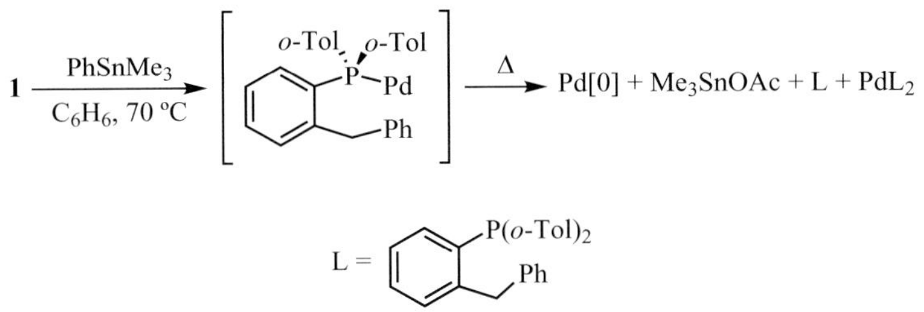

Scheme 8.6 Generation of Pd(0) species from phosphane palladacycle **1** in a Stille reaction.

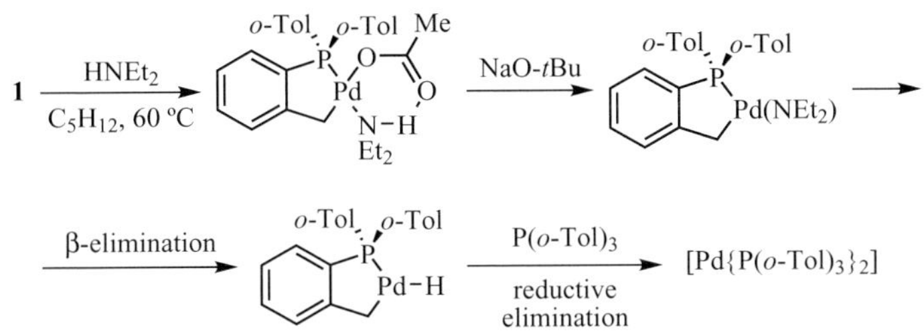

Scheme 8.7 Generation of Pd(0) species from phosphane palladacycle **1** in the amination reaction.

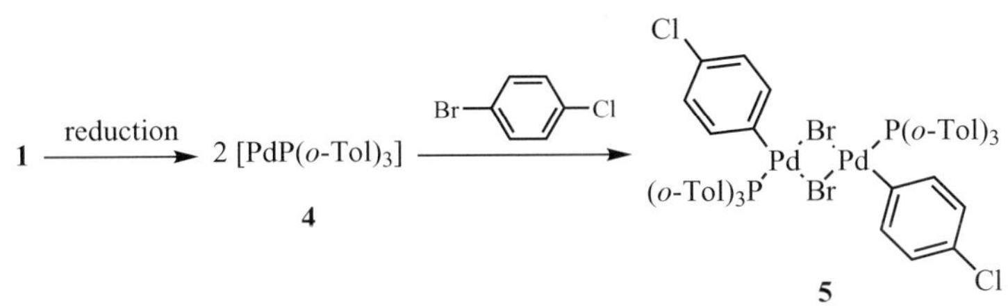

Scheme 8.8 Proposed generation of Pd(0) species from phosphane palladacycle **1** by Beller and coworkers.

Pd(0) species under the reaction conditions during their studies on the amination and Stille couplings of aryl halides [9a] (Schemes 8.6 and 8.7). The nucleophiles (organostannanes or amines) were at the origin of the formation of Pd(0)complexes at the beginning of the catalytic reaction, either by transmetallation of the acetate ligand of **1** by the organostannane followed by reductive elimination (Scheme 8.6) or by cleavage of the acetate bridge in **1** by complexation of the secondary amine, deprotonation of the latter by the base, β-hydride elimination of the resulting amide ligand, and finally reductive elimination (Scheme 8.7).

Beller and Riermeier also involved palladacycle **1** in a Pd(0)/Pd(II) catalytic cycle for the palladium-catalyzed synthesis of trisubstituted olefins [25]. The observed induction period before the production of the arylated alkene was interpreted as the time required for the *in situ* generation of the monophosphane Pd(0) complex **4** from the palladacycle via an undetailed reduction. This species would undergo oxidative addition to the bromoarene to give palladium complex **5** (Scheme 8.8).

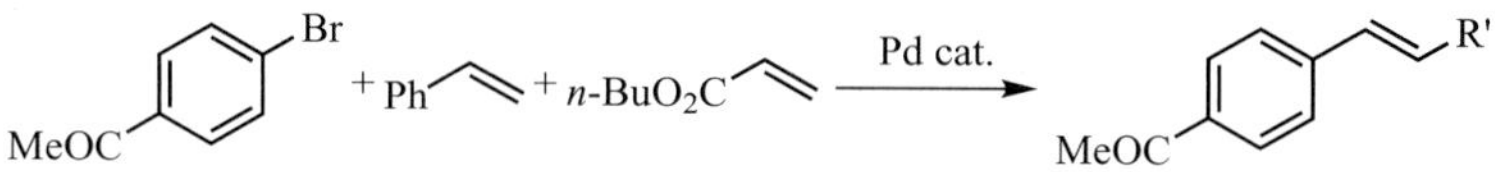

7, R' = Ph
8, R' = n-BuCO$_2$

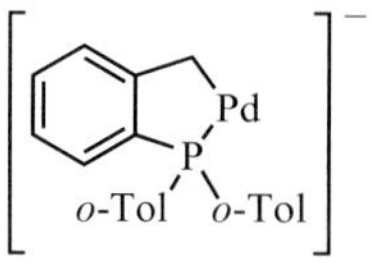

6

Pd catalyst	**7/8**
1	4/96
Pd[(o-Tol)$_3$P]$_2$	3/97
[ArPdBr(o-Tol)$_3$P]$_2$	2/98

Scheme 8.9 Competition experiments.

A cyclopalladated anionic Pd(0) species **6** was proposed by Herrmann and Böhm as a plausible active catalyst in the Heck coupling of aryl bromides with olefins catalyzed by palladacycle **1** (Scheme 8.9) [26]. A cyclometallated catalytically active species such as **6** would account for the observed high activity and stability of phosphane palladacycle catalysts in the Heck reaction. The authors arrived at the conclusion that, under similar reaction conditions, **1** and the catalytic system formed by Pd(0) and (o-Tol)$_3$P lead to different active species, although both catalytic systems were believed to involve a Pd(0)/Pd(II) cycle. This conclusion was based on a series of competition experiments, isotope effect measurements and Hammett studies. For instance, as depicted in Scheme 8.9, under pseudo-first order conditions with respect to the aryl bromide the reaction of *para*-substituted aryl bromides with styrene or n-butyl acrylate in a competition experiment afforded the same ratio of products despite the precatalyst employed, which pointed to the presence of the same Pd(0) catalytic species.

The above-mentioned proposals for the actual role of palladacycles as source of Pd(0) species were not proved in detail due to the extremely difficult characterization of any of the solution species at the very low catalyst concentrations used. However, already at this point some very important conclusions were obtained from those interesting studies. Palladacycles are not stable under the conditions used in the Heck reaction and undergo full or partial disassembly of the dimeric precursor complex, reduction of Pd(II) to Pd(0) and ligand dissociation. Thus, palladacycles should definitively be considered as "reservoirs" of low-ligated highly active Pd(0) species and/or colloidal Pd(0) that are slowly released to the reaction medium. By themselves cyclopalladated compounds such as **1** do not participate in the real catalytic cycle of the Heck reaction. This scenario was supported by Nowotny and coworkers for nitrogen palladacycles from the observation that the polystyrene immobilized imine palladacycle **9** [27], recovered by filtration from an initial catalytic run between iodobenzene and styrene at 140 °C, was completely inactive in a consecutive run while the filtrate of the initial run exhibited an undiminished level of activity upon addition of another equivalent of substrates and base (Scheme 8.10) [28]. This result again is consistent with **9** being a slow source of active colloidal Pd particles. Other attempts to create recyclable palladacycles led to similar conclusions, as in the case of fluorous imine based palladacycles **10**

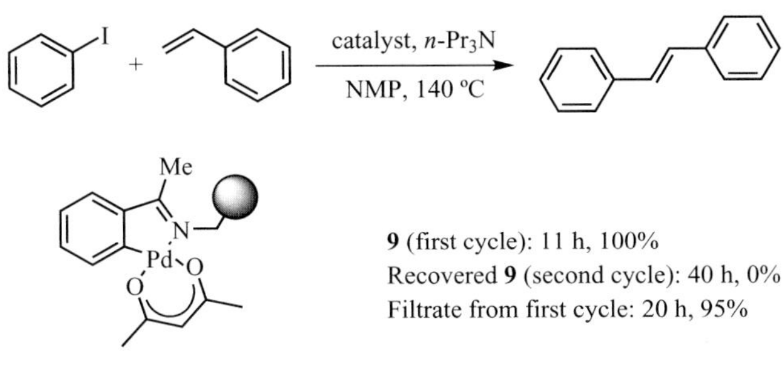

9

Scheme 8.10 Loss of activity of an immobilized palladacycle.

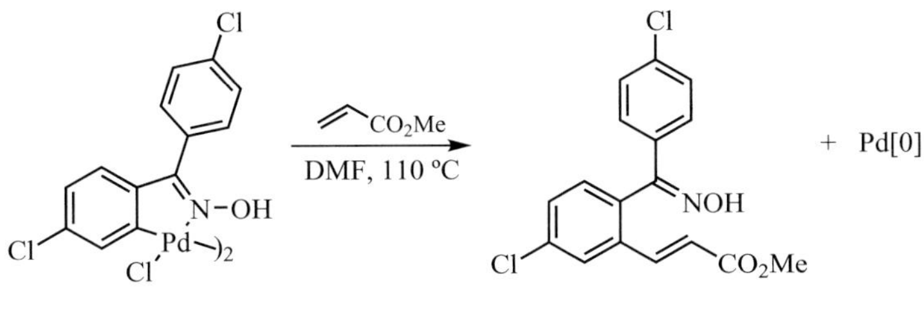

12

Scheme 8.11 Generation of Pd(0) by olefin insertion in Nájera's palladacycle.

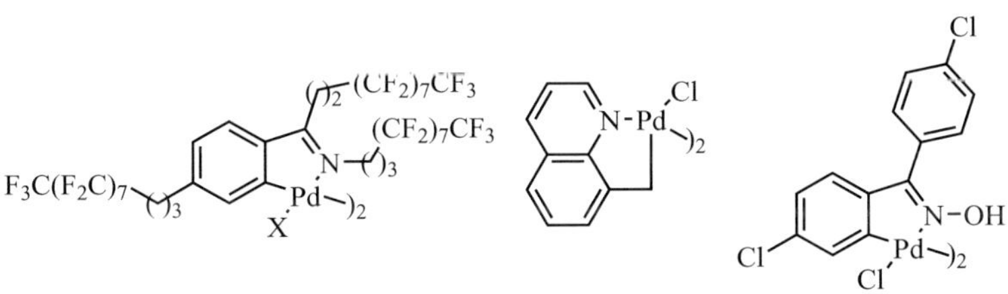

10 (X = I, Cl, OAc) **11** **12**

Figure 8.4 Nitrogen-derived palladacycles.

(Figure 8.4), which can transfer catalytic activity to the non-fluorous phase that contains the responsible palladium colloids, which were also detected by transmission electron microscopy (TEM) [29].

Stoichiometric olefin insertion in the Pd–C bond of the palladacycle [30] has been proposed by Beletskaya and coworkers as a plausible mechanism for the Pd(0) generation from nitrogen-containing palladacycles such as **11**, based on the observed induction periods and sigmoidal kinetics curves [31]. A heavy precipitation of Pd(0) black was observed when oxime-derived catalyst **12** reacted with methyl acrylate in DMF at 110 °C due to Pd–C bond cleavage with C–C bond-forming reductive elimination (Scheme 8.11). The absence of ligands to inhibit

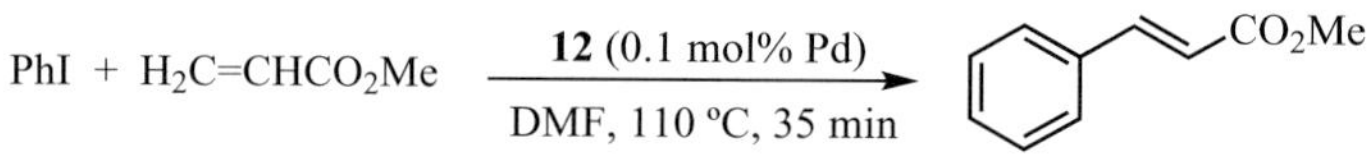

Activation of **12**	Induction period	Yield (%)
—	20 min	35
Heating in DMF at 110 °C	—	99

Scheme 8.12 Activation of oxime palladacycle **12**.

palladium nucleation and growth, stabilizing the initially formed Pd(0) colloids, was the most probable reason for this observation (D. A. Alonso, *et al.* unpublished results).

The Heck reaction between phenyl iodide and methyl acrylate in DMF at 110 °C catalyzed by catalyst **12** showed an induction period of about 20 min (Scheme 8.12) [32a]. This induction stage disappeared when the dimer was previously heated in DMF at 110 °C, leading to a very high yield and a shorter reaction time. This is consistent with the slow formation at high temperatures of Pd(0) nanoparticles. Indeed, TEM analysis of aliquots taken from the reaction showed the presence of Pd nanoparticles (0.9–1.2 nm in average size) and, to much less extent, the presence of Pd nanoparticle colloids with a 10 nm average size [32b]. XPS analysis of the crude reaction mixture showed that, at least in the solid surface, Pd(II) was the most abundant oxidation state, supporting once more a slow thermal decomposition of the precatalyst in the reaction media [32].

Herrmann and coworkers proposed in their initial studies that the acetate-bridged phosphapalladacycle **1** becomes catalytically active at about 80 °C [23a]. Although the pathway that palladacycle precatalysts take to arrive to the active Pd(0) species is unclear, some insights have been reported recently. For instance, high dilution (0.5 mM) EXAFS experiments performed with Herrmann's catalyst **1** in NMP solution at room temperature have detected an equilibrium between the dimer and monomer **13**; notably, NMP interactions are minimal or do not exist under the tested high dilution conditions (0.5 mM) (Scheme 8.13) [33]. In contrast, Jutand and coworkers found, during their cyclic voltammetry and ^{31}P NMR studies with **1** in DMF at higher concentrations (1–2 mM) and in the absence of additives, an equilibrium between **1** and the isomeric neutral monomers **14** preceding the electron transfer (Scheme 8.13). This would rule out the existence of ionic species such as **13** at higher concentrations [34]. Interestingly, the electrochemical reduction of both **1** and **14** leads to the anionic Pd(0) complex **6** initially proposed by Herrmann and coworkers (Scheme 8.9), which after protonation and reductive elimination yields the complex Pd(0){P(*o*-Tol)$_3$}$_2$.

In the same study Jutand and coworkers demonstrated that in the absence of any reducing agent the Pd(0) complex **15a** is generated through an endergonic equilibrium from phosphapalladacycle **1** in DMF at 80 °C, probably via a reductive elimination process between the acetate ligand and the *o*-benzyl moiety of the ligand (Scheme 8.14) [34]. Owing to the endergonic character of the equilibrium a fast backward intramolecular oxidative addition of the monophosphane Pd(0)

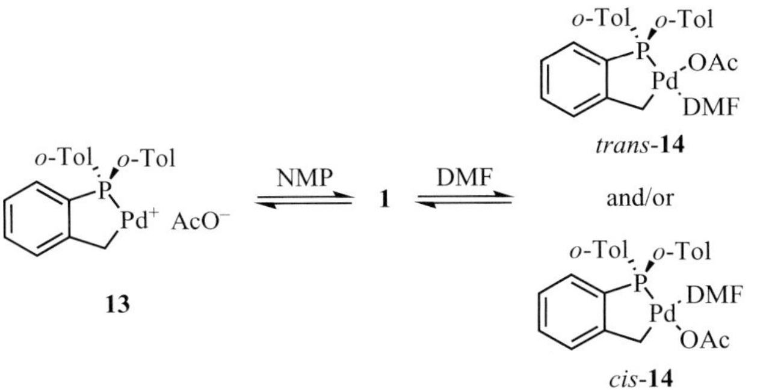

Scheme 8.13 Monomer/dimer equilibria for Herrmann's palladacycle **1**.

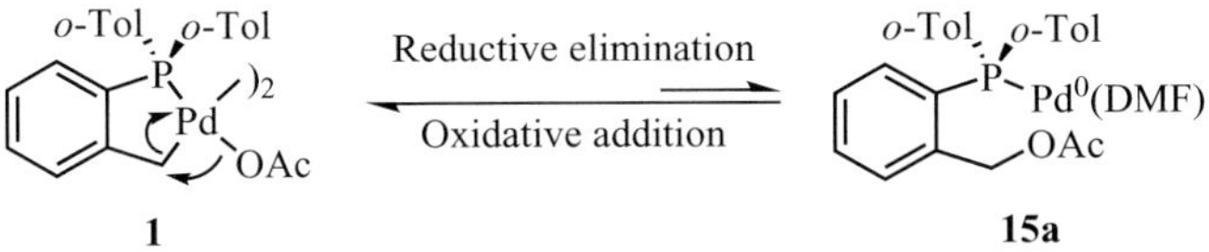

Scheme 8.14 Generation of Pd(0) from Herrmann's palladacycle **1**.

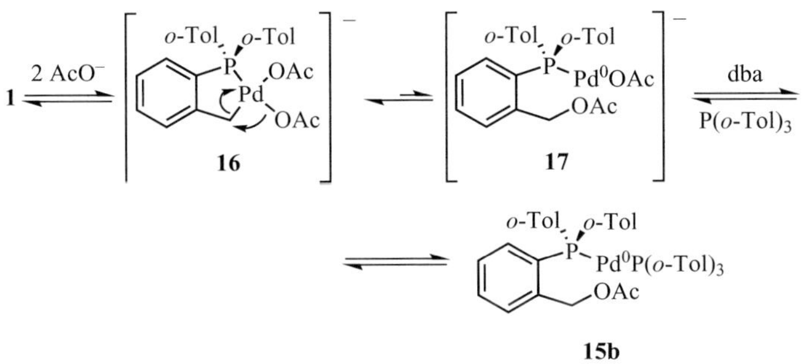

Scheme 8.15 Generation of Pd(0) complex **15b** from **1** in the presence of acetate ions.

complex into the benzyl-acetate bond ensures the observed stability of the CP palladacycle structure under the usually harsh reaction conditions employed in the Heck coupling.

The Pd(0) complex **15b** was trapped and stabilized by addition of dibenzylideneacetone (dba) and P(*o*-Tol)$_3$ and was also detected in cyclic voltammetry (Scheme 8.15). Its generation is favored by acetate anions (often used as base in Heck reactions) via the formation of a monomeric anionic CP palladacycle **16**, which would suffer the reductive elimination process (Scheme 8.15). In the absence of acetate ions, Pd(0) complex **15b** would be formed via reductive elimination over *cis*-**14** in a similar fashion as described for **16** in Scheme 8.15 [34].

Regarding the mechanism followed by pincer palladacycles such as **3** (Figure 8.3), many researchers still support the idea of a Pd(II)/Pd(IV) mechanism due to

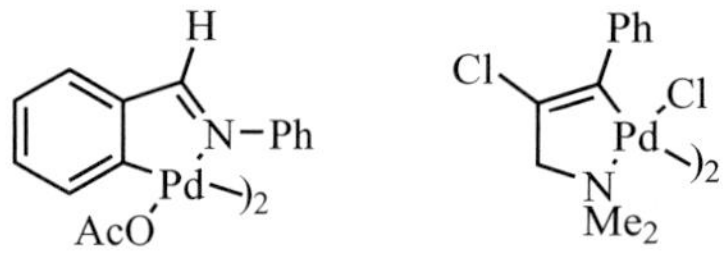

18 19

Figure 8.5 Nitrogen-derived palladacycles used in kinetic studies.

the high stability, probably conferred by the tridentate pincer ligand, observed for these complexes. So far, it has not been possible to rule out beyond doubt catalysis by the intact pincer complex occurring in parallel with catalysis by leached highly active Pd(0) species. However, numerous studies [35] have demonstrated that pincer palladacycles indeed release catalytically active Pd(0) and they do it much more slowly than do four-electron donor palladacycles. This situation would confer an apparent stability to the tridentate systems and the possibility to recover and reuse them as reported by several authors [35a, 35d, 35e].

The different activities observed in the Heck reaction among the palladacycle precursors are usually related to the kinetics of the catalyst preactivation step, which depends on the reagent used, the reaction conditions and the structure of the reactants. Detailed kinetic investigations, kinetic modeling, and experimental studies carried out by Blackmond, Pfaltz and coworkers [36] and Dupont and coworkers [37] on the Heck reaction with PC palladacycle **1** [36] and the NC palladacycles **18** [36] and **19** [37] (Figure 8.5) have shown that the formation of active catalytic species from palladacycle reservoirs gives rise to complicated kinetics.

Blackmond's kinetic model for the Heck olefination of bromobenzaldehyde with *n*-butyl acrylate using dimeric palladacycles **1** and **18** explains experimental observations such as the existence of an induction period and that the catalyst efficiency increases at lower catalyst loadings [36]. This latter point has been also detected by de Vries and coworkers, who observed a better turnover number (TON) for $Pd(OAc)_2$ the lower the concentration, as a consequence of the formation, at higher concentrations of Pd, of inactive Pd(0) black [19]. In the same study, de Vries and coworkers found very similar kinetic behavior for homeopathic $Pd(OAc)_2$ and **1**, indicating that the same catalytically active Pd species should be involved for both systems. This has been corroborated by Blackmond and coworkers, who have demonstrated that the resting state and the rate-limiting step within the catalytic cycle are the same for phosphapalladacycle **1**, azapalladacycles **18** and ligandless $Pd(OAc)_2$, being the olefin addition/insertion process [36]. The half-order dependence of the reaction rate on palladium concentration was used to test the mechanism depicted in Scheme 8.16, where the oxidative addition step was found to be relatively fast (reaction between bromobenzaldehyde and *n*-butyl acrylate), with the dominant species in the catalytic cycle turn being the oxidative addition product **20**. Intermediate **20** is in equilibrium with the halide-bridged dimer **21**, which exists outside the catalytic cycle and is the major species present under the reaction

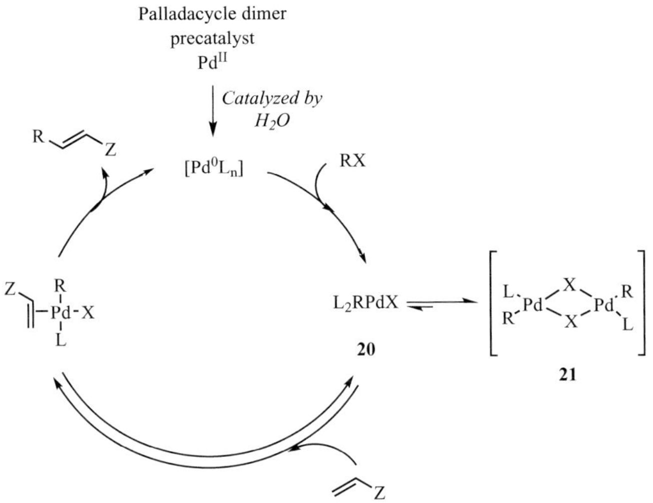

Scheme 8.16 Blackmond's proposed mechanism for the Heck reaction with palladacycle dimers.

conditions studied [36]. This type of halide-bridged dimeric palladacycle has been isolated after Heck coupling reactions of aryl bromides employing palladacycle precursor **1** [38]. The presence of most of the palladium in the form of a halide-bridged complex is also consistent with EXAFS studies carried out on a Heck reaction between iodobenzene and 2-methylprop-2-en-1-ol in NMP catalyzed by homeopathic amounts of $Pd(OAc)_2$, where complexes such as $[Pd_2I_6]^{2-}$ and $[Ph_2Pd_2I_4]^{2-}$ have been detected and isolated [39].

Blackmond's studies also demonstrated a profound effect of the amount of water present in the reaction on the rate at which the active monomeric Pd species is produced from the dimeric palladacycle precursor, decreasing or even suppressing the induction period (Scheme 8.16). This behavior has been further demonstrated in different studies with other palladacycles, such as in the microwave-promoted double arylation of vinyl ethers catalyzed by Herrmann's palladacycle [40].

Dupont and coworkers have recently performed a comprehensive study of the Heck reaction between aryl halides and *n*-butyl acrylate catalyzed by the NC palladacycle **19** [37]. The authors determined not only the nature of the catalytically active species involved but also the kinetics of the reaction, presenting a plausible mechanism of the process. According to their conclusions, palladacycle **19** would act merely as a reservoir of catalytically active Pd(0) species in the form of Pd colloids or highly active forms of low ligated Pd(0) species stabilized by anions and/or solvent molecules that undergo oxidative addition of the aryl halide on the surface with subsequent detachment, generating homogeneous Pd(II) species. Scheme 8.17 depicts the *in situ* transformations of palladium proposed by Dupont and coworkers. A first possible decomposition pathway (A, Scheme 8.17) takes place

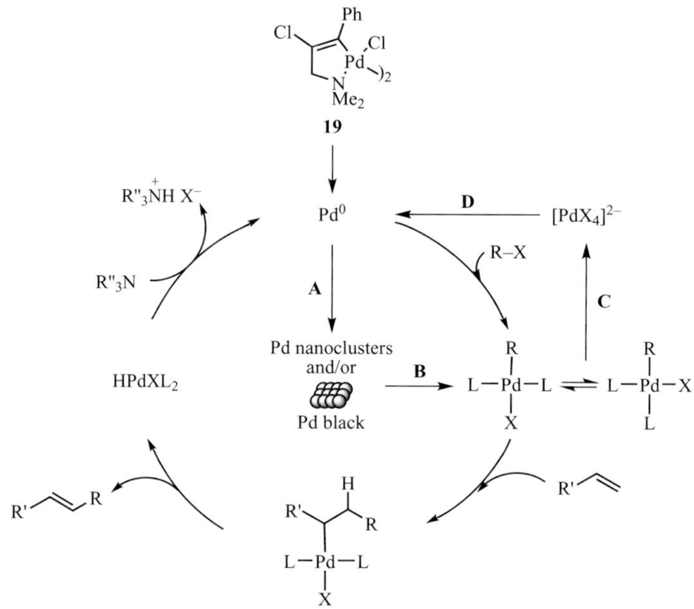

Scheme 8.17 Dupont's proposed transformations of palladium in a Heck reaction catalyzed by an NC palladacycle.

via formation of Pd nanoparticles and eventually inactive Pd metal through autocatalytic agglomeration of Pd(0). According to the kinetic model proposed by the authors, a slight excess of the alkene relative to the haloarene would lead to a rapid rise of Pd(0) concentration that could explain the observed palladium black deposition. When ligands and/or stabilization agents are present in the reaction mixture they exert a kinetic control on the nucleation step, enabling a regeneration process (B, Scheme 8.17) through oxidative addition of the haloarene to form soluble Pd(II) species. This regeneration process depends on halide concentration, with the palladium nucleation rate increasing when the halide concentration decreases [41]. A second decomposition pathway (C, Scheme 8.17) consists of the formation of Pd(II) halide species such as PdX_2, PdX_3^-, PdX_4^{2-} or $Pd_2X_6^{2-}$ together with the production of biaryls or haloarene reduction products from the oxidative addition intermediate. This would occur in the presence of a slight excess of iodobenzene relative to the alkene, a scenario in which the oxidative addition product is the resting state of the catalytic cycle and reaction deactivation by formation of a Pd(II) halide species PdX_n^{m-} is highly probable.

A second regeneration step of the catalytically active species (D, Scheme 8.17) occurs through reduction of the Pd(II) halides promoted by the base.

To conclude this section, all the collected data from the palladacycle precatalysts point to the view that such complexes are stable reservoirs of inactive Pd that slowly release low-ligated catalytically active Pd(0) species to the Pd(0)/Pd(II) catalytic cycle, helping to counterbalance deactivation processes.

8.1.3
Catalysts

The Mizoroki–Heck reaction has become a key step in many syntheses of organic chemicals, natural products and new materials. Practically all forms of palladium can be used as precatalysts for the most reactive substrates such as aryl iodides and activated aryl bromides. However, highly active palladium catalysts are required for the oxidative addition of unactivated substrates such as aryl chlorides [22f, 42] and alkyl halides [43]. During the past few years significant advances have been made in this direction, with most of the success due to the development of new, very active and, at the same time, more stable palladium catalysts [1, 4]. A promising new class of highly active catalysts is palladium complexes containing σ-donating electron-rich and bulky ancillary ligands such as phosphanes [1a, 22f, 44] and N-heterocyclic carbenes (NHC) [1a, 22f, 45]. These ligands provide, through sterically-driven dissociation, highly coordinatively unsaturated electron-rich Pd(0) species that readily undergo oxidative addition reactions with unreactive substrates such as aryl chlorides and aryl tosylates [46]. In addition, the high σ-donicity can stabilize monoligated Pd(0) species, which function as highly reactive yet relatively stable catalytic intermediates. Owing to the large cone angle of the ancillary ligands, the Pd(II) complexes easily suffer reductive elimination as a consequence of the steric relief [47]. On the other hand, ligand-free palladium-catalyzed Heck reactions have attracted increasing attention due to easier work-up and reduced costs. A wide variety of studies have fully confirmed the ability of simple Pd salts such as $Pd(OAc)_2$, $PdCl_2$ and $PdCl_2(SEt_2)_2$ to catalyze Heck reactions of aryl bromides and chlorides at high temperatures. Under ligand-free conditions a competition between oxidative addition and Pd inactivation usually takes place. Thus, it is usually necessary to work under particular reactions conditions such as very low catalyst loadings and in the presence of certain bases [48] and additives such as ammonium salts [49] or nitrogen ligands [21] to avoid Pd inactivation due to aggregation.

In the last decade palladacycles have emerged as a very promising family of organometallic catalyst precursors [22]. Since the cyclopalladation of azobenzene by Cope [50] in 1965, palladacycles have been proposed as intermediate species in many palladium promoted reactions, leading to complex molecular architectures [51]. Their synthesis, structural properties and applications in stoichiometric organic synthesis have been also extensively studied. Twenty years ago the catalytic activity of cyclopalladated complexes was investigated by Lewis in the hydrogenation of alkenes and alkynes [52]. However, the field of palladacycles as catalysts in cross-coupling reactions truly began in 1995 with the introduction by Herrmann and Beller of the new cyclopalladated tri-*o*-tolyl phosphane **1** for the palladium-catalyzed Heck [23a] and Suzuki–Miyaura [53] reactions, using unreactive aryl bromides as well as activated aryl chlorides with unprecedented TONs. Rapid enhancements in activity and stability observed on orthometallation led to a wide variety of phosphorous-, nitrogen-, sulfur-, and oxygen-derived palladacycles as well as cyclopalladated pincer complexes of the type PCP, PCS and NCN, with

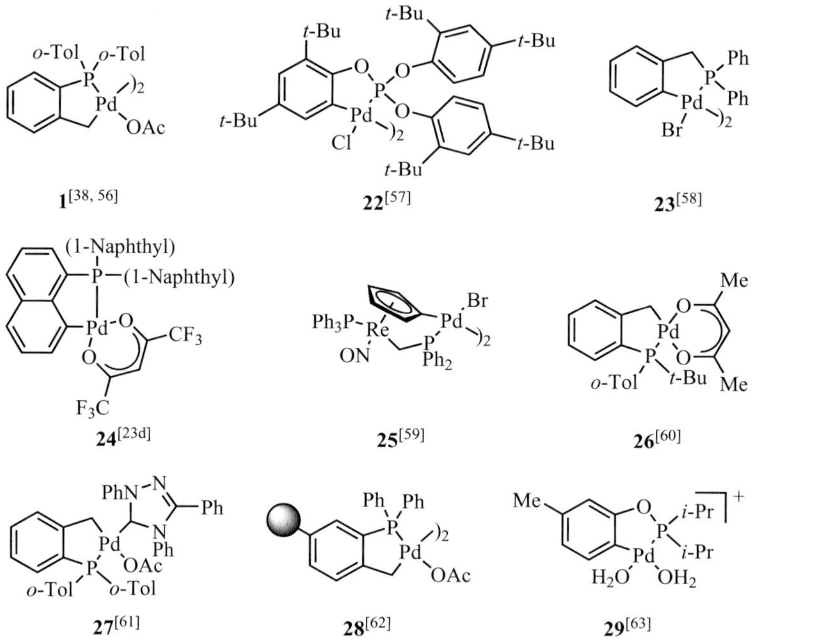

Figure 8.6 Phosphane palladacycles.

reports of extremely high TONs in different C—C and C—heteroatom bond forming reactions.

As already pointed out, since the introduction of Herrmann's phosphane palladacycle **1** as a very effective promoter for the Heck reaction [23a, 54], different phosphorous-derived palladacycles [22g, 55] such as **22–29** have been successfully used for the arylation of alkenes (Figure 8.6). Almost any phosphorous-derived palladacycle can promote the coupling of aryl iodides and bromides with alkenes at high temperatures. However, few of them can perform the Heck coupling of deactivated aryl bromides and aryl chlorides. Table 8.1 depicts representative results. As seen, Herrmann's palladacycle **(1)** is an excellent catalyst for the olefination of deactivated aryl bromides such as 4-bromoanisol in DMAc at 140 °C [38]. Palladacycle **1** is also very active in the Heck reaction of activated aryl chlorides, with better yields in the presence of substoichiometric amounts of soluble bromides such as TBAB [38]. Improved Herrmann's catalyst efficiencies, compared to the organic solvents, have been observed for reactions of bromoarenes and chloroarenes by the employment of non-aqueous ionic liquids such as TBAB as solvents at 130 °C [56].

Some close analogues of **1** such as palladacycles **22, 23, 26** and **27** also show good activity, especially for the Heck reaction of aryl bromides (Table 8.1). Herrmann's palladacycle also shows very good activity when supported in a polystyrene matrix, as in **28** [62], or entrapped in zeolites NaY [64].

The water-soluble palladacycle aqua complex **29** has been prepared recently and tested in the Heck reaction of 3-bromobenzoic acid with 4-vinylbenzoic acid in

Table 8.1 Heck reactions of aryl bromides and aryl chlorides catalyzed by phosphane-derived palladacycles.

Catalyst (mol.% Pd)	RX	Alkene	Conditions	Yield (%)	Ref.
1 (1)	*p*-BrC_6H_4OMe	CO_2n-Bu	NaOAc, DMAc, 140 °C, 48 h	87	[38]
1 (0.1)	*p*-BrC_6H_4OMe	Ph	NaOAc, DMAc, 140 °C, 30 h	69	[38]
1 (0.2)	*p*-ClC_6H_4CHO	CO_2n-Bu	NaOAc, DMAc, TBAB (20 mol.%), 140 °C, 24 h	81	[38]
1 (0.1)	*p*-ClC_6H_4COMe	Ph	NaOAc, DMAc, 140 °C, 54 h	69	[38]
1 (0.1)	*p*-BrC_6H_4OMe	Ph	NaOAc, TBAB, 120 °C, 17 h	79	[56]
1 (2)	PhCl	Ph	NaOAc, TBAB, $[AsPh_4]Cl$ (20 mol.%), 150 °C, 16 h	96	[56]
22 (0.2)	*p*-BrC_6H_4OMe	CO_2n-Bu	K_2CO_3, DMAc, 160 °C, 18 h	88	[57]
23 (0.004)	*p*-BrC_6H_4OMe	CO_2n-Bu	NaOAc, DMAc, 130 °C, 175 h	68	[58]
26 (0.1)	*p*-BrC_6H_4OMe	Ph	NaOAc, NMP, 130 °C, 18 h	84	[60]
27 (1)	*p*-BrC_6H_4OMe	Ph	NaOAc, DMAc, TBAB (20 mol.%), 130 °C, 14 h	92[a]	[61]
27 (0.1)	*p*-ClC_6H_4COMe	Ph	Cs_2CO_3, DMAc, TBAB (20 mol.%), 130 °C, 14 h	73	[61]

a *E*:*Z* = 88:12.

water [63]. The reaction, performed under reflux conditions and employing 0.02 mol.% of Pd, is pH dependent, with optimum catalytic activity at pH 11.5. Under these conditions, the corresponding stilbene derivative is obtained in a modest 51% yield after 3 h.

Catalyst **1** is very useful for the Heck reaction between 1,1-disubstituted olefins such as α-methylstyrene and *n*-butyl methacrylate and various aryl bromides [25]. The nature of the base significantly influences the regiochemistry of the reaction, with internal olefins being synthesized with high selectivities with organic bases such as Bu_3N or diisopropylethylamine (DIPEA) (Scheme 8.18).

The stable phosphapalladacycle **1** has become a standard precatalyst, and has been widely used in many types of Heck reactions. For instance, a very interesting Heck arylation of acrolein with various aryl and heteroaryl bromides has been pos-

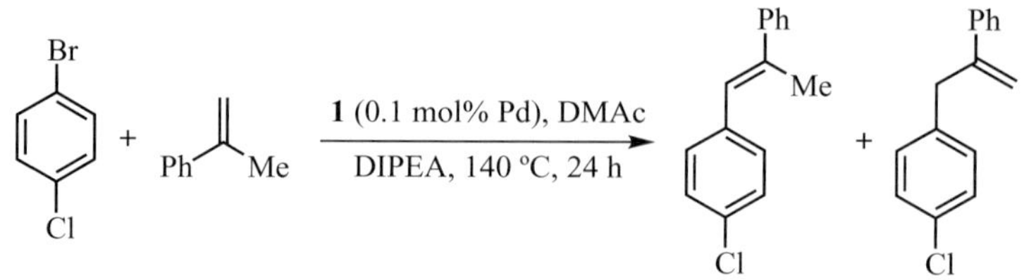

95% 5%

Scheme 8.18 Heck reaction of α-methylstyrene catalyzed by **1**.

Table 8.2 Heck arylation of acrolein with Herrmann's catalyst **1** [65].

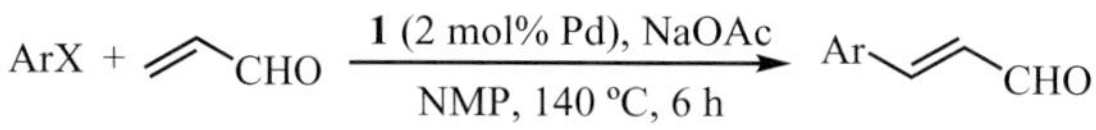

ArX	Yield (%)
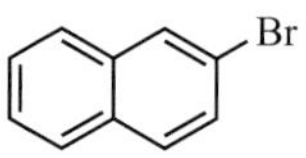	87
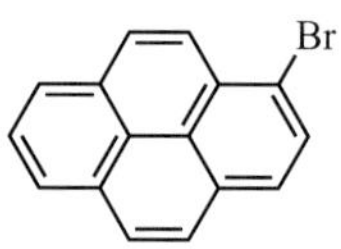	67
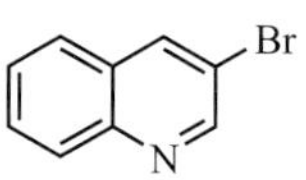	82
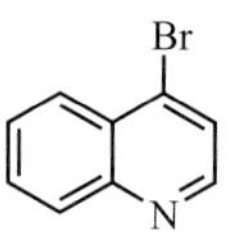	83
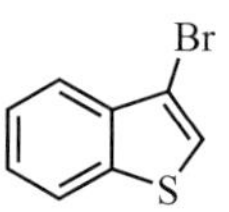	40

sible using catalyst **1**, with yields of up to 87% (Table 8.2) [65]. The results shown in Table 8.2 are remarkable since very poor yields have been observed in the preparation of cinnamaldehyde derivatives directly from acrolein when other palladium catalytic systems are employed due to the high propensity of acrolein to polymerize. In those cases acrolein acetals have to be used to avoid polymerization [66]. For palladacycle **1**, when the reaction was carried out with the diethyl acetal of acrolein the arylated ethyl ester was obtained as a consequence of internal stabilization between the Pd(II) center and the aromatic ring in the carbopalladated intermediate **30** (Scheme 8.19). This interaction prevents internal rotation and leads to a *syn* β-hydrogen elimination via the available H *gem* to the acetal group.

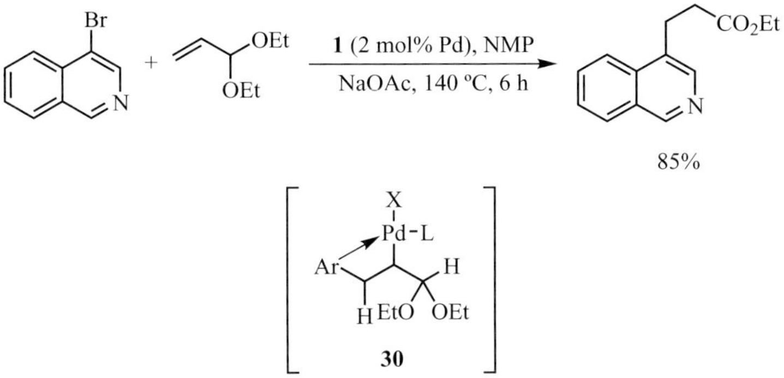

Scheme 8.19 Palladacycle-catalyzed Heck arylation of acrolein diethyl acetal.

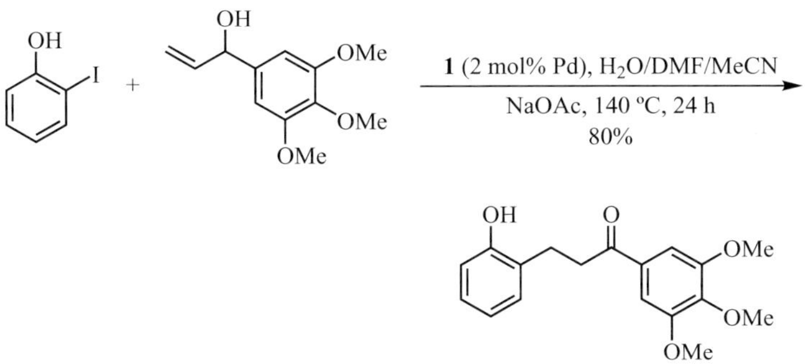

Scheme 8.20 Synthesis of dihydrochalcones catalyzed by **1**.

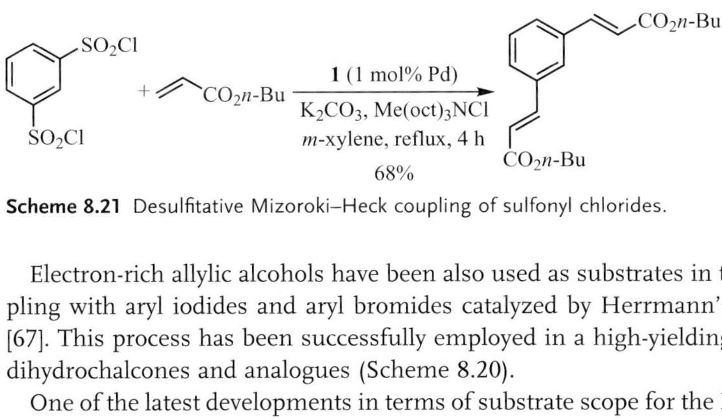

Scheme 8.21 Desulfitative Mizoroki–Heck coupling of sulfonyl chlorides.

Electron-rich allylic alcohols have been also used as substrates in the Heck coupling with aryl iodides and aryl bromides catalyzed by Herrmann's palladacycle [67]. This process has been successfully employed in a high-yielding synthesis of dihydrochalcones and analogues (Scheme 8.20).

One of the latest developments in terms of substrate scope for the Heck reaction is the employment of sulfonyl chlorides. Palladacycle **1** is a very active catalyst in the desulfitative Heck-type cross-coupling reaction of sulfonyl chlorides with mono- and disubstituted olefins at 140 °C in the presence of bulky tri-*n*-octylmethylammonium chloride as phase-transfer catalyst (Scheme 8.21) [68]. Interestingly, other palladium catalysts such as $[PdCl_2(PhCN)_2]$ and Pd_2dba_3 in the presence of bulky electron-rich phosphane or carbene ligands have shown modest activity in this process.

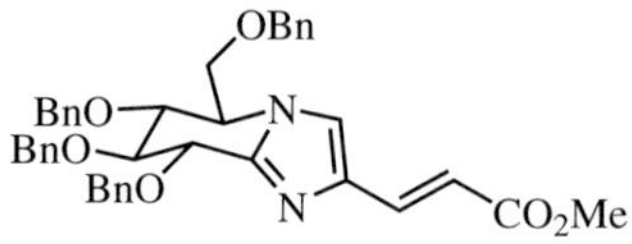

31

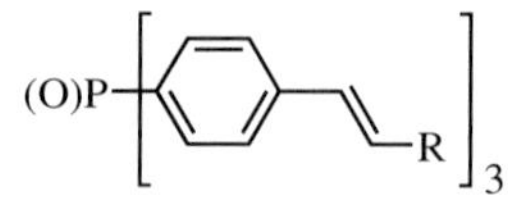

32a, R = CO_2n-Bu, 98%
32b, R = CN, 74%
32c, R = n-C_4H_9, 92%
32d, R = Ph, 91%
32e, R = $CH_2CH(CH_2Rf_8)_2$, 84%

$Pd(OAc)_2$ (12 mol% Pd), Ph_3P, K_2CO_3, DMF, 80 °C, 8 h: 63%
1 (5 mol% Pd), K_2CO_3, DMF, 80 °C, 2 h: 92%

1 (0.25-0.9 mol% Pd), NaOAc, DMF, 125-130 °C

Figure 8.7 Vinylation products of Heck reactions catalyzed by Herrmann's palladacycle **(1)**.

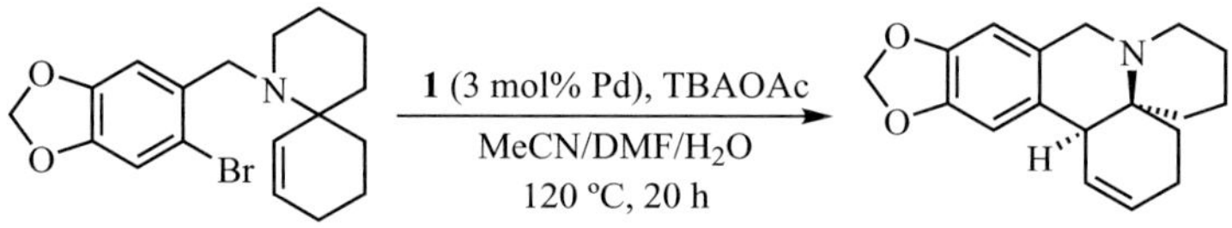

75%

Scheme 8.22 Intramolecular Heck reactions catalyzed by palladacycle **1** *en route* to a cephalotaxine analog.

Improved yields for the Heck vinylation have been obtained when using catalyst **1** in the synthesis of imidazole derivatives such as **31** [69], the vinylation of 2-iodo-*p*-carborane [70] and in the synthesis of functionalized arylphosphine ligands such as **32**, substrates with different steric and electronic properties that can be attached to solid surfaces for immobilization or for use as ponytails and split ponytails in fluorinated solvents [71] (Figure 8.7).

Tietze and coworkers have employed phosphane palladacycle **1** in various intramolecular Heck reactions *en route* to the synthesis of complex alkaloids such as cephalotaxine analogues (Scheme 8.22) and steroid derivatives [72]. Moreover, an efficient access to novel enantiomerically pure steroidal δ-amino acids has been reported recently by de Meijere and coworkers through an intermolecular Heck coupling of vinyl bromide derivatives with *tert*-butyl acrylate catalyzed by **1** [73]. A double Heck inter- and intramolecular coupling of bridged *o,o′*-dibromobiaryls with ethyl acrylate has also been performed in the presence of Herrmann–Beller palladacycle [74].

A very interesting domino-Heck double cyclization process employing allylsilanes catalyzed by **1** has allowed efficient and selective preparation of complex polycyclic structures containing a tetrasubstituted double bond of defined configuration (Scheme 8.23) [75]. The final vinyl silanes with *(E)*-configuration were obtained after oxidative addition of the aromatic bromide to Pd(0) and double *syn*-addition to the alkyne and alkene moieties.

Sulfur and oxygen palladacycles (Figure 8.8) have been much less studied and have always showed lower catalytic activity than the phosphorous or nitrogen pal-

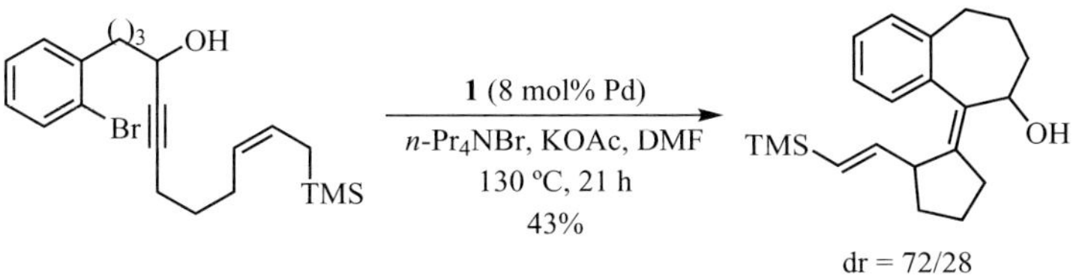

Scheme 8.23 Domino Heck double cyclization catalyzed by palladacycle **1**.

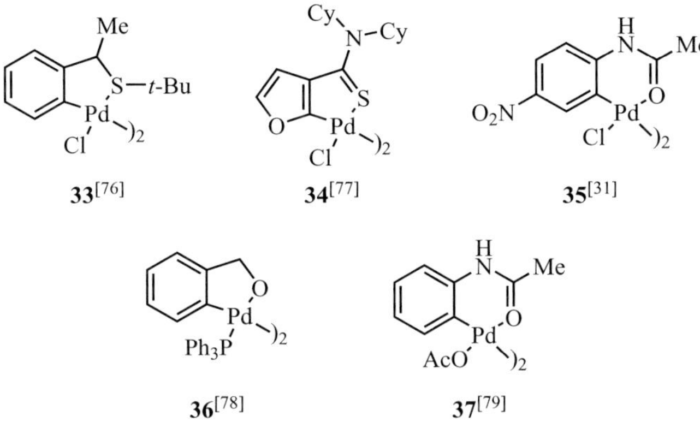

Figure 8.8 Sulfur and oxygen palladacycles.

Table 8.3 Heck reactions of aryl bromides and aryl chlorides catalyzed by sulfur-derived palladacycles.

Catalyst (mol.% Pd)	RX	Alkene	Conditions	Yield (%)	Ref.
33 (4×10^{-3})	*p*-BrC_6H_4OMe	$CH_2{=}CHCO_2$*n*-Bu	NaOAc, DMAc, TBAB, 170 °C, 6 h	61	[76]
33 (0.2)	*p*-$ClC_6H_4NO_2$	$CH_2{=}CHCO_2$*n*-Bu	NaOAc, DMAc, TBAB, 170 °C, 2 h	49	[76]
33 (0.2)	*p*-$ClC_6H_4NO_2$	$CH_2{=}CHPh$	NaOAc, DMAc, TBAB, 170 °C, 5 h	41	[76]
34 (0.5)	*p*-BrC_6H_4OMe	$CH_2{=}CHCO_2$*n*-Bu	NaOAc, DMAc, TBAB, 120 °C, 15 h	58	[77]
34 (0.5)	*p*-BrC_6H_4OMe	$CH_2{=}CHPh$	NaOAc, DMAc TBAB, 120 °C, 15 h	75	[77]

ladacycles. In the absence of external ligands, only the benzylsulfide palladacycle **33** [76] and the furan-carbothiamide-based complex **34** [77] have been reported to be able to catalyze Heck reactions of non-activated aryl bromides and activated aryl chlorides (only **33**) with synthetically useful yields (Table 8.3). However, in the

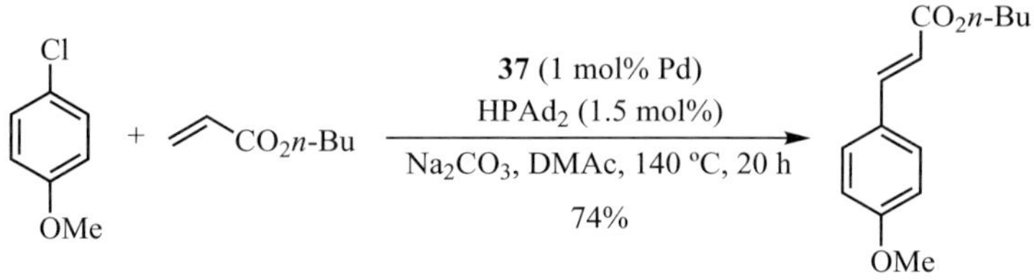

Scheme 8.24 Heck coupling catalyzed by acetamide-derived palladacycle **37**.

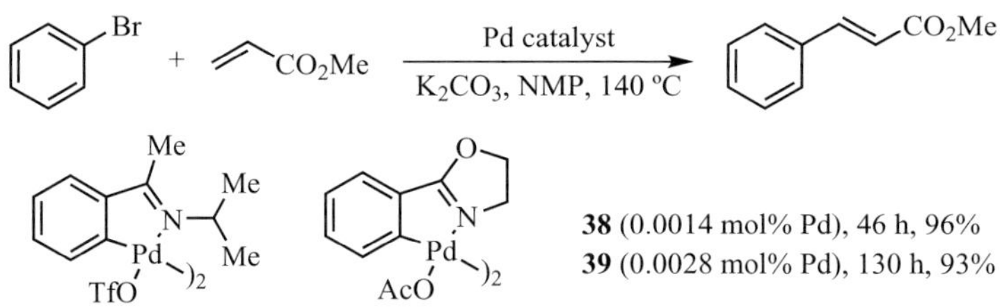

Scheme 8.25 Heck coupling of bromobenzene catalyzed by Milstein's imine palladacycles.

presence of the highly donating sterically demanding secondary phosphane bis(adamantyl)phosphane, Indolese and Studer have presented acetamide-derived palladacycle **37** as a very active catalyst for the Heck reaction of deactivated 4-chloroanisole with *n*-butyl acrylate at 140 °C (Scheme 8.24) [79]. This catalyst combines the stability induced by the presence of a palladacycle framework with the high activity commonly associated with palladium/phosphane complexes. Palladacycles alone failed to catalyze the reaction with chloroarenes. The catalytic activity in the Heck reaction of bromobenzene with styrene of a new aliphatic oxygen palladacycle prepared in the pores of 3-hydroxypropyl-triethoxysilane functionalized MCM-41 has also been studied [80].

By far the largest number of structurally different nitrogen palladacycles tested in the Mizoroki–Heck reaction are cyclopalladated compounds. In pioneering work on the Heck coupling employing nitrogen palladacycles Milstein and coworkers introduced imine palladacycles **38** and **39** as exceptional catalysts for the coupling of bromobenzene and methyl acrylate in NMP at 140 °C (Scheme 8.25) [27a]. Kinetic measurements performed by Blackmond and coworkers by precision calorimetry showed that imine palladacycles such as **18** (Figure 8.5) are indeed much more active than Herrmann's phosphane palladacycle **1** in the Heck coupling of aryl halides with olefins [81].

In further studies, several dimeric and monomeric imine and amine analogues of Milstein's palladacycles have been reported (Figure 8.9), showing in general very good activity in the Heck reaction of aryl bromides (including non-activated ones such as 4-bromoanisole) and poor activity for activated aryl chlorides even at high temperatures (Table 8.4). Of special interest are the results for dicyclohexylphosphane-derived monomeric palladacycle **46** developed by Indolese

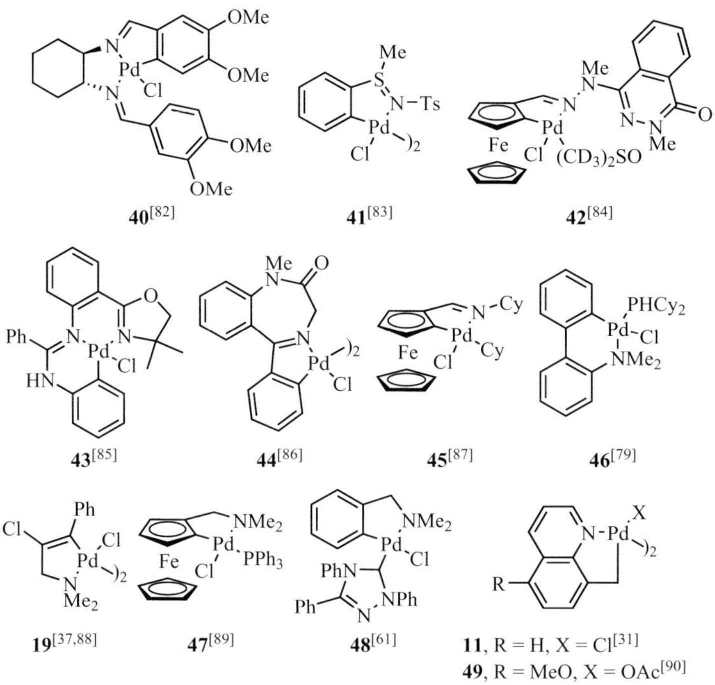

Figure 8.9 Imine and amine-derived palladacycles.

and coworkers [79]. This catalyst, which can be either isolated or prepared *in situ* starting from the corresponding palladacycle bridge dimer and the secondary phosphane, is a highly active palladium catalyst with broad applicability for various coupling reactions employing deactivated chloroarenes as starting materials. In particular, as depicted in Table 8.4, palladacycle **46** affords an excellent yield for the Heck coupling of 4-chloroanisole with *n*-butyl acrylate at 140 °C under low loading conditions [79]. The high efficiency of **46** containing a secondary phosphane is surprising since secondary phosphanes have a much lower basicity and are less sterically hindered than other tertiary derivatives such as P(*t*-Bu_3).

A promising new class of highly active cyclopalladated catalyst has been presented recently by Dupont and coworkers [88]. Palladacycle **19**, derived from the chloropalladation of 3-(dimethylamino)-1-phenyl-1-propyne, promotes the arylation of olefins such as *n*-butyl acrylate and styrene with deactivated aryl bromides and activated aryl chlorides at high reaction temperatures in good yields (Table 8.4). Moreover, catalyst **19** affords very good yields in the room temperature Heck coupling of aryl iodides and activated aryl bromides [88]. The lability of this aliphatic complex, which can suffer under the reaction conditions retrochloropalladation to afford catalytically active species of Pd(0), seems to be related to the possibility of lowering the reaction temperature. Then, the stability of a palladacycle catalyst should not be an unconditional requirement to achieve high yields in a Heck process.

Table 8.4 Heck reactions of aryl bromides and aryl chlorides catalyzed by imine and amine-derived palladacycles.

Catalyst (mol.% Pd)	RX	Alkene	Conditions	Yield (%)	Ref.
40 (2)	PhBr	$=\!\!\diagup CO_2Me$	Et_3N, DMAc, reflux, 24h	68	[82]
41 (5×10^{-4})	PhBr	$=\!\!\diagup CO_2Me$	Et_3N, NMP, 140°C, 12h	21	[83]
43 (1)	*p*-BrC_6H_4OMe	$=\!\!\diagup Ph$	Cs_2CO_3, toluene, 110°C, 8h	84	[85]
43 (2)	*p*-ClC_6H_4Ac	$=\!\!\diagup Ph$	Cs_2CO_3, DMAc, 135°C, 24h	80	[85]
44 (0.1)	*p*-BrC_6H_4OMe	$=\!\!\diagup CO_2n\text{-}Bu$	NaOAc, DMAc, TBAB, 120°C, 24h	70	[86]
44 (0.1)	*p*-ClC_6H_4Ac	$=\!\!\diagup CO_2n\text{-}Bu$	NaOAc, DMAc, TBAB, 120°C, 24h	31	[86]
45 (0.1)	*p*-BrC_6H_4OMe	$=\!\!\diagup CO_2Et$	Et_3N, DMF, 140°C, 16h	54	[87]
45 (1)	*m*-$ClC_6H_4NO_2$	$=\!\!\diagup CO_2Et$	Et_3N, DMF, 140°C, 24h	47	[87]
46 (0.25)	*p*-ClC_6H_4OMe	$=\!\!\diagup CO_2n\text{-}Bu$	Na_2CO_3, DMAc, $HPAd_2$, 140°C, 20h	89	[79]
19 (0.01)	*p*-BrC_6H_4OMe	$=\!\!\diagup CO_2n\text{-}Bu$	NaOAc, DMAc, TBAB, 150°C, 24h	72	[88]
19 (0.1)	*p*-$ClC_6H_4NO_2$	$=\!\!\diagup CO_2n\text{-}Bu$	NaOAc, DMAc, TBAB, 150°C, 24h	91	[88]
48 (1)	*p*-BrC_6H_4OMe	$=\!\!\diagup Ph$	Cs_2CO_3, DMAc, TBAB, 130°C, 14h	81	[61]
48 (1)	*p*-ClC_6H_4Ac	$=\!\!\diagup Ph$	$Ca(OH)_2$, DMAc, 130°C, 14h	97[a]	[61]
11 (0.2)	PhBr	$=\!\!\diagup CO_2Et$	n-Bu_3N, DMAc, TBAB, 140°C, 24h	72	[31]
49 (2×10^{-4})	PhBr	$=\!\!\diagup CO_2n\text{-}Bu$	CsOAc, DMAc, TBAB, 140°C, 48h	96	[90]

a $E:Z = 96:4$.

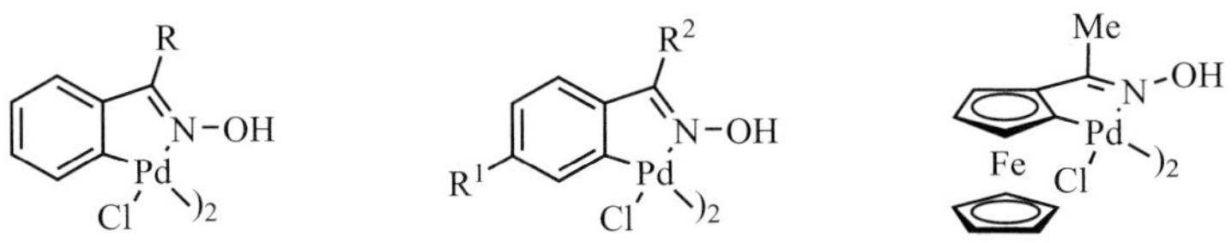

50, R = H[92]
51, R = Me[32,92]
52, R = Ph[32,92]

53, R^1 = OH, R^2 = Me[94]
54, R^1 = OH, R^2 = p-HOC_6H_4[94]
55, R^1 = OMe, R^2 = p-$MeOC_6H_4$[32]
12, R^1 = Cl, R^2 = p-ClC_6H_4[32]

56[32,95]

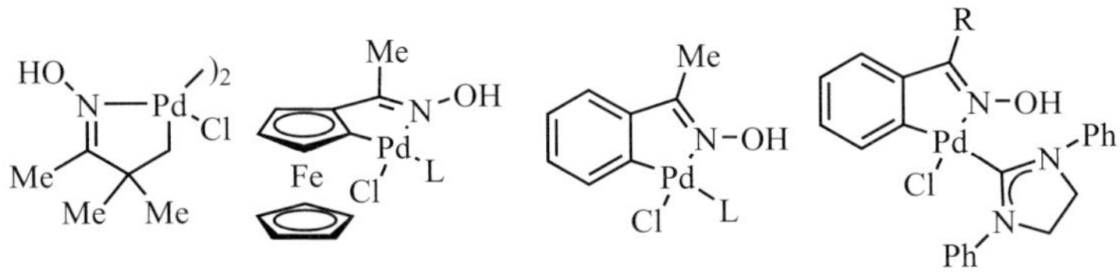

57[32]

58, L = PPh_3[92b]
59, L = $P(OEt)_3$[92b]

60, L = PPh_3[97]
61, L = $P(OEt)_3$[97]

62, R = Me[92b]
63, R = Ph[92b]

Figure 8.10 Oxime palladacycles.

The potential of oxime-derived palladacycles as catalysts in C—C bond forming reactions has been also demonstrated by Nájera and coworkers. These catalysts are very stable, easily prepared and not sensitive to oxygen or moisture [91]. Several groups have prepared various oxime-derived palladacycles in a straightforward synthesis via aromatic metallation of the corresponding oximes with Li_2PdCl_4 in MeOH at r.t. in the presence of NaOAc as base. Various dimeric chloro-bridged palladacycles such as **12** and **50–57** (Figure 8.10), derived from aromatic and aliphatic oximes from benzaldehyde [92], acetophenone [32, 92, 93], benzophenone [32, 92], 4-hydroxyacetophenone [94], 4,4′-dihydroxybenzophenone [94], 4,4′-dimethoxybenzophenone [32], 4,4′-dichlorobenzophenone [32], acetylferrocene [32, 95], and pinacolone [32], have been described and their catalytic activity checked in Heck and different cross-coupling processes such as Suzuki, Ullmann, Stille, Glaser and alkyne acylation [32b, 96]. Furthermore, an interesting approach in palladacycle catalysis consists in combining oxime-palladacycles with bulky and electron-rich ligands such as phosphanes and carbenes. Particularly interesting is the use of preformed hybrids of the palladacycle with potent monodentate phosphorous, nitrogen, or carbene ligands such as **58–63** (Figure 8.10). These monomeric complexes are prepared by reaction of the corresponding chloro-bridged cyclopalladated compounds with the desired ligand.

Pioneering studies performed by Nájera and coworkers on the use of oxime palladacycles as catalysts for the Mizoroki–Heck reaction were carried out in organic solvents, demonstrating that these complexes can be efficiently used in the Heck coupling of aryl bromides and chlorides with olefins [32, 93]. Of oxime catalysts tested, complex **12**, derived from 4,4′-dichlorobenzophenone oxime, was the most efficient catalyst precursor for the Mizoroki–Heck vinylation of aryl halides employing DMF or NMP as solvents, K_2CO_3 as base and TBAB as additive (Table 8.5) [32]. Under high temperature conditions (130–160 °C), palladacycle **12**

Table 8.5 Heck reactions of aryl bromides and aryl chlorides catalyzed by oxime catalyst **12** [32].

Catalyst (mol.% Pd)	RX	Alkene	Conditions	Yield (%)
12 (0.5)	*p*-BrC_6H_4OMe	$CH_2{=}CH{-}CO_2Me$	K_2CO_3, DMF, TBAB, 130 °C, 9 h	57
12 (0.001)	*p*-BrC_6H_4OMe	$CH_2{=}CH{-}Ph$	K_2CO_3, NMP, TBAB, 160 °C, 16 h	85
12 (0.1)	*p*-$ClC_6H_4NO_2$	$CH_2{=}CH{-}CO_2n\text{-}Bu$	K_2CO_3, DMF, TBAB, 160 °C, 4.5 h	92
12 (0.5)	*p*-$ClC_6H_4NO_2$	$CH_2{=}CH{-}Ph$	K_2CO_3, DMF, TBAB, 130 °C, 8 h	70
12 (0.2)	*p*-ClC_6H_4OMe	$CH_2{=}CH{-}Ph$	K_2CO_3, NMP, LiBr, 160 °C, 24 h	22

is a very effective catalyst under aerobic conditions for the Heck coupling of deactivated aryl bromides and activated aryl chlorides with a wide variety of olefins, affording lower yields in the case of less reactive deactivated substrates such as 4-chloroanisole (Table 8.5). Excellent yields are also observed when polybromobenzenes such as 1,2-dibromobenzene and 1,3,5-tribromobenzene react with an excess of methyl acrylate, affording the corresponding aromatic di- and triacrylic methyl esters [32]. In the same way, a highly efficient **12**-catalyzed multiple vinylation of polybromo tribenzotriquinacenes and fenestrindanes with styrene or methyl acrylate has been presented, employing typical conditions for Heck coupling of aryl bromides (Scheme 8.26) [98]. Under similar reaction conditions other different palladium sources such as $Pd(OAc)_2$, $PdCl_2$ and $Pd(PPh_3)_4$ did not give satisfactory results, yielding a complex mixture containing partial cross-coupling products.

Palladacycle **12** is also a very efficient precatalyst for the palladium-catalyzed annulation of internal alkynes with *o*-halobenzaldehydes and *o*-haloanilines, developed by Larock and coworkers, for the synthesis of indenones and indoles [32]. The annulation reaction works with 2-bromobenzaldehyde, 2-chlorobenzaldehyde and 2-iodoaniline, employing only 0.5 mol.% of **12** with yields of 47–98% (Scheme 8.27), whereas 5 mol.% of $Pd(OAc)_2$ is used under Larock's conditions.

Oxime-derived monomeric complexes **58–63** (Figure 8.10) bearing an extra bulky electron-rich extra ligand such as PPh_3, $P(OEt)_3$, or a N-heterocyclic carbene show comparable reactivity to the corresponding dimeric chlorobridge counterparts in Heck catalysis when coupling aryl bromides [92b, 97]. Only phosphite adduct **59** possess, to some extent, better activity than **56** in the coupling of activated aryl chlorides such as 4-chloroacetophenone and 4-chlorobenzonitrile with styrene [92b].

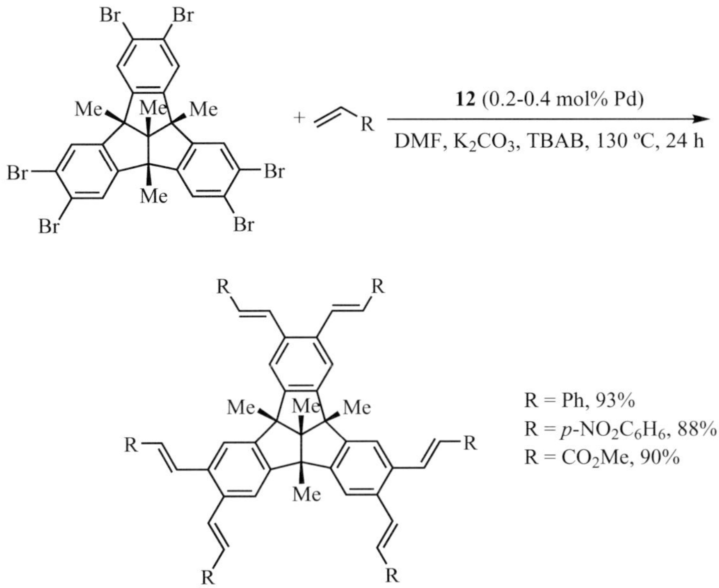

R = Ph, 93%
R = p-$NO_2C_6H_6$, 88%
R = CO_2Me, 90%

Scheme 8.26 Multiple Heck vinylation catalyzed by oxime palladacycle **12**.

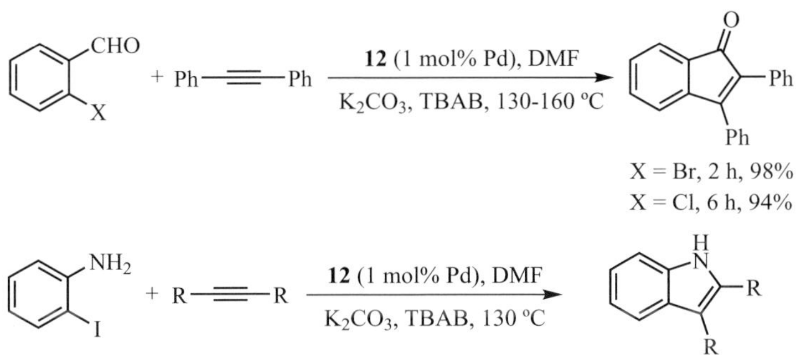

X = Br, 2 h, 98%
X = Cl, 6 h, 94%

R = n-C_3H_7, 9 h, 93%
R = Ph, 10 h, 98%
R = TMS, 4 h, 47%

Scheme 8.27 Annulation of internal alkynes catalyzed by oxime palladacycle **12**.

4-Hydroxyacetophenone-derived palladacycle **53** (Figure 8.10) has been shown as one of the most active and versatile palladium catalysts for the Suzuki reaction of aryl bromides and chlorides in organic and aqueous solvents [94, 96]. With regard to Heck olefination, catalyst **53** is an efficient precatalyst for the mono- and diarylation of olefins under aqueous conditions (Table 8.6, Scheme 8.28) [99]. Unsubstituted and substituted α,β-unsaturated carbonyl compounds are efficiently monoarylated using (dicyclohexyl)methylamine (Cy_2NMe) as base under thermal conditions or microwave irradiation (Table 8.6). The rate enhancement observed

Table 8.6 Monoarylation of α,β-unsaturated carbonyl compounds catalyzed by oxime catalyst **53** [99].

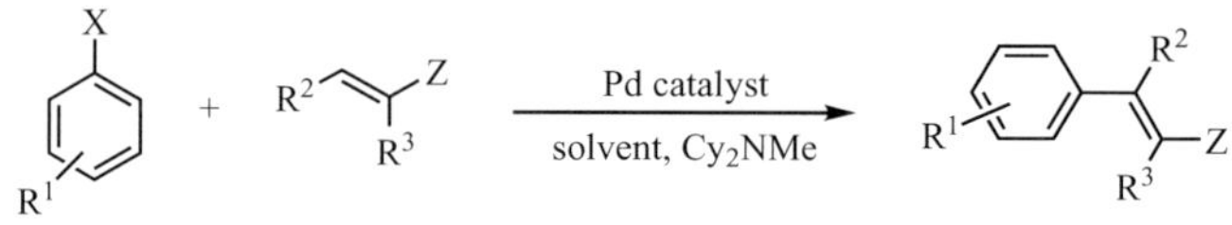

Catalyst (mol.% Pd)	X	R^1	R^2	R^3	Z	Conditions	Yield (%)
53 (0.01)	I	*p*-Cl	H	H	CO_2t-Bu	H_2O, 120 °C, 3 h	94
$Pd(OAc)_2$ (0.01)	I	*p*-Cl	H	H	CO_2t-Bu	H_2O, 120 °C, 3 h	83
53 (0.01)	I	*p*-Cl	H	H	CO_2t-Bu	H_2O, 120 W, 120 °C, 10 min	87
$Pd(OAc)_2$ (0.01)	I	*p*-Cl	H	H	CO_2t-Bu	H_2O, 120 W, 120 °C, 10 min	88
53 (0.01)	Br	*p*-Cl	H	H	CO_2t-Bu	DMAc/H_2O, TBAB, 130 °C, 14 h	85
$Pd(OAc)_2$ (0.01)	Br	*p*-Cl	H	H	CO_2t-Bu	DMAc/H_2O, TBAB, 130 °C, 14 h	47
53 (0.1)	Br	*p*-MeO	H	H	CO_2t-Bu	DMAc/H_2O, TBAB, 130 °C, 14 h	40
53 (1)	Br	*p*-MeO	Me	H	CO_2t-Bu	DMAc/H_2O, TBAB, 130 °C, 14 h	54[a]
53 (1)	Cl	*p*-Ac	Ph	H	CO_2Et	DMAc/H_2O, TBAB, 140 °C, 24 h	38[b]

a *E*:*Z* = 97:3.
b *E*:*Z* = 61:39.

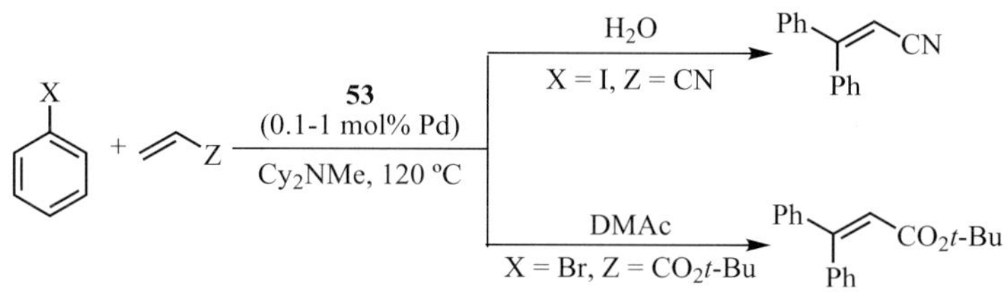

Scheme 8.28 Diarylation of α,β-unsaturated carbonyl compounds and derivatives catalyzed by oxime palladacycle **53**.

when using Cy_2NMe as base is due to the more rapid conversion of the intermediate [$L_2Pd(HX)$], since the base, which is needed to remove HX, is already bound to the palladium center [100]. In addition, the use of an organic base avoids the hydrolysis of acrylic esters in aqueous media. Under these conditions, regioselective monoarylation of unsubstituted α,β-unsaturated carbonyl compounds takes place with aryl iodides at 120 °C in water employing either **53** or $Pd(OAc)_2$ as catalysts. On the other hand, aqueous DMAc in the presence of TBAB as additive provide the most convenient conditions for the **53**-catalyzed monoarylation with aryl bromides. In this latter case, $Pd(OAc)_2$ is much less effective than oxime palladacycle **53**, and the coupling does not take place, or is much less effective, under microwave irradiation. For the monoarylation of substituted α,β-unsaturated carbonyl compounds, good E-stereoselectivities are obtained for crotonates, whereas cinnamic acid derivatives afford lower stereoselectivities. Very low yields are observed when performing the arylation with activated aryl chlorides.

The β,β-diarylation of acrylic systems can be achieved using 2 equivalents of the aryl halide and higher catalyst loadings of **53** either in water at 120 °C for aryl iodides or in DMAc for aryl bromides (Scheme 8.28) [99]. Notably, in these particular reactions the use of $Pd(OAc)_2$ as catalyst is unproductive. This methodology was applied to the stereoselective synthesis of methoxylated (*E*)-stilbene derivatives, which are important chemotherapeutic agents for cancer treatment [101]. A Heck reaction between less reactive aryl halides and styrenes, using oxime palladacycle **53** or $Pd(OAc)_2$ as catalysts, can be performed using Cy_2NMe in aqueous DMAc or in water with TBAB as additive or in DMAc with TEA as base [102]. The former reaction conditions allow the coupling between 3,5-dimethoxyiodobenzene and styrenes with the best regioselectivity.

The synthetic applications of oxime-derived palladacycles have been expanded to the chemoselective synthesis of cinnamaldehyde derivatives and ethyl 3-arylpropanoates through a chemoselective arylation of acrolein diethyl acetal [66] catalyzed by palladacycle **53** (Scheme 8.29) [103]. The preparation of cinnamaldehyde derivatives is performed by Heck reaction of acrolein diethyl acetal with iodo-, bromo- and chloroarenes in DMAc using K_2CO_3 as base at 120 °C and TBAB or KCl as additives. For the preparation of 3-aryl propanoate esters, the arylation of acrolein diethyl acetal with iodoarenes is performed at 90 °C in aqueous DMAc using Cy_2NMe as base. Aryl bromides require TBAB as additive and higher temperatures (120 °C). The reaction can be performed under thermal or microwave irradiation conditions with shorter reaction times. Complex **53** is again more efficient than other palladium sources such as $Pd(OAc)_2$ in this process.

Several supported oxime palladacycles have been designed to combine the advantages of both homogeneous and heterogeneous catalysts [104]. These systems have been successfully used in different cross-coupling reactions such as Heck, Suzuki and Sonogashira couplings. With respect to the Heck reaction, supported catalysts **64–66** (Figure 8.11) have been prepared and tested as promoters for the coupling of aryl bromides with olefins. Among them, only Kaiser oxime resin derived palladacycles **65** are efficient precatalyst for the reaction in DMF or aqueous solvents under relatively moderate temperatures (110–120 °C) [104e]. Although supported catalysts **65a** and **66** can be reused several times after recycling, low to moderate leaching of Pd(0) is always observed, which is in agreement with the palladacycles acting as a reservoir of the truly catalytically active Pd(0) species. In

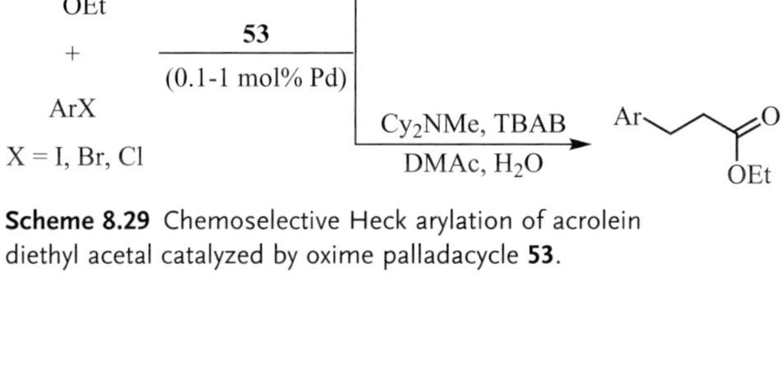

Scheme 8.29 Chemoselective Heck arylation of acrolein diethyl acetal catalyzed by oxime palladacycle **53**.

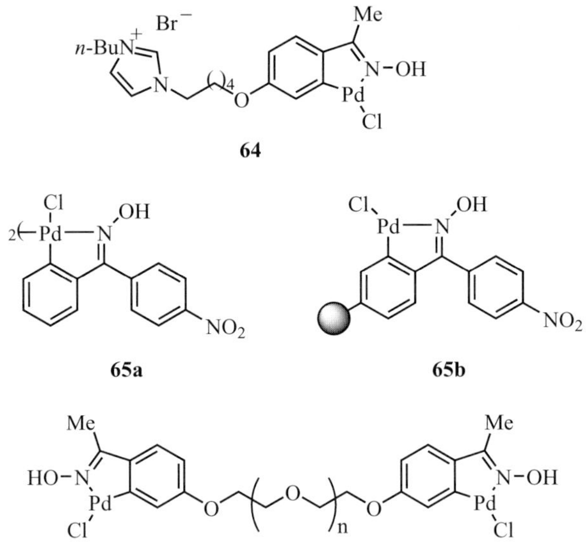

Figure 8.11 Supported oxime palladacycles employed in the Heck reaction.

particular, poisoning with Hg and XPS studies over palladacycle **65a**, which is easily recovered in water, have clearly demonstrated the role this polymer has as a source of highly active Pd(0) species [104e].

Milstein and coworkers carried out pioneering work on the employment of pincer palladacycles [24] as efficient catalysts for the Heck reaction [23b]. The first pincer palladacycles studied, PCP-complexes **67** and **68**, were presented as very effective catalysts for the Heck reaction of aryl bromides (catalyst **68**) and iodides (**67** and **68**) with styrene or acrylates, generally at 140 °C in NMP as solvent, affording the corresponding coupled products in very high yields (Scheme 8.30) [23b].

After Milstein's studies, various phosphane- and phosphinito-derived PCP-pincer complexes have been presented as efficient catalysts for the Heck reaction, although only a few of them show good activity for the coupling of less activated substrates such as aryl bromides and aryl chlorides (Figure 8.12, Table 8.7). Among them, phophinito complex **3**, developed by Jensen and coworkers, is the most active since it catalyzes efficiently the olefination of a broad scope of aryl chlorides in excellent yields (Table 8.7) [23e]. Palladacycle **3** does not require additional co-catalyst such as TBAB, LiBr and so on to achieve high catalytic activity for both electron-rich and sterically hindered aryl chlorides (Table 8.7) and although it requires long reaction times for completion (5 days), a significant reduction to 1 day is achieved by increasing the temperature to 180 °C [23b]. Palladacycle **3** is also a very effective catalyst for the synthesis of 1,2-disubstituted and trisubstituted alkenes [107]. 1,2-Disubstituted alkenes are obtained in good yields as a mixture of isomers from the catalytic coupling of styrene and aryl iodides in DMF

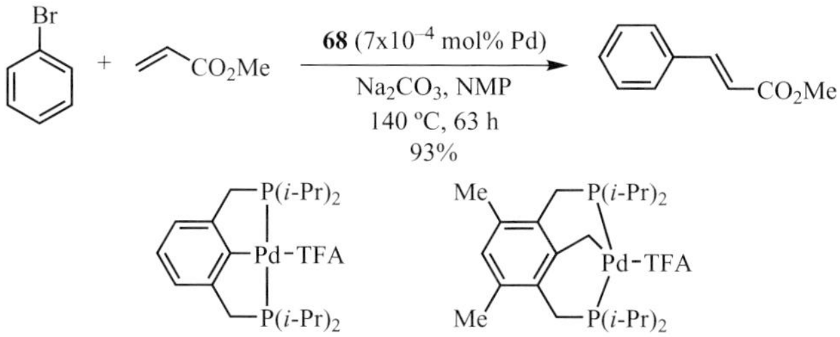

67 **68**

Scheme 8.30 Milstein's pincer palladacycles employed in the Heck reaction.

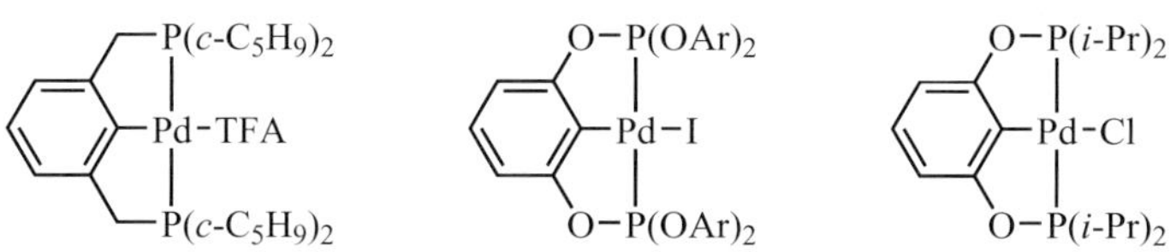

69[105] **70** (Ar = p-MeOC$_6$H$_4$)[106] **3**[23e]

Figure 8.12 Phosphane- and phosphinito-derived pincer palladacycles.

at 180 °C. Under similar reaction conditions, the reaction of aryl bromides with styrene or 1,1-disubstituted alkenes such as α-methylstyrene and *n*-butyl acrylate affords the corresponding trisubstituted alkenes with high regio- and diastereoselectivity.

Catalytic activity in the Mizoroki–Heck reaction has also been demonstrated by various pincer nitrogen palladacycles. Figure 8.13 depicts those that display reasonable cross-coupling yields with neutral and deactivated aryl bromides as well as activated aryl chlorides. These Pd(II) pincer complexes are composed of two six-membered fused palladacycles, which reduces the bond angle strain around the metal center. This feature seems to be crucial to displaying good activity since five-membered palladacycle analogues are catalytically active only with aryl iodides or electron-poor aryl bromides [111].

Pincer sulfur palladacycles catalyze the Heck reaction of iodoarenes with various olefins such as acrylates, acrylonitrile, styrene, vinyl pyridines and enol ethers in high yields [112]. However, these types of catalysts are not very active with less activated substrates – very few examples, such as palladacycles **74** [113] (Figure 8.14), have been reported to efficiently perform the reaction with bromoarenes. Interestingly, the selenium pincer analog **75** (Figure 8.14) has been shown to perform the Heck reaction of deactivated aryl and heterocyclic bromides in the absence of additives in moderate, for electron-rich aryl bromides, to excellent yields [114].

Finally, some interesting studies have been carried out with non-symmetrical pincer palladacycles such as NCP and NCS catalyst precursors **76** [115] and **77** [116] (Figure 8.14), which have also demonstrated good activity in the reaction with non-activated aryl bromides and aryl chlorides at high temperatures.

Table 8.7 Heck reactions of aryl bromides and aryl chlorides catalyzed by pincer palladacycles.

Catalyst (mol.% Pd)	RX	Alkene	Conditions	Yield (%)	Ref.
69 (0.5)	PhBr	CO_2Me	Na_2CO_3, NMP, 135 °C, 19 h	100	[105]
3 (0.7)	PhCl	Ph	CsOAc, dioxane, 120 °C, 120 h	>99	[23e]
3 (0.7)	*p*-ClC_6H_4Ac	Ph	CsOAc, dioxane, 120 °C, 120 h	>99	[23e]
3 (0.7)	*p*-ClC_6H_4CHO	Ph	CsOAc, dioxane, 120 °C, 120 h	81	[23e]
3 (0.7)	*p*-ClC_6H_4OMe	Ph	CsOAc, dioxane, 120 °C, 120 h	86	[23e]
3 (0.7)	*o*-MeC_6H_4Cl	Ph	CsOAc, dioxane, 120 °C, 120 h	83	[23e]

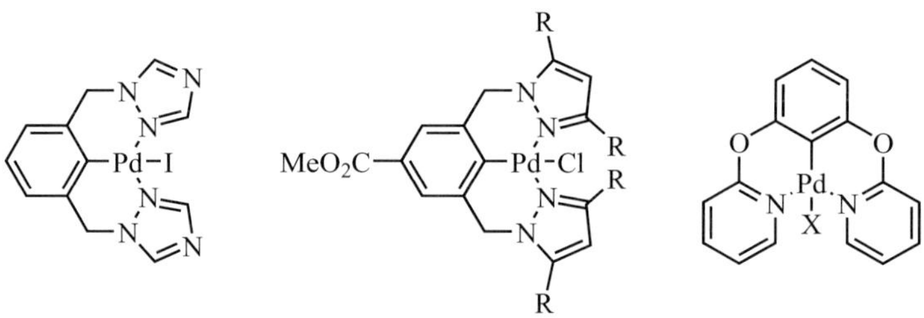

71[108] **72** (R = H, Me)[109] **73** (X = OTf)[110]

Figure 8.13 Nitrogen-derived pincer palladacycles.

8.2 Sonogashira Reaction

8.2.1 Introduction

The palladium-catalyzed Csp^2–Csp cross-coupling reaction between aryl or alkenyl halides or triflates and terminal alkynes has become the most important method to prepare conjugated acetylenic derivatives, compounds that are valuable intermediates for the synthesis of natural products, pharmaceuticals and molecular organic materials [117]. In 1975, Cassar [118], Heck [119], and Sonogashira [120]

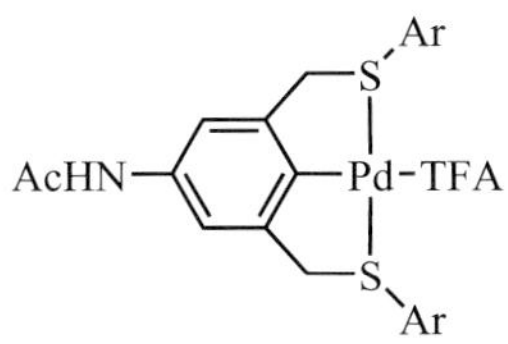

74a (Ar = *p*-AcHNC$_6$H$_4$)[113]
74b (Ar = 2,4-(MeO)$_2$C$_6$H$_3$)[113]

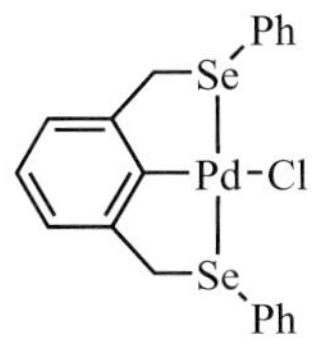

75[114]

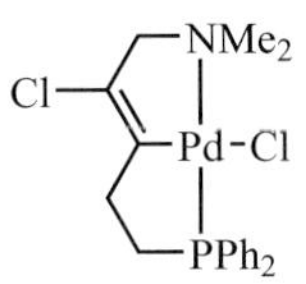

76[115]

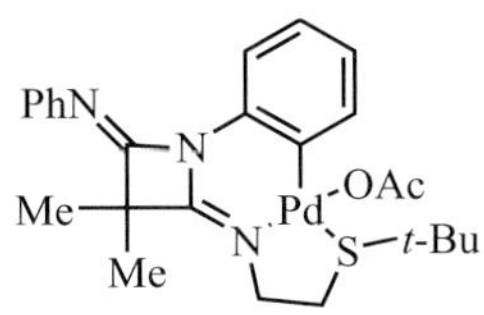

77[116]

Figure 8.14 Sulfur-, selenium and non-symmetrical pincer palladacycles.

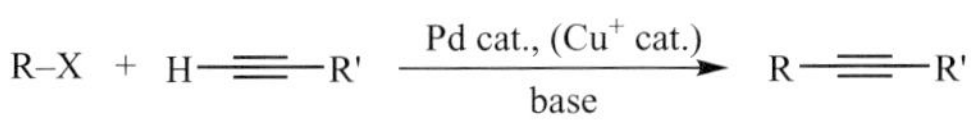

R = aryl, vinyl
R' = aryl, alkenyl, alkyl, SiR$_3$
X = Cl, Br, I, OTf

Scheme 8.31 Cassar–Heck–Sonogashira alkynylation.

independently reported the first studies on this reaction. The former two methods were developed as an extension of the Mizoroki–Heck palladium-catalyzed arylation or alkenylation of alkenes employing a phosphane-palladium complex at high temperatures (Scheme 8.31). On the other hand, Sonogashira, Tohda and Hagihara were able to perform the cross-coupling reaction at room temperature employing a catalytic amount of copper(I) iodide as co-catalyst, a result directly connected with the Stephens–Castro cross-coupling reaction [121].

An outstanding number of studies followed the initial work performed by these groups, mostly devoted to finding more efficient catalysts, milder reaction conditions, expanding the substrate scope of the reaction and understanding the mechanism of the process. In this way, aryl chlorides [42], primary alkyl bromides or iodides [122], and secondary alkyl bromides [123] have been successfully alkynylated using a Sonogashira protocol. Copper salts often promote the homocoupling of the terminal alkyne (Glaser coupling [124]) when working under aerobic conditions so significant efforts have also been dedicated to developed efficient and mild amine- and copper-free (Cassar–Heck alkynylation coupling) procedures by increasing the activity of the catalytic system. During the past few years important advances have been made in this direction with most of the success owed to the development of new, very active and, at the same time, more stable palladium catalysts [4a]. Among them, palladacycles [22] display an important role since they behave as a source of very active Pd catalytic species, allowing for mild reaction conditions.

8.2.2
Mechanism

Sonogashira and Hagihara originally proposed a mechanism for the copper co-catalyzed reaction [120] based on the discovery of the CuI-catalyzed transmetallation in an amine as solvent [125], which still today is generally accepted. As shown in Scheme 8.32, the reaction takes place through two independent catalytic cycles, which involve the normal addition–reductive elimination process common to the Pd-catalyzed C—C bond-forming reactions. However, the exact mechanism for the reaction is not known since the structure of the truly catalytically active species and the role of the copper co-catalyst remain unclear. The first step of the process is the generation, under the employed reaction conditions, of the active Pd(0) species from the Pd(II) precatalyst. Neutral coordinatively unsaturated 12- [Pd(0)L] [126] or 14-electron [$Pd(0)L_2$] [120] complexes as well as anionic [$(Pd(0)L_2X)^-$] [6b, 127] active Pd(0) species have been postulated. A fast oxidative addition of the electrophile to the real Pd(0) catalyst and a rate-determining transmetallation from the copper acetylide formed in the Cu-cycle generate the corresponding alkynylpalladium(II) derivatives, which finally collapse to give the coupled product after trans/cis isomerization and reductive elimination, regenerating the active Pd(0) species.

The amine plays a very important role in the Sonogashira reaction since it is involved in the generation of the copper acetylide. This intermediate is implicated in the production of the active Pd(0) species and in the transmetallation step and it is formed by a Cu^+-assisted deprotonation of the acetylenic C—H bond (Scheme 8.32). The usually employed bases in the Sonogashira reaction are not basic enough to deprotonate the acetylene proton, so the copper(I) salt acts as a Lewis acid by formation of a transient π-alkyne-Cu complex [128]. A similar π-alkyne-Ag complex has been detected recently by NMR techniques after generation of silver acetylides in silver co-catalyzed Sonogashira couplings [129]–something that theoretically could be extended to the typical copper co-catalyzed reaction.

Anions and halides coming from the palladium precatalyst, the copper co-catalyst, or formed during the catalytic cycle have been postulated to play a crucial role in the Sonogashira reaction by Amatore and Jutand [6b, 127]. In fact, anionic Pd(0) species such as [$(PdL_2X)^-$] in which the Pd(0) is ligated by an anion could be the truly active catalyst instead of the coordinatively unsaturated $Pd(0)L_2$. As depicted in Scheme 8.33, new pentacoordinate Pd(II) anionic complexes are formed during the catalytic cycle where the anion previously ligated to the Pd(0) remains ligated to the Pd(II) intermediates, thus conditioning their stability and reactivity.

As mentioned earlier, the Pd-catalyzed cross-coupling reaction between aryl or vinyl halides and triflates with terminal acetylenes also proceeds without copper co-catalyst. In this case, specific amines such as piperidine, morpholine and diisopropylamine are usually employed in large excess or as the solvent. Despite much speculation over the years and indirect evidence, a clear mechanistic picture, sup-

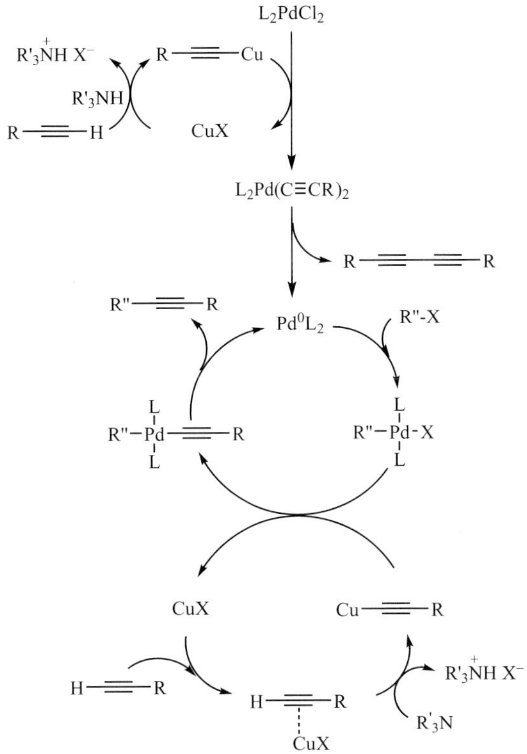

Scheme 8.32 Mechanism of the Sonogashira reaction.

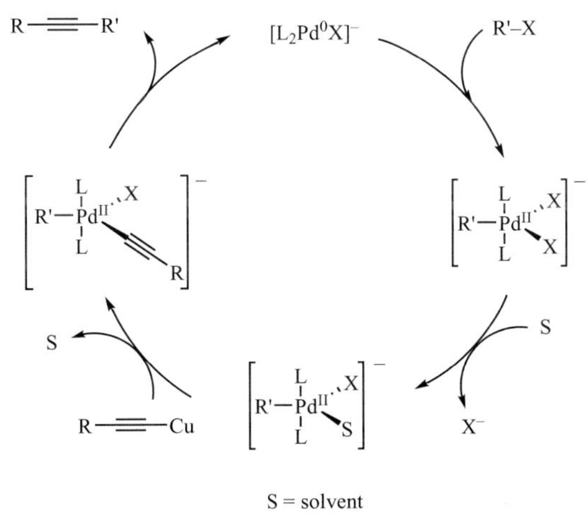

Scheme 8.33 Mechanism of the Sonogashira reaction in the presence of halides.

ported by definitive experimental evidence, is still lacking. A plausible catalytic cycle has been proposed (Cycle A, Scheme 8.34) [130]. As previously commented, the amines generally employed in the process are not usually able to deprotonate the alkyne for the reaction with the *trans*-R′PdXL$_2$ complex formed after oxidative

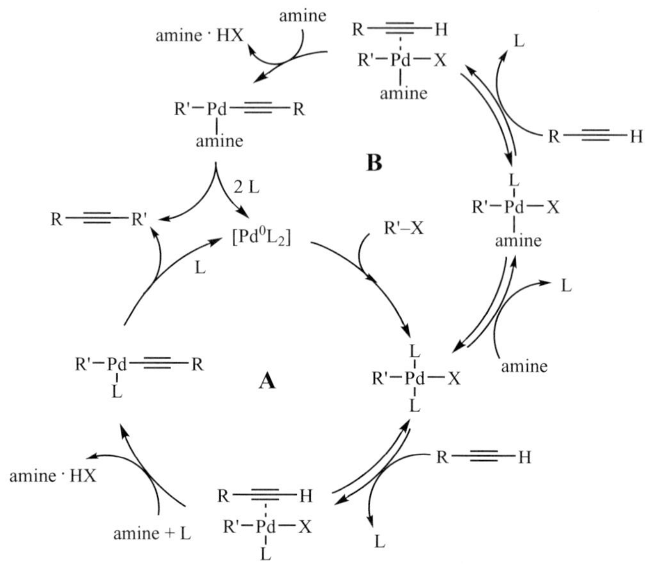

Scheme 8.34 Mechanism of the palladium-catalyzed alkynylation reaction.

addition. Therefore, complexation of the alkyne to this Pd(II) complex is supposed to proceed first with displacement of one ligand to give intermediate complex [η^2—RC≡CH]R′PdXL. Deprotonation of the ligated alkyne by the base leads to the new complex R′Pd(—C≡CR)L_2 which gives the coupling product R—C≡C—R′ by reductive elimination (Cycle A, Scheme 8.34). A similar mechanistic scenario has been proposed recently by Dupont and coworkers for the Heck alkynylation of iodobenzene with phenylacetylene, employing cyclopalladated allylamines **19** and **78** as catalysts, using *N,N*-dimethylacetamide (DMAc) as solvent and tetra-*n*-butylammonium acetate (TBAOAc) as base (Figure 8.15) [131]. These authors have suggested the involvement of soluble Pd(0) species as the most probable catalytically active species involved in the reaction (generated from the cyclopalladated precursor).

Very recently, Jutand and coworkers have demonstrated that the amine base may play a multiple role in the catalytic cycle that is not just limited to deprotonation of the alkyne [132]. First, the oxidative addition step is faster when performed in the presence of amines, the formation of the more reactive complex [Pd(0)L(amine)] probably being responsible for this effect. However, depending on the neutral ligand (L) on the catalyst precursor and the amine employed, a different catalytic cycle (Cycle B, Scheme 8.34) could operate where the *trans*-R′PdXL_2 complex suffers substitution of one ligand (L) by the amine via the dimer [Pd(μ-I)R′L_2] to generate the R′PdX(L)(amine) complex in a reversible reaction whose equilibrium constant depends on R′, X, the basicity and the steric hindrance of the amine. Therefore, depending on the rate of competition between the amine and the alkyne in the substitution of one L group in *trans*-R′PdXL_2, two different mechanisms (A or B) may operate (Scheme 8.34).

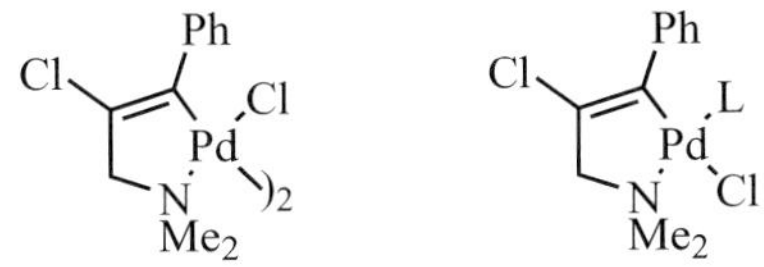

19 78a, L = $PH(t\text{-}Bu)_2$
78b, L = $(p\text{-}CF_3C_6H_4)_3P$

Figure 8.15 Cyclopalladated allyl amine catalysts.

The terminal alkynes employed in the coupling reaction can also play an unexpected role in the catalytic cycle since they influence the rate of the oxidative addition before their implication in the transmetallation step [133]. The carbon–carbon triple bond decreases the rate of the oxidative addition reaction by coordination to the Pd(0) active complex, forming the low reacting complex $(\eta^2\text{—RC}{\equiv}\text{CH})Pd(0)L_2$. The stationary regime of a catalytic cycle is more easily reached if the reaction rates of all the elemental steps are as close as possible to each other. This can be achieved by accelerating the rate-determining step (i.e. destabilizing stable intermediate complexes) or decelerating the fast reactions by stabilization of high-energy species [6]. Thus, decreasing the rate of a fast oxidative addition (i.e. aryl iodides) may favor the efficiency of the catalytic cycle by bringing its rate closer to the slower transmetallation step. However, if the oxidative addition is slower than the transmetallation (i.e. aryl chlorides or deactivated aryl bromides) and is therefore the rate-determining step of the catalytic cycle, it will be even slower in the presence of the nucleophilic alkyne and the catalytic reaction would be less efficient–any technique that allows maintaining a low concentration of the alkyne (i.e. slow addition) being beneficial for the efficiency of the catalytic reaction [133]. Complexation of the active Pd(0) complex by the some final acetylenic reaction products may explain why some catalytic reactions stop before total conversion of the reagents.

Finally, the recent finding that some commercially available palladium salts such as $PdCl_2$ or $Pd(OAc)_2$ contain small amounts of copper [134] brings reasonable doubts about the copper-free Csp^2—Csp cross-coupling reactions.

8.2.3
Catalysts

At present, the most employed reaction conditions for the Sonogashira coupling still involve standard palladium-phosphane complexes such as $[Pd(PPh_3)_2Cl_2]$, $[Pd(PPh_3)_4]$, $[Pd(dppe)Cl_2]$, $[Pd(dppp)Cl_2]$ and $[Pd(dppf)Cl_2]$ as catalysts, Cu(I) salts as co-catalyst and in the presence of an amine in large excess or as a solvent [117h]. Under these conditions excellent results are usually obtained, though, most frequently, high catalyst and co-catalyst loadings are required, showing very low activity in more demanding situations as in the case of non-activated substrates such as aryl chlorides. The high price of palladium catalysts, and the increasing interest showed by the industry to use palladium-mediated Csp^2–Csp cross-coupling reactions, has shifted the attention of the research community towards making these reactions more economical and practical [4a]. Recent research in this area has focused on the development of highly active catalysts that could also circumvent the use of copper and the excess of base.

An important breakthrough in the area of high turnover palladium catalysts has been the use of electron-rich and/or bulky ligands such as phosphanes [22a, 22f, 44] and phosphane-free N-heterocyclic carbenes (NHC) [22a, 22f, 44, 45]. Electron-rich ligands have been shown to perform an easier oxidative addition to aryl halides, which is ideal when deactivated bromoarenes or chloroarenes are employed. Furthermore, a ligand with steric demand promotes an easier dissociation from the $Pd(0)L_2$, which is necessary prior to the oxidative addition step [135]. Table 8.8 illustrates representative examples of the Sonogashira reaction between aryl or alkyl halides with terminal alkynes catalyzed by newly developed catalytic systems using bulky phosphanes or carbenes as ligands. The reactions are most often performed by generating the precatalyst *in situ*, by combining the palladium source (typically in the range 0.4–5 mol.% Pd) and the ligand, with good results being obtained for aryl bromides and chlorides.

Catalytic activity in the Sonogashira reaction has also been demonstrated by various carbene-derived palladium complexes. Although preliminary research showed that elevated temperatures and activated aryl halides were required, recent studies have demonstrated the efficiency of these systems in the Sonogashira-type reaction of unactivated primary and secondary alkyl halides in the presence of CuI, usually employing high catalyst loadings (4–5 mol.% Pd) but under mild reaction conditions [122, 123] (Table 8.8).

A few examples of the Heck alkynylation employing palladium-nitrogen complexes have been reported. Pyridines [134, 142], pyrimidines [143], imidazoles [144], oxazolines [145] and pyrazoles [146] have shown good complexation properties for palladium and have been employed in the formation of efficient catalysts in the Sonogashira reaction. In this context, just $PdCl_2$ and monomeric and polymeric di-2-pyridylmethylamine-based palladium complexes **79** and **80** (Figure 8.16) have been successfully used in the copper-free Sonogashira and sila-Sonogashira [147] coupling of aryl iodides and bromides in water and in NMP under homogeneous and heterogeneous conditions [134, 142a–d].

The use of simple palladium salts as catalysts is of major interest from an industrial and academic standpoint since several useful ligands are considered to

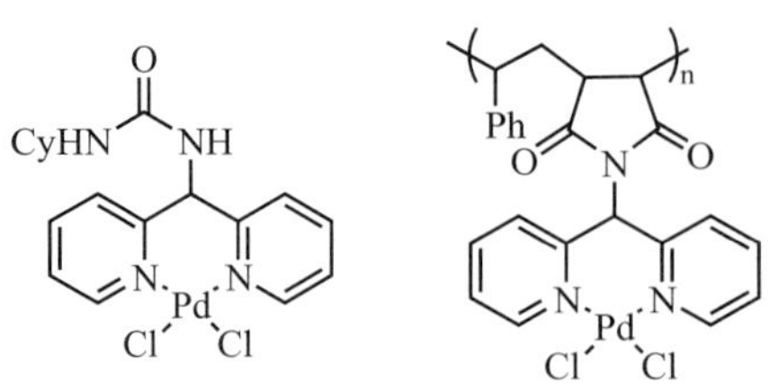

79 **80**

Figure 8.16 Highly efficient pyridine-derived palladium catalysts.

Table 8.8 Examples of Sonogashira reactions using phosphane or carbene ligands.

Catalyst	RX	Alkyne	Conditions	Yield (%)	Ref.
[Pd(PhCN)$_2$Cl$_2$] P(*t*-Bu)$_3$	*p*-BrC$_6$H$_4$OMe	HC≡CPh	i-Pr$_2$NH, CuI, dioxane, r.t., 1 h	94	[136]
[Pd$_2$(dba)$_3$] P(*t*-Bu)$_3$	*p*-BrC$_6$H$_4$OMe	HC≡CPh	Et$_3$N, r.t., 20 h	51	[137]
[Pd(PCy$_3$)$_2$Cl$_2$]	PhCl	HC≡CPh	Cs$_2$CO$_3$, DMSO, 150 °C, 12 h	81	[138]
[Pd(MeCN)$_2$Cl$_2$], 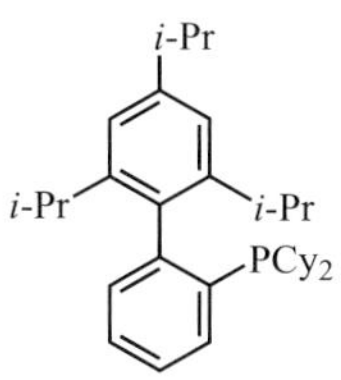	*o*-ClC$_6$H$_4$OMe	HC≡CPh	Cs$_2$CO$_3$, MeCN, 97 °C, 3 h	95	[139]
[Pd(η$_3$-C$_3$H$_5$)Cl$_2$], 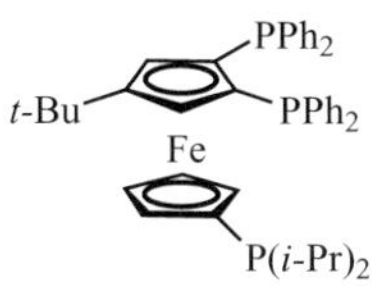	*p*-ClC$_6$H$_4$CN	HC≡CPh	K$_2$CO$_3$, DMF, 130 °C, 20 h	86	[140]
[Pd(η$_3$-C$_3$H$_5$)Cl$_2$], 	*p*-ClC$_6$H$_4$OMe	HC≡CPh	K$_2$CO$_3$, DMF, 140 °C, 20 h	96	[141]
[Pd(π-allyl)Cl]$_2$, 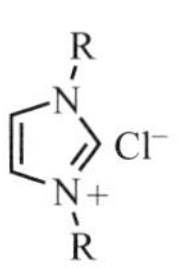	CN(CH$_2$)$_3$Br	HC≡Cn-Hex	Cs$_2$CO$_3$, CuI, DMF/Et$_2$O, 45 °C, 16 h	79	[122]
R = 1-adamantyl 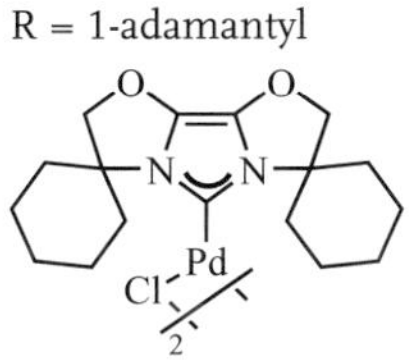	Ph(CH$_2$)$_3$Br	HC≡Cn-Oct	Cs$_2$CO$_3$, CuI, DMF/DME, 60 °C, 16 h	66	[123]

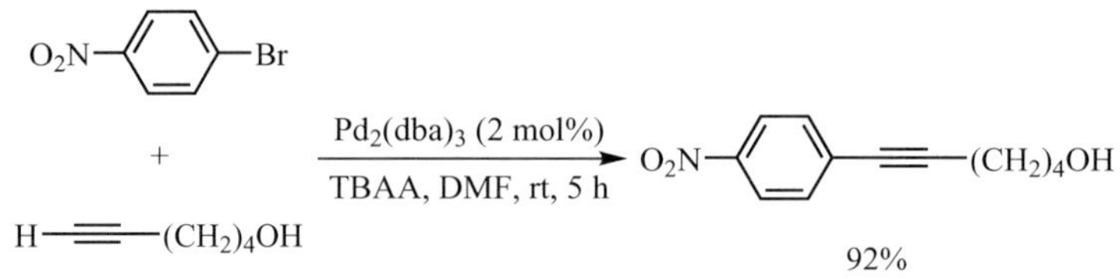

Scheme 8.35 Ligand-, amine- and copper-free alkynylation reaction.

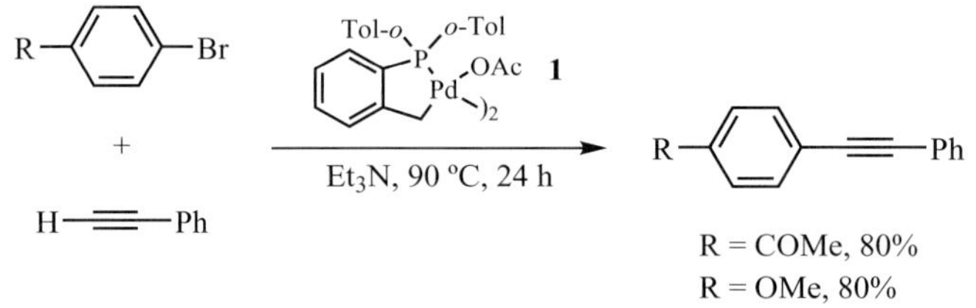

Scheme 8.36 Heck alkynylation of aryl bromides catalyzed by **1**.

be expensive, sensitive and difficult to obtain. Successful studies on ligand-free versions of the Cassar–Heck–Sonogashira alkynylation reactions are scarce and have appeared only quite recently [148]. For instance, conditions for an efficient ligand-, copper-, and amine-free palladium-catalyzed alkynylation reaction of aryl iodides and activated aryl bromides have been developed, employing $Pd_2(db_2)_3$ as catalyst at room temperature in the presence of TBAA as basic additive (Scheme 8.35) [148a]. Since carboxylated *n*-tetrabutylammonium salts are known to promote the reduction of $Pd(OAc)_2$ to catalytically active Pd(0) species [149], the involvement of highly reactive palladium nanoparticles in the process is very probable.

Palladacycles have proven to be useful and versatile catalytic precursors for the Sonogashira and sila-Sonogashira reactions under homogeneous and heterogeneous conditions. Herrmann and coworkers first reported on the use of palladacycles for the Heck alkynylation of aryl bromides with terminal alkynes [22a, 150]. Phosphapalladacycle **1** performs the copper-free coupling of bromoarenes with phenyl acetylene in triethylamine at 90 °C, achieving very high yields for activated and non-activated aryl bromides, respectively (Scheme 8.36). Neither aryl chlorides nor alkyl acetylenes give satisfactory results.

Herrmann's palladacycle in the presence of CuI as co-catalyst has been employed efficiently in a catalytic traceless solid-phase approach to the synthesis of 2,6,9-trisubstituted purines from resin bound 6-thiopurines (Scheme 8.37) [151].

Several nitrogen-derived palladacycles have been presented that show good activity in the Sonogashira reaction under different reaction conditions. Initial studies were carried out by Nájera and coworkers with oxime-derived palladacycles [93]. Preliminary studies with several benzo- and acetophenone-derived palladacycles showed these precatalysts as promising systems for the conventional copper-cocatalyzed Sonogashira coupling of iodobenzene and phenylacetylene in pyrrolidine at 90 °C for catalyst **55** (0.5 mol.% Pd) (Table 8.9) [93]. However, a subsequent

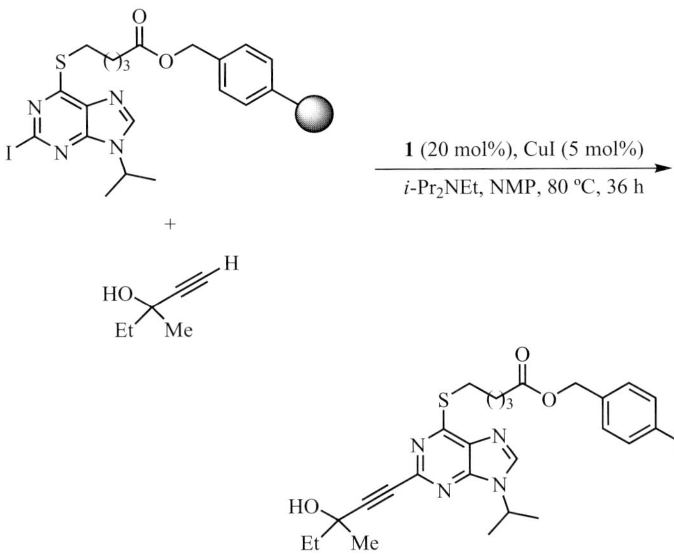

Scheme 8.37 Solid-phase Sonogashira cross-coupling catalyzed by Herrmann's palladacycle **1**.

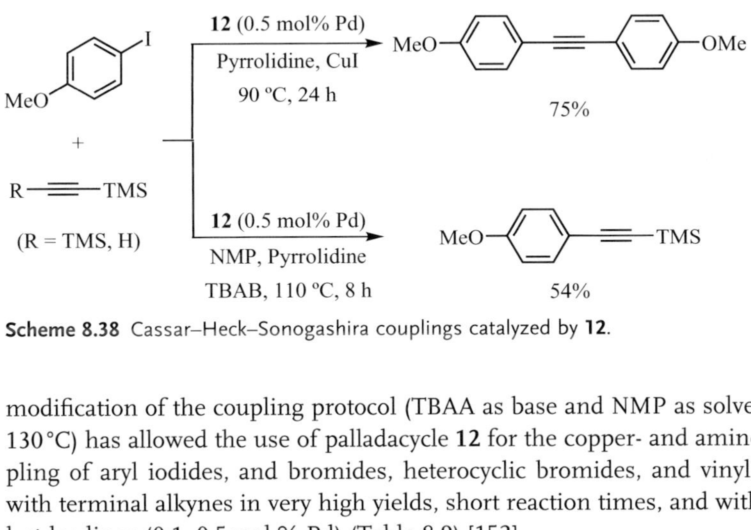

Scheme 8.38 Cassar–Heck–Sonogashira couplings catalyzed by **12**.

modification of the coupling protocol (TBAA as base and NMP as solvent at 110–130 °C) has allowed the use of palladacycle **12** for the copper- and amine-free coupling of aryl iodides, and bromides, heterocyclic bromides, and vinyl bromides with terminal alkynes in very high yields, short reaction times, and with low catalyst loadings (0.1–0.5 mol.% Pd) (Table 8.9) [152].

Catalyst **12** is also an effective promoter of the sila-Sonogashira [147] coupling between alkynyl silanes and aryl iodides and bromides in the presence of CuI or TBAB as co-catalysts [152b]. Fine-tuning the reaction conditions allows one to control the reaction outcome to obtain either diarylated alkynes or silylated mono-arylated alkynes using mono- or bis(trimethylsilyl)acetylene (Scheme 8.38). Diarylation is observed when the reaction was carried out with bis(trimethylsilyl)acetylene in the presence of CuI as co-catalyst and pyrrolidine as solvent at 90 °C. However, when the same reaction is carried out in NMP as solvent, and in the presence of

Table 8.9 Examples of Sonogashira reactions catalyzed by oxime palladacycles.

Catalyst	RX	Alkyne	Conditions	Yield (%)	Ref.
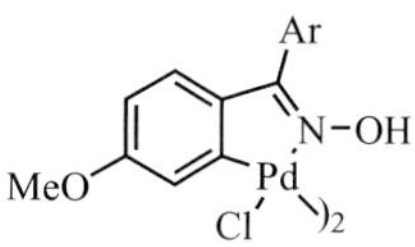 **55** (Ar = *p*-MeOC$_6$H$_4$)	PhI	HC≡CPh	Pyrrolidine, CuI, dioxane, 90 °C, 5.5 h	75	[93]
	p-ClC$_6$H$_4$I	HC≡CPh	TBAOAc, NMP, 110 °C, 24 h	72	[152]
	p-ClC$_6$H$_4$Br	HC≡CPh	TBAOAc, NMP, 110 °C, 7.5 h	90	[152]
	p-ClC$_6$H$_4$Br	HC≡Cn-Hex	TBAOAc, NMP, 110 °C, 2 h	78	[152]
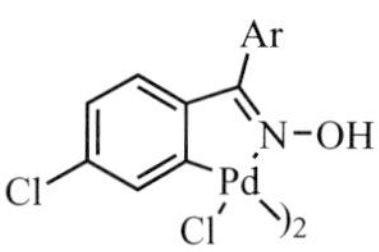 **12** (Ar = *p*-ClC$_6$H$_4$)	*o*-BrC$_{10}$H$_7$	HC≡Cn-Hex	TBAOAc NMP, 110 °C, 1 h	75	[152]
	p-MeOC$_6$H$_4$Br	HC≡CTIPS	TBAOAc, NMP, 130 °C, 1 h	85	[152]
	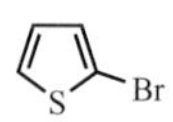	HC≡Cn-Hex	TBAOAc NMP, 110 °C, 3 h	62	[152]
	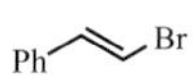	HC≡CTIPS	TBAOAc, NMP, 110 °C, 1 h	63	[152]

TBAB and pyrrolidine at 110 °C, the silylated alkyne is the main product. This methodology has also been used for the synthesis of asymmetrically substituted alkynes via sila-Sonogashira coupling of silylated alkynes.

Dichlorobenzophenone oxime palladacycle catalyst **12** has been employed as a copper-free promoter for the acylation of terminal alkynes with different carboxylic acid chlorides in toluene in the presence of TEA as base, giving the corresponding ynones in good yields (Table 8.10) [153]. The reaction coupling can normally be performed under air but an inert atmosphere is necessary when using very low catalyst loadings or sensitive carboxylic acid chlorides. The protocol permits the synthesis of ynones at 110 °C, at r.t., or under microwave irradiation conditions, with good yields being obtained for aromatic and aliphatic carboxylic acid chlorides and different acetylenes. Pd(OAc)$_2$ also catalyzes the ligandless cross-coupling process but usually working under higher loading conditions and giving lower yields (Table 8.10).

Besides oxime-derived palladacycles, other new types of active Sonogashira nitrogen cyclopalladated catalysts have been developed (Figure 8.17). For instance, sulfinimine palladacycle **41** performs less efficiently than cyclopalladated oximes when employed in the coupling of aryl iodides in triethylamine at 80 °C, showing

Table 8.10 Examples of Sonogashira reactions using oxime palladacycle **12** [153]

Catalyst	RX	Alkyne	Conditions	Yield (%)
12	PhCOCl	HC≡CPh	Toluene, Et_3N, 25 °C, 23 h	96
12	PhCOCl	HC≡CTIPS	Toluene, Et_3N, 110 °C, 7 h	82
$Pd(OAc)_2$	PhCOCl	HC≡CTIPS	Toluene, Et_3N, 110 °C, 7 h	56
12	*p*-ClC_6H_4Cl	HC≡CPh	Toluene, Et_3N, 110 °C, 2.5 h	72
12	*p*-$MeOC_6H_4Cl$	HC≡CPh	Toluene, Et_3N, 110 °C, 4 h	70
12	*o*-$MeOC_6H_4Cl$	HC≡CPh	Toluene, Et_3N, 110 °C, 13.5 h	99
12	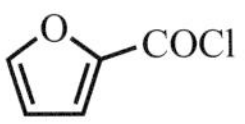	HC≡CPh	Toluene, Et_3N, 110 °C, 4 h	50
12	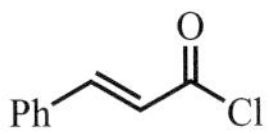	HC≡CPh	Toluene, Et_3N, 110 °C, 3 h	77
12	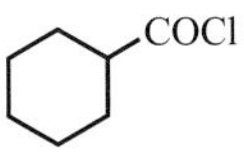	HC≡CPh	Toluene, Et_3N, 110 °C, 4 h	60
12	*t*-BuCOCl	HC≡CPh	Toluene, Et_3N, 110 °C, 4 h	99

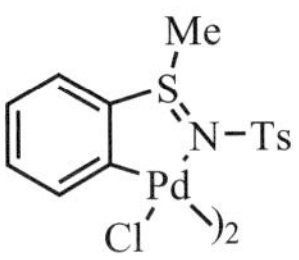

41

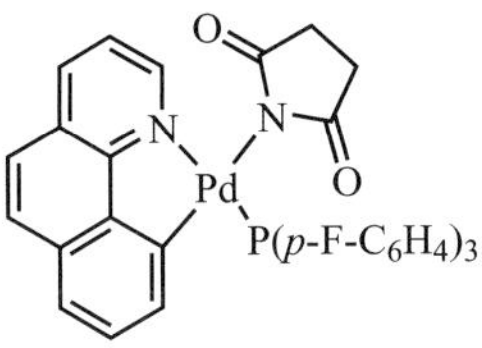

81

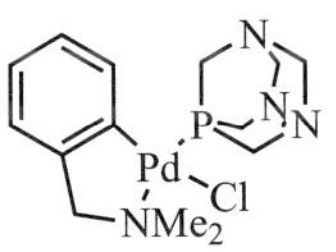

82

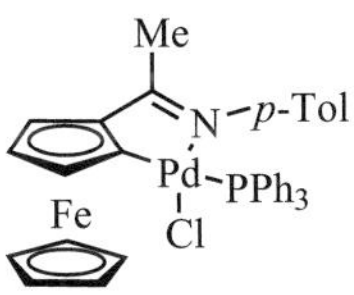

83

Figure 8.17 Nitrogen palladacycles for the Sonogashira reaction.

good yields [83]. The coupling of bromo- and, especially, chlorobenzene affords very low yields under similar conditions (25% and 10%, respectively). In addition, palladacycle **81**, derived from benzoquinoline (Figure 8.17), has been found as the most effective from a series of related nitrogen complexes in the Sonogashira reaction of *p*-bromoacetophenone and phenylacetylene, using triethylamine as co-solvent at 100 °C, although the presence of Cu(I) is essential to enhance the reaction rate [154].

Table 8.11 Examples of Sonogashira reactions of aryl bromides and chlorides catalyzed by nitrogen-derived palladacycles.

Catalyst	RX	Alkyne	Conditions	Yield (%)	Ref.
82	p-$MeOC_6H_4Br$	HC≡CPh	Cs_2CO_3, MeCN, 80 °C, 24 h	60	[155]
$[Pd(OAc)_2]$ PTA	p-$MeOC_6H_4Br$	HC≡CPh	Cs_2CO_3, MeCN, 80 °C, 24 h	70	[155]
83	p-$MeOC_6H_4Br$	HC≡CPh	KOAc, DMAc, TBAB, 80 °C, 20 h	42	[156]
82	p-$MeOC_6H_4Br$	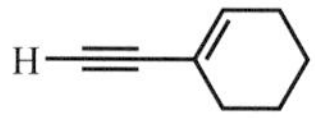	Cs_2CO_3, MeCN, 80 °C, 24 h	80	[155]
$[Pd(OAc)_2]$ PTA	p-$MeOC_6H_4Br$	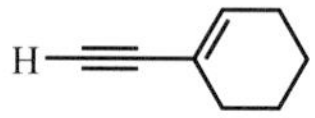	Cs_2CO_3, MeCN, 80 °C, 24 h	92	[155]
82	PhCl	HC≡C*t*-Bu	Cs_2CO_3, TBAB, MeCN, 80 °C, 24 h	70	[155]
$[Pd(OAc)_2]$ PTA	PhCl	HC≡C*t*-Bu	Cs_2CO_3, TBAB, MeCN, 80 °C, 24 h	72	[155]
82	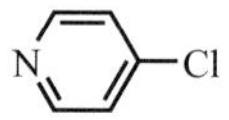	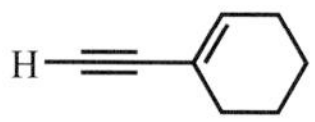	Cs_2CO_3, TBAB, MeCN, 80 °C, 24 h	100	[155]
$[Pd(OAc)_2]$ PTA	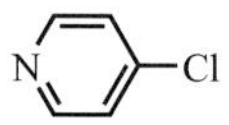	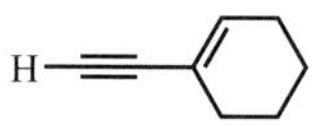	Cs_2CO_3, TBAB, MeCN, 80 °C, 24 h	100	[155]
83	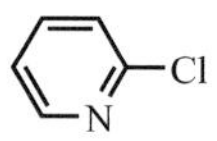	HC≡CPh	KOAc, DMAc, TBAB, 110 °C, 24 h	87	[156]

Monomeric phosphane-derived cyclopalladated complexes **82** [155] and **83** [156] are of special interest since they efficiently perform the Sonogashira cross-coupling reaction of aryl iodides, aryl bromides and activated aryl chlorides with aliphatic and aromatic terminal alkynes under amine- and copper-free conditions, typically employing 2.5–3 mol.% of Pd. 1,3,5-Triaza-7-phosphaadamantane (PTA) phosphane-derived complex **82** performs the coupling with excellent results in acetonitrile as solvent at 80 °C, with the presence of TBAB as additive being necessary for the coupling of aryl chlorides (Table 8.11) [155]. Interestingly, the presence of a copper co-catalyst such as CuI has a deleterious effect on the reaction, a result that had been previously found by Buchwald and coworkers [139]. In the same study, the authors also showed that $Pd(OAc)_2$/PTA is an equally catalytic system for the alkynylation reaction of aryl bromides and chlorides under the same reaction conditions (Table 8.11) [155]. The use of TBAB as additive is also essential for a good performance when employing monomeric ferrocenylimine-derived palladacycle **83** (Table 8.11) [156].

There are also examples of the use of bis-chelated palladacycles, usually referred to as "pincer" complexes [24], in Sonogashira couplings, such as the PCP pincer complex **3** (Figure 8.18), which is reactive enough to cross-couple a wide range of activated and non-activated aryl chlorides with phenylacetylene using cesium carbonate as base, although $ZnCl_2$ (10 or 100 mol.%) has to be added as additive and the reaction is performed under harsh conditions (DMSO, 160 °C) [157]. On the other hand, the N-heterocyclic NCN-pincer palladium complexes **72** (Figure 8.18)

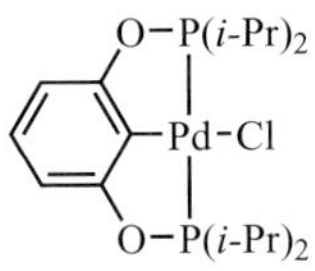

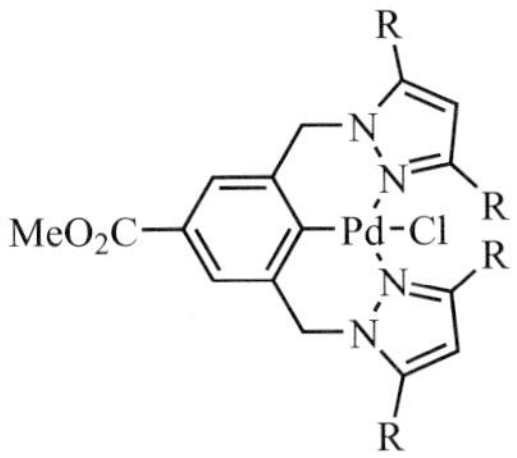

3 **72** (R = H, Me)

Figure 8.18 Cyclopalladated pincer catalysts for the Sonogashira reaction.

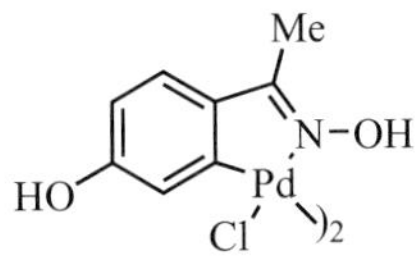

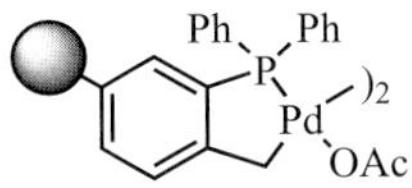

53 **28**

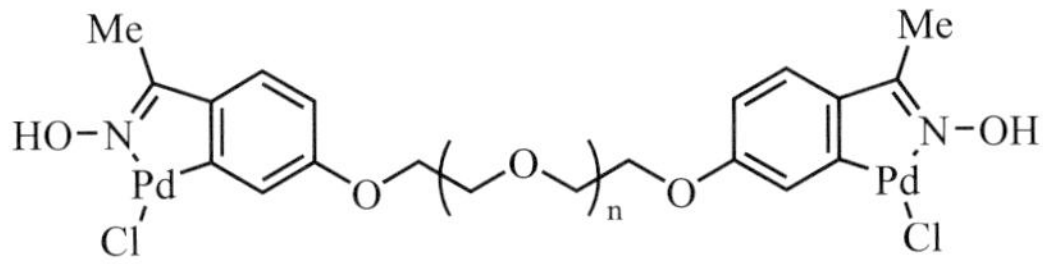

66

Figure 8.19 Supported oxime and phosphane palladacycles.

have been employed recently in the coupling of aryl and naphthyl iodides and terminal alkynes (0.1 mol.% catalyst loading) in pyrrolidine as solvent at 100 °C [109].

Considering the high activity of cyclopalladated compounds as catalyst in the palladium-catalyzed Heck alkynylation, much interest been shown in converting such Pd catalysts into recoverable and reusable catalytic systems. Thus, with the aim of developing a homogeneous and reusable palladium catalytic system for the Sonogashira reaction, the activity and stability of the oxime-derived palladacycle **53** (Figure 8.19) has been tested in the alkynylation reaction using ionic liquids or PEG as recyclable solvents [158]. Under copper-free conditions and using cesium acetate as base, heating in ionic liquids at 120 °C, palladacycle **53** generally suffers extensive decomposition [158]. This problem does not occurs upon prolonged heating in PEG, with decomposition instead occurring under the real reaction conditions, giving place to PEG-stabilized active nanoparticles in a homogeneous recyclable system.

On the other hand, the soluble linear polystyrene-supported phosphapalladacycle **28** (Figure 8.19) (0.2 mol.%) has been used in a low turnover frequency (TOF) copper-free Sonogashira coupling of 4-bromoacetophenone and phenylacetylene in triethylamine at 90 °C [62]. This polymeric catalyst is precipitated by addition of diethyl ether and has been reused up to four times, maintaining conversions of >90%, although the amount of recycled catalyst has to be increased to 5 mol.%; no palladium leaching studies were performed. In addition, the oxime palladacycle **53** has been anchored to soluble PEG and the resulting polymer **66** has been used as a catalyst solubilized in PEG for a copper-free Sonogashira reaction using cesium acetate as base at 150 °C [104d]. The catalyst is effective in the coupling of a substrate such as 4-bromoacetophenone and phenylacetylene and can be reused after precipitation of the PEG in ether. This PEG-anchored carbopalladacycle **66** mostly decomposes during the first catalytic cycle, forming palladium nanoparticles stabilized by PEG, thus retaining its catalytic properties and avoiding palladium leaching from the PEG phase.

8.3 Conclusions

Without doubt palladacycles have become one of the most important catalytic systems for the Mizoroki–Heck and Sonogashira reactions, due to their facile synthesis, stability, ease of handling and, of course, high activity. Moreover, palladacycles have served as a very useful mechanistic tool. The catalytic potential and versatility of cyclopalladated complexes is very similar to that of ligand-stabilized palladium chemistry and, in many cases, superior to ligandless processes. Different ligands such as phosphanes and N-heterocyclic carbenes can be also combined with palladacycles to prepare hybrid monomeric catalysts that, while still being very stable, are reactive enough to perform cross-coupling reactions of non-activated substrates such as aryl chlorides. Palladacycles provide sources of more active Pd catalytic species than classical palladium sources. This very often has allowed the reaction to be performed under very mild reaction conditions, such as the employment of aqueous solvents, room temperature couplings, and the possibility of carrying out Sonogashira couplings under copper-free conditions.

Thus, the chemistry of palladacycles is still a very attractive area and definitely still has much to offer.

References

1 (a) Tsuji, J. (2004) *Palladium Reagents and Catalysts*, Wiley Sons, Ltd, Chichester.
(b) Cornils, B. and Herrmann, W.A. (1996) *Applied Homogeneous Catalysis with Organometallic Compounds*, Wiley-VCH Verlag GmbH, Weinheim.
(c) Beller, M. and Stang, P.J. (1998) *Metal–Catalyzed Cross–Coupling Reactions*, Wiley-VCH Verlag GmbH, Weinheim.
(d) Negishi, E. (2002) *Handbook of Organopalladium Chemistry for Organic Synthesis*, Willey–Interscience, New York.
(e) Bräse, S., Kirchhoff, J.H. and Köbberling, J. (2003) *Tetrahedron*, **59**, 885–939.

(f) Diederich, F. and de Meijere, A. (2004) *Metal–Catalyzed Cross–Coupling Reactions*, 2nd edn, Wiley-VCH Verlag GmbH, Weinheim.
(g) Beller, M. and Bolm, C. (2004) *Transition Metals for Organic Synthesis, Building Block and Fine Chemicals*, 2nd edn, Wiley-VCH Verlag GmbH, Weinheim.
(h) Nicolau, K.C., Bulger, P.G. and Sarlah, D. (2005) *Angewandte Chemie–International Edition*, **44**, 4442–89.
(i) Corbet, J.-P. and Mignani, G. (2006) *Chemical Reviews*, **106**, 2651–710.
(j) Yin, L. and Liebscher, J. (2007) *Chemical Reviews*, **107**, 133–73.

2 For recent reviews, see: (a) de Vries, J.G. (2001) *Canadian Journal of Chemistry–Revue Canadienne de Chimie*, **79**, 1086–92.
(b) Beletskaya, I.P. and Cheprakov, A.V. (2000) *Chemical Reviews*, **100**, 3009–66. (c) Whitcombe, N.J., Hii, K.K. and Gibson, S.E. (2001) *Tetrahedron*, **57**, 7449–76.
(d) Alonso, F., Beletskaya, I.P. and Yus, M. (2005) *Tetrahedron*, **61**, 11771–835.

3 (a) Mizoroki, T., Mori, K. and Ozaki, A. (1971) *Bulletin of the Chemical Society of Japan*, **44**, 581.
(b) Heck, R.F. and Nolley, J.P. (1972) *Journal of Organic Chemistry*, **37**, 2320–2.

4 (a) Farina, V. (2004) *Advanced Synthesis Catalysis*, **346**, 1553–82.
(b) Blaser, H.-U., Indolese, A., Naud, F., *et al.* (2004) *Advanced Synthesis Catalysis*, **346**, 1583–98.
(c) Fairlamb, I.J.S. (2005) *Tetrahedron*, **61**, 9661–62.

5 Heck, R.F. (1979) *Accounts of Chemical Research*, **12**, 146–51.

6 (a) Amatore, C. and Jutand, A. (1999) *Journal of Organometallic Chemistry*, **576**, 254–78.
(b) Amatore, C. and Jutand, A. (2000) *Accounts of Chemical Research*, **33**, 314–21.

7 (a) Amatore, C., Jutand, A. and M'Barki, M.A. (1992) *Organometallics*, **11**, 3009–13.
(b) Ozawa, F., Kubo, A. and Hayashi, T. (1992) *Chemistry Letters*, 2177–80.

8 (a) Jutand, A. and Mosleh, A. (1995) *Organometallics*, **14**, 1810–17.
(b) Shmidt, A.F. and Smirnov, V.V. (2002) *Kinetics Catalysis*, **43**, 195–8.

9 (a) Louie, J. and Hartwig, J.F. (1996) *Angewandte Chemie–International Edition in English*, **35**, 2359–61.
(b) Benhaddou, R., Czernecki, S., Ville, G. and Zegar, A. (1988) *Organometallics*, **7**, 2435–2439.

10 (a) Amatore, C. and Plüger, F. (1990) *Organometallics*, **9**, 2276–82.
(b) Jutand, A. and Mosleh, A. (1995) *Organometallics*, **14**, 1810–17.
(c) Cianfriglia, P., Narducci, V., Sterzo, C.L. *et al.* (1996) *Organometallics*, **15**, 5220–30.
(d) Amatore, C., Carré, E., Jutand, A. and Medjour, Y. (2002) *Organometallics*, **21**, 4540–5.

11 (a) Brumbaugh, J.S., Whittle, R.R., Parvez, M. and Sen, A. (1990) *Organometallics*, **9**, 1735–47.
(b) Cabri, W. and Candiani, I. (1995) *Accounts of Chemical Research*, **28**, 2–7.
(c) Amatore, C., Godin, B., Jutand, A. and Lemaître, F. (2007) *Organometallics*, **26**, 1757–61.

12 Sabino, A.A., Machado, A.H.L., Correia, C.R.D. and Eberlin, M.N. (2004) *Angewandte Chemie–International Edition*, **43**, 2514–18.

13 Brown, J.M., Pérez-Torrente, J.J., Alcock, N.W. and Clase, H.J. (1995) *Organometallics*, **14**, 207–13.

14 Cabri, W., Candiani, I., Bedeschi, A. and Santi, R. (1992) *Journal of Organic Chemistry*, **57**, 3558–63.

15 (a) Pei, W., Mo, J. and Xiao, J. (2005) *Journal of Organometallic Chemistry*, **690**, 3546–51.
(b) Mo, J. and Xiao, J. (2006) *Angewandte Chemie–International Edition*, **45**, 4152–7.
(c) Hyder, Z., Mo, J. and Xiao, J. (2006) *Advanced Synthesis Catalysis*, **348**, 1699–704.
(d) Liu, S., Berry, N., Thomson, N., *et al.* (2006) *Journal of Organic Chemistry*, **71**, 7467–70.

16 Cabri, W., Candiani, I., Bedeschi, A., *et al.* (1992) *Journal of Organic Chemistry*, **57**, 1481–6.

17 De Vries, J.G. (2006) *Dalton Transactions*, 421–9.

18 (a) Scheeren, C.W., Machado, G., Dupont, J., *et al.* (2003) *Inorganic Chemistry*, **42**, 4738–42.
(b) Cassol, C.C., Umpierre, A.P., Machado, G., *et al.* (2005) *Journal of the American Chemical Society*, **127**, 3298–9.

19 De Vries, A.H.H., Mulders, J.M.C.A., Mommers, J.H.M., *et al.* (2003) *Organic Letters*, **5**, 3285–8.

20 (a) Jeffery, T. (a) (1984) *Journal of the Chemical Society D – Chemical Communications*, 1287–9.
(b) Jeffery, T. (1985) *Tetrahedron Letters*, **26**, 2667–70.
(c) Reetz, M.T. and Westermann, E. (2000) *Angewandte Chemie – International Edition*, **39**, 165–8.

21 Reetz, M.T., Westermann, E., Lohmer, R. and Lohmer, G. (1998) *Tetrahedron Letters*, **39**, 8449–52.

22 (a) Herrmann, W.A., Böhm, V.P.W. and Reisinger, C.-P. (1999) *Journal of Organometallic Chemistry*, **576**, 23–41.
(b) Dupont, J., Pfeffer, M. and Spencer, J. (2001) *European Journal of Inorganic Chemistry*, 1917–27.
(c) Bedford, R.B. (2003) *Chemical Communications*, 1787–96.
(d) Singleton, J.T. (2003) *Tetrahedron*, **59**, 1837–57.
(e) Bedford, R.B., Cazin, C.S.J. and Holder, D. (2004) *Coordination Chemistry Reviews*, **248**, 2283–321.
(f) Beletskaya, I.P. and Cheprakov, A.V. (2004) *Journal of Organometallic Chemistry*, **689**, 4055–82.
(g) Dunina, V.V. and Gorunova, O.N. (2004) *Russian Chemical Reviews*, **73**, 309–50.
(h) Dupont, J., Consorti, C.S. and Spencer, J. (2005) *Chemical Reviews*, **105**, 2527–71.

23 (a) Herrmann, W.A., Brossmer, C., Öfele, K., *et al.* (1995) *Angewandte Chemie – International Edition*, **34**, 1844–8.
(b) Ohff, M., Ohff, A., Van der Boom, M.E. and Milstein, D. (1997) *Journal of the American Chemical Society*, **119**, 11687–8.
(c) Shaw, B.L. (1998) *New Journal of Chemistry*, 77–9.
(d) Shaw, B.L., Perera, S.D. and Staley, E.A. (1998) *Chemical Communications*, 1361–2.
(e) Morales-Morales, D., Redón, R., Yung, C. and Jensen, C.M. (2000) *Chemical Communications*, 1619–20.
(f) Sundermann, A., Uzan, O. and Martin, J.M.L. (2001) *Chemistry – A European Journal*, **7**, 1703–11.

24 For reviews on palladium pincer complexes, see: (a) ref 22d.
(b) van der Boom, M.E. and Milstein, D. (2003) *Chemical Reviews*, **103**, 1759–92.

25 (a) Beller, M. and Riermeier, T.H. (1996) *Tetrahedron Letters*, **37**, 6535–8.
(b) Beller, M. and Riermeier, T.H. (1998) *European Journal of Inorganic Chemistry*, 29–35.

26 Böhm, V.P.W. and Herrmann, W.A. (2001) *Chemistry – A European Journal*, **7**, 4191–7.

27 For the homogeneous version of this catalyst in Heck and Suzuki reactions, see: (a) Ohff, M., Ohff, A. and Milstein, D. (1999) *Chemical Communications*, 357–8.
(b) Weissman, H. and Milstein, D.D. (1999) *Chemical Communications*, 1901–2.

28 Nowotny, M., Hanefeld, U., van Koningsveld, H. and Maschmeyer, T. (2000) *Chemical Communications*, 1877–8.

29 (a) Rocaboy, C. and Gladysz, J.A. (2002) *Organic Letters*, **4**, 1993–6.
(b) Rocaboy, C. and Gladysz, J.A. (2003) *New Journal of Chemistry*, **27**, 39–49.

30 (a) Tsuji, J. (1969) *Accounts of Chemical Research*, **2**, 144–52.
(b) Moritani, I. and Fujiwara, Y. (1973) *Synthesis*, 524–33.
(c) Ryabov, A.D. (1985) *Synthesis*, 233–52.

31 Beletskaya, I.P., Kashin, A.N., Karlstedt, N.B., *et al.* (2001) *Journal of Organometallic Chemistry*, **622**, 89–96.

32 (a) Alonso, D.A., Nájera, C. and Pacheco, M.C. (2002) *Advanced Synthesis Catalysis*, **344**, 172–83.
(b) Alacid, E., Alonso, D.A., Botella, L., *et al.* (2006) *The Chemical Record*, **6**, 117–32.

33 Fiddy, S.G., Evans, J., Newton, M.A., *et al.* (2003) *Chemical Communications*, 2682–3.

34 D'Orlyé, F. and Jutand, A. (2005) *Tetrahedron*, **61**, 9670–8.

35 (a) Chanthateyanonth, R. and Alper, H. (2003) *Journal of Molecular Catalysis A-Chemical*, **201**, 23–31.
(b) Eberhard, M.R. (2004) *Organic Letters*, **6**, 2125–8.
(c) Nilsson, P. and Wendt, O.F. (2005) *Journal of Organometallic Chemistry*, **690**, 4197–202.
(d) Sommer, W.J., Yu, K., Sears, J.S., *et al.* (2005) *Organometallics*, **24**, 4351–61.
(e) Weck, M. and Jones, C.W. (2007) *Inorganic Chemistry*, **46**, 1865–75.
(f) Yu, K., Sommer, W., Richardson, J.M., Weck, M., Jones, C.W. (2005) *Advanced Synthesis Catelysis*, **347**, 161–71.

36 (a) Rosner, T., Le Bars, J., Pfaltz, A. and Blackmond, D.G. (2001) *Journal of the American Chemical Society*, **123**, 1848–55.
(b) Rosner, T., Pfaltz, A. and Blackmond, D.G. (2001) *Journal of the American Chemical Society*, **123**, 4621–2.
(c) Mathew, J.S., Klussmann, M., Iwamura, H., *et al.* (2006) *Journal of Organic Chemistry*, **71**, 4711–22.
(d) Blackmond, D.G., Schultz, T., Mathew, J.S., *et al.* (2006) *Synlett*, 3135–9.

37 Consorti, C.S., Flores, F.R. and Dupont, J. (2005) *Journal of the American Chemical Society*, **127**, 12054–65.

38 Herrmann, W.A., Brossmer, C., Reisinger, C.-P., *et al.* (1997) *Chemistry-A European Journal*, **3**, 1357–64.

39 Evans, J., O'Neill, L., Kambhampati, V.L., *et al.* (2002) *Journal of the Chemical Society-Dalton Transactions*, 2207–12.

40 Svennebring, A., Nilsson, P. and Larhed, M. (2004) *Journal of Organic Chemistry*, **69**, 3345–9.

41 Shmidt, A.F. and Khalaika, A. (1998) *Kinetics Catalysis*, **39**, 803–7.

42 Littke, A.F. and Fu, G.C. (2002) *Angewandte Chemie – International Edition*, **41**, 4176–211.

43 (a) Luh, T.-Y., Leung, M.-K. and Wong, K.-T. (2000) *Chemical Reviews*, **100**, 3187–204.
(b) Cárdenas, D.J. (2003) *Angewandte Chemie-International Edition*, **42**, 384–7.

44 Christmann, U. and Vilar, R. (2005) *Angewandte Chemie-International Edition*, **44**, 366–74.

45 (a) Hillier, A.C., Grasa, G.A., Viciu, M.S., *et al.* (2002) *Journal of Organometallic Chemistry*, **653**, 69–82.
(b) Peris, E. and Crabtree, R.H. (2004) *Coordination Chemistry Reviews*, **248**, 2239–46.
(c) Scott, N.M. and Nolan, S.P. (2005) *European Journal of Inorganic Chemistry*, 1815–28.
(d) Kantchev, E.A.B., O'Brien, C.J. and Organ, M.G. (2006) *Aldrichimica Acta*, **39**, 97–111.

46 (a) Van Strijdonck, G.P.F., Boele, M.D.K., Kamer, P.C.J., *et al.* (1999) *European Journal of Inorganic Chemistry*, 1073–6.
(b) Galardon, E., Ramdeehul, S., Brown, J.M., *et al.* (2002) *Angewandte Chemie-International Edition*, **41**, 1760–3.

47 (a) Mann, G., Shelby, Q., Roy, A.H. and Hartwig, J.F. (2003) *Organometallics*, **22**, 2775–89.
(b) Culkin, D.A. and Hartwig, J.F. (2004) *Organometallics*, **23**, 3398–416.
(c) Yamashita, M. and Hartwig, J.F. (2004) *Journal of the American Chemical Society*, **126**, 5344–5.
(d) Hartwig, J.F. (2007) *Inorganic Chemistry*, **46**, 1936–47.

48 Gruber, A.S., Pozebon, D., Monteiro, A.L. and Dupont, J. (2001) *Tetrahedron Letters*, **42**, 7345–8.

49 Yao, Q., Kinney, E.P. and Yang, Z. (2003) *Journal of Organic Chemistry*, **68**, 7528–31.

50 (a) Cope, A.C. and Siekman, R.W. (1965) *Journal of the American Chemical Society*, **87**, 3272–3.
(b) Cope, A.C. and Friedrich, E.C. (1968) *Journal of the American Chemical Society*, **90**, 909–13.

51 (a) Dyker, G. (1997) *Chemische Berichte*, **130**, 1567–78.
(b) Catellani, M. (2003) *Synlett*, 298–313.
(c) Echavarren, A.M., Gómez-Lor, B., González, J.J. and de Frutos, O. (2003) *Synlett*, 585–97.

52 Lewis, L.N. (1986) *Journal of the American Chemical Society*, **108**, 743–9.

53 Beller, M., Fischer, H., Herrmann, W.A., *et al.* (1995) *Angewandte Chemie – International Edition*, **34**, 1848–9.

54 Zapf, A. and Beller, M. (2005) *Chemical Communications*, 431–40.

55 Dunina, V.V. and Gorunova, O.N. (2005) *Russian Chemical Reviews*, **74**, 871–913.

56 (a) Böhm, V.P.W. and Herrmann, W.A. (2000) *Chemistry – A European Journal*, **6**, 1017–25.
(b) Herrmann, W.A. and Böhm, V.P.W. (1999) *Journal of Organometallic Chemistry*, **572**, 141–5.

57 Albisson, D.A., Bedford, R.B. and Noelle Scully, P. (1998) *Tetrahedron Letters*, **39**, 9793–6.

58 Gibson, S., Foster, D.F., Eastham, G.R., *et al.* (2001) *Chemical Communications*, 779–80.

59 Friedlein, F.K., Kromm, K., Hampel, F. and Gladysz, J.A. (2006) *Chemistry – A European Journal*, **12**, 5267–81.

60 Frey, G.D., Reisinger, C.-P., Herdtweck, E. and Herrmann, W.A. (2005) *Journal of Organometallic Chemistry*, **690**, 3193–201.

61 Frey, G.D., Schütz, J., Herdtweck, E. and Herrmann, W.A. (2005) *Organometallics*, **24**, 4416–26.

62 (a) Lin, C.-A. and Luo, F.-T. (2003) *Tetrahedron Letters*, **44**, 7565–8.
(b) Luo, F.-T., Xue, C., Ko, S.-L., *et al.* (2005) *Tetrahedron*, **61**, 6040–5.

63 Ogo, S., Takebe, Y., Uehara, K., *et al.* (2006) *Organometallics*, **25**, 331–8.

64 Djakovitch, L., Heise, H. and Köhler, K. (1999) *Journal of Organometallic Chemistry*, **584**, 16–26.

65 (a) Nejjar, A., Pinel, C. and Djakovitch, L. (2003) *Advanced Synthesis Catalysis*, **345**, 612–19.
(b) Noël, S., Djakovitch, L. and Pinel, C. (2006) *Tetrahedron Letters*, **47**, 3839–42.

66 (a) Battistuzzi, G., Cacchi, S. and Fabrizi, G. (2003) *Organic Letters*, **5**, 777–80.
(b) Battistuzzi, G., Cacchi, S., Fabrizi, G. and Bernini, R. (2003) *Synlett*, 1133–6.

67 Briot, A., Baehr, C., Brouillard, R., *et al.* (2004) *Journal of Organic Chemistry*, **69**, 1374–7.

68 Dubbaka, S.R. and Vogel, P. (2005) *Chemistry – A European Journal*, **11**, 2633–41.

69 Panday, N., Canac, Y. and Vasella, A. (2000) *Helvetica Chimica Acta*, **83**, 58–79.

70 Eriksson, L., Winberg, K.J., Claro, R.T. and Sjöberg, S. (2003) *Journal of Organic Chemistry*, **68**, 3569–73.

71 (a) Chen, W., Xu, L. and Xiao, J. (2001) *Tetrahedron Letters*, **42**, 4275–8.
(b) Chen, W., Xu, L., Hu, Y., *et al.* (2002) *Tetrahedron*, **58**, 3889–99.
(c) Wende, M., Seidel, F. and Gladysz, J.A. (2003) *Journal of Fluorine Chemistry*, **124**, 45–54.
(d) Xu, L., Mo, J., Baillie, C. and Xiao, J. (2003) *Journal of Organometallic Chemistry*, **687**, 301–12.

72 (a) Tietze, L.F., Schirok, H., Wöhrmann, M. and Schrader, K. (2000) *European Journal of Organic Chemistry*, 2433–44. (b) Tietze, L.F. and Krahnert, W.-R. (2001) *Synlett*, 560–2.
(c) Tietze, L.F. and Petersen, S. (2001) *European Journal of Organic Chemistry*, 1619–24.
(d) Tietze, L.F., Sommer, K.M., Schneider, G., *et al.* (2003) *Synlett*, 1494–6.

73 Sünnemann, H.W., Hofmeister, A., Magull, J. and de Meijere, A. (2006) *Chemistry – A European Journal*, **12**, 8336–44.

74 Prashad, M., Liu, Y., Mak, X.Y., *et al.* (2002) *Tetrahedron Letters*, **43**, 8559–62.

75 Tietze, L.F., Kahle, K. and Raschke, T. (2002) *Chemistry – A European Journal*, **8**, 401–7.

76 Gruber, A.S., Zim, D.Z., Ebeling, G., *et al.* (2000) *Organic Letters*, **2**, 1287–90.

77 Xiong, Z., Wang, N., Dai, M., *et al.* (2004) *Organic Letters*, **6**, 3337–40.

78 Muñoz, M.P., Martín-Matute, B., Fernández-Rivas, C., *et al.* (2001) *Advanced Synthesis Catalysis*, **343**, 338–42.

79 Schnyder, A., Indolese, A.F., Studer, M. and Blaser, H.-U. (2002) *Angewandte Chemie – International Edition*, **41**, 3668–71.

80 Venkatesan, C. and Singh, A.P. (2004) *Journal of Catalysis*, **227**, 148–63.

81 Blackmond, D.G., Rosner, T. and Pfaltz, A. (1999) *Organic Process Research and Development*, **3**, 275–80.

82 Bravo, J., Cativiela, C., Navarro, R. and Urriolabeitia, E.P. (2002) *Journal of Organometallic Chemistry*, **650**, 157–72.

83 Thakur, V.V., Kumar, N.S.C.R. and Sudalai, A. (2004) *Tetrahedron Letters*, **45**, 2915–18.

84 Nagy, T.Z., Csámpai, A. and Kotschy, A. (2005) *Tetrahedron*, **61**, 9767–74.

85 Wu, K.-M., Huang, C.-A., Peng, K.-F. and Chen, C.-T. (2005) *Tetrahedron*, **61**, 9679–87.

86 Spencer, J., Sharratt, D.P., Dupont, J., *et al.* (2005) *Organometallics*, **24**, 5665–72.

87 Xu, C., Gong, J.-F., Yue, S.-F., *et al.* (2006) *Dalton Transactions*, 4730–9.

88 Consorti, C.S., Zanini, M.L., Leal, S., *et al.* (2003) *Organic Letters*, **5**, 983–6.

89 Yang, F., Zhang, Y., Zheng, R., *et al.* (2002) *Journal of Organometallic Chemistry*, **651**, 146–8.

90 Evans, P., Hogg, P., Grigg, R., *et al.* (2005) *Tetrahedron*, **61**, 9696–704.

91 (a) Onoue, H., Minami, K. and Nakagawa, K. (1970) *Bulletin of the Chemical Society of Japan*, **43**, 3480–5. (b) Constable, A.G., McDonald, W.S., Sawkins, L.C. and Shaw, B.L. (1978) *Journal of the Chemical Society D–Chemical Communications*, 1061–2. (c) Constable, A.G., McDonald, W.S., Sawkins, L.C. and Shaw, B.L. (1980) *Journal of the Chemical Society–Dalton Transactions*, 1992–2000. (d) Nielson, A.J. (1981) *Journal of the Chemical Society–Dalton Transactions*, 205–11. (e) Ryabov, A.D., Kazankov, G.M., Yatsimirsky, A.K., *et al.* (1992) *Inorganic Chemistry*, **31**, 3083–90.

92 (a) Iyer, S. and Ramesh, C. (2000) *Tetrahedron Letters*, **41**, 8981–4. (b) Iyer, S., Kulkarni, G.M. and Ramesh, C. (2004) *Tetrahedron*, **60**, 2163–72.

93 Alonso, D.A., Nájera, C. and Pacheco, M.C. (2000) *Organic Letters*, **2**, 1823–6.

94 Botella, L. and Nájera, C. (2002) *Angewandte Chemie–International Edition*, **41**, 179–81.

95 Iyer, S. and Jayanthi, A. (2001) *Tetrahedron Letters*, **42**, 7877–8.

96 Alonso, D.A., Botella, L., Nájera, C. and Pacheco, M.C. (2004) *Synthesis*, 1713–18.

97 Iyer, S. and Jayanthi, A. (2003) *Synlett*, 1125–8.

98 Cao, X.-P., Barth, D. and Kuck, D. (2005) *European Journal of Organic Chemistry*, 3482–8.

99 (a) Botella, L. and Nájera, C. (2004) *Tetrahedron Letters*, **45**, 1833–6. (b) Botella, L. and Nájera, C. (2005) *Journal of Organic Chemistry*, **70**, 4360–9.

100 Gürtler, C. and Buchwald, S.L. (1999) *Chemistry–A European Journal*, **5**, 3107–12.

101 Kim, S., Ko, H., Park, J.E., *et al.* (2002) *Journal of Medicinal Chemistry*, **45**, 160–4.

102 Botella, L. and Nájera, C. (2004) *Tetrahedron*, **60**, 5563–70.

103 Nájera, C. and Botella, L. (2005) *Tetrahedron*, **61**, 9688–95.

104 (a) Baleizão, C., Corma, A., García, H. and Leyva, A. (2003) *Chemical Communications*, 606–7. (b) Baleizão, C., Corma, A., García, H. and Leyva, A. (2004) *Journal of Organic Chemistry*, **69**, 439–46. (c) Corma, A., García, H. and Leyva, A. (2004) *Tetrahedron*, **60**, 8553–60. (d) Corma, A., García, H. and Leyva, A. (2006) *Journal of Catalysis*, **240**, 87–99. (e) Alacid, E. and Nájera, C. (2006) *Synlett*, 2959–64.

105 Kiewel, K., Liu, Y., Bergbreiter, D.E. and Sulikowski, G.A. (1999) *Tetrahedron Letters*, **40**, 8945–8.

106 Miyazaki, F., Yamaguchi, K. and Shibasaki, M. (1999) *Tetrahedron Letters*, **40**, 7379–83.

107 Morales-Morales, D., Grause, C., Kasaoka, K., *et al.* (2000) *Inorganica Chimica Acta*, **300–2**, 958–63.

108 Díez-Barra, E., Guerra, J., Hornillos, V., *et al.* (2003) *Organometallics*, **22**, 4610–12.

109 Churruca, F., SanMartín, R., Tellitu, I. and Domínguez, E. (2005) *Synlett*, 3116–20.

110 Yoon, M.S., Ryu, D., Kim, J. and Ahn, K.H. (2006) *Organometallics*, **25**, 2409–11.

111 (a) Jung, I.G., Son, S.U., Park, K.H., *et al.* (2003) *Organometallics*, **22**, 4715–20. (b) Takenaka, K. and Uozumi, Y. (2004) *Advanced Synthesis Catalysis*, **346**, 1693–6. (c) Takenada, K., Minakawa, M. and

Uozumi, Y. (2005) *Journal of the American Chemical Society*, **127**, 12273–81.
(d) Vicente, J., Abad, J.-A., López-Serrano, J., *et al.* (2005) *Organometallics*, **24**, 5044–57.
(e) Soro, B., Stoccoro, S., Minghetti, G., *et al.* (2006) *Inorganica Chimica Acta*, **359**, 1879–88.

112 For examples, see: (a) ref. 22d.
(b) Bergbreiter, D.E., Osburn, P.L. and Liu, Y.-S. (1999) *Journal of the American Chemical Society*, **121**, 9531–8.
(c) Bergbreiter, D.E., Osburn, P.L., Wilson, A. and Sink, E.M. (2000) *Journal of the American Chemical Society*, **122**, 9058–64.
(d) Hossain, M.A., Lucarini, S., Powell, D. and Bowman-James, K. (2004) *Inorganic Chemistry*, **43**, 7275–7.

113 Bergbreiter, D.E., Osburn, P.L. and Frels, J.D. (2005) *Advanced Synthesis Catalysis*, **347**, 172–84.

114 Yao, Q., Kinney, E.P. and Zheng, C. (2004) *Organic Letters*, **6**, 2997–9.

115 Consorti, C.S., Ebeling, G., Flores, F.R., *et al.* (2004) *Advanced Synthesis Catalysis*, **346**, 617–24.

116 Chen, M.-T., Huang, C.-A. and Chen, C.-T. (2006) *European Journal of Inorganic Chemistry*, 4642–8.

117 (a) Sonogashira, K. (2002) *Handbook of Organopalladium Chemistry for Organic Synthesis* (ed. E. Negishi), John Wiley & Sons, Inc., New York, pp. 493–529.
(b) Sonogashira, K. (2002) *Journal of Organometallic Chemistry*, **653**, 46–9.
(c) Negishi, E. and Anastasia, L. (2003) *Chemical Reviews*, **103**, 1979–2017.
(d) Tykwinski, R.R. (2003) *Angewandte Chemie – International Edition*, **42**, 1566–8.
(e) Brandsma, L. (2004) *Synthesis of Acetylenes, Allenes and Cumulenes: Methods and Techniques*, Elsevier, Oxford, p. 293.
(f) Sonogashira, K. (2004) *Metal–Catalyzed Cross–Coupling Reactions*, 2nd edn, Vol. **1** (eds F. Diederich and A. de Meijere), Wiley-VCH Verlag GmbH, Weinheim, pp. 319–45.
(g) Doucet, H. and Hierso, J.–C. (2007) *Angewandte Chemie – International Edition*, **46**, 834–71.
(h) Chinchilla, R. and Nájera, C. (2007) *Chemical Reviews*, **107**, 874–922.

118 Cassar, L. (1975) *Journal of Organometallic Chemistry*, **93**, 253–7.

119 Dieck, H.A. and Heck, F.R. (1975) *Journal of Organometallic Chemistry*, **93**, 259–63.

120 Sonogashira, K., Tohda, Y. and Hagihara, N. (1975) *Tetrahedron Letters*, 4467–70.

121 (a) Castro, C.E. and Stephens, R.D. (1963) *Journal of Organic Chemistry*, **28**, 2163.
(b) Stephens, R.D. and Castro, C.E. (1963) *Journal of Organic Chemistry*, **28**, 3313–15.

122 Eckhardt, M. and Fu, G.C. (2003) *Journal of the American Chemical Society*, **125**, 13642–3.

123 Altenhoff, G., Würtz, S. and Glorius, F. (2006) *Tetrahedron Letters*, **47**, 2925–8.

124 (a) Siemsen, P., Livingston, R.C. and Diederich, F. (2000) *Angewandte Chemie – International Edition*, **39**, 2632–57.
(b) Kotora, M. and Takahashi, T. (2002) *Handbook of Organopalladium Chemistry for Organic Synthesis* (ed. E. Negishi), John Wiley & Sons, Inc., New York, pp. 973–93.

125 Sonogashira, K., Yatake, T., Tohda, Y., *et al.* (1977) *Journal of the Chemical Society D – Chemical Communications*, 291–2.

126 Stambuli, J.P., Bühl, M. and Hartwig, J.F. (2002) *Journal of the American Chemical Society*, **124**, 9346–7.

127 (a) Amatore, C., Jutand, A., Khalil, F., *et al.* (1993) *Organometallics*, **12**, 3168–78.
(b) Grosshenny, V., Romero, F.M. and Ziessel, R. (1997) *Journal of Organic Chemistry*, **62**, 1491–500.
(c) Amatore, C. and Jutand, A. (2002) *Handbook of Organopalladium Chemistry for Organic Synthesis* (ed. E. Negishi), John Wiley & Sons, Inc., New York, pp. 943–72.

128 Bertus, P., Fécourt, F., Bauder, C. and Pale, P. (2004) *New Journal of Chemistry*, **28**, 12–14.

129 Létinois-Halbes, U., Pale, P. and Berger, S. (2005) *Journal of Organic Chemistry*, **70**, 9185–90.

130 Soheili, A., Albaneze-Walker, J., Murry, J.A., *et al.* (2003) *Organic Letters*, **5**, 4191–4.

131 Consorti, C.S., Flores, F.R., Rominger, F. and Dupont, J. (2006) *Advanced Synthesis Catalysis*, **348**, 133–41.

132 Tougerti, A., Negri, S. and Jutand, A. (2007) *Chemistry–A European Journal*, **13**, 666–76.

133 (a) Jutand, A. (2004) *Pure and Applied Chemistry*, **76**, 565–76.
(b) Amatore, C., Bensalem, S., Ghalem, S., *et al.* (2004) *European Journal of Organic Chemistry*, 366–71.

134 Gil-Moltó, J. and Nájera, C. (2006) *Advanced Synthesis Catalysis*, **348**, 1874–82.

135 Barrios-Landeros, F. and Hartwig, J.F. (2005) *Journal of the American Chemical Society*, **127**, 6944–5.

136 Hundertmark, T., Littke, A.F., Buchwald, S.L. and Fu, G.C. (2000) *Organic Letters*, **2**, 1729–31.

137 Böhm, V.P.W. and Herrmann, W.A. (2000) *European Journal of Organic Chemistry*, 3679–81.

138 Yi, C. and Hua, R. (2006) *Journal of Organic Chemistry*, **71**, 2535–7.

139 Gelman, D. and Buchwald, S.L. (2003) *Angewandte Chemie–International Edition*, **42**, 5993–6.

140 Hierso, J.-C., Fihri, A., Amardeil, R., *et al.* (2004) *Organic Letters*, **6**, 3473–6.

141 Feuerstein, M., Doucet, H. and Santelli, M. (2004) *Tetrahedron Letters*, **45**, 8443–6.

142 (a) Nájera, C., Gil-Moltó, J., Karlström, S. and Falvello, L.R. (2003) *Organic Letters*, **5**, 1451–4.
(b) Gil-Moltó, J. and Nájera, C. (2005) *European Journal of Organic Chemistry*, 4073–81.
(c) Gil-Moltó, J., Karström, S. and Nájera, C. (2005) *Tetrahedron*, **61**, 12168–76.
(d) Li, J.-H., Zhang, X.-D. and Xie, Y.-X. (2005) *European Journal of Organic Chemistry*, 4256–9.
(e) Chouzier, S., Gruber, M. and Djakovitch, L. (2004) *Journal of Molecular Catalysis A–Chemical*, **212**, 43–52.

143 Buchmeiser, M.R., Schareina, T., Kempe, R. and Wurst, K. (2001) *Journal of Organometallic Chemistry*, **634**, 39–46.

144 Park, S.B. and Alper, H. (2004) *Chemical Communications*, 1306–7.

145 (a) Gossage, R.A., Jenkins, H.A. and Yadav, P.N. (2004) *Tetrahedron Letters*, **45**, 7689–91. (b) Eisnor, C.R., Gossage, R.A. and Yadav, P.N. (2006) *Tetrahedron*, **62**, 3395–401.

146 Wang, R., Piekarski, M.M. and Shreeve, J.M. (2006) *Organic and Biomolecular Chemistry*, **4**, 1878–86.

147 Nishihara, Y., Ikegashira, K., Mori, A. and Hiyama, T. (1997) *Chemistry Letters*, 1233–4.

148 (a) Urgaonkar, S. and Verkade, J.G. (2004) *Journal of Organic Chemistry*, **69**, 5752–5.
(b) Liang, B., Dai, M., Chen, J. and Yang, Z. (2005) *Journal of Organic Chemistry*, **70**, 391–3.
(c) Li, J.-H., Zhang, X.-D. and Xie, Y.-X. (2005) *Synthesis*, 804–8.
(d) Li, J.-H., Liang, Y. and Xie, Y.-X. (2005) *Journal of Organic Chemistry*, **70**, 4393–6.
(e) Li, J.-H., Hu, X.-C., Liang, Y. and Xie, Y.-X. (2006) *Tetrahedron*, **62**, 31–8.

149 Reetz, M.T. and Maase, M. (1999) *Advanced Materials*, **11**, 773–7.

150 Herrmann, W.A., Reisinger, C.-P., Öfele, K., *et al.* (1996) *Journal of Molecular Catalysis A–Chemical*, **108**, 51–6.

151 Brun, V., Legraverend, M. and Grierson, D.S. (2002) *Tetrahedron*, **58**, 7911–23.

152 (a) Alonso, D.A., Nájera, C. and Pacheco, M.C. (2002) *Tetrahedron Letters*, **43**, 9365–8.
(b) Alonso, D.A., Nájera, C. and Pacheco, M.C. (2003) *Advanced Synthesis Catalysis*, **345**, 1146–58.

153 Alonso, D.A., Nájera, C. and Pacheco, M.C. (2004) *Journal of Organic Chemistry*, **69**, 1615–19.

154 Fairlamb, I.J.S., Kapdi, A.R., Lee, A.F., *et al.* (2004) *Dalton Transactions*, 3970–81.

155 Ruiz, J., Cutillas, N., López, F., *et al.* (2006) *Organometallics*, **25**, 5768–73.

156 Yang, F., Cui, X., Li, Y., *et al.* (2007) *Tetrahedron*, **63**, 1963–9.

157 Eberhard, M.R., Wang, Z. and Jensen, C.M. (2002) *Chemical Communications*, 818–19.

158 Corma, A., García, H. and Leyva, A. (2005) *Tetrahedron*, **61**, 9848–54.

9
Palladacyclic Pre-Catalysts for Suzuki Coupling, Buchwald–Hartwig Amination and Related Reactions

Robin B. Bedford

9.1
Introduction

Palladium-catalyzed coupling of organoboron compounds with electrophilic coupling partners such as aryl halides is known as the Suzuki or Suzuki–Miyaura reaction [1]. A generalized equation for the reaction is shown in Scheme 9.1, whilst Scheme 9.2 depicts a highly simplified version of the generally accepted catalytic cycle.

The palladium catalysts traditionally employed include $[Pd(PPh_3)_4]$ or mixtures of triarylphosphines and appropriate palladium (II) or (0) precursors such as palladium acetate or dipalladium tris(dibenzylideneacetone), respectively. The primary function of the base is to increase the nucleophilicity of the aryl boronic acid, which it does by forming a boronate complex $[ArBX(OH)_2]^-$, where X is the anion from the base or hydroxide formed from adventitious water.

The nucleophilic coupling partner is typically an arylboronic acid; however, boronic esters can sometimes be employed. Other examples of C-sp^2-based organoboron species that can be used include vinyl boronic acids and heterocyclic boronic acids. The exploitation of *alkyl* boronic acids and esters in the Suzuki reaction is a topical field of research, but these substrates tend to require very exacting catalysts and reaction conditions. As for the electrophilic coupling partners, aryl halides are most often used but halide surrogates such as aryl triflates and aryldiazonium salts can be employed. Aryl chlorides are the substrates of choice in terms of both cost and commercial availability but, unfortunately, they are the least easily used since the high C—Cl bond strength compared with C—Br and C—I bonds disfavors oxidative addition, the first step in the catalytic cycle, and makes the coupling of such substrates far more challenging [2].

The Suzuki reaction has matured into a very powerful technique for the formation of new carbon–carbon bonds and is routinely used in fine chemicals research and development and pharmaceutical discovery laboratories. In several instances it has been scaled to commercial operation, for instance in the production of Valsartan, an angiotensin II inhibitor produced by Novartis for the treatment of

Palladacycles: Synthesis, Characterization and Applications. Edited by Jairton Dupont and Michel Pfeffer

ISBN: 978-3-527-31781-3

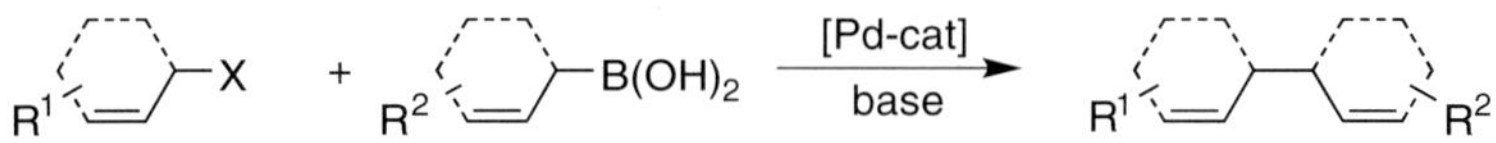

Scheme 9.1 The Suzuki or Suzuki–Miyaura coupling reaction.

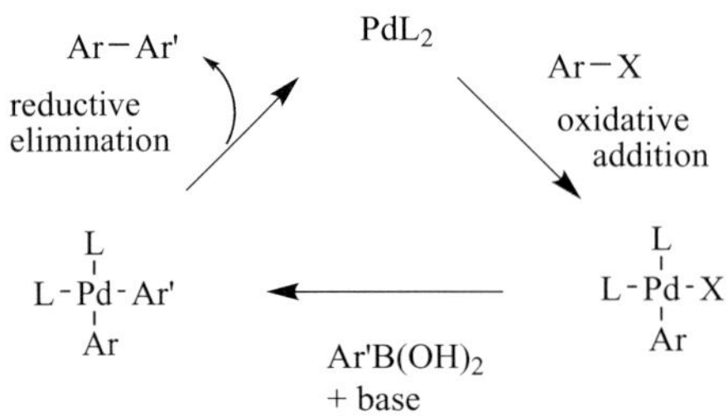

Scheme 9.2 Highly simplified version of the catalytic cycle for the Suzuki reaction.

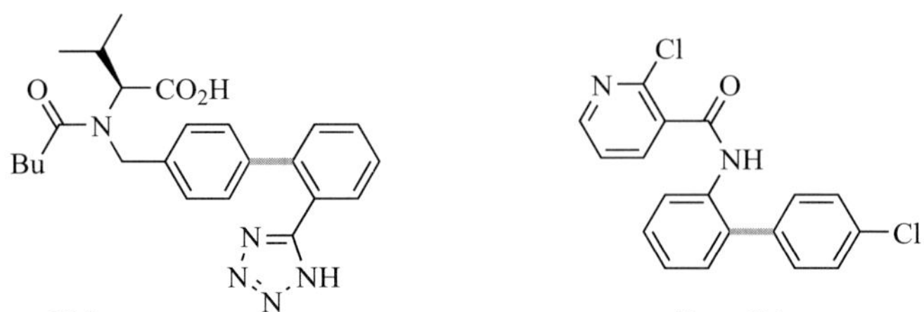

Figure 9.1 Examples of commercial products that employ Suzuki coupling in their manufacture. The C—C bonds produced in the Suzuki step are highlighted.

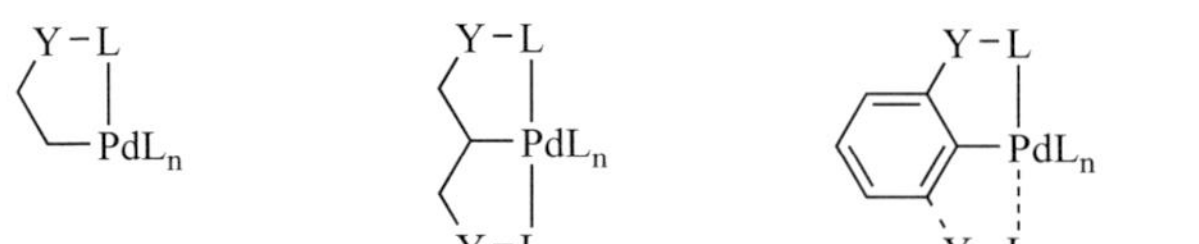

Figure 9.2 Generic structures of palladacycles (**I**) and pincer complexes (**II, III**).

hypertension and congestive heart failure, and of Boscalid, a fungicide produced by BASF for crop protection (Figure 9.1).

The growth in the use palladacycles (**I**, Figure 9.2) and related pincer complexes **(II)** (typically with an aryl group incorporated into the backbone; structure **III**) as pre-catalysts in the Suzuki coupling reaction has been driven by two main considerations. The first is that often they can be used in very low loadings, which is important because the use of palladium-catalyzed couplings in the production of fine chemicals and pharmaceuticals can be hampered by the need to remove heavy metal contamination down to the ppm level. Therefore, catalysts that show good

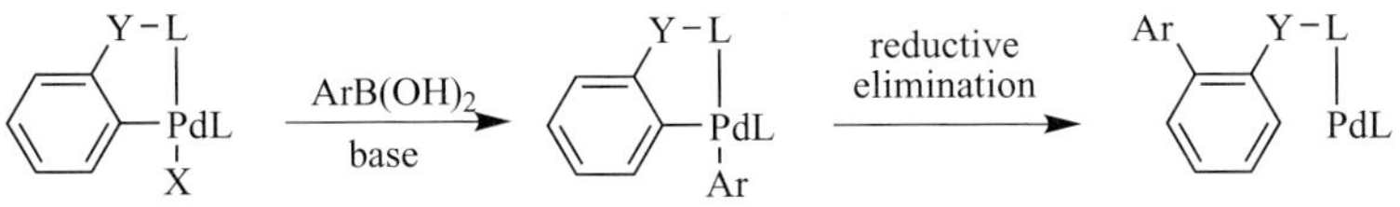

Scheme 9.3 Arylboronic acid mediated reduction of Pd(II) palladacyclic pre-catalysts to active Pd(0) species.

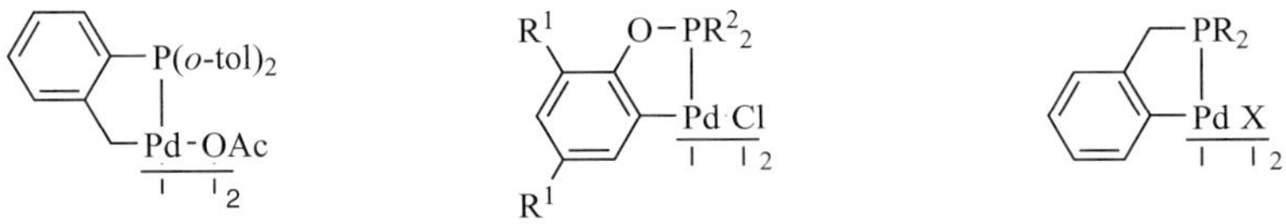

1

2a: R^1 = ^{t}Bu; R^2 = OC_6H_3-2,4-$^tBu_2$
b: R^1 = H; R^2 = Ph
c: R^1 = ^{t}Bu; R^2 = Ph
d: R^1 = ^{t}Bu, R^2 = ^{i}Pr

3a: R = Ph; X = Cl
b: R = Ph; X = Br
c: R = ^{t}Bu; X = Br

Figure 9.3 Examples of phosphorus-based palladacyclic catalysts.

activity at very low loadings negate the need for subsequent removal of palladium from the product stream, making them attractive for commercial application. The second driving consideration is that relatively simple modifications to the pre-catalysts can give good to excellent activity with challenging aryl chloride substrates.

Much of the success enjoyed by palladacycles in the Suzuki reaction can be traced to the fact that they act as well-defined, easily handled, stable precursors to highly active Pd(0) catalysts. The conversion of the palladacyclic pre-catalysts into active Pd(0) species typically involves their reaction with arylboronic acids (activated by the base present in the reaction mixture) followed by reductive elimination of the resultant aryl-ligated palladacycle (Scheme 9.3).

9.2
Phosphorus-Based Palladacycles and Pincer Complexes

The first report to reveal the potential promise of palladacyclic pre-catalysts in Suzuki coupling reactions was published by Beller, Hermann and coworkers in 1995 [3]. They showed that complex **1** (Figure 9.3), formed by reaction of palladium acetate with tri(*o*-tolyl)phosphine [4], shows good to excellent activity in the coupling of arylboronic acids with a range of aryl bromide substrates and can even be exploited in the coupling of electronically activated (electron-deficient) aryl chlorides. Thus, the coupling of phenylboronic acid with 4-bromoacetophenone, 4-bromoanisole and 4-chloroacetophenone gave turnover numbers (TON, mol product per mol catalyst) of 74 000, 7600 and 2100, respectively. Notably, though, it was later recognized that highly electronically activated substrates such 4-bromoacetophenone do not provide a particularly useful yardstick with which to gauge catalyst performance as high TONs can be obtained with this substrate using palladium acetate in the absence of added ligands [5]. Beyond its good activity, other advantages of using the palladacyclic complex **1** compared to more classical catalysts include that it is air-stable and thus easily handled and that the phosphine to palladium ratio is maintained at 1:1, saving on the cost of the ligand.

The orthometallated triarylphosphite complex **2a** is considerably more active in the Suzuki coupling of aryl bromides than complex **1**; the coupling of phenylboronic acid with 4-bromoacetophenone gives TONs of up to 1 000 000 [6]. With the more electronically challenging substrate, 4-bromoanisole, the maximum TON obtained is 30 000. Complex **2a** can also be used in the coupling of aryl bromides with *alkyl* boronic acids. This increased activity arises despite the fact that the palladium center in complex **2a** is more electron-deficient than in complex **1** due to the greater π-acidity of the phosphorus donor atom and the lower σ-basicity of the orthometallated carbon donor in the former complex. At that time it was usually assumed that oxidative addition of the aryl bromide would be the rate-limiting step in the catalytic cycle; however, this means that triarylphosphites should be relatively poor ligands.

In view of these findings a more systematic study was conducted on the effect of "tuning" the electronic and steric profile of the orthometallated ligand in complexes of type **2** [7]. In general it appears that increasing the size of the substituents on the orthometallated ring leads to increased activity. The electronic effects are more subtle; in the series of orthometallated complexes with ligands of the type κ^2-*P*,*C*-$Y_2P(OAr)$ the more electron donating phosphinite ligands ($Y = {}^iPr > Ph$) show better activity than the phosphite ligands ($Y = OAr$). Thus when complex **2d** is employed, with an extra equivalent of the phosphinite ligand added, TONs of up to 475 million and 8.75 million are observed with the substrates 4-bromoacetophenone and 4-bromoanisole, respectively. These data suggest that *increasing* the electron density on the palladium center is beneficial; however, this can only be pushed so far. This is demonstrated by the fact that the orthopalladated benzylphosphine complex **3a**, which is sterically essentially identical to **2b**, but is more electron rich at the palladium center, shows less than half the TON in the coupling of phenylboronic acid with 4-bromoanisole [8].

Despite the very high activity shown in the coupling of aryl bromide substrates, the complex **2d** shows only modest activity with aryl chloride substrates; the coupling of phenyl boronic acid with the non-activated and activated aryl chlorides 4-chlorotoluene and 4-chlorobenzaldehyde gives 76% and 100% conversion, respectively, at 1.0 mol% Pd catalyst loading [8]. Cole-Hamilton and coworkers showed that more electron-rich palladacycle **3c** gives a good TON of 2700 with 4-chlorobenzaldehyde, but at the expense of selectivity; the product is contaminated with substantial amounts of 1-(4-chlorophenyl)-1-phenylmethanol and 1,4-biphenyl-1-phenylmethanol [9]. In contrast with the more electron-deficient system **2d**, complex **3c** shows no activity with activated substrate 4-chloroacetophenone, again indicating that relatively subtle differences in the electronic properties of the palladium center can have significant effects on the rate of catalysis.

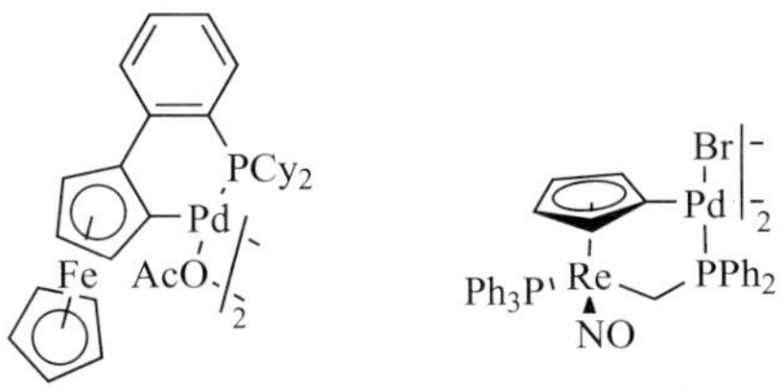

4 **5**

Figure 9.4 Palladacyclic pre-catalyst incorporating cyclopentadienyl-metal complexes.

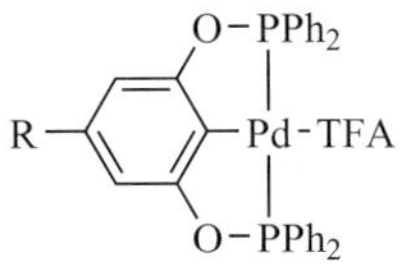

6: R = H, Me

Figure 9.5 "PCP"-pincer pre-catalysts.

While some of the palladacyclic complexes described above can activate electron-deficient aryl chloride substrates, they are not usually able to participate in the efficient oxidative addition of electronically deactivated examples. The ferrocene-based palladacyclic system **4** is sufficiently electron-rich at palladium to be able to couple both electronically deactivated and sterically hindered aryl chlorides such as 4-chloroanisole and 2-chloro-*m*-xylene with phenylboronic acid, even at room temperature [10]. When the reaction temperature is increased to 60 °C then the coupling of the non-activated substrate 4-chlorotoluene proceeds with TONs of over 9500 (Figure 9.4).

An alternative approach to the incorporation of a cyclopentadienyl-metal fragment into the backbone of a palladacyclic pre-catalyst was adopted by Gladysz and coworkers, who showed the Cp-rhenium based complex **5** can show reasonably good activity in the coupling of aryl bromides, with TONs of up to nearly 100 000 with 4-bromoanisole [11].

As well as simple κ^2-*P*,*C*-palladacycles, κ^3-*P*,*C*,*P*-pincer complexes have been employed to good effect in Suzuki coupling reactions. Not only do the complexes **6** show reasonable activity in the coupling of electronically deactivated aryl bromides, they can also be used with activated aryl chloride substrates (Figure 9.5) [12]. The TONs obtained with aryl bromides using **6** are typically 1–3 orders of magnitude lower than those obtained with the notionally related phosphinite-based palladacycles **2b–d**. Substitution of the PPh_2 phenyl groups in complexes of type **6** with other functions proved to be deleterious; however, the study was limited to the use of the easy-to-couple substrate 4-bromoacteophenone [13].

9.3
Nitrogen-Based Palladacycles

The use of palladacyclic pre-catalysts is not limited to phosphorus-containing systems and there has been a substantial amount of research undertaken on the

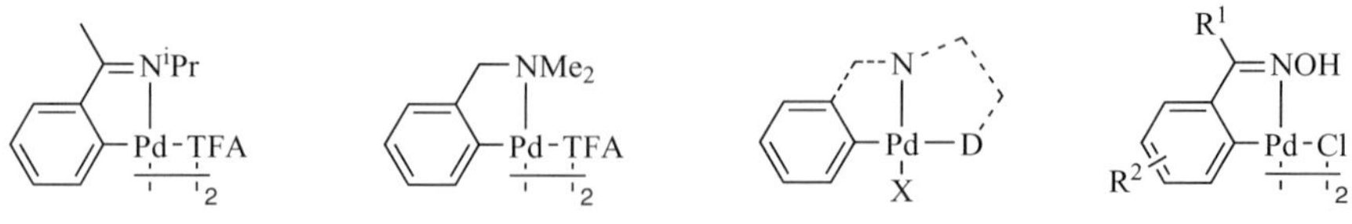

Figure 9.6 Examples of N,C-based complexes used in the Suzuki reaction.

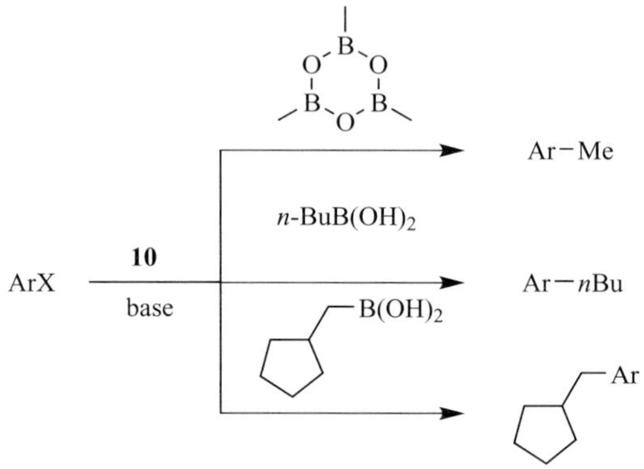

Scheme 9.4 Suzuki coupling of aryl halide with alkyl boron reagents catalyzed by oxime-based palladacycle **10**.

use of N,C-based complexes in the Suzuki reaction. This was initiated by Milstein and coworkers, who showed that the imine-based palladacycle **7** could give TONs of up to 840 000 and 136 000 for the coupling of phenylboronic acid with 4-bromoacetophenone and 4-bromoanisole respectively [14]. The related benzylamine-based palladacycle **8** shows slightly lower activity [15]. Sulfilimine- [16], benzodiazepine- [17] and arylurea-based palladacycles [18] have all been shown to give good activity with aryl bromides. Where activity is observed in the coupling of aryl chlorides using N,C-palladacycles it is supported by the addition of either tetrabutylammonium bromide (TBAB) or alkylphosphines, whose probable roles are discussed below. Further synthetic elaboration led to the production of imine-based palladacycles with pendant donor functions (general structure **9**, D = N, S), all of which have been applied to the Suzuki reaction [19–21] (Figure 9.6).

Nájera and coworkers have performed extensive studies on the use of oxime-based palladacycles of the type **10** in a range of Suzuki coupling reactions [22]. For instance, reasonable TONs of up to 500 000 are seen with the electronically activated substrate 4-bromoacetophenone [23]. The complexes can also be used in the coupling of aryl chlorides, again provided that TBAB is added to the system. Usefully, the catalysis can be performed under aqueous conditions [24, 25]. These complexes can also catalyze the coupling of aryl halides with *alkyl* boronic acids and anhydrides – a topical area of research (Scheme 9.4). Furthermore, they can be employed in the coupling of aryl boronic acids with allylic chlorides and acetates as well as benzylic chlorides in organic or aqueous media [23, 25].

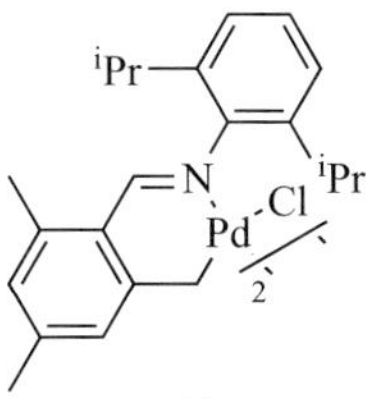

11

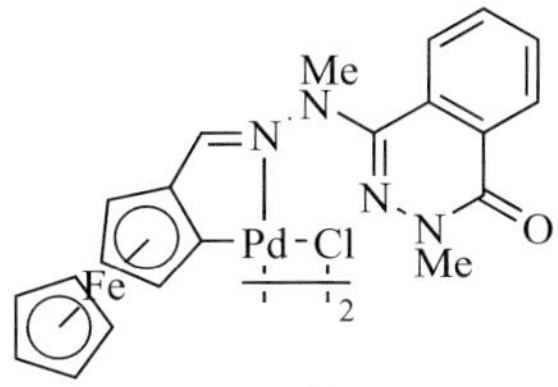

12

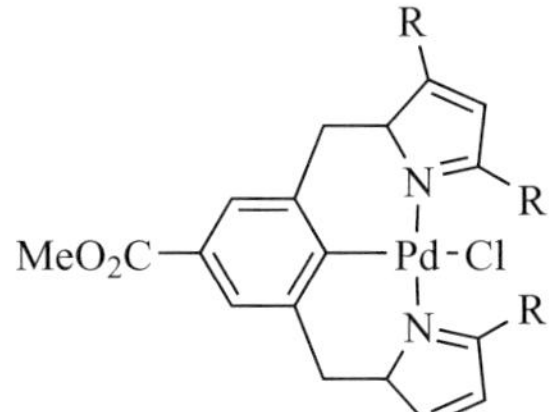

13: R = H, Me

Figure 9.7 Some imine-based palladacycles.

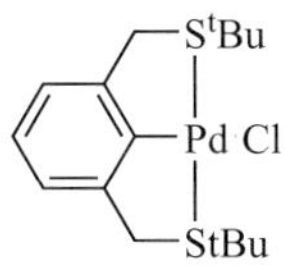

14

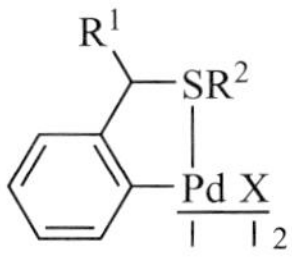

15a: R^1 = Me; R^2 = ^{t}Bu, X = Cl
b: R^1 = Me; R^2 = Me, X = Cl
c: R^1 = H; R^2 = ^{t}Bu, X = Cl
d: R^1 = Me; R^2 = ^{t}Bu, X = OAc

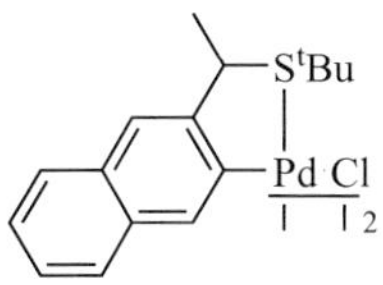

16

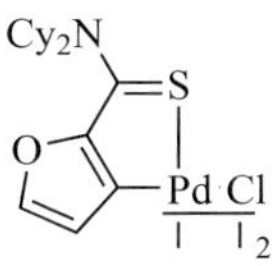

17

Figure 9.8 Sulfur-based palladacycles used in Suzuki couplings.

Liu and coworkers synthesized the imine-based C(sp^3)-palladacycle **11**, which gives TONs of up to 1 000 000 with 4-bromoanisole under air in EtOH and reasonable activity with activated and non-activated aryl chlorides in water with TBAB under air (Figure 9.7) [26]. As with P,C-based palladacycles, an N,C-palladacycle incorporating a ferrocene moiety in the orthometallated ligand has been examined as a pre-catalyst in the Suzuki reaction; however, only modest activity was observed with aryl bromide substrates [27]. Again activity is not limited to simple palladacycles; for instance, the N,C,N-pincer complex **13** shows very good activity in the coupling of aryl bromide substrates in water, giving TONs of up to 8.6 million with bromobenzene [28].

9.4 Sulfur-Based Palladacycles

Sulfur-based palladacycles have also been exploited in the Suzuki reaction. Dupont and Monteiro showed that the complex **14** could be used to good effect in coupling the electronically deactivated aryl bromide substrate 4-bromoanisole in Suzuki coupling reactions (Figure 9.8) [29]. Similarly, complexes **15** and **16** display good activity; for example, catalyst **15a** even shows some, albeit very limited, activity with the deactivated aryl chloride substrate 4-chloroanisole. This catalyst also shows good activity in the sterically challenging coupling of bromomesitylene with 2-tolylboronic acid. The best activities are seen when TBAB is employed as an additive. The furancarbothioamide-based palladacycle **17** shows activity in the coupling of aryl bromide substrates in aqueous DMF under aerobic conditions and this advantage has been exploited in the synthesis of a small library of biaryl intermediates to chlorobenzothiophene-containing "Hedgehog" pathway agonists [30].

9.5 Phosphine and Carbene Adducts of Palladacycles

The simple palladacyclic complexes discussed so far can show good to excellent activity in the coupling of aryl bromide or iodide substrates, some giving very high turnover numbers indeed. However, most of these complexes show at best only limited activity with more desirable aryl chlorides and in these instances it is also usually necessary to add tetrabutylammonium bromide. The addition of phosphines and related carbene ligands to dimeric palladacycles leads to the breaking of the dimer and the formation of adducts (Scheme 9.5) that often show enhanced catalytic activity, particularly with aryl chloride substrates.

This was first demonstrated with the complexes **18** and **19a**, which show excellent activity in the coupling of electronically deactivated aryl chlorides, giving TONs of up to 8000 in the coupling of 4-chloroanisole with phenylboronic acid (Figure 9.9) [31]. This is interesting as tricyclohexylphosphine is not typically the phosphine of choice in the Suzuki coupling of aryl chlorides when using other palladium precursors: better activity is usually observed with other ligands such as tri-*tert*-butylphosphine or PCy_2(*o*-biphenyl). Yet the adducts of these phosphines, complexes **19b** and **19c**, both show reduced activity compared with **19a.** This highlights the importance of the palladium source; indeed, when palladium acetate is used as the palladium precursor the activity with respect to the ligands is PCy_2(*o*-biphenyl) > P^tBu_3 > PCy_3 [32]. Tricyclohexylphosphine is the ligand of choice due to its lower cost and, compared with P^tBu_3, its lower air-sensitivity. With both PCy_3 and P^tBu_3 the activity is strongly influenced by the amount of added phosphine, with optimum activity observed at a P : Pd ratio of 2 : 1; increasing the level of added phosphine further is highly deleterious. By contrast, little effect is seen on adding extra equivalents of PCy_2(*o*-biphenyl) to **19c.** The related complexes **20**, containing secondary or tertiary alkylphosphine, can also be used to good effect with aryl chloride substrates [33], as can the ferrocenyl-based system **21** [34]. Dupont,

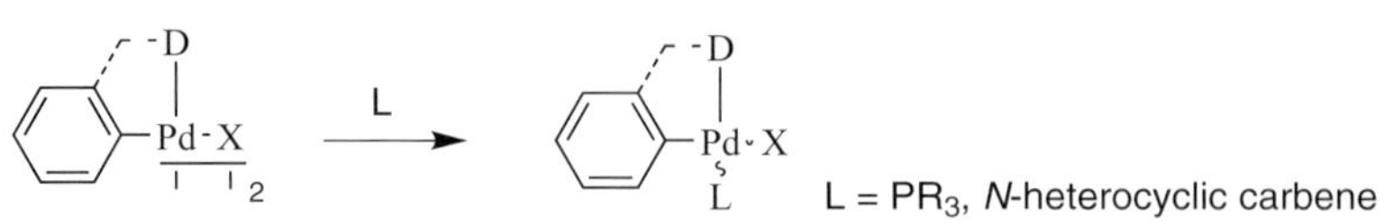

Scheme 9.5 Formation of adducts of palladacycles.

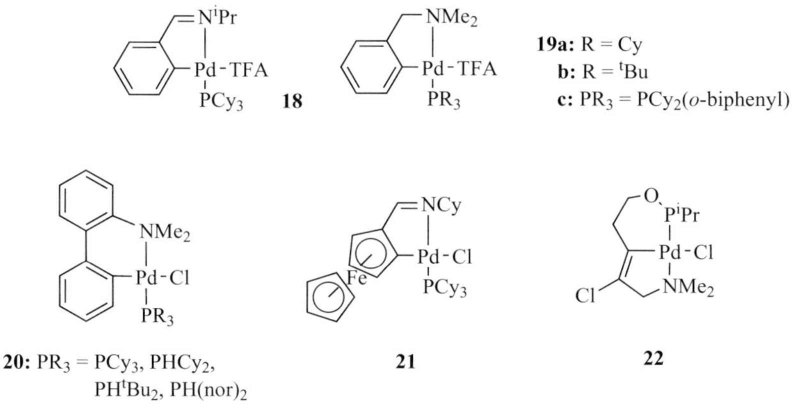

Figure 9.9 Phosphine adducts of N,C-palladacycles.

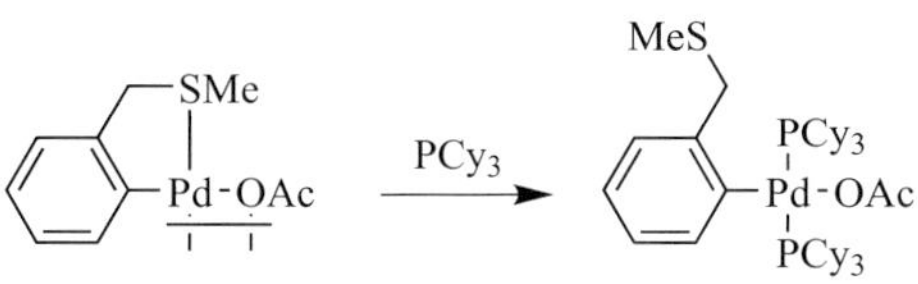

23 **24**

Scheme 9.6 Reaction of an S-based palladacycle with PCy$_3$.

Monteiro and coworkers showed that the unusual tridentate vinyl-based "*NCP*"-palladacycle **22**, which is notionally related to the phosphine adducts of N,C-palladacycles, can also be used to couple aryl chloride substrates [35].

Notably, the use of highly electron-donating phosphines is not always beneficial; adducts of the type **18** and **19** show no enhancement in activity compared to the dimeric parent palladacycles when aryl bromides are used as substrates, presumably because in this instance the rate-determining step is probably not the oxidative addition of the aryl halide, but rather lies later in the catalytic cycle and, therefore, is favored by *lower* electron density on the metal. Indeed, when the PCy$_3$ is replaced by more π-acidic arylphosphines or arsines an enhancement in activity is observed compared with the dimers [15].

It is not necessary to use a nitrogen-based palladacycle, catalysts formed *in situ* from the sulfur-based palladacycle **23** and PCy$_3$, P^{t}Bu$_3$ or PCy$_2$(*o*-biphenyl) give good activity with aryl chloride substrates (Scheme 9.6) [36]. Again, the same general trends are observed on increasing the P : Pd ratio as described above. In this case the adduct formation is not clean; the addition of PCy$_3$ gives several species, the major one with increasing amounts of added ligand is complex **24**, in which the SMe donor has been displaced by a second equivalent of the phosphine ligand.

In all cases described so far, the active catalysts produced on reduction of the pre-catalysts according to Scheme 9.3 would contain only one P-donor on the resultant palladium(0) complex; furthermore, the reductively eliminated N- or

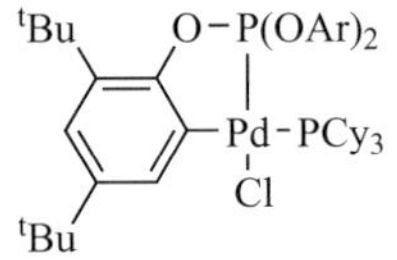

25

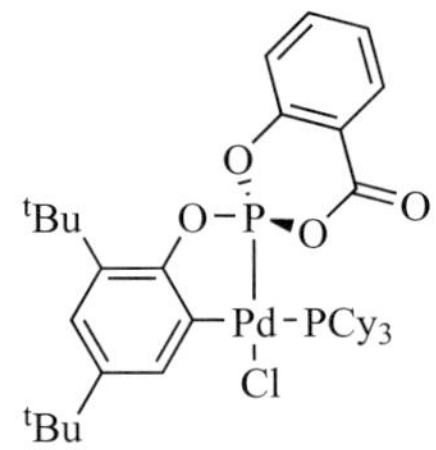

26

Figure 9.10 Phosphine adducts of phosphite-based palladacycles, typically prepared *in situ*.

S-donor ligand is likely to be fairly labile. By contrast, the reductive activation of a tricyclohexylphosphine adduct of a triarylphosphite-based palladacycle would give both a trialkylphosphine and a triarylphosphite in the coordination sphere of the resultant Pd(0) complex. The palladium center in this mixed trialkylphosphine–triarylphosphite complex would not be as electron-rich as a simple trialkylphosphine-containing species; therefore, oxidative addition, the rate-determining step with most aryl chlorides, would be slower and consequently overall activity may be expected to be significantly lower with these substrates. However, when a PCy_3 adduct of complex **2a** (complex **25**, formed *in situ*) is employed then it proves to be extremely active compared with the amine-based palladacycle **19a** for a range of activated and deactivated aryl chloride substrates (Figure 9.10) [37]. The vastly increased TONs observed with **25** is *not* accounted for by an increase in the rate of catalysis, but rather by substantially enhanced catalyst longevity; catalyst **19a** is only active for approximately 1/2 h whereas the phosphite-based adduct remains active after 24 h. This enhancement in catalyst productivity by increased longevity is, presumably, attributable to stabilization of the catalyst's resting state. In reactions where oxidative addition is rate determining then the resting state will be a Pd(0) complex – such species are relatively well stabilized by π-acidic phosphite ligands. If this is true, then longevity should vary with the π-acidity of the ortho-metallated co-ligand. This indeed turns out to be the case, as catalysts formed *in situ* from tricyclohexylphosphine and the palladacycles **2a**, **c** and **d** show maximum rates in the order **2d** > **c** > **a**, that is to say that rate falls off with increasing π-acidity, but the opposite trend is observed for total TONs. The salicylaldehyde-based phosphite ligand in the palladacyclic complex **26** is even more electron-withdrawing and this adduct gives very high TONs in the coupling of aryl chlorides – up to 128 000 and 2 000 000 with 4-chloroanisole and 4-chloronitrobenzene respectively [38].

N-Heterocyclic carbenes derived from imidazolium salts are excellent σ-donors and very poor π-acids that give strong bonds with Pd(II). The electronic profile of these ligands means that they give electron-rich Pd(0) complexes that undergo facile oxidative addition reactions. Thus, unsurprisingly, the carbene adduct of an *NC*-based palladacycle, complex **27**, is highly active in the Suzuki coupling of aryl chloride substrates, even at room temperature [39, 40]. This system is particularly useful for the synthesis of hindered biaryls. What is more surprising, given the activity of the PCy_3 adducts of phosphite-based discussed above, is that the analo-

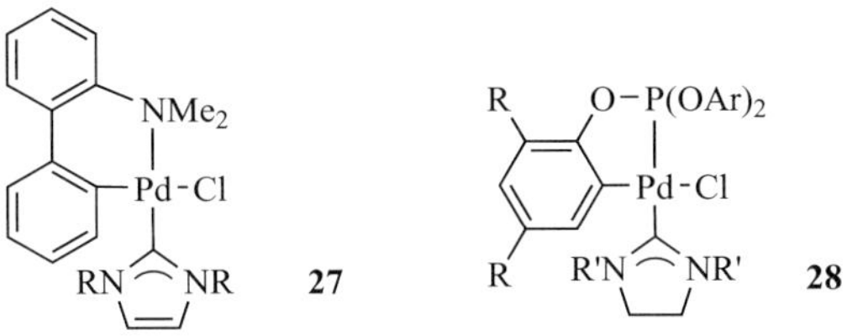

Figure 9.11 Carbene adducts of palladacycles.

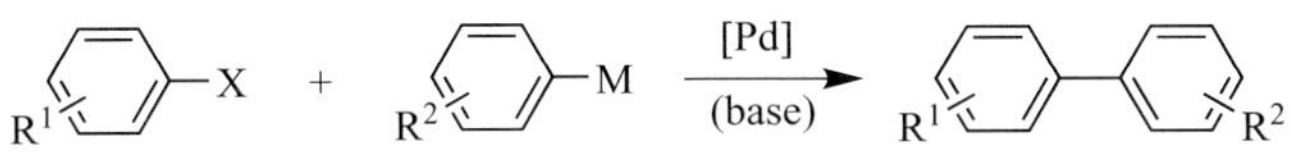

Scheme 9.7 The Stille (M = SnR_3), Kumada (M = MgX) and Negishi (M = ZnX) coupling reactions.

gous carbene adducts **28** show at best only modest activity with activated and non-activated aryl chloride substrates [41] (Figure 9.11).

9.6 Palladacyclic Catalysts for Other Cross-Coupling Reactions

Palladacyclic complex **1** has been exploited in a range of other biaryl coupling reactions, namely the Stille coupling of aryl bromide substrates [42, 43] and the Kumada and Negishi reactions, (Scheme 9.7) [44]. Nájera and coworkers showed that oxime-based palladacycles can also be used for the Stille coupling of aryl bromide substrates [44]. Similarly, complex **2a** gives high TONs in the Stille coupling of aryl bromides [6], while the tricyclohexylphosphine adduct complex **25**, can be used with aryl chlorides [45]. Interestingly, the use of the phosphite-based palladacycle as a pre-catalyst shows no benefit over $Pd(OAc)_2/PCy_3$ mixtures in this instance. This is readily explained by the observation that the rate-determining step in the Stille reaction with aryl chlorides is not the oxidative addition of the chloride but rather the transmetallation step. Therefore, the resting state in the catalytic cycle would be a Pd(II) species rather than a Pd(0) complex. The inclusion of the phosphite ligand would offer little or no extra benefit in stabilizing this Pd(II) species compared with PCy_3 alone.

9.7 Palladacyclic Catalysts for Buchwald–Hartwig Amination

The Buchwald–Hartwig amination of aryl halides (Scheme 9.8) is a very powerful method for the formation of new C–N bonds and is widely exploited in organic synthesis. The first application of a palladacycle as a pre-catalyst in this reaction was reported by Hartwig and Louie, who showed that complex **1** could be used with aryl bromide substrates [42]. Subsequently, Beller and coworkers found that this pre-catalyst could be used with activated aryl chlorides [46].

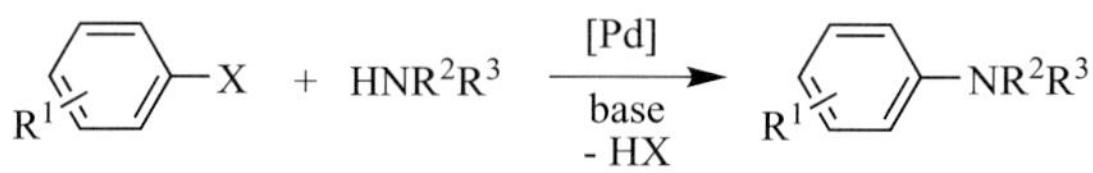

Scheme 9.8 Buchwald–Hartwig amination of aryl halides.

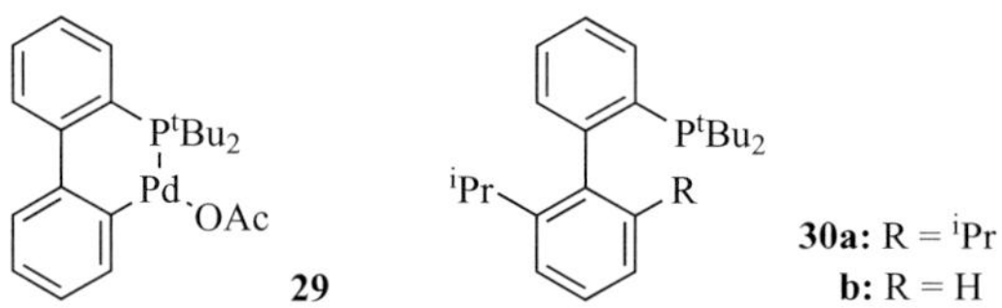

Figure 9.12 A palladacycle **(29)** and ligands **(30)** employed in amination reactions.

As observed in the Suzuki reaction, alkylphosphine adducts of N,C-palladacycles show substantially enhanced activity in the Buchwald–Hartwig amination of aryl chlorides and can be used in the coupling of electronically deactivated aryl chlorides [32, 33]. For instance, complex **19b**, formed *in situ* from complex **8** and P^tBu_3, gives a TON of 920 in the coupling of 4-chloroanisole with morpholine, whilst the use of palladium acetate and P^tBu_3 gives a TON of only 10 under the same conditions, highlighting the importance of choosing the correct palladium precursor [32]. The phosphite- and phosphinite-based palladacycles can also be exploited as palladium sources for the *in situ* formation of active catalysts from tris(*tert*-butyl)phosphine [47].

Carbene adducts of palladacycles also show excellent activity; complex **27** (R = 2,6-$^iPr_2C_6H_4$) can be used for the amination of both aryl chlorides and aryl triflates under mild conditions with good selectivity for mono-arylated products when mono-substituted amines are used as substrates [48].

Buchwald and coworkers have produced the palladacyclic complex **29** as a single source, air-stable, easily handled catalyst for the amination of aryl chlorides (Figure 9.12) [49]. The same group showed that formation of palladacycles *in situ* is actually *deleterious* to catalyst performance and that the use of ligand **30a**, which does not undergo C—H activation during catalytic amination reactions, gives much higher rates than **30b** which can [50].

9.8
What Are the True Active Catalysts?

Early on in the use of palladacycles in cross-coupling and amination reactions there was a debate as to whether they were reduced under reaction conditions to give palladium(0) complexes that then entered a classical Pd(0)/Pd(II) catalytic pathway or whether oxidative addition of the aryl halide substrate to the palladacycle occurred, leading to a Pd(II)/Pd(IV) manifold. The vast majority of the data produced subsequently points to facile reductive activation pathways for the palladacycles in all cases. In Suzuki and related cross-coupling reactions catalyzed by several distinct types of palladacycles, it has been demonstrated that this occurs by the activation pathway outlined above in Scheme 9.3, that is to say that the arylboronic acid attacks the palladium center to generate a palladacyclic complex with an additional aryl ligand. This aryl group and the orthometallated aryl function then undergo reductive elimination to yield a palladium(0) complex. One obvious consequence of this activation pathway is that the new biaryl-containing ligand formed should be traceable and that its presence helps confirm this type of activation pathway. This was first discussed by Louie and Hartwig when they showed that the palladacycle **1** undergoes reaction with $PhSnMe_3$ (Scheme 9.9) to yield the new ligand **31** [42]. Subsequently, this type of activation pathway has been supported by the presence of eliminated biaryl ligands for imine- [51], amine- [32], phosphinite- [8], and thioether-based palladacycles on reaction with arylboronic acids [37]. Therefore, it is tempting to conclude that *all* classes of palladacycles could undergo this process with ease in cross-coupling reactions.

Louie and Hartwig also showed that palladacycle **1** could react with diethylamine to give the amine adduct **32**, which is deprotonated in the presence of NaOtBu to give the palladium(0) complex **33** via β-elimination of the amide ligand (Scheme 9.9) [42]. The activities of **1, 32** and **33** are identical in the amination of 4-bromobenzophenone with *N*-methylaniline, indicating a common Pd(0) catalyst. When β-elimination is not possible, for instance when diphenylamine is used as the coupling partner, then complex **1** is inert, whereas the pre-formed Pd(0) catalyst **33** is active.

A similar hydridic β-elimination/reductive elimination sequence has been proposed by Nolan in the Suzuki coupling of aryl chlorides using the palladacycles **27** in aqueous isopropanol solution. In this case isopropoxide replaces the diethylamide ligand as the hydride source [40]. The P^tBu_3 adduct of complex **2a** shows

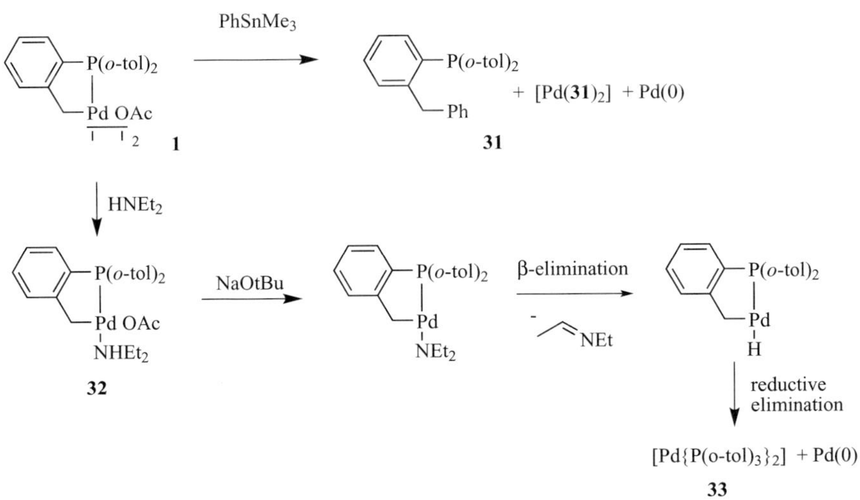

Scheme 9.9 Formation of palladium(0) species from palladacycle **1** in Stille cross-coupling and Buchwald–Hartwig amination reactions.

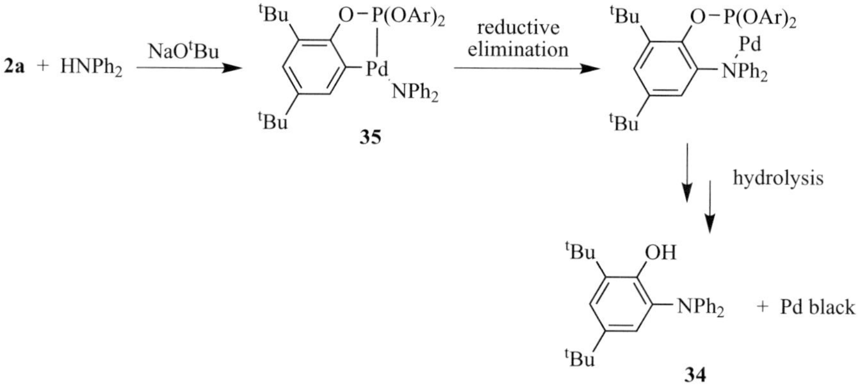

Scheme 9.10 Possible mechanism for the formation of palladium(0) in the coupling of diphenylamine with aryl chlorides catalyzed by mixtures of complex **2a** and P^tBu_3.

good activity in amination reactions with diphenylamine. In this instance β-elimination is not possible and an alternative reduction pathway must be invoked. GC/MS analysis of the reaction mixture indicates the presence of small amounts of the aminophenol **34**, which is presumably formed by the sequence shown in Scheme 9.10, and indicated that the putative amido intermediate **35** can itself undergo reductive elimination with the orthometallated aryl fragment [49].

Having established the existence of facile reduction pathways for palladacycles in Suzuki and related cross-coupling reactions and in Buchwald–Hartwig amination reactions, the question remains as to what are the true active catalysts. If the product of the reductive elimination is an insufficiently strong monodentate ligand to stabilize the resultant complex then rapid palladium aggregation can occur, particularly at higher catalyst loadings, which ultimately leads to early palladium precipitation. In this scenario the orthometallated ligand in the pre-catalyst plays a purely sacrificial role and over-elaboration of the palladacyclic architecture is counter-productive. The precipitation of palladium proceeds via a series of equilibria through the intermediate formation of soluble clusters then colloidal nanoparticles and finally precipitation of palladium black [52]. The soluble clusters and nanoparticles can be stabilized either by high dilution of the catalyst or by the addition of stabilizer. Thus, many of the high TONs observed with palladacyclic catalysts in the coupling of more easily activated aryl halides is probably a consequence of their low dilution, leading to retardation of palladium precipitation. As the ease of oxidative addition of the aryl halide decreases then competitive catalyst precipitation becomes more of an issue. This can be countered by stabilizing the clusters or colloids with added stabilizers, such as tetrabutylammonium bromide. Interestingly, in this regard, many of the catalysts described above require the addition of TBAB to function with aryl chloride substrates. Indeed it has been shown that even electronically deactivated substrates such as 4-chloroanisole can be coupled with phenylboronic acid using *only* palladium acetate in TBAB/water mixtures in the absence of any added ligands [53].

In the examples where trialkylphosphine or N-heterocyclic carbene adducts of palladacyclic complexes are used as the pre-catalysts, then the reductive activation leads to low-coordinate mono-phosphine or carbene complexes. Such mono-ligand species have been identified as being the likely active catalysts in a range of aryl chloride coupling reactions and in these instances the function of the palladacyclic pre-catalyst is to act as a very effective, conveniently handled source for their clean production.

9.9 Summary

In conclusion, palladacyclic pre-catalysts can show very high activity in Suzuki coupling, Buchwald–Hartwig amination and related coupling reactions. This is typically seen as high turnover numbers at elevated reaction temperatures but can be manifest as good activities under mild conditions, for instance at room temperature or in water. Their ease of handling and low air-sensitivity adds to their general appeal. It has become clear that the palladacycles are not the active catalysts in these reactions, but rather serve as precursors to highly active palladium(0) species. These are typically soluble clusters or colloids, often stabilized by surfactants such as TBAB or highly active, low-coordinate complexes of electron-rich phosphines or carbenes. In both cases the role of the palladacycle is essentially sacrificial. It is tempting to conclude, therefore, that palladacycles should play no further role in catalyst development since surely all they do is fall apart. However, the situation is more subtle than that; there is a fine balance at play between the relative rates of active catalyst production and decomposition and it seems likely that in the many examples were palladacyclic precursors outperform more classical palladium sources the balance seems to be tipped in favor of the former, particularly at high catalyst dilution. This, coupled with the observation that in some cases the reductively eliminated ligands can still have a crucial role to play in stabilizing catalyst resting state, will ensure that research in the field remains active for some time to come and may yet lead to new developments in catalyst design.

References

1 For selected reviews see: (a) Miyaura, N. and Suzuki, A. (1995) *Chemical Reviews*, **95**, 2457.
(b) Stanforth, S.P. (1998) *Tetrahedron*, **54**, 263.
(c) Suzuki, A. (1999) *Journal of Organometallic Chemistry*, **576**, 147.

2 For a discussion see: Grushin, V.V. and Alper, H. (1994) *Chemical Reviews*, **94**, 1047.

3 Beller, M., Fischer, H., Herrmann, W.A., *et al.* (1995) *Angewandte Chemie, International Edition in English*, **34**, 1848.

4 Herrmann, W.A., Broßmer, C., Öfele, K., *et al.* (1995) *Angewandte Chemie, International Edition in English*, **34**, 1844.

5 Wolfe, J.P., Singer, R.A., Yang, B.H. and Buchwald, S.L. (1999) *Journal of the American Chemical Society*, **121**, 9550.

6 Albisson, D.A., Bedford, R.B., Lawrence, S.E. and Scully, P.N. (1998) *Chemical Communications*, 2095.

7 Bedford, R.B. and Welch, S.L. (2001) *Chemical Communications*, 129.

8 Bedford, R.B., Hazelwood, S.L., Horton, P.N. and Hursthouse, M.B. (2003) *Dalton Transactions*, 4164.

9 Gibson, S., Foster, D.F., Eastham, G.R., *et al.* (2001) *Chemical Communications*, 779.

10 Roca, F.X. and Richards, C.J. (2003) *Chemical Communications*, 3002.

11 Friedlein, F.K., Kromm, K., Hampel, F. and Gladysz, J.A. (2006) *Chemistry–A European Journal*, **12**, 5267.

12 Bedford, R.B., Draper, S.M., Scully, P.N. and Welch, S.L. (2000) *New Journal of Chemistry*, **24**, 745.

13 Kimura, T. and Uozumi, Y. (2006) *Organometallics*, **25**, 4883.

14 Weissman, H. and Milstein, D. (1999) *Chemical Communications*, 1901.

15 Bedford, R.B., Cazin, C.S.J., Coles, S.J., *et al.* (2003) *Dalton Transactions*, 3350.

16 Thakur, V.V., Kumar, N.S.C.R. and Sudalai, A. (2004) *Tetrahedron Letters*, **45**, 2915.

17 Spencer, J., Sharratt, D.P., Dupont, J., *et al.* (2005) *Organometallics*, **24**, 5665.

18 Vicente, J., Abad, J.-A., López-Serrano, J., *et al.* (2005) *Organometallics*, **24**, 5044.

19 Bianchini, C., Lenoble, G., Oberhauser, W., *et al.* (2005) *European Journal of Inorganic Chemistry*, 4794.

20 Chen, M.-T., Huang, C.-A. and Chen, C.-T. (2006) *European Journal of Inorganic Chemistry*, 4642.

21 Wu, K.-M., Huang, C.-A., Peng, K.-F. and Chen, C.-T. (2005) *Tetrahedron*, **61**, 9679.

22 Alacid, E., Alonso, D.A., Botella, L., *et al.* (2006) *The Chemical Record*, **6**, 117.

23 Alonso, D.A., Nájera, C. and Pacheco, M.C. (2002) *The Journal of Organic Chemistry*, **67**, 5588.

24 Botella, L. and Nájera, C. (2002) *Angewandte Chemie, International Edition*, **41**, 179.

25 Botella, L. and Nájera, C. (2002) *Journal of Organometallic Chemistry*, **663**, 46.

26 Chen, C.-L., Liu, Y.-H., Peng, S.-M. and Liu, S.-T. (2005) *Tetrahedron Letters*, **46**, 521.

27 Nagy, T.Z., Csámpai, A. and Kotschy, A. (2005) *Tetrahedron*, **61**, 9767.

28 Churruca, F., SanMartin, R., Tellitu, I. and Domínguez, E. (2005) *Synlett*, 3116.

29 Zim, D., Gruber, A.S., Ebeling, G., *et al.* (2000) *Organic Letters*, **2**, 2881.

30 Xiong, Z., Wang, N., Dai, M., *et al.* (2004) *Organic Letters*, **6**, 3339.

31 Bedford, R.B. and Cazin, C.S.J. (2001) *Chemical Communications*, 1540.

32 Bedford, R.B., Cazin, C.S.J., Coles, S.J., *et al.* (2003) *Organometallics*, **22**, 987.

33 Schnyder, A., Indolese, A.F., Studer, M. and Blaser, H.-U. (2002) *Angewandte Chemie, International Edition*, **41**, 3668.

34 Xu, C., Gong, J.-F., Yue, S.-F., *et al.* (2006) *Dalton Transactions*, 4730.

35 Rosa, G.R., Ebeling, G., Dupont, J. and Monteiro, A.L. (2003) *Synthesis*, 2894.

36 Bedford, R.B., Cazin, C.S.J., Hursthouse, M.B., *et al.* (2004) *Dalton Transactions*, 3864.

37 Bedford, R.B., Cazin, C.S.J. and Hazelwood, S.L. (2002) *Angewandte Chemie, International Edition*, **41**, 4120.

38 Bedford, R.B., Hazelwood, S.L. and Limmert, M.E. (2002) *Chemical Communications*, 2610.

39 Navarro, O., Kelly, R.A. and Nolan, S.P. (2003) *Journal of the American Chemical Society*, **125**, 16194.

40 Navarro, O., Marion, N., Oonishi, Y., *et al.* (2006) *The Journal of Organic Chemistry*, **71**, 685.

41 Bedford, R.B., Betham, M., Blake, M.E., *et al.* (2005) *Dalton Transactions*, 2774.

42 Louie, J. and Hartwig, J.F. (1996) *Angewandte Chemie, International Edition*, **35**, 2359.

43 Herrman, W.A., Böhm, V.P.W. and Reisinger, C.-P. (1999) *Journal of Organometallic Chemistry*, **576**, 23.

44 Alonso, D.A., Nájera, C. and Pacheco, M.C. (2000) *Organic Letters*, **2**, 1823.

45 Bedford, R.B., Cazin, C.S.J. and Hazelwood, S.L. (2002) *Chemical Communications*, 2608.

46 Beller, M., Riermeier, T.H., Reisinger, C.-P. and Herrmann, W.A. (1997) *Tetrahedron Letters*, **38**, 2073.

47 Bedford, R.B. and Blake, M.E. (2003) *Advanced Synthesis and Catalysis*, **345**, 1107.

48 Viciu, M.S., Kelly, R.A., Stevens, E.D., *et al.* (2003) *Organic Letters*, **5**, 1479.

49 Zim, D. and Buchwald, S.L. (2003) *Organic Letters*, **5**, 2413.

50 Strieter, E.R. and Buchwald, S.L. (2006) *Angewandte Chemie, International Edition*, **45**, 925.

51 Bedford, R.B., Cazin, C.S.J., Hursthouse, M.B., *et al.* (2001) *Journal of Organometallic Chemistry*, **633**, 173.

52 For an excellent overview of nanoparticles in coupling see: de Vries, J.G. (2006) *Dalton Transactions*, 421.

53 Bedford, R.B., Blake, M.E., Butts, C.P. and Holder, D. (2003) *Chemical Communications*, 466.

10
Other Uses of Palladacycles in Synthesis

John Spencer

10.1
Introduction

Besides Heck, Suzuki and related transformations, palladacycles have applications in many areas of organic synthesis [1]. The present chapter describes such miscellaneous uses of palladacycles, including their employment as chiral catalysts for C—C bond-forming reactions such as those of the aldol-type and sigmatropic rearrangements or Heck-type reactions, cyclopropanations and related C—C bond formations. The stereochemical outcome of these reactions ranges from product that is racemic, despite the employment of a homochiral non-racemic palladacycle, to product that is highly enriched in one enantiomeric form. The mechanistic rationale for these reactions will be presented hereafter. In the latter part of this chapter, the role of palladacycles as intermediates in synthetically useful oxidations will be presented. Many of these reactions employ catalytic C—H activations and enable the formation of oxygenated or halogenated products with high degrees of chemo and regioselectivities and show great potential for the elaboration of molecules of high value to the medicinal and fine chemical industries.

10.2
Chiral Palladacycles in Aldol and Related Transformations

When using a chiral palladacycle for asymmetric synthesis, examples of which are given in Scheme 10.1, two distinct outcomes can be observed, depending on whether the palladacycle acts as a chiral Lewis acid, where highly enantioselective reactions are often observed, or if there is a prior change in Pd oxidation state. In the second case, such as in attempted enantioselective Heck-type reactions employing homochiral palladacycles [Pd(II)-containing precatalysts and precursors to Pd(0)], racemic product is usually obtained since the generated Pd(0) is no longer attached to the chiral ligand. Hence, Heck reactions [2], hydroarylations [3] and cyclopropanations [4] catalyzed by homochiral palladacycles **1–4** respectively, yield racemic product (Scheme 10.1).

Palladacycles: Synthesis, Characterization and Applications. Edited by Jairton Dupont and Michel Pfeffer

ISBN: 978-3-527-31781-3

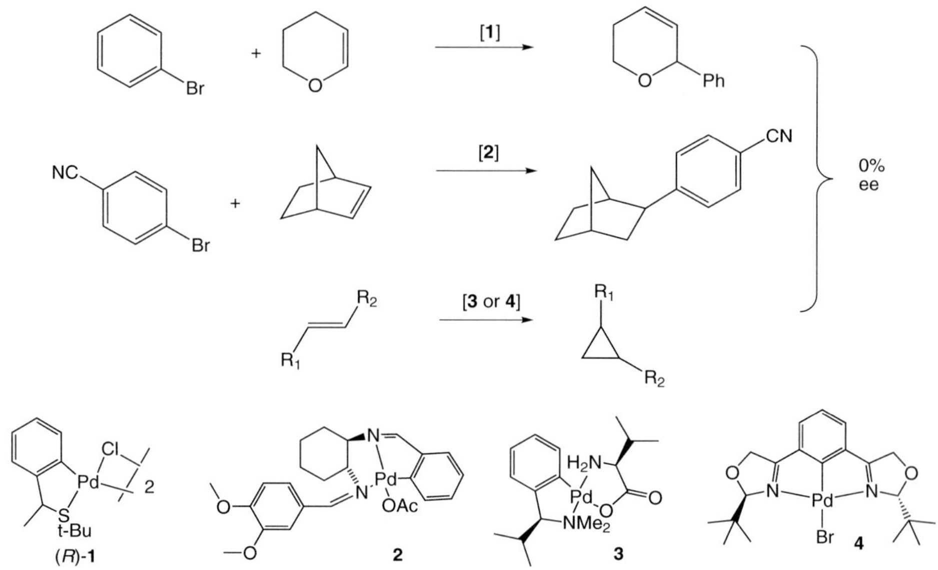

Scheme 10.1 Poor enantioselectivity in Pd-catalyzed C—C bond formations.

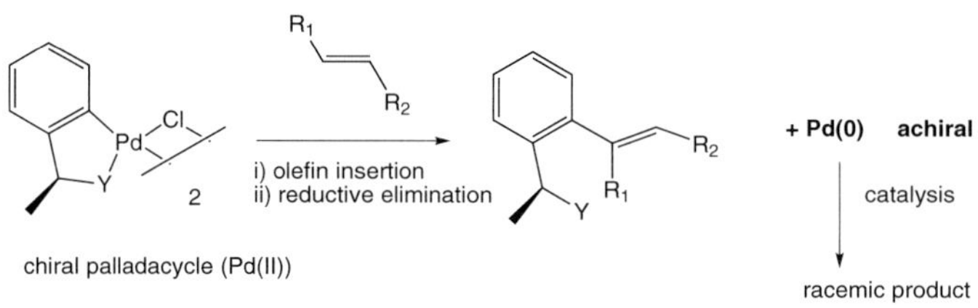

Scheme 10.2 A Heck reaction yielding racemic product catalyzed by a homochiral palladacycle.

In the case of the Heck reaction, the palladacycle, upon activation (olefin insertion in a Pd–C bond followed by β elimination), gives rise to (achiral) Pd(0), which is no longer bound to the chiral ligand (Scheme 10.2).

10.3 Catalytic Allylic Rearrangements

Palladacycles, such as the planar chiral **5–8** (Figure 10.1), have been successfully employed in enantioselective catalysis as in [3,3]-sigmatropic rearrangements of allyl amidates and trichloroacetimidates. Here, the palladacycle acts as a chiral Lewis acid and retains its stereochemical integrity in the Pd(II) state [5] (Scheme 10.3).

Intramolecular aminopalladations, with e.e.s >90%, were observed with oxazoline-containing palladacycles such as **5**, opening up the possibility of

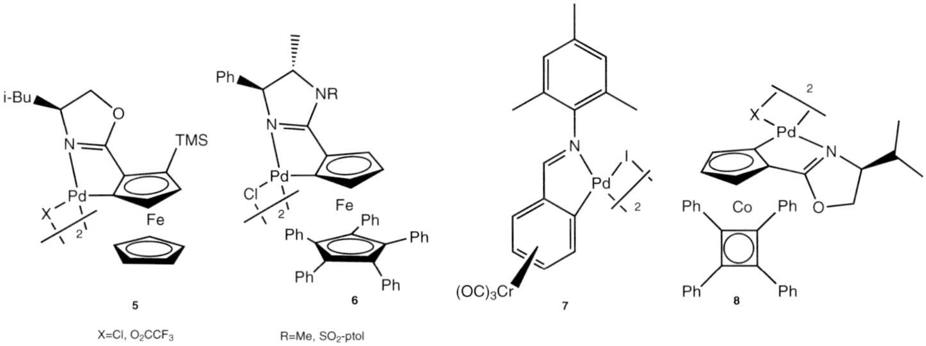

Figure 10.1 Planar chiral palladacycles used in enantioselective allylic rearrangements.

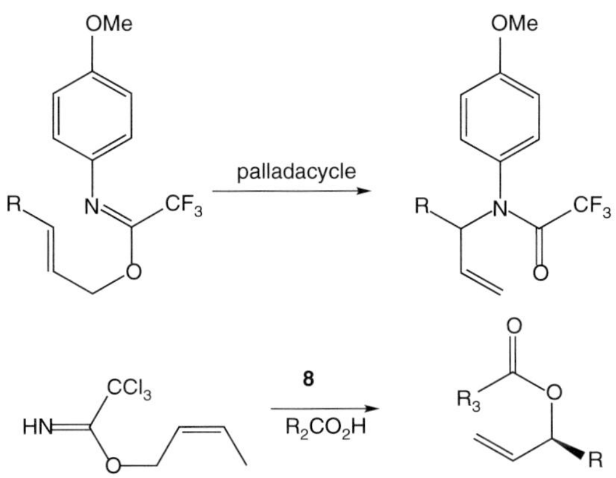

Scheme 10.3 Allylic rearrangements catalyzed by palladacycles.

obtaining high value end-products such as enantioenriched heterocycles [6] (Scheme 10.4).

10.4
Catalytic C—C Bond-Forming Reactions

Pincer-type palladacycles have also been employed as chiral Lewis acids for aldol and Michael-type chemistry in the formation of heterocyclic and acyclic compounds. Moderate to good degrees of success in terms of asymmetric induction were observed. The 34% e.e. observed for the formation of the acyclic Michael product is in fact an encouraging result considering that a quaternary carbon centre has been formed (Scheme 10.5) [7].

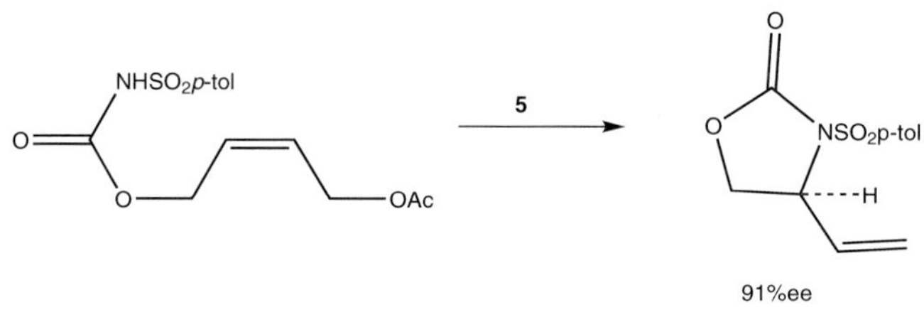

Scheme 10.4 Oxazoline-containing palladacycle catalysts for enantioselective intramolecular aminopalladations.

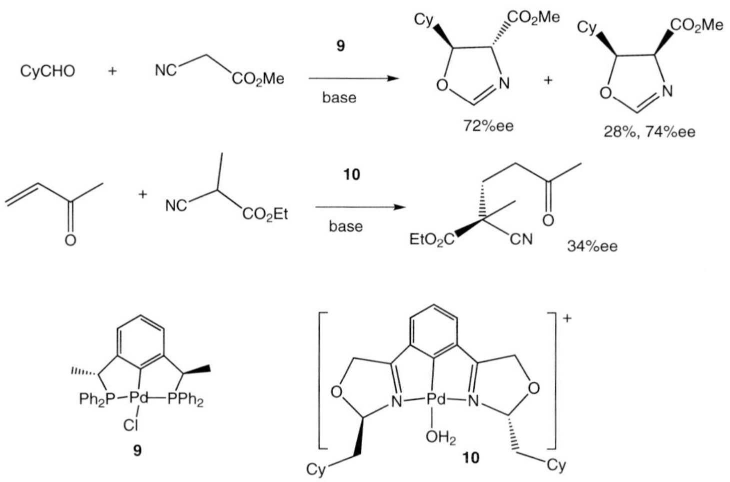

Scheme 10.5 Asymmetric aldol and Michael reactions catalyzed by palladacycles.

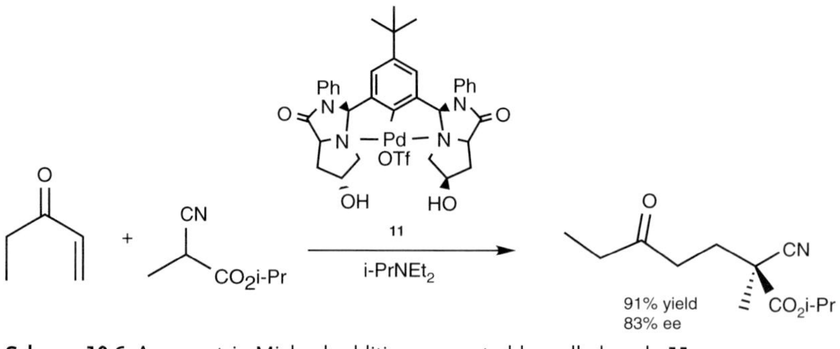

Scheme 10.6 Asymmetric Michael addition promoted by palladacycle **11**.

A related Michael addition employed the elaborate pincer complex **11** as catalyst, whose synthesis was surprisingly straightforward. Yields for the formation of the quaternary carbon centre were impressive, coupled with excellent enantioselectivities that attained 83% (Scheme 10.6) [8].

PCP pincer complexes catalyze the addition of allylstannanes to aldehydes and imines. The robust pincer system ensures that the allylic intermediate is

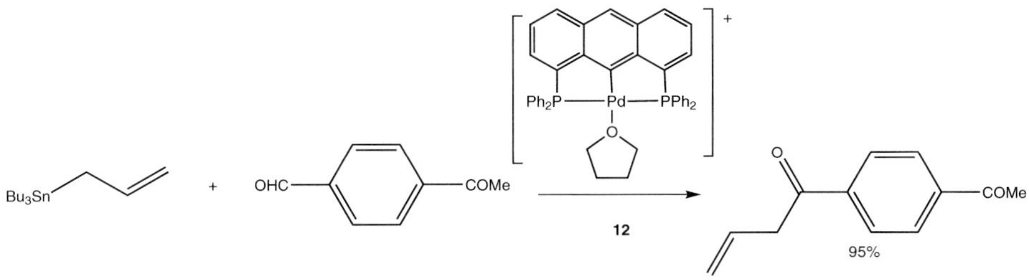

Scheme 10.7 Addition of allylstannanes to aldehydes.

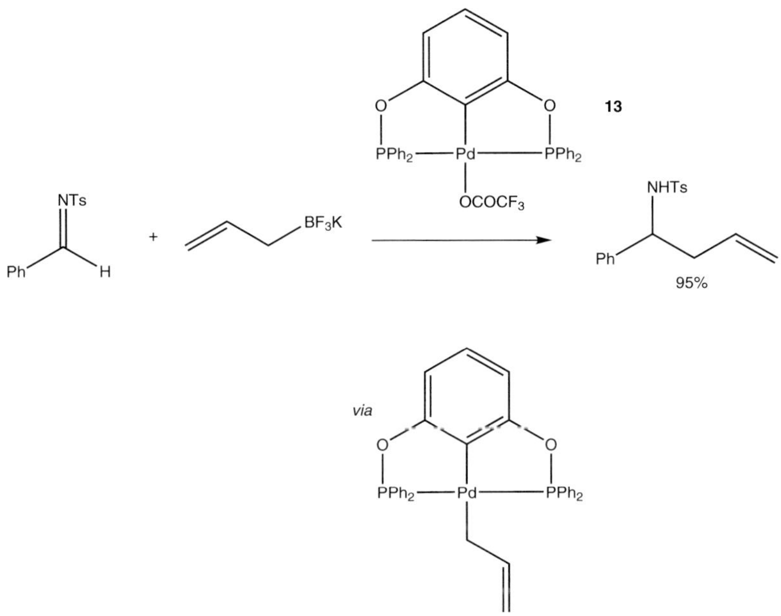

Scheme 10.8 Allylation of tosylimines promoted by palladacycle **13**.

nucleophilic, η^1-bound to the metal and monomeric; the example shown in Scheme 10.7 illustrates the high chemoselectivity of this process since it differentiates an aldehyde functionality from a ketone [9].

The toxic tin substrates described in the former scheme can be replaced with more eco-friendly allylborates. For example, the palladium PCP pincer complex **13** catalyzes the allylation of tosylimines with borate salts, yielding racemic product. The purported η^1-allylic intermediate is shown for the mechanistic rationalization of this reaction [10] (Scheme 10.8).

A natural extension to this nucleophilic allylic addition chemistry would obviously aim to achieve enantioselective versions of the above reactions. This is feasible since such processes are thought to involve a monomeric Pd(II) complex as opposed to (achiral) Pd(0). Hence, a chiral variant of this reaction has been

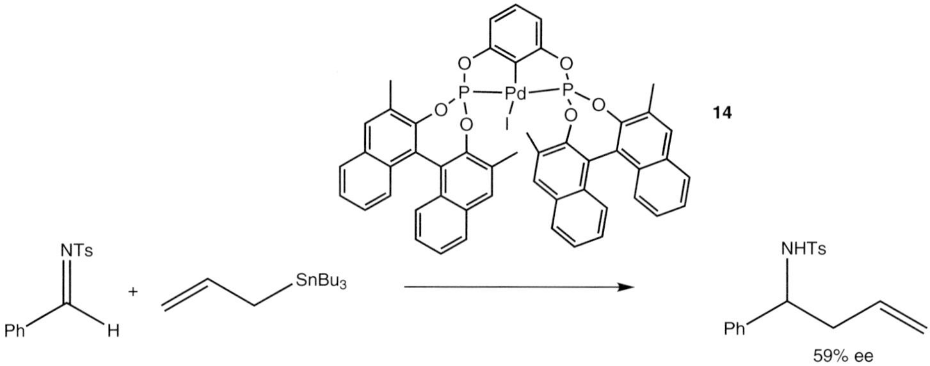

Scheme 10.9 A chiral allylation reaction catalyzed by a chiral palladacycle.

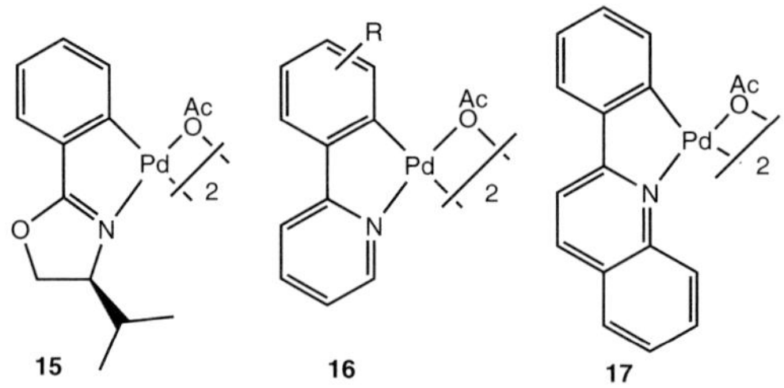

Figure 10.2 Palladacycles employed as catalyst precursors to oxidation reactions.

disclosed with acceptable e.e.s when employing the BINOL-substituted palladacycle **14** [11] (Scheme 10.9).

10.5 Oxidations Involving Palladacycles

Palladium-based catalysts can be used for environmentally friendly oxidations employing oxygen at atmospheric pressure. Palladacycles derived from oxazoles, 2-phenylpyridine and quinoline (**15–17** respectively, Figure 10.2) have been reported as effective catalysts for the oxidation of primary and secondary alcohols to the corresponding aldehydes and ketones under an atmospheric pressure of air without the addition of any other re-oxidants [12].

We now discuss reactions where palladacycles are likely to be (non-isolated) intermediates in oxidation processes in order to explain the very high degree of regioselectivities observed. Regioselective Pd-catalyzed C–H bond activation can be used for the conversion of Csp^2 and Csp^3 C–H bonds into esters, ethers and aryl-halides. The reaction proceeds via a cyclopalladated intermediate (Scheme 10.10) [13].

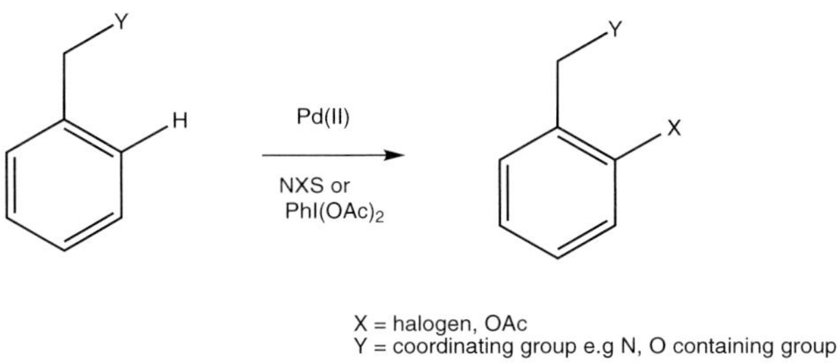

Scheme 10.10 Orthometallation/halogenation of arenes via palladacycles.

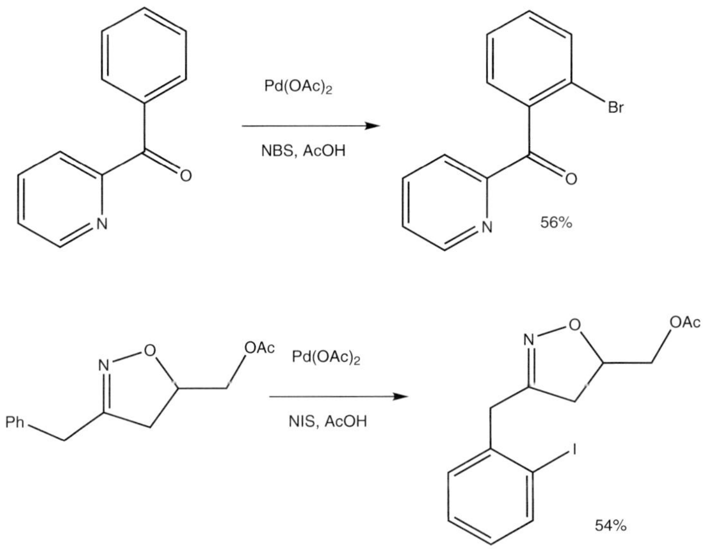

Scheme 10.11 Catalytic orthohalogenations.

This is a powerful method for the halogenation of various aryl-containing molecules (Scheme 10.11) [14].

In some cases the direction of the halogenation using palladium catalysis is different to that using conventional chemistry (Scheme 10.12) [14, 15]. The ortho-substituted product resulting from palladium catalysis strongly suggests the involvement of a palladacycle intermediate.

Acetanilides can also be ortho-halogenated using similar chemistry although copper salts were also required (Scheme 10.13) [15, 16].

The fluorination of carbon–hydrogen bonds has also been disclosed and both aliphatic and aromatic C—H bonds can be substituted. The fluorination of aromatic C—H bonds may have applications in pharmaceutical chemistry, where a C—F bond is often used for blocking metabolism. For accelerated synthesis, microwave-mediated reactions were also performed (Scheme 10.14) [17].

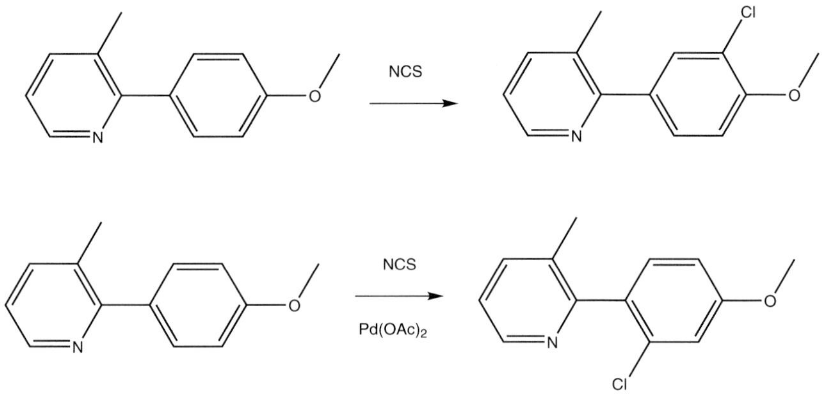

Scheme 10.12 Palladium-directed versus classical electrophilic chlorination.

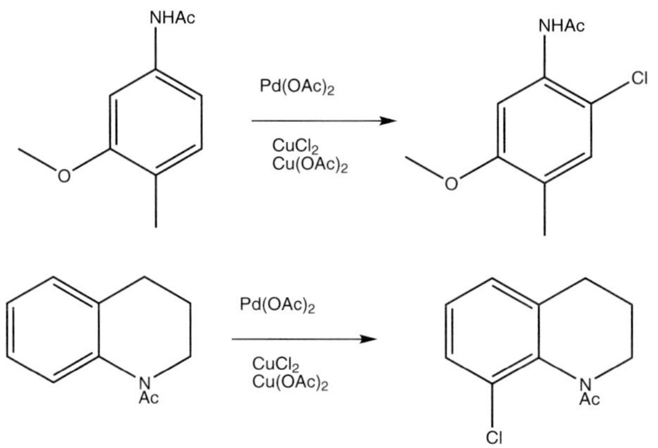

Scheme 10.13 Catalytic orthohalogenations of acetanilides.

The oxygenation of unactivated sp^3 C—H bonds of various O-methyl oximes can occur via five-membered ring palladacycles (Scheme 10.15) [18].

Related sp^3 and sp^2 C—H oxidations have also been reported using $PhI(OAc)_2$ as the oxidant and source of acetate. High levels of ortho regioselectivity were observed, and in certain cases double oxidations are observed, which appear to be avoided when a meta substituent is present (Scheme 10.16) [19]. For improved work-up procedures, supported oxidants can also be used [20].

Oxone can also be used as a cleaner, environmentally safer oxidant, and both acetate and ether products can be synthesized, the latter when methanol is used as solvent (Scheme 10.17) [21].

A mechanism involving palladacycle intermediates and a Pd(II)–Pd(IV) catalytic cycle has been proposed (Scheme 10.18) [17].

Indeed, Pd(IV) complexes have been formed and shown to undergo reductive elimination in impressive NMR experiments [22] (Scheme 10.19).

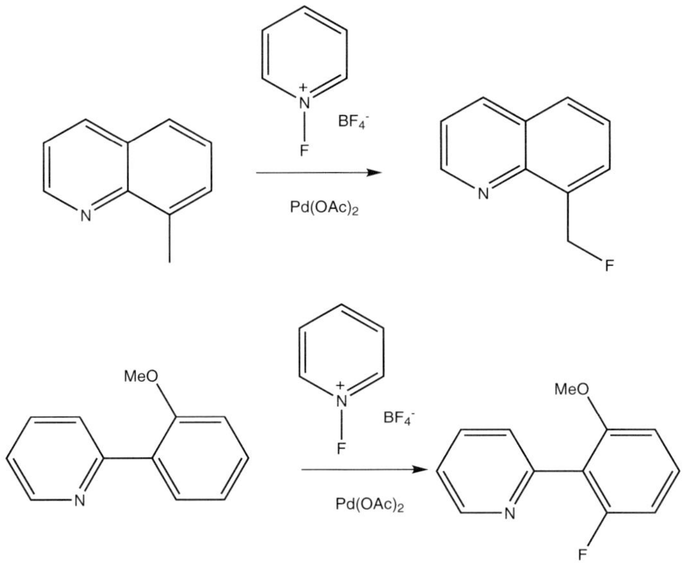

Scheme 10.14 Catalytic fluorination reactions.

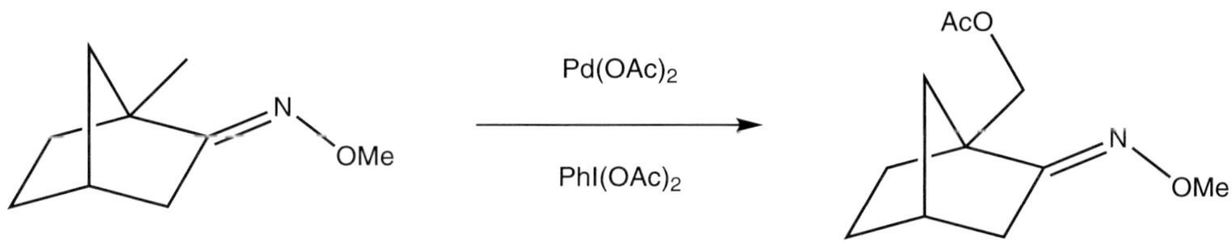

Scheme 10.15 Catalytic sp^3 C—H bond activation/oxidation.

10.6 Conclusion

Palladacycles have numerous applications beyond the usual scope of Heck and Suzuki and related chemistries. A greater understanding of the mechanisms involved in palladacycle-mediated reactions has led to the development of reactions that have wide applications in organic synthesis. For example, we now know that many redox processes involving palladacycles will lead to racemic products. Hence, it is futile to prepare a chiral palladacycle for an attempted asymmetric Heck-type process, whereby Pd(0) is generated *in situ*, as racemic product will be formed. However, chiral palladacycles are very important catalysts for enantioselective allylic rearrangements, allylic additions and aldol chemistry, since they act as Pd(II) Lewis acids and no redox process occurs. It is now possible to selectively oxidize/halogenate, even fluorinate, aromatic and aliphatic C—H bonds, be they sp^2 or sp^3, using palladacycle catalysts. Palladacycles will no doubt continue to have

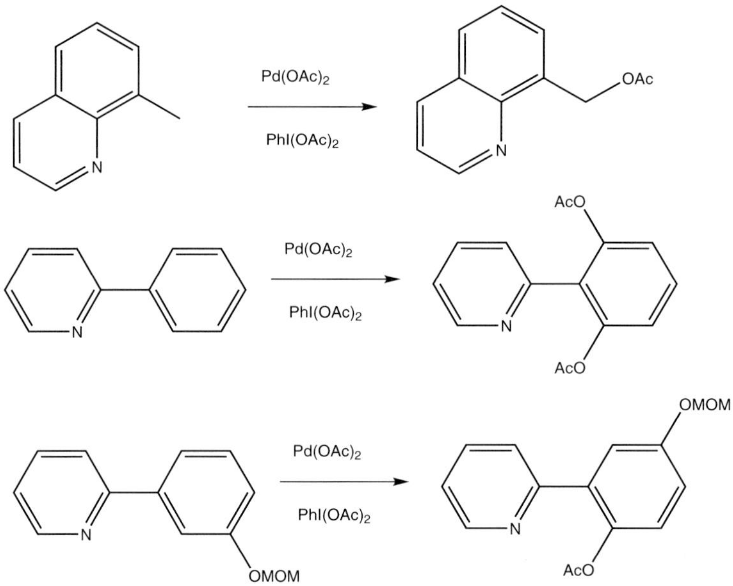

Scheme 10.16 Catalytic sp^3 and sp^2 C—H bond activation/oxidation.

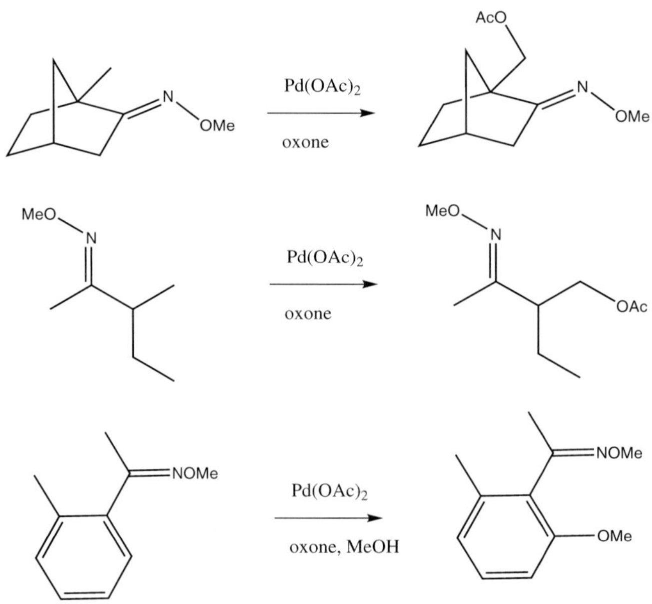

Scheme 10.17 Oxone as oxidant in Pd-catalyzed oxidations.

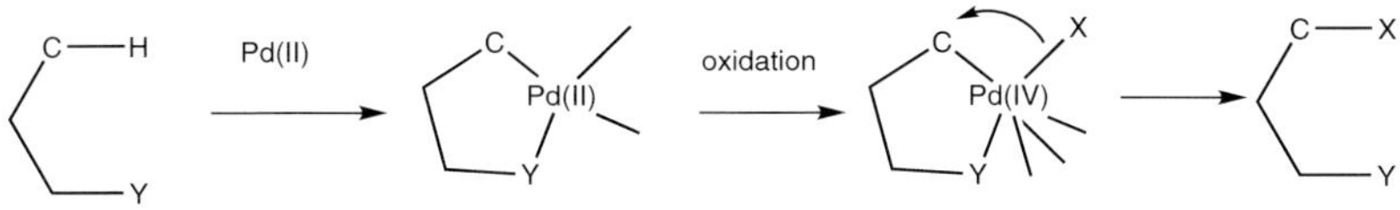

Scheme 10.18 Proposed palladacycle and Pd(IV) intermediates in oxidation reactions.

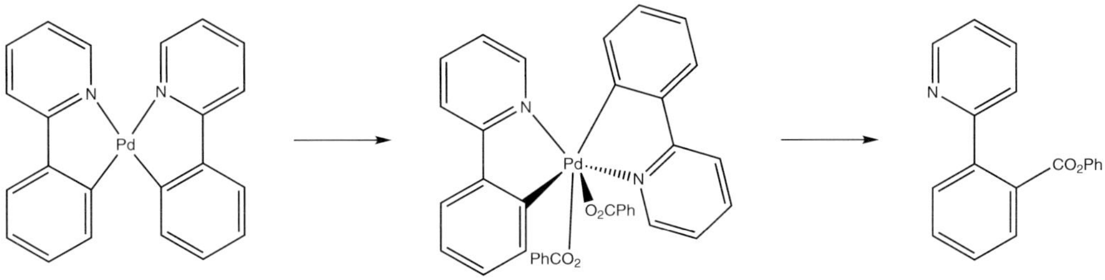

Scheme 10.19 An NMR experiment to illustrate the reductive elimination process.

a huge impact in atom economical synthesis and the area of C–H activation chemistry will continue to flourish.

References

1 Dupont, J., Consorti, C.S. and Spencer, J. (2005) *Chemical Reviews*, **105**, 2527.

2 Dupont, J., Gruber, A.S., Fonsesca, G.S., *et al.* (2001) *Organometallics*, **20**, 171.

3 Bravo, J., Cativela, C., Navarro, R. and Urriolabeitia, E.P. (2002) *Journal of Organometallic Chemistry*, **650**, 157.

4 (a) Denmark, S.E., Stavenger, R.A., Faucher, A.-M. and Edwards, J.P. (1997) *The Journal of Organic Chemistry*, **62**, 3375–89.
(b) Navarro, R., Urriolabeitia, E.P., Cativiela, C., *et al.* (1996) *Journal of Molecular Catalysis A*, **105**, 111.

5 (a) Moyano, A., Rosol, M., Moreno, R.M., *et al.* (2005) *Angewandte Chemie, International Edition*, **44**, 1865.
(b) Weiss, M.E., Fischer, D.F., Xin, Z.Q., *et al.* (2006) *Angewandte Chemie, International Edition*, **45**, 5694.
(c) Jautze, S., Seiler, P. and Peters, R. (2007) *Angewandte Chemie, International Edition*, **46**, 1260.
(d) Hollis, T.K. and Overman, L.E. (1999) *Journal of Organometallic Chemistry*, **576**, 290.
(e) Burke, B.J. and Overman, L.E. (2004) *Journal of the American Chemical Society*, **126**, 16820.
(f) Overman, L.E., Owen, C.E., Pavan, M.M. and Richards, C.J. (2003) *Organic Letters*, **5**, 1809.
(g) Prasad, R.S., Anderson, C.E., Richards, C.J. and Overman, L.E. (2005) *Organometallics*, **24**, 77.
(h) Kirsch, S.F., Overman, L.E. and Watson, M.P. (2004) *The Journal of Organic Chemistry*, **69**, 8101.
(i) Kang, J., *et al.* (2002) *Tetrahedson Letters*, **43**, 9509.

6 Overman, L.E. and Remarchuk, T.P. (2002) *Journal of the American Chemical Society*, **124**, 12–3.

7 (a) Longmire, J.M., Zhang, X. and Shang, M. (1998) *Organometallics*, **17**, 4374.
(b) Albrecht, M., Kocks, B.M., Spek, A.L. and van Koten, G. (2001) *Journal of Organometallic Chemistry*, **624**, 271.
(c) Stark, M.A. and Richards, C.J. (1997) *Tetrahedron Letters*, **38**, 5881.

(d) Stark, M.A., Jones, C.J. and Richards, C.J. (2000) *Organometallics*, **19**, 1282.

8 Takenaka, K. and Uozumi, Y. (2004) *Organic Letters*, **6**, 1833.

9 Solin, N., Kjellgren, J. and Szabó, K.J. (2004) *Journal of the American Chemical Society*, **126**, 7026.

10 (a) Solin, N., Wallner, O.A. and Szabó, K.J. (2005) *Organic Letters*, **7**, 689.
(b) Szabó, K.J. (2006) *Synlett*, 811.

11 Wallner, O.A., Olsson, V.J., Eriksson, L. and Szabó, K.J. (2006) *Inorganica Chimica Acta*, **359**, 1767.

12 (a) Hallman, K. and Moberg, C. (2001) *Advanced Synthesis and Catalysis*, **343**, 260.
(b) Paavola, S., Zetterberg, K., Privalov, T., *et al.* (2004) *Advanced Synthesis and Catalysis*, **346**, 237.
(c) Privalov, T., Linde, C., Zetterberg, K. and Moberg, C. (2005) *Organometallics*, **24**, 885.

13 Desai, L.V., Hull, K.L. and Sanford, M.S. (2004) *Journal of the American Chemical Society*, **126**, 9542–3.

14 Kalyani, D., Dick, A.R., Anani, W.Q. and Sanford, M.S. (2006) *Organic Letters*, **8**, 2523.

15 (a) Kalyani, D., Dick, A.R., Anani, W.Q. and Sanford, M.S. (2006). *Tetrahedron*, **62**, 11483–98.
(b) Dick, A.R. and Sanford, M.S. (2006) *Tetrahedron*, **62**, 2439–63.

16 Wan, X., Ma, Z., Li, B., *et al.* (2006) *Journal of the American Chemical Society*, **128**, 7416.

17 Hull, K.L., Anani, W.Q. and Sanford, M.S. (2006) *Journal of the American Chemical Society*, **128**, 7134.

18 Desai, L.V., Hull, K.L. and Sanford, M.S. (2004) *Journal of the American Chemical Society*, **126**, 9542.

19 (a) Dick, A.R., Hull, K.L. and Sanford, M.S. (2004) *Journal of the American Chemical Society*, **126**, 2300.
(b) Kalyani, D. and Sanford, M.S. (2005) *Organic Letters*, **7**, 4149.

20 Kalberer, E.W., Whitfield, S.R. and Sanford, M.S. (2006) *Journal of Molecular Catalysis A*, **251**, 108.

21 Desai, L.V., Malik, H.A. and Sanford, M.S. (2006) *Organic Letters*, **8**, 1141.

22 (a) Dick, A.R., Kampf, J.W. and Sanford, M.S. (2005) *Journal of the American Chemical Society*, **127**, 12790.
(b) Dick, A.R., Kampf, J.W. and Sanford, M.S. (2005) *Organometallics*, **24**, 482.

11
Liquid Crystalline Ortho-Palladated Complexes

Bertrand Donnio and Duncan W. Bruce

11.1
Introduction

This chapter gives an account of orthopalladated complexes that form liquid crystal mesophases and describes the way these complexes are organized within the phases. The general field of thermotropic metallomesogens began to develop rapidly in the mid-1980s, and has been covered by several reviews [1] and a book [2], which will inevitably provide the interested reader with different and, in many cases, more detailed perspectives.

11.2
Liquid Crystals

Matter is often considered as existing in one of three states – solid, liquid and gas. Yet there exists a state of matter between the solid and liquid states that possesses properties reminiscent of each, so that like a crystal it has order (molecules possess three-dimensional orientational and translational ordering), while like a liquid it is fluid. The combination of order and mobility results in anisotropy of the physical properties, which is the basis of their widespread applications. The liquid-crystalline (LC) state is generated as a function of temperature (thermotropic), that is a compound passes between the crystal, LC and liquid states by a progressive loss of order on heating, or by a solvent (lyotropic) – when the phase transitions are driven by the concentration of a LC in a solvent. The term amphotropic applies when mesomorphism can be induced independently by both methods [3].

Since these initial discoveries, liquid crystals have become a major, multidisciplinary field of research that has impacted on society in a major way following the discovery of the cyanobiphenyl liquid crystals in the early 1970s [4]. Their utilization in the twisted nematic mode display device [5] led to the birth of the liquid crystal display industry, which is now worth over 45 billion annum^{-1} and is ever increasing. This multidisciplinarity is evidenced by the pervasive nature of liquid crystal science, extending from biology (the materials in cell membranes are liquid

Palladacycles: Synthesis, Characterization and Applications. Edited by Jairton Dupont and Michel Pfeffer

ISBN: 978-3-527-31781-3

crystals) [6], through chemistry to physics, mathematics and electronic engineering [7]. It is a constantly expanding and developing field with new applications challenging the synthetic chemist and new phase types prompting physicists to see how they may be harnessed. Thus, despite being regarded as a "mature" discipline, liquid crystal research has probably never been so vibrant, and one of the significant developments in the last 20 years or so has been that of metallomesogens–a term coined in the first review of the subject by Giroud-Godquin and Maitlis [8].

11.2.1
Thermotropic Liquid Crystals

Before developing further, it is necessary define the vocabulary used in thermotropic LCs. Thus, a liquid-crystalline material is referred to as a "mesogen" and is said to exhibit "mesomorphism" (or a "mesophase"); a "liquid-crystal-like" molecule is known as "mesogenic", although it is not necessarily mesomorphic. The liquid state is referred to as the "isotropic" liquid. The temperature at which a material passes from the solid state into a mesophase is referred to as the "melting point", while the temperature at which the mesophase transforms into an isotropic fluid is called the "clearing point". Enantiotropic mesophases are found in a reversible phase sequence, whilst "monotropic" mesophases appear only on cooling (metastable phases).

Thermotropic LC mesophases are formed by molecules endowed with specific structural criteria. Usually, a mesogen has a molecular structure composed of at least two portions of contrasting structural and/or chemical character (i.e. an "amphipathic" molecule), for example a rigid anisotropic moiety attached to peripheral, flexible segments. The amphipathic character is at the origin of the multi-steps melting process [9] and phase formation and stabilization is driven and results in most cases from a phase separation process [10]. The mesophase is further stabilized by intermolecular (non-covalent) interactions (dipolar, electrostatic, hydrogen bonding, van der Waals), anisotropic dispersion forces that result from the anisometry of the molecules and repulsive forces that result from its amphipathic character [8].

For a long time, the classification of thermotropic liquid crystalline mesophases was based on the essential shape of the molecules [11]; however, as will become evident, this taxonomy does not address all issues of physical properties and phase behavior [12]. The two most common types of LC molecular structure are the rod-like (calamitic) and the disc-like (discotic). Thus, rod-like (Figure 11.1**a**: calamitic) molecules are much longer than they are broad and, hence, possess one unique, long axis. By comparison, disc-like molecules (Figure 11.1**e**: discotic) are rather flat and hence possess one unique, short axis. The rigid part (for both rods and discs) consists in the specific arrangements (e.g. linear for rods and planar for discs) of phenyl and/or heterocyclic (unsaturated or not, rigid or flexible) rings, linked together through σ, double, triple bonds, or other functional linkers such

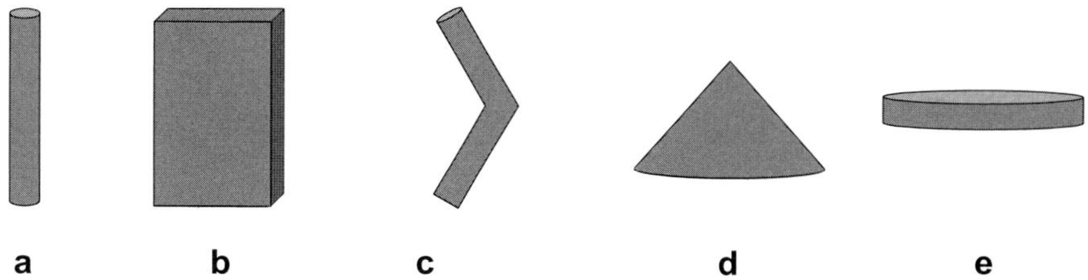

Figure 11.1 Sketches of the five main molecular anisotropies found in thermotropic LC materials: (a) calamitic, (b) sanidic, (c) bent-core, (d) bowlic and (e) discotic motifs.

as –COO–, –N=CH–, –N=N– that maintain the overall anisotropy, whereas the flexible moieties, often hydrocarbon chains, are connected at one or several extremities of the rigid part. Additionally, dipolar groups such as F, CN, NH_2 and NO_2 can be incorporated within the anisometric part subtly to modify some physical properties [7, 13]. The need to control macroscopic structures further led to the exploration of mesogens with different molecular shapes (Figure 11.1) [14]. A significant area of research deals with the so-called bent-core mesogens where a circa 120° bend angle in the molecule is effectively a requirement [15]. Other systems include sanidic mesogens (lath-like structure) [16], bowlic LCs (disk or pseudo-disc-like cores with reduced symmetry) [17], polycatenar mesogens (LCs bearing more than two peripheral chains) [18], polyphilic block molecules (small ABC-like block molecules) [19], oligomers and [20] dendrimers [21], polymers [7], amphiphiles [22], carbohydrates [23] and supramolecular hydrogen-bonded LCs (Figure 11.1) [24].

Despite the great variety of molecular shapes, thermotropic LC mesophases must be classified according to their symmetry. Mesophases are essentially subdivided into three main categories: the nematic, the smectic and the columnar phases (and their chiral modifications). Less commonly observed are the three-dimensional mesophases with cubic [25] and tetragonal symmetries [19b,c].

11.2.2
Nematic Phase

The nematic phase (N) has the simplest structure of all of the mesophases, is very fluid and is also the least ordered mesophase. The word nematic comes from the Greek *nematos* meaning thread-like–this arises from the observed optical texture of the phase between crossed polarizers. The N phase is characterized by one-dimensional orientational order of the molecules by virtue of correlations of the principal molecular axes, although note that the orientational order is not polar; there is no translational order within the N phase (Figure 11.2). This very fluid phase is commonly observed with calamitic LCs [7], whilst remaining more elusive in discotic materials (N_D) [26].

Figure 11.2 Schematic representations of the molecular arrangement in the nematic phases: (left) N from calamitics, (middle) N_D from discotics and (right) N_b; *n (m, l)* is the nematic director.

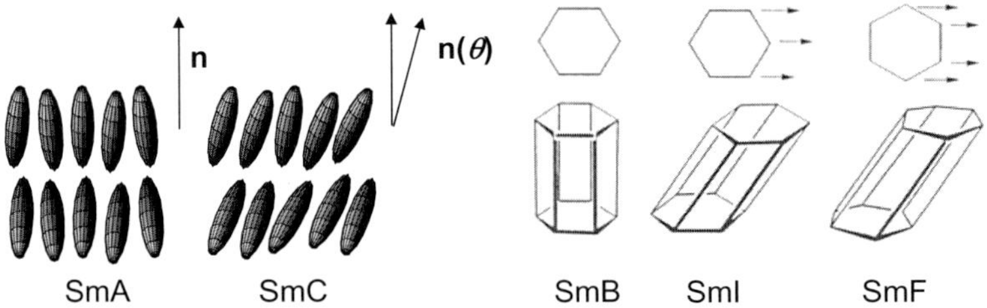

Figure 11.3 Sketches of the SmA, SmC, SmB, SmI and SmF phases (ellipsoids represent the rod mesogens).

The biaxial nematic phase (N_b) [27], has a D_{2h} symmetry compared to the $D_{\infty h}$ symmetry of its uniaxial counterpart, and is characterized by orientational correlations perpendicular to the principal director, *n* [28].

11.2.3 Smectic Phases

The true smectic phases, formed principally by calamitic mesogens (Figure 11.1a), consist of the superposition of equidistant molecular layers, and are characterized by orientational correlations of the principal axis and by partial translational ordering of the molecules within layers, but with no long-range in-plane positional order [29]. The simplest smectic phase is the smectic A phase (SmA), in which the long molecular axes are oriented on average in the same direction and parallel to the layer normal, but with the molecules loosely associated into layers (Figure 11.3). If the molecular director is tilted with respect to the layer normal, then the SmC phase is obtained (correlation of the tilt).

Hexatic smectic phases result from increasing the in-plane (short-range) positional and long-range bond orientational orders. In the SmB phase, the molecules sit at the nodes of a 2D hexagonal lattice. Two tilted variations of the SmB exist (Figure 11.3): the SmI phase (the mesogens are tilted towards a *vertex* of the hexagonal net) and the SmF phase (the mesogens are tilted towards the *edge* of the hexagonal net) [30]. In all cases, diffusion between the layers occurs readily and the phases are fluid.

The crystalline smectic phases are derived from the true smectics, and characterized by the appearance of inter-layer correlations and, in some cases, by the loss of molecular rotational freedom. Thus, the B, G and J phases are SmB, SmF and SmI phases, respectively, with inter-layer correlations, while the E, H and K phases are B, G and J phases that have lost rotational freedom. These phases possess considerable disorder and are therefore properly intermediate between both crystal and liquid states.

Smectic phases are also exhibited by sanidic and bent-shape molecules. As for bent-core systems, they self-organize into various smectic-like modifications, which are different from the true smectic phases, since various polar orderings within the layers, for example ferro-, ferri- and anti-ferroelectric [15], usually observed in chiral compounds, may be obtained. The origin of this polar order and macroscopic chirality is attributed to symmetry-breaking instabilities imposed by the polar-controlled packing, the tilt of the mesogens and the steric constraints of the intrinsic bent shape.

11.2.4 Columnar Mesophases

Columnar phases result from the stacking of disc-like molecules (or pseudo-circular molecular aggregates) into columns, which are packed parallel into 2D ordered lattices [31]. They are characterized by the symmetry of the side-to-side molecular arrangement of supramolecular columns. The common 2D lattices of the columnar phases shown in Figure 11.4, namely hexagonal (a), square (b), rectangular (c) and oblique (d), are represented as "aerial" views showing projections of the columns onto a two-dimensional plane; circles represent disks that are orthogonal within the columns, whereas ellipses represent disks that are tilted. Diffusion between and within the columns occurs readily and the phases are fluid. They are also exhibited by many non-discotic molecules such as polycatenars [18], polyphilic[16, 19] and bent-shape mesogens (polar and non-polar columnar phases) [15], with different processes of self-organization. When this 2D order is lost (e.g. by addition of a solvent), nematic arrays of infinite columnar stacks can be formed; the phase is referred to as a columnar nematic phase (N_{Col}).

11.2.5 Chiral Mesophases

Chiral modifications of N and smectic phases also exist for calamitic mesogens, for example N*, SmC*, SmI* and SmF*, either in a pure enantiomer or with a non-chiral compound doped with a small amount of chiral additive [32]. Owing to the packing constraints imposed by the materials being chiral, the director is forced to precess through the phase, describing a helix. Because of the low symmetry in the chiral smectic phases, the molecular dipoles align within the layers

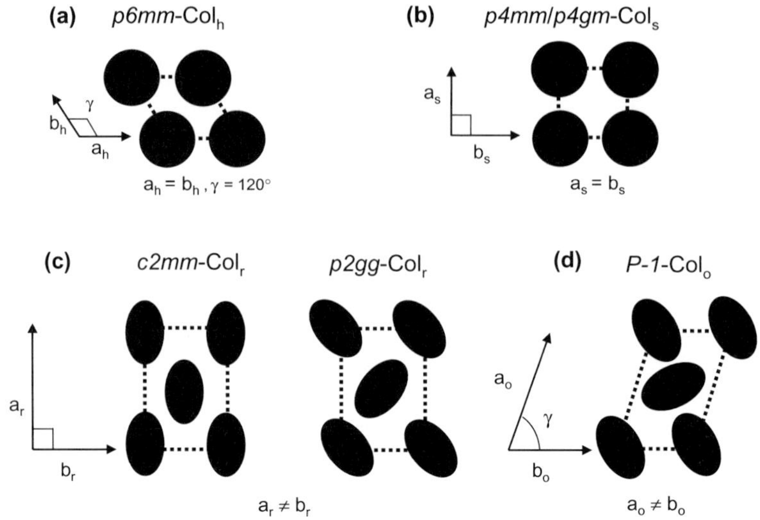

Figure 11.4 Representations of the lattices of the (a) hexagonal, (b) square, (c) rectangular, and (d) oblique columnar phases.

that are then ferroelectric. Chirality also gives rise to more exotic frustrated supramolecular structures such as the twist grain boundaries (TGB) phases [33] and the blue phases (BP) [34]. Ferroelectric columnar phases have also been claimed in discotic systems bearing chiral chains [35], and in bent-shaped and bowlic mesogens [7, 15].

11.3 Mesophase Characterization

Once a material is synthesized it is necessary to establish which mesophases it forms and at what temperatures the transitions occur. Two techniques are used routinely in all laboratories for this purpose, namely polarized optical microscopy and differential scanning calorimetry, and it is important that these are used in conjunction with one another. In addition, the technique of X-ray scattering is often used to give unequivocal phase identification when microscopy cannot so do, and also to provide additional insights into the structures adopted [1, 2].

11.4 Liquid Crystalline Ortho-Palladated Complexes

Ortho-palladated dinuclear and mononuclear complexes represent an important fraction of metallomesogens. Various types have been reported, and to facilitate the discussion and the description of the various systems the division of the subject matter in this chapter is by ligand type.

11.4.1 Ortho-Palladated Azobenzene Complexes

One of the earlier contributions to the development of metal-containing liquid crystal systems was the synthesis of the ortho-palladated complexes of mesogenic azobenzenes **(1, 2)** by Ghedini and coworkers, which represented the first systematic attempt to coordinate metals to known liquid crystal systems [36]. Initial studies [37] investigated the dipalladium complexes **1** [X = Cl; m = 1, 2; R = $C_4H_9CO_2$, $C_6H_{13}CO_2$, $CH_2{=}CH(CH_2)_8CO_2$] and the related mononuclear complexes **2** [X = Cl; L = PPh_3, pyridine (py), quinoline and aniline] [38]. With such non-symmetrical ligands, ortho-metallation was shown to occur in the more electron-rich ring. All dinuclear complexes **1** showed an enantiotropic N phase at elevated temperature (165–215 °C), depending on chain-length; the free ligands showed a low-temperature N phase (<125 °C). The role of the bridging halide was investigated (**1**: X = Cl, Br, I; m = 2; R = $C_6H_{13}CO_2$) [38] and it was found that the melting point increased in the order Cl < Br < I and that the temperature at which the N phase first appeared increased according to the same order. Clearing points, however, surprisingly followed a different trend, namely Br > I > Cl. Moreover, while none of the related mononuclear complexes with PPh_3 or aniline was mesomorphic, those with L = py and quinoline gave materials with N and smectic phases. For the py (*N,N*-cis) complex, the N phase stability was similar to that found in the parent dinuclear systems (above 200 °C), while for quinoline (*N,N*-trans) complex, transition temperatures were somewhat reduced (above 180 °C). In **2**, with respect to **1**, smectic phases are preferred to the N phase, which is likely due to the increase of the molecular aspect ratio and to more favorable lateral interactions.

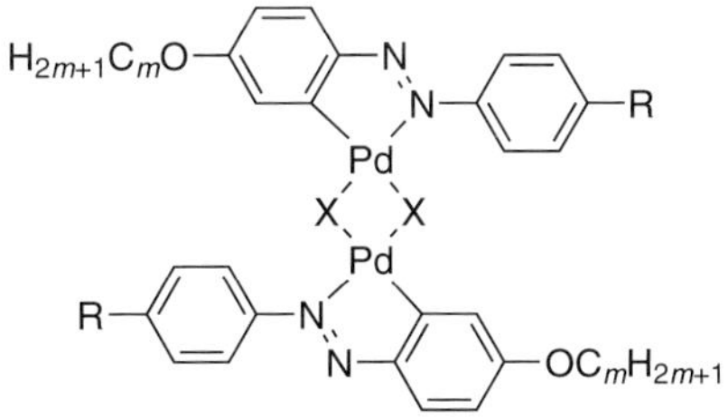

1

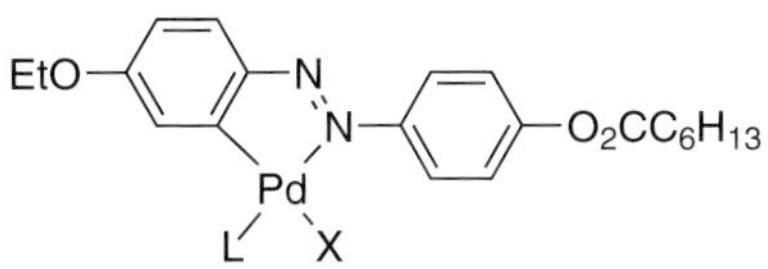

2

The fact that complexation of such systems enhances mesophase stability was further demonstrated in azopalladium(II) complexes with 4-alkoxyazobenzenes (**1**: X = Cl; R = H) and 4-alkyl-4′-alkoxyazobenzenes (**1**: X = Cl; R = C_nH_{2n+1}, $n = 1$–3) for which mesophase induction and/or stabilization were observed (m = 1, 2, 7, 12, 14, 18) [39–41]. All dinuclear complexes exist as equimolar mixtures of cis and trans isomers, and metallation occurs generally in the benzene ring bearing the alkoxy group. Mesomorphism (monotropic SmA and N phases) were induced for all complexes of 4-alkoxyazobenzenes (**1**: m = 10, 12, 14) [36, 40]. An enantiotropic N phase with, in some cases, an additional SmA phase was obtained for the analogous 4-alkyl-4′-alkoxyazobenzenes complexes; for short chain-length complexes (**1**: R = Me; m = 1, 2, 7), mesomorphism occurred above 200 °C, whilst for the derivatives with long alkoxy chains (**1**: R = Me, m = 12, 18; R = Et, Pr, m = 12) mesophases appeared between 130–180 and 160–200 °C; the SmA phase becomes enantiotropic for $n + m \geq 14$ [41].

Several azopalladium(II) complexes **1** of 4,4′-dialkoxyazobenzenes were also prepared. In one study, ortho-palladation of the 4,4′-di(tetradecyloxy)azobenzene (**1**: R = $OC_{14}H_{29}$; m = 14; X = OAc, Cl) afforded an enantiotropic SmC phase for the acetato-bridged complex, and SmB phase for the chloro-bridged analog [42]. A complete series of cyclopalladated dimers obtained from the nematogen 4,4′-di(hexyloxy)azobenzene with various bridging systems (**1**: X = Cl, Br, I, N_3, SCN, OAc; R = OC_6H_{13}; m = 6; **3**) has been prepared, and the effectiveness of the bridging group in promoting mesophases evaluated [43, 44]. From crystalline structures of homologous compounds, all the complexes, except those containing the acetato-bridged, are planar and in their trans conformation, the latter possessed a sort of "roof-shape" and existed as a cis : trans mixture [45]. All complexes but the OAc derivative exhibited high transition temperatures (ca 200 °C) and mesophases were only seen for the chloro- (N and E phases), bromo- (monotropic N), azido- (SmA), and oxalato-complexes (**3**: N, SmA). Surprisingly, for two complexes (**1**: X = Cl and **3**) the N phase is found below the more ordered smectic phases. While such re-entrant behavior is known and possible [46], the authors tentatively explained this observation by the dissociation of molecular pairs into single molecular species of different mesomorphism. With the oxalato complex, the N phase was transient and never reappeared on successive heating–cooling runs.

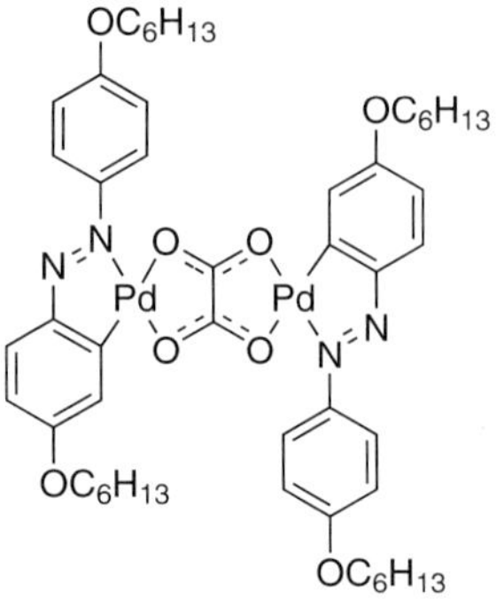

3

A low-temperature ferroelectric SmC* phase (enantiotropic with small temperature ranges or monotropic) and a SmA phase were induced in chiral complexes obtained either by substitution of the acetate with a chiral carboxylate bridge (**1**: R = $OC_{14}H_{29}$, m = 14, X = $CH_3ClCHCO_2$) [42] or by incorporating chiral alkoxy substituents R [**1**: M = Pd, X = Cl, I; R = OC*HMeC_6H_{13}, $O(CH_2)_2$C*HMe$(CH_2)_2$CH=C Me_2; m = 7, 10, 12, 14] [47]. For the latter systems, the free ligands with a citronellol chain exhibited N* and/or SmA phases, while it was not mesomorphic with 2-octanol.

Low-melting, mononuclear ortho-metallated metallomesogens combining 4,4′-bis(alkoxy)azobenzene and various types of simple chelating ligands, crucial for physical measurements and potential applications, were obtained from the reaction of the dinuclear, chloro-bridged complexes **1** with various anionic chelating ligands. Thus, complexes **4** (n = m = 6) with chelating ligand X–Y such as the *O,O*-monoanionic acac, 1,1,1,5,5,5-hexafluoro-2,4-pentanedionato anion (hfac), and tropolonate, the *N,O*-2-aminophenate and 2-amino-2-methyl-1-propanoate [48], and cyclopentadienyl (Cp) [49] were prepared. However, none was mesomorphic, probably due to the bulkiness of the co-ligands relative to the small anisotropy of the complex. In contrast, palladium(II)-acac complexes bearing a chiral chain [**4**: n = 7, 10, 12, 14; OC_mH_{2m+1} = *(R)*-(−)-2-octanol, *(S)*-(−)-β-citronellol] [50] or amino-acid-chelated palladated complexes (**4**: n = m = 14) [51] exhibited low-temperature N* and SmA or a SmC*, respectively. Mesomorphism (essentially N phase) is restored by increasing the overall anisotropy of the ligands [**5**, **9** (see below), M=Pd]; **5** again exists as a mixture of two isomers [48, 49].

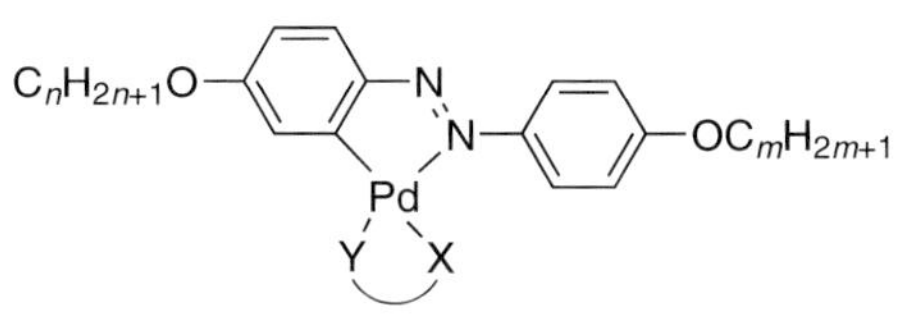

4

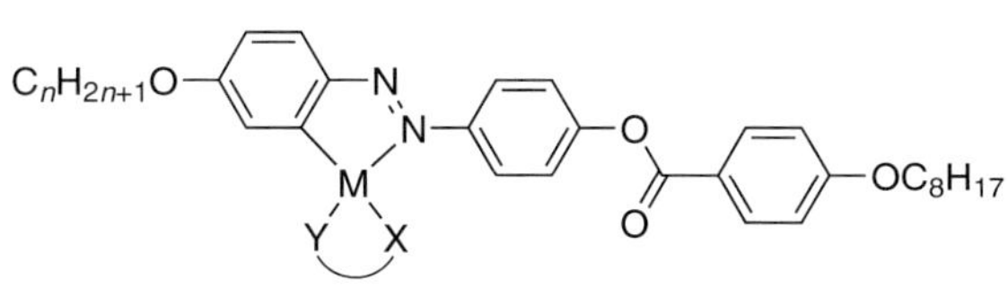

5

Mixed non-symmetrical cyclopalladated azobenzene complexes bearing mesogenic salicylaldimato co-ligands were also synthesized (**6**). The complexes [**6**: R^1 = *(R)*-(−)-2-octanol, *(S)*-(−)-β-citronellol; R^2 = C_2H_5, C_8H_{17}, *(R)*-(−)-2-octanol, *(S)*-(−)-β-citronellol] showed ferroelectric mesophases (SmC*, SmF*) and N* below 100 °C over narrow temperature ranges [52]. The one bearing two citronellyl chains [53, 54] was investigated as a possible photorefractive material [55].

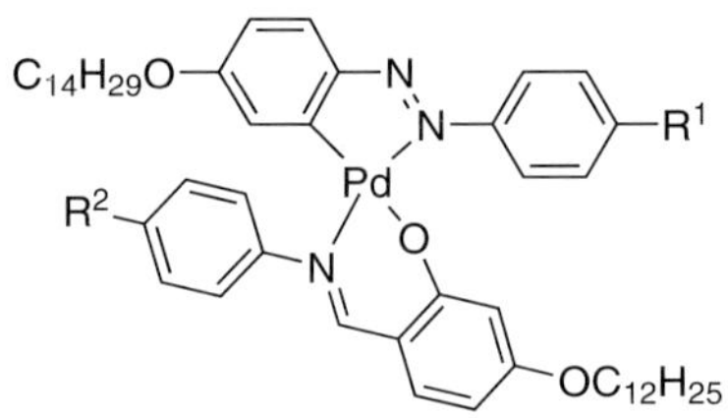

6

Dinuclear and mononuclear ortho-platinated complexes (**1**, **4**, **5**, **7–9**: M = Pt) [56, 57, 58] were also prepared for comparison. Most of the new mononuclear organometallic species **5** and **9** were found to show mainly an N phase as the homologous Pd complexes do, whilst **1** and **4** were not mesomorphic. Interestingly, some complexes derived from **5** and **9** are the first examples of mesomorphic complexes of octahedral Pt^{IV} (obtained from the oxidative addition to the square-planar Pt^{II} complexes by I_2 and MeI). A slight reduction of the transition temperatures was generally observed, and smectic mesomorphism (mostly SmC) was favored in octahedral complexes derived from **9** with respect to the square planar parent complexes. The Pt^{II} complexes usually showed higher melting and clearing temperatures than the homologous Pd^{II} complexes. For the dinuclear chloro-bridged complexes (**7**, **8**: M = Pd, Pt), a SmC phase was observed above 200 °C for **7** and above 250 °C for **8**.

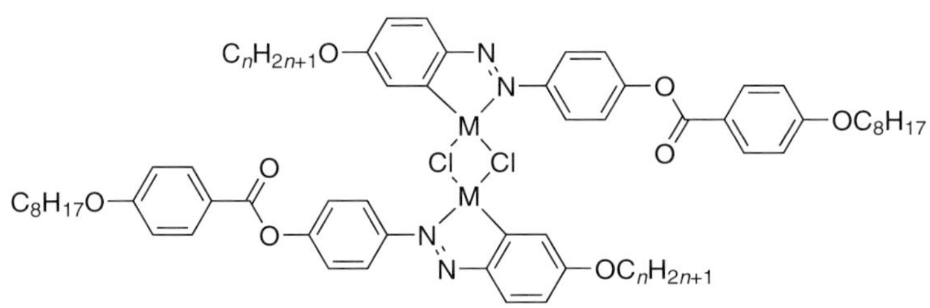

7

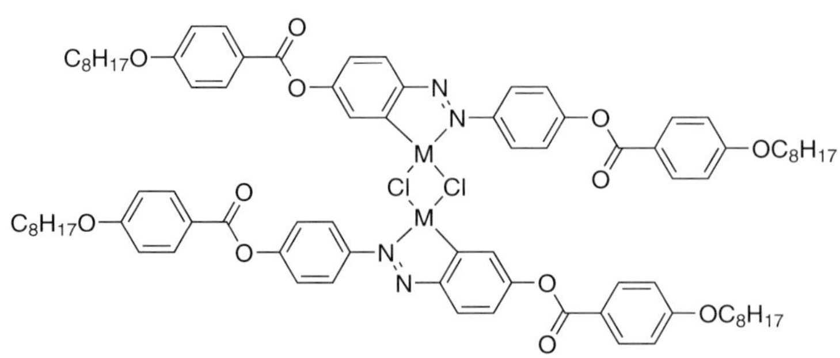

8

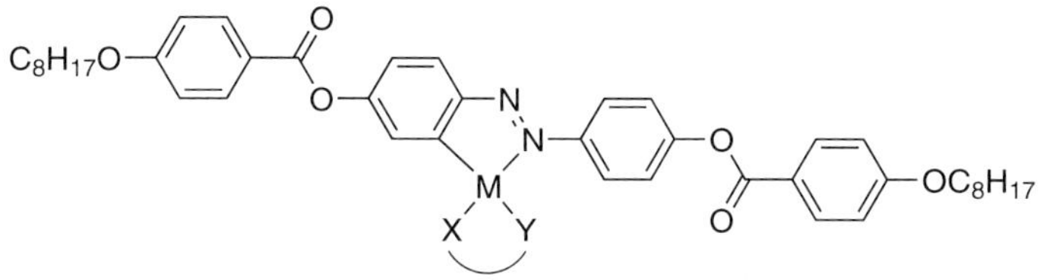

9

An early example of such ortho-metallated complexes with elongated azobenzenes was reported by Hoshino *et al.* (**10:** $n = 1$–10, 12, 14, 16, 18) [59]. In common with the parent ligands, the complexes were nematic, although melting and clearing points were raised by 60–100 and 100–140 °C, respectively, on complexation, giving the complexes a much wider nematic range.

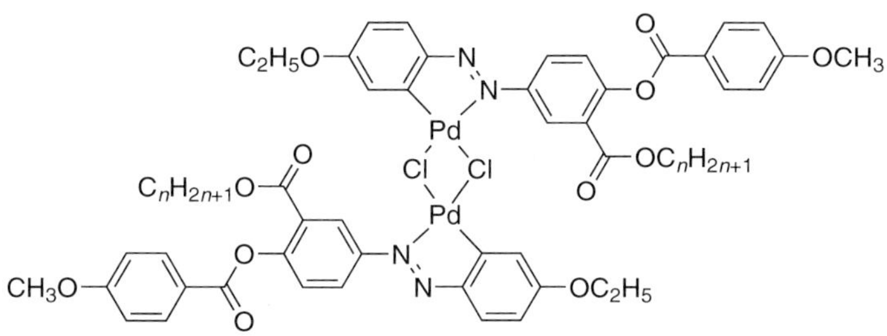

10

11.4.2
Ortho-Metallated Azoxybenzene Complexes

The dinuclear, cyclopalladated complex with 4,4′-dihexyloxyazoxybenzene was not mesomorphic, unlike its close azobenzene parent **1**, decomposing without clearing near 300 °C. Thus, mononuclear, ortho-palladated 4,4′-dihexyloxyazoxybenzene complexes bearing various co-ligands, obtained by reaction of the dinuclear complex with various chelating monoanionic ligands, were studied instead. Examples included *O,O*-monoanionic acac and tropolonate, or the *N,O*-2-aminophenate, 2-amino-2-methyl-1-propanoate, salicylialdinate and hydroxy azobenzenes. In one study, the acac derivative **11a** showed a low-temperature nematic phase between 90 and 105 °C [60], in contrast to the non-mesomorphic tropolonate, aminophenate and aminomethylpropanoate complexes [48]. Compared to the mononuclear azobenzene homologs, mesomorphism seemed slightly favored.

Other, mixed-ligand, mononuclear complexes whose molecular structure consists of two different ligands, namely 4,4′-dihexyloxyazoxybenzene with 2-hydroxy-4-alkoxy-4′-alkylazobenzene (**11b:** n = 1–4, 6, 8; m = 7, 9, 12, 14) [61], *N*-(4-dodecyloxysalicylidene)-4′-alkylaniline (**11c:** $n = 0$–4, 6, 8) [62], and the chiral alkoxy analogues [**11d:** R* = *(S)*-(–)-β-citronellyl, *(R)*-(–)-2-octyl] [63] ligands were studied (Figure 11.5). For complexes **11b**, obtained as single trans-isomers, a

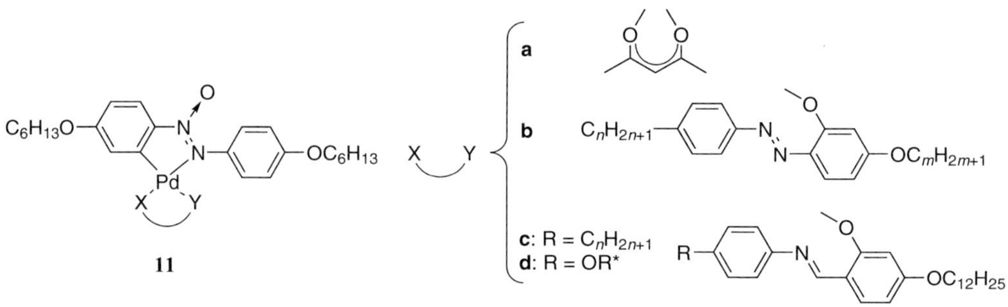

Figure 11.5 Ortho-palladated azoxybenzene compounds (**11**).

monotropic nematic phase was most commonly observed (the hydroxyazobenzene ligands exhibit N and SmC phases below ca 80 °C). However, smectic behavior seemed to be promoted in complexes **11c** and **11d**. For **11c** [62], obtained as an isomeric mixture of *N,N*-trans : *N,N*-cis in a 5 : 1 ratio [64], both SmA and N phases were observed – monotropic for $n = 0$, 2, and 4 and enantiotropic for $n = 1$, 3, 6 and 8. In most cases, the mesophases occurred between 125 and 145 °C. Incorporation of a chiral chain in the Schiff base ligand[63] led to the formation of a chiral crystal H* phase for both complexes **11d** bearing the citronellol or the 2-octanol chain (recall that both chiral imines exhibited a SmC* phase). In this case the *N,N*-trans/*N,N*-cis ratio was 9 : 1.

Using such azoxybenzene ligands bearing the chiral *(S)*-(+)-2-octyloxy group at one end and a hexyloxy or decyloxy chain on the other, Ghedini and coworkers reported the formation of ortho-metallated complexes of chloromercury(II). These complexes existed as 1 : 1 equimolar mixture of isomers due to non-selective metallation, and showed a room-temperature SmC* phase up to 58 and 63.5 °C, respectively [65]. The mesophase stability was slightly enhanced upon complexation.

11.4.3
Ortho-Palladated Benzalazine Complexes

Dinuclear complexes of symmetric benzalazines were prepared with various bridges (**12**: X = OAc, Cl, Br, SCN; $n = 10$) [66]. Only the trans isomers were observed for the halo-bridged complexes, while cis and trans isomers in a 60 : 40 ratio were observed for the thiocyanato complex [67]. In each of these examples, the complex was assumed to be planar by comparison with related structures. As for the acetato-bridged complexes, both trans and cis isomers in the ratio 3 : 1 were produced consistently. ^{1}H NMR studies went on to show that the trans isomer was optically active and hence the structure had to be that of an "open book," although in the synthesis a racemic mixture was produced. For the non-acetato-bridged dimers, the only mesophase seen was SmC, which was typically in the range 100–250 °C, while for the acetato-bridged complexes a nematic phase was seen for $6 \leq n \leq 8$ and a SmC phase for $n \geq 7$ (between ca 100 and 160 °C). Mesophase ranges were very much larger in the planar materials (lower melting points and greater mesophase stabilities).

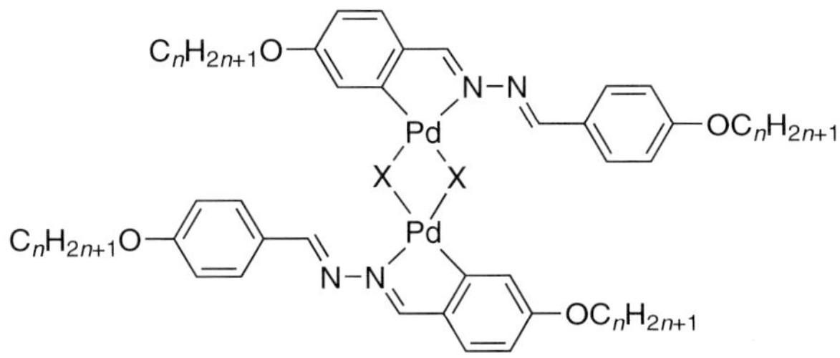

12

Using the fact that the trans isomer was intrinsically chiral, a derivative was synthesized where the bridging carboxylate group was the optically pure *(R)*-2-chloropropionate (**12:** X = $MeC^{*}ClHCO_2$; $n = 10$) [68]. Synthesized from the μ-Cl_2 species by reaction with the sodium salt of the acid, a mixture was produced, which was shown by ^{1}H NMR spectroscopy to have the composition *trans*-Λ*R,R* (34%), *trans*-Δ*R,R* (34%), *cis*-*R,R* (32%). Thus, while the trans-components described a pair of diastereoisomers, the cis isomer was optically pure by virtue of the chiral acid groups. The full series of *(S)*-chloropropionate complexes from $n = 6$ to 16 was later published by another group, revealing an important mesophase stabilization, essentially due to depression of the melting point [42, 69]. One complex with the larger, chiral carboxylate [**12:** X = *(S)*-$Me_2CHCH_2CHClCO_2$; $n = 14$] also yielded a SmC* phase, but at much lower temperatures (45–58 °C).

The effect of the length of the bridging carboxylate was also studied (**12:** X = $C_mH_{2m+1}CO_2$, m = 0–11, 13,15,17; $n = 10$) [70]. Thus, the shortest (m = 0–3) and longest ($m \geq 10$) carboxylates gave rise to a SmC phase, enantiotropic only for m = 0–2, whereas a N phase was seen for m 3, both enantiotropic ($2 \leq m \leq 6$, and m = 10, 13, 17) and monotropic ($7 \leq n \leq 9$, and m = 11, 15). The mesomorphic range decreased rapidly as the chain-length was increased, dropping from ca 40 °C to almost nothing, although the transition temperatures became stabilized at around 100 °C, revealing the important perturbation brought to the lateral molecular packing by the carboxylate [71].

Interestingly, **12** (X = OAc; $n = 14$) reacted with an amino acid (alanine) to yield a mononuclear complex exhibiting a broad SmC* phase [51, 72].

11.4.4
Ortho-Metallated Imine Complexes

As for the ortho-metallated azo compounds just described, a broad systematic study has been carried out on the imines and their corresponding mono- and dinuclear complexes, to determine the role of both the chain-length and -type (alkyl, alkoxy, perfluorinated, polyether), in addition to the role of bridging group (dinuclear complexes) and co-ligands (mononuclear complexes) in determining the mesomorphism.

Dinuclear ortho-palladated complexes, based on alkyl-alkoxy and dialkoxy imines **(13a)** and alkyl-alkoxy methyl imine ligands **(13b)** and various bridging groups, were investigated (**13a:** X = OAc, Cl, Br, SCN; R = C_nH_{2n+1}/OC_nH_{2n+1}; R′ = OC_mH_{2m+1}; *n*,*m* = 2, 6, 10; R = R′ = OC_nH_{2n+1}, *n* = 2, 4, 6, 8, 10 and **13b**: X = OAc, Cl, Br, SCN; R = $C_{10}H_{21}$; R′ = $OC_{10}H_{21}$) [73–75]. Only the benzylidene anilines are mesomorphic, showing G, SmF, SmB, SmC and SmA, N phases for the alkyl-alkoxy substitution and SmF, SmC and SmA, N phases for the dialkoxy derivatives; the transition temperatures were higher for the dialkoxy anilines. Concerning the influence of the nature of the chains and bridges, the overall trends can be summarized as follows. The longer the terminal chains, the lower are the melting points and the more ordered are the mesophases (smectic phases in place of nematic).

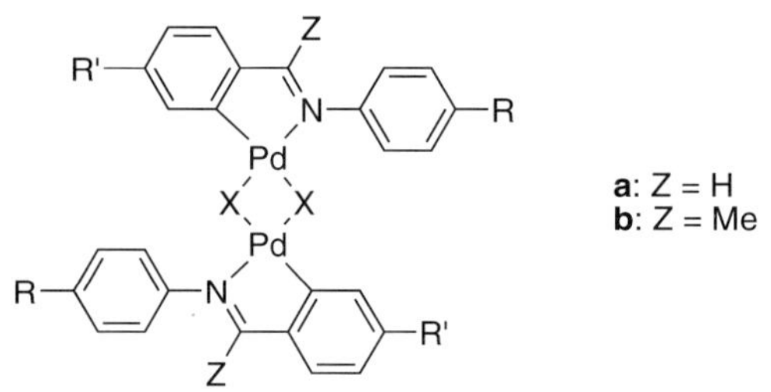

13

Complexes with solely alkoxy chains have more stable mesophases than those with both alkyl and alkoxy chains. In general, the acetate bridge was ineffective in promoting mesomorphism, in contrast with the findings reported by Zhang *et al.* [42], who observed a monotropic SmC phase for long-chain acetato-bridged palladated systems, and a SmC phase for the chloro-bridged analogues (**13a:** X = OAc, Cl; R = R′ = OC_nH_{2n+1}, *n* = 6–12, 14). Chloro-bridged derivatives exhibited SmC and SmA phases, those with bromide exhibited a SmA phase (and N for a few homologues), with the complete absence of any SmC phase. The mesophases existed at higher temperatures in the latter case (X = Br) compared to the former (X = Cl). Mesomorphic ranges were typically 80–100 °C for X = Cl or Br, with clearing temperatures near 240 °C. For X = SCN, 30 °C was the maximum range for the SmA phase, but the mesophase occurred at much higher temperatures than for the halo-bridged complexes. As with their dinuclear azo analogues, mixtures of cis/trans complexes have to be considered. Thus, while the acetato- and halo-bridged systems were found to exist as single, trans compounds, the thiocyanato compounds existed as a mixture of several isomers due to the different possibilities of coordination of the two, unsymmetric thiocyanato groups [76]. Interestingly, the lateral methyl group did not lead to any significant effects in the complexes **(13b)** even though it had suppressed mesomorphism in the parent ligand.

Studies of chloro- and acetate-bridged complexes derived from benzylidene ligands bearing polar groups either in the para-position of the aldehyde ring (**13a:**

X = OAc, Cl; R = C_8H_{17}/OC_8H_{17}; R′ = H, F, Cl, Br, CN, NO_2, Me, OMe, CF_3, OCOMe, $OCOC_6H_5$, CO_2Me) [77, 78] or aniline ring (**13a:** X = OAc, Cl; R′ = OC_8H_{17}; R = H, Cl, CN, NO_2, Me, OMe) [79] were undertaken. None of the acetato-bridged complexes were mesomorphic, whereas all the chloro-bridged derivatives but one (R = H) exhibited a high-temperature SmA phase (melting from 140 up 200 °C, at which temperature most of them started to decompose). This behavior was observed regardless of the location of the polar group. Interestingly, those with a cyano group or with the shortest chain were the only complexes to show a nematic phase.

As for complexes **1** and **12**, chirality may be introduced either by chiral peripheral chains (R and R′) or alternatively by chiral carboxylate bridging groups. Chloro-bridged complexes derived from non-mesomorphic benzylidene ligands bearing one chiral (*S*/*R*)-2-octanol chain either in the para-position of the aldehyde ring (**13a:** X = Cl; R = OC_nH_{2n+1}, n = 6, 8, 10, 14; R′ = $OC^*HMeC_6H_{13}$) or aniline ring (**13a:** X = Cl; R = $OC^*HMeC_6H_{13}$; R′ = OC_8H_{17}) or both rings (**13a:** X = Cl; R = R′ = $OC^*HMeC_6H_{13}$) [80] all displayed a SmC* phase (monotropic for the latter) along with a SmA phase. The mesophases occurred typically between 100 and 220–230 °C the mesomorphic temperature range of the SmC* increasing with elongation of the alkoxy chain. Interestingly, comparing the two octyloxy isomers, the SmC* phase was more stable by 20 °C when the chiral chain was attached to the aniline ring. The ferroelectric properties are improved when the chiral chains are fixed on the ortho-metallated ring, likely due to the hindered freedom of the chiral chains, and hence a better coupling of the molecular dipoles. The same study was carried out with the cycloplatinated homologs and the same trend observed but with slightly more stable SmC* and SmA phases [81].

Isomeric chloro- and acetato-bridged dinuclear complexes derived from elongated (three-ring) benzylidene ligands bearing one (*R*)-2-octanol chain chiral either in the para-position of the aniline ring (**13a:** X = Cl, OAc; R = $CO_2PhCO_2C^*HMeC_6H_{13}$; R′ = OC_8H_{17}) or aldehyde ring (**13a:** X = Cl, OAc; R = OC_8H_{17}, R′ = $CO_2PhCO_2C^*HMeC_6H_{13}$) were reported by Serrano and coworkers [82]. The former imine showed SmA, SmC* and a monotropic SmC_A^* phases (antiferroelectric), while the latter showed, in addition, a monotropic, antiferroelectric SmI_A^* phase. On complexation, both μ-acetato complexes showed only a SmA phase before decomposing close to 200 °C. The μ-Cl complexes also retained the SmA phase of the ligand, but in addition showed a monotropic SmC* phase or an enantiotropic SmC_A^* phase, respectively. Both complexes decomposed in their SmA phases between 240 and 250 °C. The complexes made from the ligand elongated at the aniline side have higher melting temperatures than those obtained with the other ligand (by about 20–40 °C).

Treatment of the chloro-bridged, dimeric species with potassium 2-chloropropionate [83] or, alternatively, the acetato-bridged analogues with 2-bromopropionic acid [74] led to new carboxylato-bridged dimers (**13a:** X = $CH_3CHClCO_2$, $CH_3CHBrCO_2$; R = R′ = OC_nH_{2n+1}, n = 2, 4, 6, 8, 10) showing a broad-temperature-range SmA phase (ca 100–150 °C), with some stable up to nearly 250 °C. Only small differences were observed in the transition temperatures between the chloro- and bromo-propionato derivatives, probably due to the existence of the complexes as rich isomeric mixtures. Indeed, due to the chiral carboxylate group, in addition to the cis and trans isomers resulting from the arrangement of the two imines in the dimer, the chirality of the trans isomer leads to two enantiomers, Δ and Λ, the cis not being chiral. For both series, the thermal behavior was slightly complicated and most of the complexes exhibited double clearing (or melting) behavior, that is the Cr-SmA-I′ (or Cr′)-SmA-I phase sequence was observed systematically. One dipalladium complex with tetradecyloxy chains and a bulkier carboxylate bridge (**13a**: X = $Me_2CHCH_2C^*HClCO_2$) was reported to show a SmC* [42].

Another interesting and rather original part of this work was concerned by the study of dinuclear cyclopalladated complexes with mixed bridges (**14a–c**) [84]. The reaction of the dichloro-bridged complexes **13a** with silver thiolate ($AgSC_mH_{2m+1}$) leads to dinuclear derivatives with mixed bridges μ-Cl/μ-SC_mH_{2m+1} (**14a**: n = 6, 8, 10; m = 6, 8, 10, 18). Likewise, treatment of the di-μ-acetato complexes **13a** with thiols (HSC_mH_{2m+1}) or, alternatively, dinuclear **14a** with silver acetate affords the mixed-bridge complexes μ-OAc/μ-SC_mH_{2m+1} (**14b**: n = 6, 8, 10; m = 6, 8, 10, 18). Notably, in addition to their mesomorphism, among the three possible isomers (one in trans and two in cis-geometry), such mixed-bridged complexes adopted a cis-geometry, as confirmed by NMR and X-ray studies (**14** as depicted). All the complexes **14a** displayed both SmC and SmA phases, at lower temperatures than their dichloro-predecessors, that is, between 100 and 200 °C with little influence of the chain length; the SmA phase existed over most of this temperature range. For complexes **14b**, the predominant phase was the SmA (N phase for n = 6 only); the transition temperatures were also reduced, decreasing with m (from 160 to 70 °C for the melting point, and 170 to 140 °C for the clearing point). Mixed-bridge complexes obtained initially by the reaction of alkylthiols to a chloropropionate bridge dinuclear complexes (**14c**: R* = MeCHCl; n = 2, 6; m = 6, 10, 18) were the first metallomesogens to show an N* phase [83, 85].. Dinuclear thiolato-bridged complexes **14d** exhibited a broad SmA phase above 150 °C.

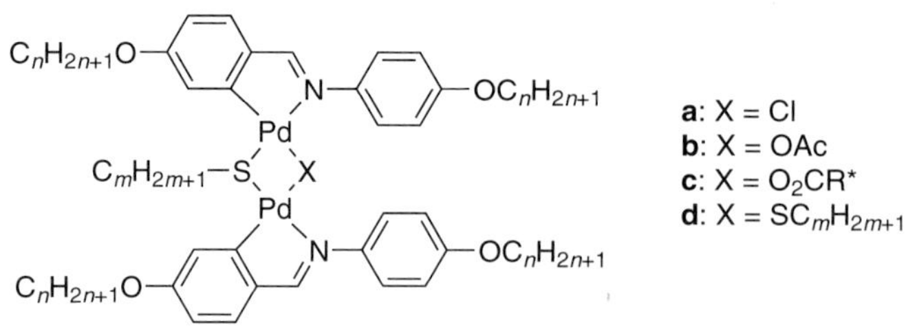

14

Espinet and coworkers [86] also undertook an extensive investigation of the effect of oligo(ethylene oxide) terminal chains in a series of complexes (**15**, Figure 11.6), in each case retaining the same overall chain length in the two related series **15-i** to **15-iv** and **15-vi** to **15-viii**. None of the precursor imine ligands containing ethylene oxide groups were mesomorphic, neither were any of the parent μ-acetato dimers. However, the μ-Cl and μ-Br dimers did show liquid-crystalline properties,

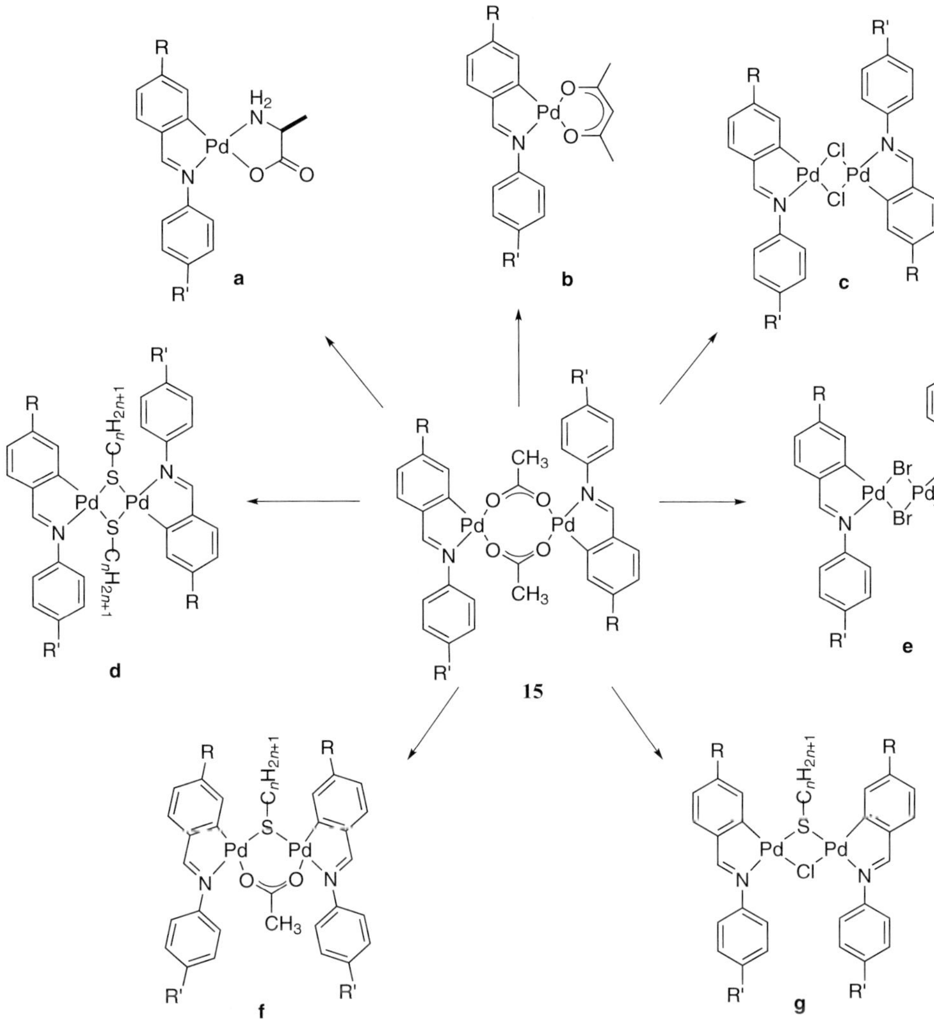

i R = R' = OC_8H_{17}	**v** R = R' = $OC_{12}H_{25}$
ii R = OC_8H_{17}, R' = $O(CH_2CH_2O)_2Et$	**vi** R = $OC_{12}H_{25}$, R' = $O(CH_2CH_2O)_3Et$
iii R = $O(CH_2CH_2O)_2Et$, R' = OC_8H_{17}	**vii** R = $O(CH_2CH_2O)_3Et$, R' = $OC_{12}H_{25}$
iv R = R' = $O(CH_2CH_2O)_2Et$	**viii** R = R' = $O(CH_2CH_2O)_3Et$

Figure 11.6 Structure of the Pd complexes **15a–g** obtained from **13a**.

with all the chloro complexes (**15c-i** to **15c-viii**) showing enantiotropic SmC and SmA phases, while for the longer-chain bromo complexes studied (**15e-v** to **15e-viii**) both phases were again seen, although the SmC phase was monotropic in the two, unsymmetric derivatives (**vi** and **vii**). The thermal behavior of the mixed chains complexes is intermediate between the tetraalkoxy and tetraethyleneoxy complexes. In general, the clearing points of the unsymmetrically substituted complexes were similar to one another and lower than those of the symmetric alkoxy materials **(i)**, while materials substituted symmetrically with ethylene oxide chains were much lower again. This effect was more pronounced in the longer-chain complexes **(v–viii)** and also in bromo rather than chloro complexes. Using octylthiolato chains, **15d-iv** were found to melt directly to the isotropic liquid – the use of chiral 2-thiooctyl led to a monotropic SmC* phase. In the μ-thiol-μ-acetato system, **15f-i** showed a SmA phase, **15f-iv** a monotropic SmC phase, while use of chiral thiolate in the latter derivative suppressed mesomorphism totally. Similarly, complexes **15g-iv** exhibited either a monotropic SmC (with octylthiolate) or an enantiotropic SmC* phase (with a chiral 2-thiooctyl thiolate chain).

Similarly, orthopalladation of crown-derivatized 4-alkoxybenzilidenes also afforded dinuclear derivatives with dichloro and diacetate bridges [87]. Whilst neither the ligands nor the acetato-bridges complexes were mesomorphic, the dichloro complexes exhibited a high-temperature SmA phase (170–220 and 215–235 °C with increasing alkoxy chain-length). To take further advantage of the ethylene oxide chains or crown ethers, experiments were undertaken to see if sodium/potassium picrate could usefully be extracted from aqueous medium by these complexes. The transport observed was rather modest, but was improved for the dipalladium fitted with crown ethers.

The reactivity of di-μ-hydroxo complexes of ortho-palladated imines, obtained by the treatment of the di- μ-acetato complexes (**13a:** X = OAc) with NaOH, towards protic substrates such amines, alkylthiols, carboxylic acids and amines in the presence of CS_2 provided a versatile entry to new metallomesogens. Thermotropic, air-stable, μ-amido-μ-hydroxo (**16a**), μ-anilido-μ-hydroxo (**16b**), bis-μ-amido (**16c**), μ-amido-μ-thiolato (**16d**), and μ-amido-μ-carboxylato (**16e**) complexes, as well as the mononuclear complexes with mixed imine and *N,N*-dialkyldithiocarbamato ligands (**16f**), were prepared (**16**, Figure 11.7) [88].

Most of the complexes **16a** were liquid crystals, showing a nematic phase with short chain-lengths at elevated temperatures ($n = 2$; $m = 14, 18$), and a SmA phase at longer chain-length ($n = 6, 10$; $m = 1, 6, 10, 14, 18$) from 60–120 °C up to 160–180 °C where they decomposed. Complexes **16b** behaved similarly to **16a** (slight increase in transition temperatures), with a nematic phase for $n = 2$ and a SmA phase for $n = 6$ and 10. Replacement of the hydroxo bridge by an alkylthiol led to complexes **16d**, which showed mainly a SmA phase for $n = 6$ and 10 above 100 °C on average, but decomposed at the clearing point. Only a monotropic SmA phase was observed for the compounds with long amido and thiolato chains ($m, p \geq 10$), and none of the complexes with $n = 2$ was mesomorphic. Replacement of the μ-OH by μ-carboxylato gave slightly better results; a stable SmA phase was seen between 76 and 107.5 °C for **16e** with $n = 10$. Complexes **16c** could not be studied because of their insolubility in organic solvents, and consequently it was impossible to isolate them as pure materials. The μ-dihydroxo complex reacts with dialkylamine in CS_2 to the give mononuclear species **16f**, which showed an enantiotropic SmA at very low temperatures for derivatives with $n = 10$ and $q = 2$ and 8. As is often

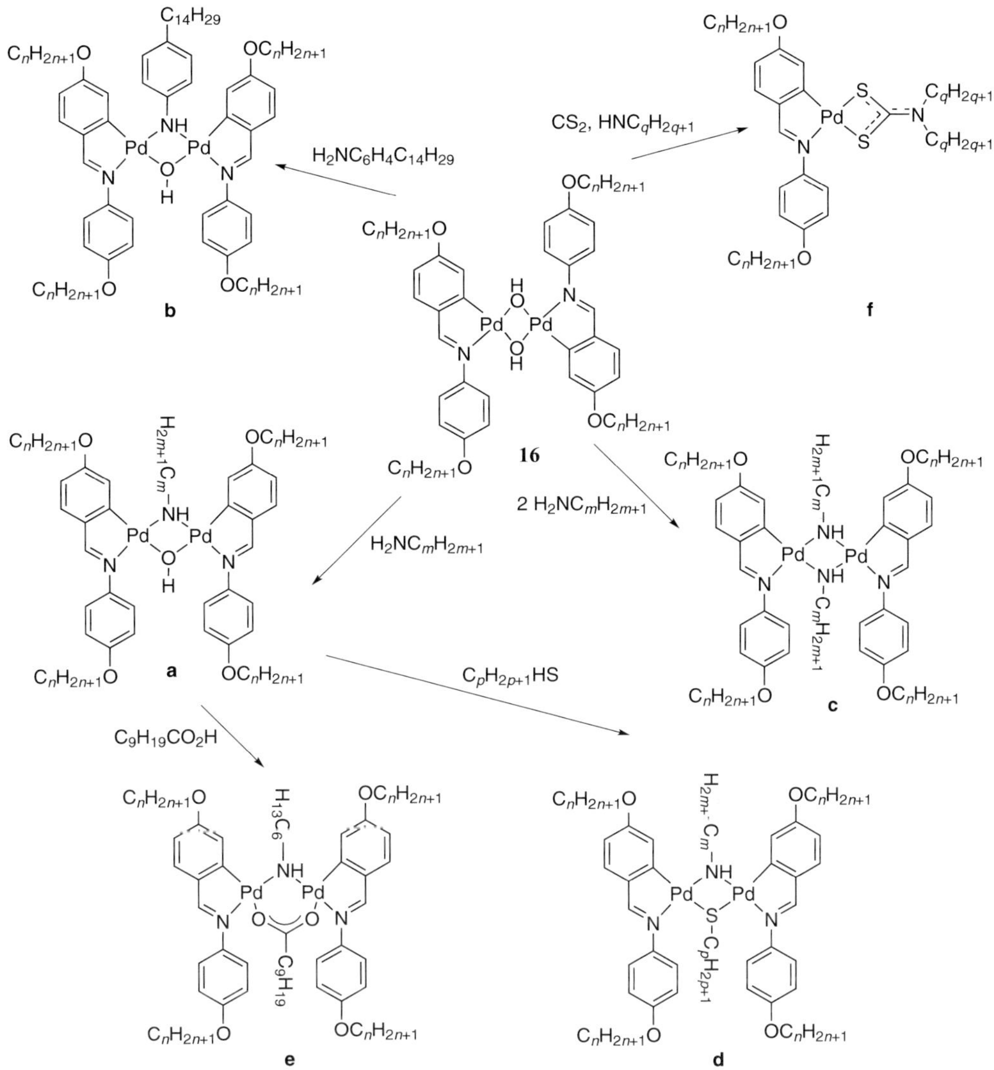

Figure 11.7 Structures of the various mixed dinuclear complexes **16a–e**, and of the mononuclear complex **16f** obtained from **13a** (X = OH): *n* = 2, 6, 10; *m* = 1, 6, 10, 14, 18; *p* = 4, 10, 18; *q* = 2, 8.

the case, the change in molecular shape from dinuclear to mononuclear metallomesogens produces a substantial lowering of the transition temperatures.

Similarly, reaction of the binuclear μ-hydroxo complex with various carboxylic acids has been greatly exploited and has led to an extensive and systematic range of complexes with different types of carboxylato bridges (**17**, Figure 11.8) [89]. Complex **17a** showed a SmA phase at high temperature (>230 °C for both values of *n*). However, when replaced by an alkylcarboxylate bridge (**17b:** R = C_mH_{2m+1},

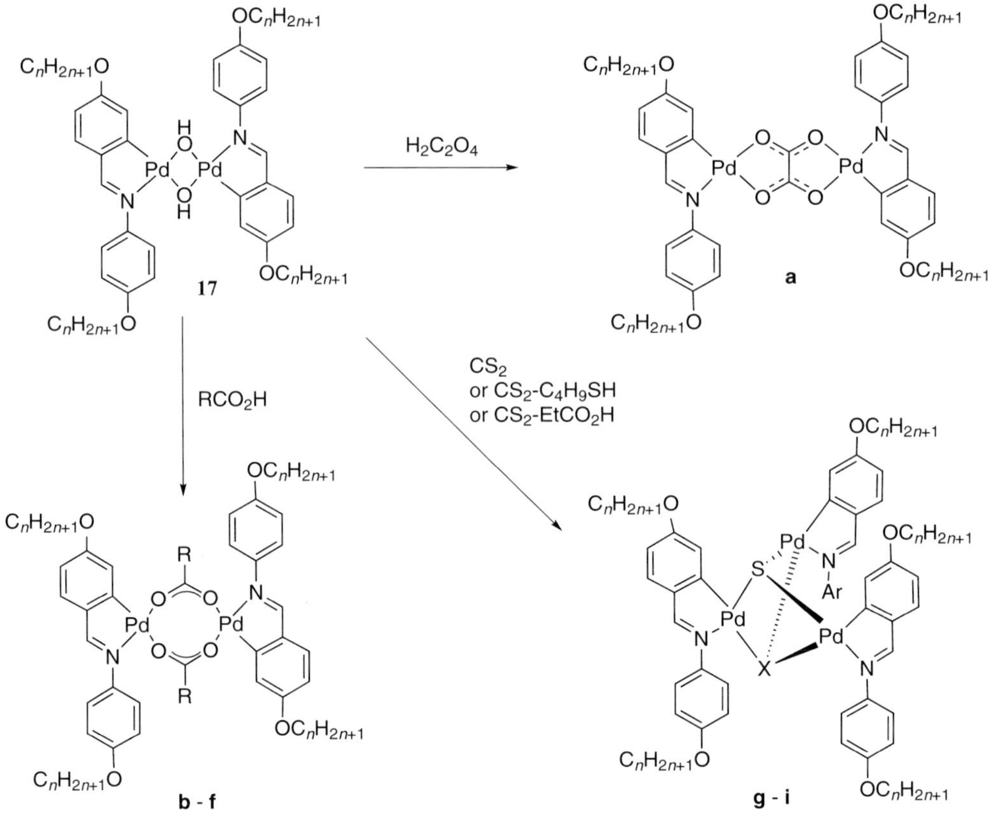

Figure 11.8 Structures of various dinuclear and trinuclear complexes **17** (n = 2, 6, 10).

m = 1, 3, 5, 7, 9, 11, 13, 15, 17), strong substituent dependence was observed so that for n = 6 only long-chain bridging ligands ($m \geq 11$) led to mesomorphism (SmA and N below 120 °C), while for n = 10 only the μ-acetato (m = 1) complex did not show a mesophase – the others showed SmA (mainly) and SmC phases monotropically except for m = 5 and 7 where mesomorphism was enantiotropic. In complex **17c** [R = $CH_2(OCH_2CH_2)_pOCH_3$, p = 1, 2], for both n = 6 and 10 and for p = 1 and 2, an enantiotropic SmA phase was seen that in most cases had a wide range (ca 50 °C, clearing between 100 and 145 °C). Use of the bridging alkoxybenzoate ligand (**17d**: R = $C_6H_4OC_rH_{2r+1}$, r = 4, 10) was not very productive, and for n = 6 and r = 4 and 10 neither was mesomorphic, while for n = 10 the complex with r = 4 gave enantiotropic SmC and SmA phases, and, for r = 10, a monotropic SmA phase was found. When the bridging ligand was the alkoxyphenylacetate (**17e**: R = $CH_2C_6H_4OC_qH_{2q+1}$, q = 2, 4, 6, 8, 10, 12), all but two homologs (n = 6, q = 2, 4) showed both a SmA and SmC phase with a total mesomorphic range of around 40 °C in most cases. Both complexes (n = 6, 10) with the bridging lactato group (**17f**: R = HC*MeOH) showed a SmA phase.

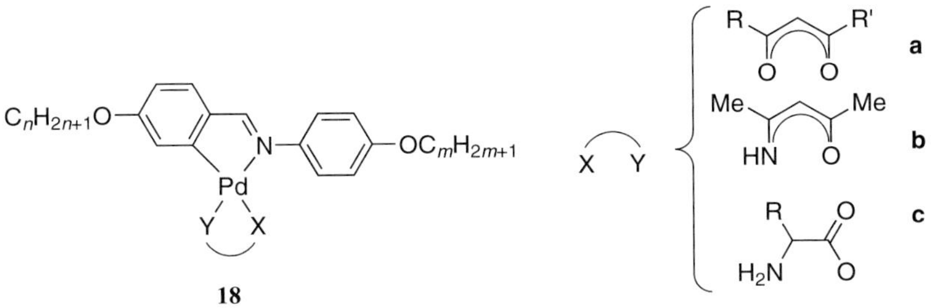

18

Figure 11.9 Structures of various mononuclear complexes **(18)**.

Notably, trinuclear, orthopalladated imine complexes with unasymmetrical (μ_3-S)–(μ_3-X) bridges (**17g–i**, Figure 11.8) were formed by the direct reaction of the di-hydroxo binuclear parent complex (**16:** n = 2, 6, 10) and CS_2 (**17g:** X = OH), followed by treatment with C_4H_9SH (**17h:** X = SC_4H_9) or $EtCO_2H$ (**17i:** X = O_2CEt). The three ligands are all in a cis-arrangement relative to the plane formed by the three metallic ions, and the μ_3-S bridge pointing toward the benzaldehyde ring. Unfortunately, no information was given about their thermal behavior [90].

Cleavage of the dinuclear (μ-Cl)$_2$ complexes (**13a:** X = Cl) with β-diketones or β-enaminoketones also led to various series of mononuclear derivatives (**18**, Figure 11.9). Thus, complexes with an acetylacetonato co-ligand [74, 91], (**18a:** R = R′ = Me; $n = m$ = 2, 4, 6, 8, 10; n = 10 and m = 2, 6, 10) were mesomorphic at much lower temperatures than their dichlorodipalladium precursors but, more interestingly, the mesophases were more accessible and more stable than those of the ligands. A monotropic ($n = m$ = 2, 4) or enantiotropic (n = 10; m = 2) nematic phase was observed for the short chain-length compounds, both enantiotropic SmA and N phases at intermediate chain-length ($n = m$ = 6, 8; n = 10, m = 6), and only the SmA phase for the derivative with two decyloxy chains. Long-chain complexes exhibit a double-melting behavior.

Similarly, mononuclear β-diketonato derivatives **(15b)** made from chains containing polyether groups[86] showed both SmA and N phases **(15b-i** to **–iii)**, while **15b-iv** gave only a monotropic nematic phase; clearing points dropped from ca 125 °C in **15b-i** to ca 21 °C in **15b-iv**. In the longer-chain derivatives, **15b-v** showed a nematic phase (monotropic) while **15b-vi** and **15b-vii** showed a SmA phase, the latter having a melting point below ambient temperature; **15b-viii** was isotropic at room temperature. The corresponding complexes bearing a crown ether showed a single SmA phase between 60–90 and 115–135 °C [87].

The lowering of the symmetry of the complex appeared therefore to be an excellent strategy for reducing the transition temperatures and simultaneously preserving a large mesomorphic range. However, the mesophase stability of these complexes was very sensitive to small, structural changes. Thus, the nematic phase became destabilized when the β-diketone used contained trifluoroacetylacetonate, with T_{NI} decreasing as R = R′ = Me > R = Me, R′ = CF_3 > R = R′ = CF_3 (**18a:** $n = m$ = 6) being a virtual transition in the last of these, probably due to the steric hindrance introduced by the CF_3 group (the dihexyloxy hemolog melted at higher

Table 11.1 Family of mononuclear complexes 18d. I: X = Y = O; II: X = –, Y = O; III: X = O, Y = –; IV: X = Y = –.

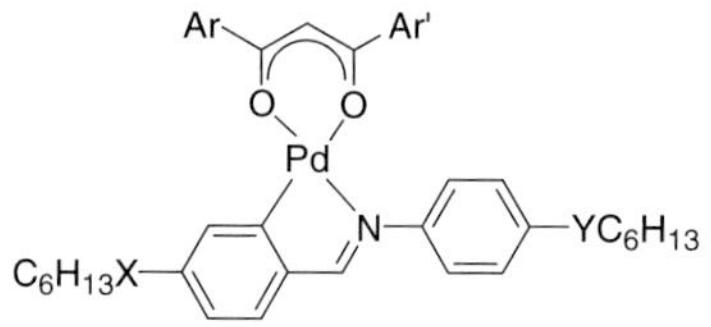

18d	i	ii	iii	iv	v	vi	vii	viii	ix	x
Ar	Ph-	4-NC-Ph-	4-Me-Ph-	4-F-Ph-	3-F-Ph-	4-CF3-Ph-	Ph-	Ph-	Ph-	Ph-
Ar	-Ph	-Ph-4-CN	-Ph-4-Me	-Ph-4-F	-Ph-3-F	-Ph-4-CF3	-Ph-4-F	-Ph-3-F	-Ph-4-CF_3	-Ph-4-Me

temperature than the hexyl-hexyloxy homolog) [27b, 92]. Mesomorphism disappeared totally for complexes **18a** with R = R′ = C_2F_5, R = C_2F_5 and R′ = Et, R = C_3F_7 and R′ = Pr, and R = R′ = C_3F_7 [93, 94]. The use of aliphatic diketones [**18a:** R = R′ = $C_{10}H_{21}$; n, m = 8*, 10–8* is (*R*)-2-methylheptyl] was correlated with a strong reduction in mesomorphic character with a monotropic SmA phase for the one-chiral chain systems or total loss of mesomorphism for the two-chiral chains complex [95]. Their mixtures with **19** (all decyloxy chains) yield monotropic N* and SmA phases.

A family of complexes **18a** with $1 \leq n = m \leq 6$ was prepared for R = Me, R′ = Ph; these complexes existed as a 2 : 1 ratio of isomers, the more predominant being that with the β-diketonato phenyl group anti to the Pd—N bond [93, 94]. For $n = m = 1$–5, a monotropic N phase is observed, whereas for $n = m = 6$ nematic and SmA phases are observed. No great changes were observed in the related mixed hexyl/hexyloxy or dihexyl homologs [94].

A family of complexes **18d** was prepared with various β-diphenylketonate ligands substituted by polar groups as shown in Table 11.1 and with imines bearing hexyloxy (**I**), hexyl-hexyloxy (**II, III**) and hexyl (**IV**) chains to evaluate the structural requirements for promoting the N phase [94]. Clearly, the presence of polar groups tends to stabilize the SmA phase at the expense of the nematic phase (as was previously observed with some derivatives of **13** bearing polar groups), and the largest temperature ranges are obtained for the complexes of the dihexyloxy series (**I**).

The mononuclear complexes with β-aminoenonate (**18b:** $n = m = 2, 4, 6, 8, 10$), obtained as the *N,N*-trans isomer only, also yielded an accessible N phase (and a SmA for longer chains), over slightly narrower temperature ranges than in the acac derivatives (**18a:** R = R′ = Me) [74]. Moreover, when the chain attached onto the fixed benzaldehyde ring was chiral (e.g. 2-octanol), complex **18b** (n = 8*; m = 8) was found to exhibit an N* phase at reasonably low temperatures [85]. Using a similar approach, a SmC* phase (**18c:** R = Me; $n = m = 14$) and a broad

Table 11.2 Structure of complexes **19i–xii**.

	$R^{1a,b}$	$R^{2a,b}$	$R^{3a,b}$	$R^{4a,b}$
i	OC_{10}	OC_{10}	OC_{10}	OC_{10}
ii	OC^*_8	OC_{10}	OC_{10}	OC_{10}
iii	OC_{10}	OC^*_8	OC_{10}	OC_{10}
iv[c]	OC_{10}	OC_{10}	OC^*_8	OC_{10}
v	OC^*_8	OC^*_8	OC_{10}	OC_{10}
vi[c]	OC^*_8	OC_{10}	OC^*_8	OC_{10}
vii[c]	OC_{10}	OC^*_8	OC^*_8	OC_{10}
viii	OC_{10}	OC_{10}	OC^*_8	OC^*_8
ix[c]	OC^*_8	OC^*_8	OC^*_8	OC_{10}
x	OC^*_8	OC_{10}	OC^*_8	OC^*_8
xi	OC_{10}	OC^*_8	OC^*_8	OC^*_8
xii	OC^*_8	OC^*_8	OC^*_8	OC^*_8

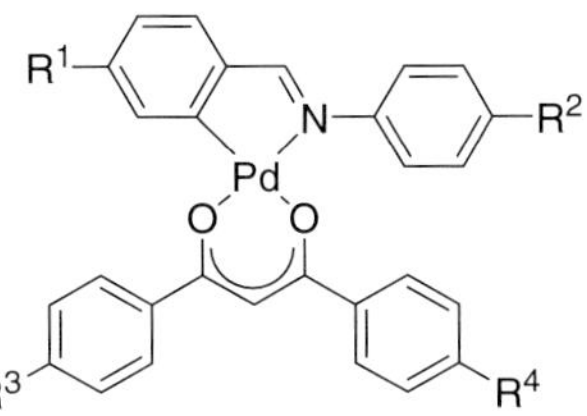

a $OC_{10} = OC_{10}H_{21}$.
b $OC^*_8 = OC^*HMeC_6H_{13}$.
c 1 : 1 mixture of cis : trans stereoisomers.

temperature-range SmA phase (**18c:** R = $C_6H_5CH_2$, Me_2CH, Me_2CHCH_2; $n = m = 14$) were obtained in related mixed complexes with chiral amino acids, again at very accessible temperatures [51]; the mesomorphic properties were lost in **18c** (R = Me) bearing one or two ethyleneoxy chains [86].

Espinet and coworkers investigated in detail chiral derivatives of ortho-palladated complexes incorporating a dialkoxybenzylidene and a di-4,4′-alkoxy-phenyl-β-diketonato ligand [96], obtained from the cleavage of the dinuclear bridge **13a** by the thallium(I) complex of the β-diketonate (**19:** $R^1 = OC^*HMeC_6H_{13}$; $R^2 = OC_{10}H_{21}$, $OC_{14}H_{29}$; R^3, $R^4 = OC_{10}H_{21}$). With a view to optimizing the system and finding an enantiotropic SmC* phase, this study was extended to evaluate the effects of the position and number of chiral chains on the ferroelectric behavior of this system, and a series of 12 new materials was prepared (**19:** R^1, R^2, R^3, R^4 = $OC^*HMeC_6H_{13}/OC_{10}H_{21}$, Table 11.2) [97]. Recall that the compounds with unsymmetric diketones exist as a cis/trans isomeric mixture. As the number of chiral chains increased, the clearing transition temperatures decreased from 155 to 115 °C (melting oscillated between 65 and 95 °C), and compounds having a chiral chain on the imine ligand exhibited monotropic behavior, whereas when this chain was on the diketone, the behavior was enantiotropic. All compounds with two chiral chains exhibited only monotropic phases, the complex having the two chiral chains on the β-diketonato ligand being devoid of mesomorphism; none of the complexes with three and four chiral chains was liquid crystalline. In addition, the number and position of these chains influenced the ferroelectric properties drastically, and particularly the spontaneous polarization, as well as the nonlinear optical responses [98].

Such a complete and systematic study was also carried out with the platinum derivatives, and dinuclear cycloplatinated complexes of 4,4′-dialkoxybenzylidene with symmetric bridges μ-dichloro (**13a**: X = Cl), μ-diacetato (**13a**: X = OAc), μ-dithiolato (**14d**), and μ-dichloropropionato (derived from **17b**) or mixed bridges μ-chloro-μ-thiolato (**14a**), μ-acetato-μ-thiolato (**14b**), and μ-chloropropionato–μ-thiolato (derived from **14c**) [99, 100]. Mononuclear species with acetylacetonato and phenyldiketonato [101, 102] co-ligands were also prepared. All the platinum complexes were mesomorphic, except the acetato-bridged material. They exhibited, in general, more ordered mesophases than their palladium analogues (e.g. a SmA phase was induced in place of the N phases) and, overall, the transition temperatures, and particularly the clearing temperatures, were slightly higher for Pt than for Pd. The substitution of palladium by platinum thus resulted in an overall increase in mesophase stability. Note that in the platinum complexes existing as isomeric mixtures, the composition was different to that in the palladium congeners, explaining partly some discrepancy between the two series.

Some of the chloro-bridged palladium complexes, **13a** (R = C_nH_{2n+1}; R′ = OC_mH_{2m+1}; n, m = 6, 10), a mixed bridged μ-acetato-μ-thiolato **14b** ($n = m = 6$) and a mononuclear complex **19** (R^1, R^2, R^3, R^4 = $OC_{10}H_{21}$), discussed above were also found to form lyotropic mesophases in contact with apolar organic solvents such as linear alkanes (octane, decane, dodecane and pentadecane), cycloocta-1,5-diene, and the chiral limonene [103, 104]. A lyotropic lamellar phase was induced for **13a** with symmetrical chain length ($n = m$ = 6, 10) in linear alkanes. Induction of a lyotropic N phase, with the complete destabilization of the thermotropic smectic phases, was observed for **14b** and **19** in alkanes; in the case of **14b**, the nematic range increased concomitantly with the length of the solvent chain. While mixtures with cyclooctadiene did not yield a mesophase for any of the complexes, mixtures with limonene resulted in suppression or destabilization of mesomorphism for complexes **13a**, and the induction of an N* phase for **14b** and **19**. This is the first case of chiral induction in binary systems between calamitic complexes and a chiral, apolar solvent.

Praefcke and coworkers reported a series of disc-shaped, dinuclear ortho-palladated benzalimine complexes **20**, which were the first example of organometallic complexes showing the nematic phase of disc-like molecules, N_D [105].

The flat, dinuclear halogeno- and thiocyanato-bridged complexes (**20**: M = Pd; X = Cl, Br, I, SCN; R = R′ = OC_6H_{13}) exhibited a monotropic N_D phase, whereas the acetato-bridged complex was not mesomorphic, a consequence of its open-book structure. The peculiar clearing process of the thiocyanato complex was caused by the composition of the nematogen, in that the thiocyanate moieties can be bridged parallel or anti-parallel to one another, leading to two structural isomers in the ratio 17:83. For structurally related platinum complexes (**20**: M = Pt; X = Cl, I, SCN), only the dichloro-bridged complex showed the monotropic N_D phase [106–108]. Interestingly, unlike the chloro-bridged palladium complexes, the platinum complex existed as an isomeric mixture (syn/anti) in solution, and attempts to separate the two isomers were unsuccessful because of decomposition processes; the thiocyanato platinum complex was obtained as a single, antiparallel isomer and the iodo mostly as the anti. The molecular geometry and the number of peripheral chains thus appeared crucial in determining the type of mesophase observed, since the related ortho-metallated imine complexes with four alkoxy chains exhibited exclusively smectic mesomorphism, mostly SmA phases (see above).

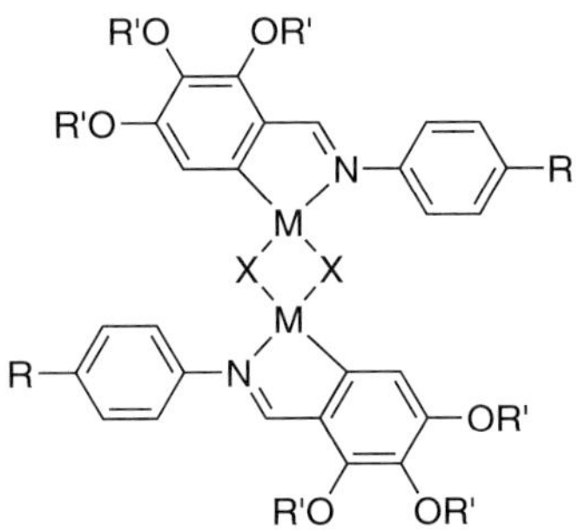

20

All these compounds formed charge-transfer (CT) complexes when doped with strong electron acceptors such as 2,4,7-trinitrofluorenone (TNF) [109, 110]. The bridging group was found to influence strongly the type of induced mesophases. Thus, enantiotropic Col_h phases were induced in the binary mixtures of chloro- and bromo-bridged complexes with TNF, with the suppression of the N_D phase above 10% of TNF. The iodo-bridged palladium complex showed both the Col_h and N_D phases, but at various TNF concentrations – above 45 mol.% TNF, a monotropic N_D phase was induced. The N_D phase became stabilized for the thiocyanato-bridged complex and, once more, the acetato-bridged complex did not show an induced mesophase in such mixtures. Contact preparations of the chloro-bridged platinum complex with TNF also resulted in an induced Col_h phase, with a higher thermal stability than its palladium analog, and an induced N_D phase for the thiocyanato compound. The structure of the different mesophases resulted from intercalation of TNF molecules between successive planar complexes in the columnar phases, while no such stacking was evidenced for the N_D phase. The differences in the mesomorphism observed for the pure compounds and in the binary mixtures were explained by unequal core dimensions caused by the bridging groups, as well as space-filling (steric) and electronic effects.

Four chiral homologous complexes were also prepared [111]. None of the palladium complexes showed mesomorphic properties, whereas a monotropic chiral N_D phase was observed for the platinum complex. All of them form CT complexes with TNF: a Col_h phase was induced for the two halo-bridged complexes, whilst, at low TNF content, a N_D^* phase was stabilized for the thiocyanato-bridged compound along with a non-chiral N_D phase at higher concentration.

Substitution of the alkoxy groups by semifluorinated chains [**20:** M = Pd; X = Cl; R = C_6H_{13}; R′ = $C_{10}H_{21}$, $(CH_2)_6C_4F_9$, $(CH_2)_4C_6F_{13}$] led to stabilization of the mesophases (enantiotropic phases, increase in the clearing temperature), and to a change in the mesophase type (the N_D^* is transformed into Col_h or SmA phases depending on the number of fluorinated chains) due to enhancement of the microsegregation of the various molecular parts [112]. Mesomorphism is enhanced in binary mixture with TNF. Mononuclear palladium and platinum complexes **21**, combining hydrocarbon and fluorocarbon chains, were obtained by ligand exchange reactions with the appropriate diketonates and dinuclear complex derived from **20** [113]. Depending on the total number of chains (5, 6, 7), the fluorocarbon : hydrocarbon chain ratio (3 : 3, 3 : 4, 2 : 3), the degree of chain fluorination [R = $(CH_2)_6C_4F_9/(CH_2)_4C_6F_{13}$; R′ = OR, H; R″ = H,$OC_{10}H_{21}$] and, to a lesser extent, the metal (M = Pd, Pt), smectic [(SmC) and SmA for the pentasubstituted], Col_h (for the hexa- and heptasubstituted) and a nematic (for the penta- and hexasubstituted) phases were observed. Interestingly, the change from the observation of smectic phases to columnar phases as a function of the number of chains is rather abrupt and does not involve some complexes that show both phases.

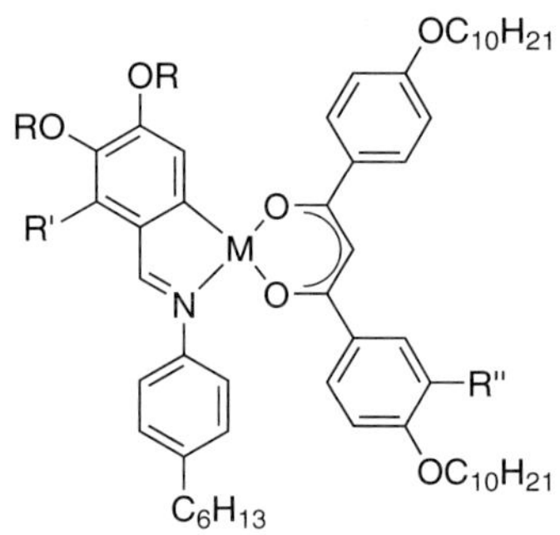

21

Rourke and coworkers reported mesomorphic mononuclear, ortho-cyclopalladated of 4-alkoxy-*N*-(4-alkoxybiphenyl)benzylidenes with various types of co-ligand (**22**, Figure 11.10), such as cyclopentadienyl **(22ai-iv)** [114], β-diketones (**22bi-iv:** *p*, *q* = 1, 4, 6, 8) [115] and amino acids [**22ci-v:** R = Me (alanine), ^{i}Pr (valine), ^{i}Bu (leucine), ^{s}Bu (isoleucine)] [116].

The free ligands displayed SmF, SmC and N phases between 150 and 250 °C, the N phase occurring at ca 200 °C. Derivatives containing the cyclopentadienyl ring, **22a**, exhibited mainly a nematic phase, while that with the longer chain **(22aiv)** showed an additional SmA phase. In this series, an important depression in transition temperatures with respect to free ligand was obtained, with the

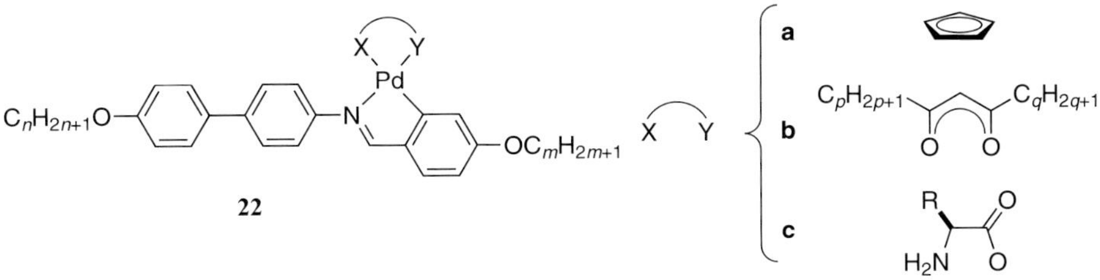

22

Figure 11.10 Structures of various mononuclear complexes, **22** (**i:** $n = m = 4$; **ii:** $n = 4$, $m = 7$; **iii:** $n = 7$, $m = 4$; **iv:** $n = m = 7$; **v:** $n = m = 10$).

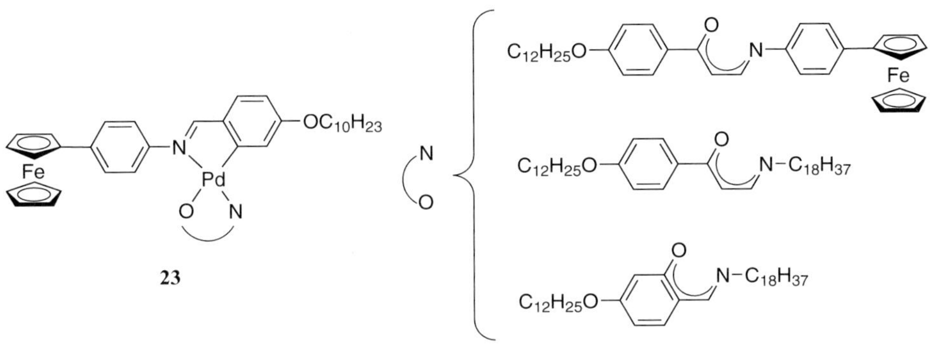

23

Figure 11.11 Structures of the mixed dinuclear and trinuclear complexes **23**.

mesophases occurring between ca 100 and 180 °C, above which temperature they decomposed. In the series with the amino acids (**22c**) the SmA phase was the only mesophase observed, but at temperatures greater than 200 °C where extensive decomposition took place. The thermal behavior of the β-diketonato complexes **22b** was intermediate between those of **22a** and **22c**. All the complexes were mesomorphic, with most of them showing both SmA and N phases, typically in the range 70–250 °C, depending on the chain lengths *n*, *m* *p*, and *q*. The transition temperatures of the acetylacetonate derivatives (**22b:** $p = q = 1$) mirrored perfectly those of the free ligands. The derivatives **22bi**, except where $p = q = 1$, all showed similar melting points at ca 130 °C. The nematic phase disappeared at the expense of the SmA phase when both *p* and *q* increased.

Enantiotropic SmA mesophases were obtained in mixed heteropolynuclear complexes (**23**, Figure 11.11) containing enaminoketone and ortho-palladated imine groups; transition temperatures and temperature ranges depended strongly on the nature of the enaminoketone derivative [117].

Double cyclopalladation of bisimine ligands, followed by reaction with various β-diketones, led to dinuclear complexes showing essentially a nematic phase [118]. However, due to the elevated transition temperatures (>200 °C), most of the acetylacetonate derivatives (**24:** R = Me; n = 4–8) decomposed in the mesophase or in the isotropic liquid. However, a great reduction in transition temperatures, without loss of the nematic phase was achieved by increasing the lateral chain length.

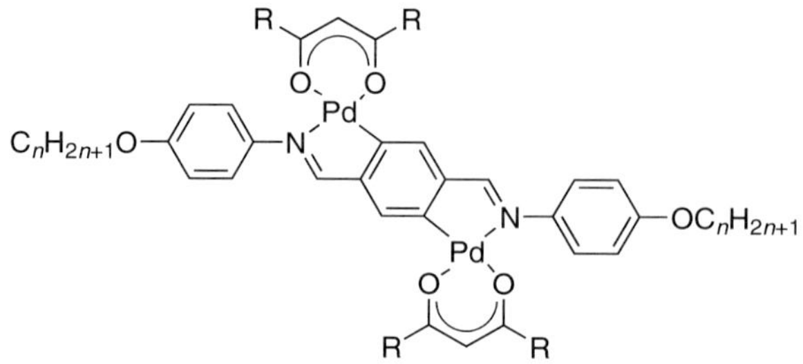

24

Orthometallation of imine-based ligands was not limited to Pd^{II} and Pt^{II}, and Bruce and coworkers demonstrated mesomorphism in benzylideneaniline complexes bound to octahedral Mn^{I} and Re^{I} ([$MMe(CO)_5$], M = Mn, Re), providing that the imine ligand was sufficiently anisotropic [119]. The parent ligands showed smectic and N phases at temperatures up to 300 °C whereas on complexation to Mn^{I}, only the nematic phase was seen for **25** (**25:** M = Mn; $n = 5, 7$) and **26** (**26:** M = Mn; X = Y = H; $n = m = 8$), which cleared below 190 °C with decomposition [120]. The related Re^{I} complexes yielded materials with very similar transition temperatures and with enhanced thermal stability, so that decomposition was not observed at the clearing point [121].

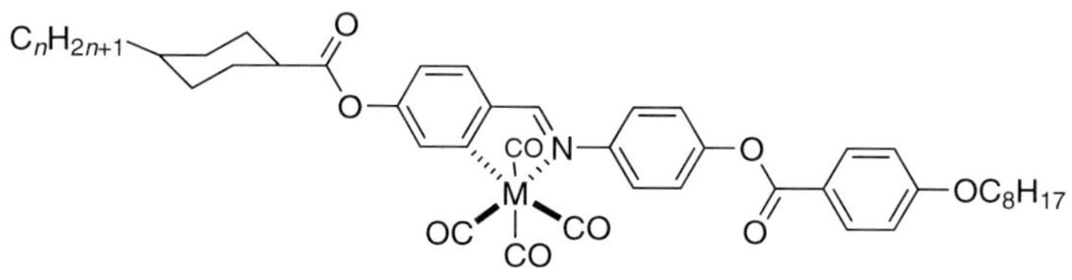

25

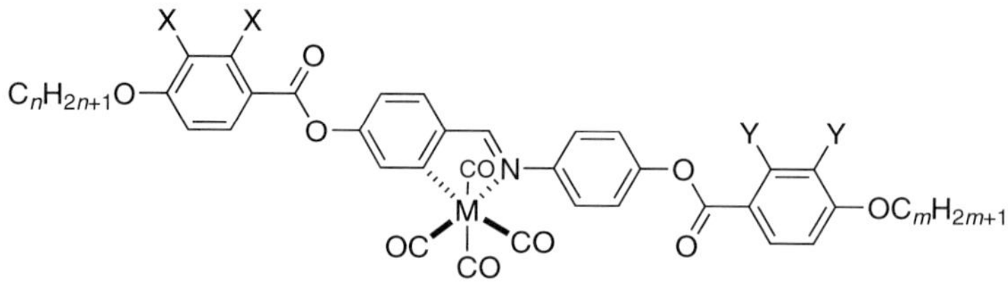

26

Both terminal chain lengths have been varied systematically [122], and lateral ligand fluorination was employed **(26)** [123]. Generally, the mesomorphism of the rhenium complexes was the same, and almost not influenced by chain length, with melting into the nematic phase between ca 130 and 155 °C and clearing between 140 and 200 °C. Fluorination, however, greatly effected the mesophase stability (**26:** M = Re; X, Y = H, F). Thus, while the nematic phase remained, its mesophase stability was considerably reduced with increasing fluorine substitution. This systematic study further revealed that complexes based on two-ring ligands were not mesomorphic, whilst those based on three-ring ligands exhibited a monotropic nematic phase [124].

The effect of the nature of the terminal groups and the position of the imine link has also been investigated. Thus, when hexyl chains are substituted at each end of the rhenium complex having the same motif as **26**, a nematic phase was observed. Moreover, when one of the two hexyl chains were replaced by one or two perfluorinated chains, the mesophase was changed to SmA phase, and occurred at higher temperatures, with decomposition taking place in the mesophase [125, 126]. Similar structural modifications using chiral aliphatic chains (citronellyloxy and its hydrogenated analogue) were performed [127], and yielded complexes with a chiral nematic phase, though when *both* terminal chains were chiral, mesomorphism was suppressed. The phases typically occurred between 120 and 160 °C. Siting the bulky rhenium fragment in one extremity of the molecule resulted in destabilization of the mesomorphism, but not its suppression. Four- and five-ring system diimine ligands, and their corresponding dinuclear rhenium complexes, were also prepared [128]. However, while the diimines exhibited smectic and nematic phases between 100 and 400 °C, none of the dirhenium complexes was mesomorphic.

Mesomorphic macroheterocyclic tetrapalladium [105, 129, 130] and tetraplatinum [108, 131] complexes, $[M_4(\mu_2\text{-}X)_4L_2]$, **27** and **28** (**27:** M = Pd, Pt; *n* = 6, 8, 10, 12, 14, 16, 18; X = OAc, Cl, Br, I, SCN, N_3; **28:** X = Cl, Br, I, OAc, SCN), derived from polycatenar bis(imine)phenylene and the nematogenic [132] bis(imine)stilbenylene ligands, respectively, were reported by Praefcke.

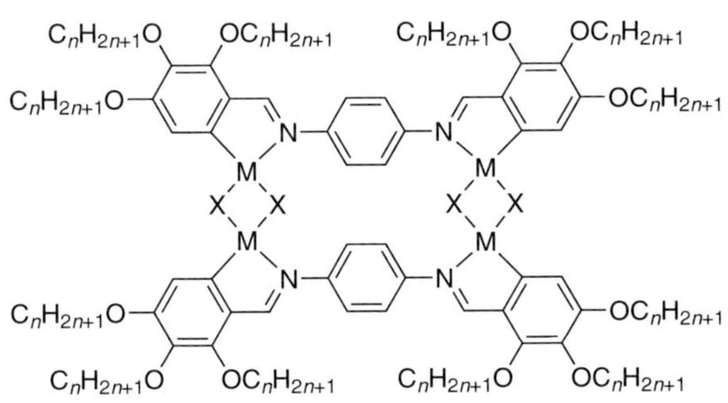

27

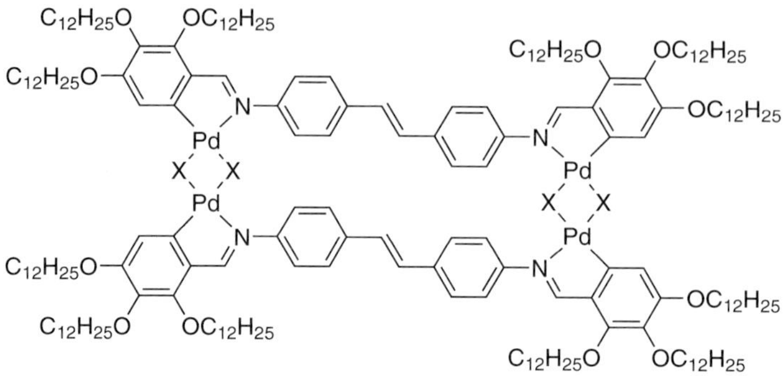

28

These large lipophilic tetrametallaorganyls, bearing three alkoxy chains in the 2-, 3- and 4-positions, were thermotropic mesogens showing broad temperature-range columnar mesophases, as expected from their molecular shape (ca 50–100 up to 250–300 °C); these columnar stacks are mainly self-organized into either rectangular or oblique two-dimensional lattices [133, 134]. The nature of the bridging group had some influence on the transition temperatures and phases symmetry, as did the chain length and the length of the rigid spacer (decrease in thermal stability from **27** → **28**), whereas the metal ion, Pd^{II} or Pt^{II}, appeared to have only minor effects. For the benzylidene derivatives **(27)**, the number of chains was reduced to eight by substituting each of the four aromatic rings at the corners in either positions 2- and 3- (n = 12, 18) or 2- and 4- (n = 12) [133]. Whilst the mesomorphism remained columnar, important modifications in the transition temperatures and phase sequences were observed. The phase transformation between the two rectangular phases of *c2mm* and *p2gg* symmetries was not clearly defined, and both mesophases appeared to co-exist over several degrees (first order transitions with slow transformation kinetics).

In addition to this thermotropic mesomorphism, the complexes formed lyotropic mesophases when dissolved in linear, lipophilic solvents such as alkanes, and in chloroform, benzene, octanol, octadecanol and stearic acid [135–138] In alkanes, the mesophases were stable over wide ranges of temperature and concentration. In general, at high complex concentration, a columnar phase was observed, while at lower concentration a nematic phase was usually induced. Here again, the mesophase behavior depended strongly on molecular intrinsic structural parameters, such as the type of bridges, the length of the chains (complexes and alkanes) and the nature of the metal. Thus, while no mesophase was induced or stabilized in the μ-acetato complex **27** (M = Pd; n = 12), both the chloro- and bromo-bridged complexes exhibited two lyotropic nematic phases (N_1, N_2) and a Col_h phase. A single nematic phase was induced for the thiocyanato-bridged complex along with a Col_h phase, while only a Col_h phase was seen for the iodo- and azido-bridged analogues. Increasing the chain length of the complex (**27**: n = 6–14) and that of the alkanes seemed to favor the occurrence of Col_h and N phases. The homologous platinum complexes in alkanes behaved more or less similarly. Extension of the rigid spacer **(27** → **28)** clearly diminished the tendency for nematic phase induction.

The transition between two nematic phases, being a unique case, was investigated thoroughly for complexes **27** (M = Pd; X = Cl; n = 6, 10, 12, 14) in various alkanes. While the high-temperature nematic phase, N_2, was present in all mixtures, the appearance of the lower-temperature nematic phase, N_1, seemed to depend on the chain-length of the alkane solvent and that of the terminal chains on the complex. A columnar structure was proposed for these two nematic phases [139], with the solvent located between columnar aggregates rather than between the complexes (N_{Col}). Due to swelling, the columns were arranged with only weak inter-columnar order. The correlation length of the columns was not given, but there was no long-range intramolecular ordering. The large, flat metallaorganyls were stacked on top of each other, and arranged perpendicular to the axis of the columns forming the N_2 phase. In the N_1 phase, however, the complexes were tilted with respect to the columnar axis.

Thus, extension of the aliphatic crown in these large metallomesogens by the incorporation of apolar aliphatic solvents enhanced the mesomorphic range by stabilizing the existing columnar phase, and favored the formation of more disordered mesophases such as the N_{Col} phase. The dependence of the mesomorphic properties on a combination of complex chain length and solvent chain length generated the idea of "internal" (that is complex chains) and "external" solvent (that is solvent chains) [129], representing an interesting way of conceptualizing mesophase formation.

Formation of inclusion complexes caused by intercalation of small, electron-acceptor molecules, such as TNF and TAPA [2′-(2,4,5,7-tetranitro-9-fluorenyliden eaminooxy)propionic acid], between large, flat electron-donor molecules has proved to be an effective means for mesophase induction, stabilization and modification [140]. The stability of such CT complexes is connected to strong electrostatic interactions between the donor and acceptor molecules, and has been shown to stabilize smectic phases by enhancing lateral interactions in calamitic systems and columnar phases by improving columnar stacking in disc-like molecules. The tetrametallated organyls **27** and **28**, except those with an acetato-bridge, also formed CT complexes with TNF, and gave rise to a viscous type of columnar phase on heating (transformation of oblique/rectangular symmetry into a hexagonal lattice). The stability of the lyotropic nematic was enhanced (higher clearing temperatures) in all systems compared to the behavior of the pure complexes in pentadecane. The columnar phase was still present in all cases. A large temperature-range, chiral lyotropic nematic phase was induced in a binary system composed of equimolar amounts of **27** and the chiral π-electron acceptor TAPA, in heptane, pentadecane and eicosane [141].

11.4.5
Ortho-Metallated Pyrimidine Complexes

Here, initially, dimeric products **29** were obtained, which were reacted further with species able to cleave the dichloro bridge to give a series of mononuclear derivatives **(30)** [142]. The trans-dinuclear complexes were studied systematically with various bridges (**29:** X = Cl, Br, I, OAc) and ligand chain-lengths (**29i-29iv. i:** n = 6, m = 1; **ii:** n = 9, m = 1; **iii:** n = 6, m = 11; **iv:** n = m = 9) [143]. None of the complexes with X = OAc and no derivatives **29i** were mesomorphic; other non-mesomorphic combinations were **29ii** with X = I and **29iv** with X = Cl. Of the remaining materials, all had a broad SmA phase (the parent ligands showed a nematic phase only), typically between 100 and 200 °C, while two materials (**29ii** with X = Cl and **29iv** with X = I) were reported to have another smectic phase (SmX) above the SmA phase. Later studies [144] showed that what had been identified as SmX was in fact SmA, while the phase identified as SmA was in fact an ordered smectic phase.

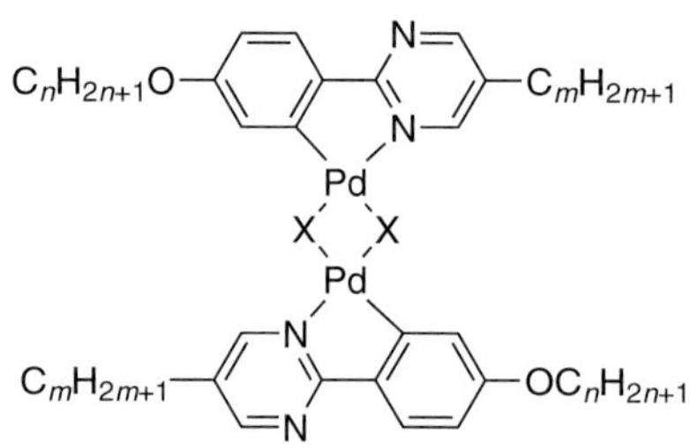

29

In another report by Guang *et al.* [145], the nature of the carboxylato bridging group was investigated (**29:** X = $MeCO_2$, $ClCH_2CO_2$, $BrCH_2CO_2$, $CH_3CHBrCO_2$, $BrCH_2CH_2CO_2$; n = 6; m = 6, 10). For m = 6, the μ-acetato complex was reported as non-mesomorphic, while all other derivatives showed a SmC phase, with the μ-chloroacetato complex showing a SmA phase, too. Clearing points varied widely, but the bromo-substituted bridges consistently gave the lowest values. For the μ-chloro- and bromo-acetato complexes, several derivatives were prepared by varying m ($6 \leq m \leq 12$). As stated already, the chloroacetate with m = 6 gave SmC and SmA phases, but then the SmC phase was suppressed at longer chain length, reappearing for m = 11 and 12. For the bromoacetates, there was a marked odd–even effect so that for m = 7, 9 and 11 only SmA was seen, while for m = 8, 10 and 12 both SmC and SmA phases were observed; m = 6 gave only a SmC phase.

Mononuclear complexes with X–Y = 8-hydroxyquinolato (**30:** x = 0) and X–Y = 1,10-phenanthroline (**30:** x = 1) were non-mesomorphic. However, when X–Y = β-diketonate (**30:** x = 0), only the acac material was mesomorphic (monotropic SmA) and when X–Y = 2,2′-bipyridine (**30:** x = 1) a material with an enantiotropic nematic phase was produced; related complexes with PF_6 or SbF_6 anions were non-mesomorphic. These are further, rare examples of ionic materials showing a thermotropic N phase [1, 2]. Other ionic palladium complexes of the type **30**, with bidentate chelating 2,2′-bipyridine ligands disubstituted in the 4,4′-positions by lateral chains such as $-CO_2C_{22}H_{45}$ or $-CH_2OH$, behaved differently and were found to show SmC (at 50–80 °C and ca 150 °C with BF_4 and ClO_4 counter-anions, respectively) or SmA phases respectively (decomposing rapidly in the mesophase above 200 °C) [146].

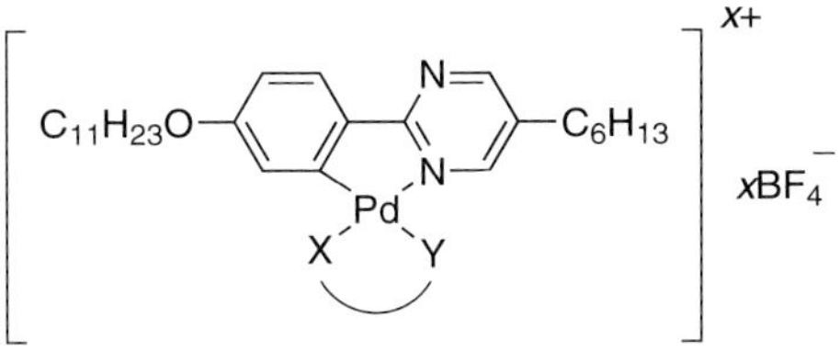

30

Tschierske *et al.* reported binuclear cyclopalladated cyclophane complexes derived from macrocyclic 2-phenylpyrimidine derivatives [147]. Mesophase stabilization or induction was observed upon cyclometallation of the macrocyclic ligands, and enantiotropic SmA and N phases were observed in the dipalladium complex **31** with X = Y = O, while only a nematic (monotropic) phase was seen for the other dinuclear complex (**31**: X = Y = CH_2O).

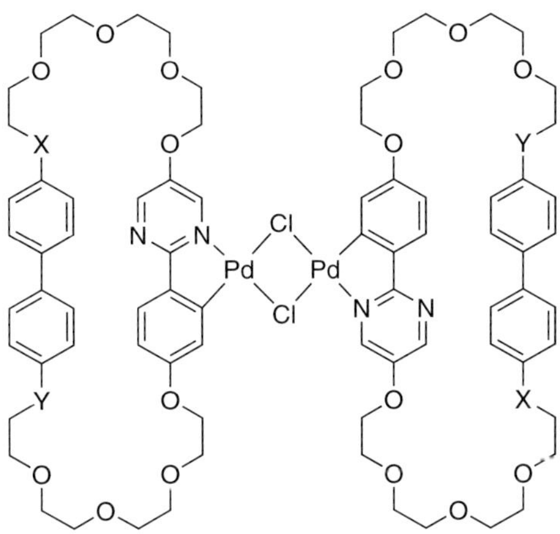

31

As for the cyclometallated azobenzenes **(4)**, azoxybenzenes **(11)** and imines **(19)**, mixed mononuclear ortho-palladated phenylpyrimidine-imine and β-diketonates **(32)** were synthesized. The mononuclear complex combining a chiral imine as in **11d** with a phenylpyrimidine ligand as in **29**, all exhibited a SmA phase between ca 100 and 130 °C losing the SmC* of the free ligand (see above, **11d**); in solution, only the *N,N*-trans complexes were formed [63].

Mesogenic molecules resulting from the combination of the calamitic 2-phenylpyrimidine sub-unit and the half-discotic 1,3-diketonato moiety (**32**, Figure 11.12) were of interest as they offered a possible insight into the "transition" from observations of smectic phases to columnar phases as a function of shape. The number of side-chains on the diketonato fragment was increased stepwise (from four to eight) so that the overall molecular structure changed continuously from a rod- to a disc-like molecular shape (**32a**: M = Pd, Pt; R_1, R_2, R_3, R_4 = $H/OC_{10}H_{21}$) [148, 149]. This type of study is important in that, with such molecular structures,

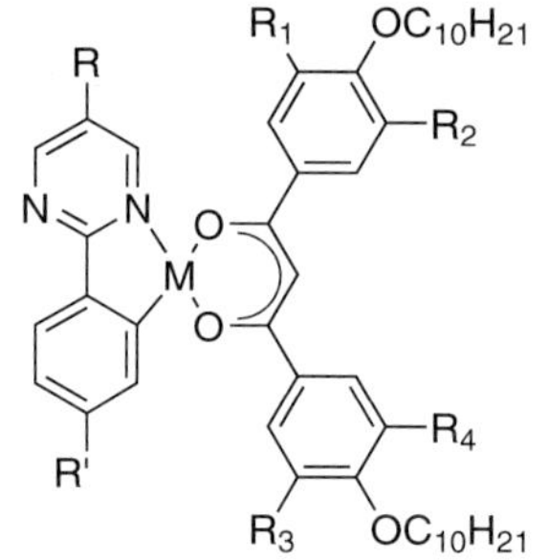

32

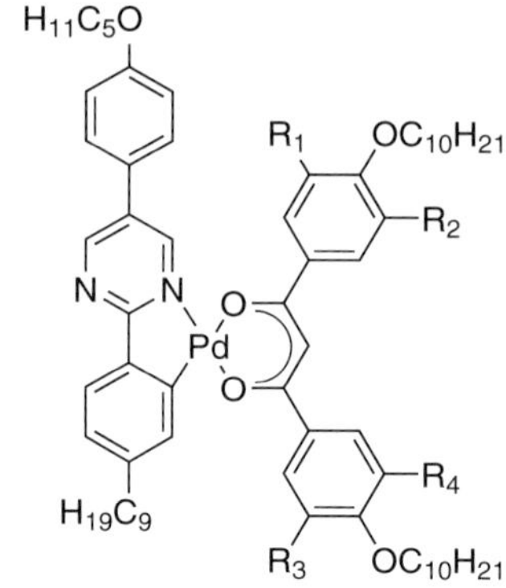

a: R = C_7H_{15}, R' = $OC_{10}H_{21}$ (M = Pd, Pt)
b: R = $C_{10}H_{21}$, R' = OC_8H17 (M = Pd)

c

Figure 11.12 Structures of mononuclear complexes **32** bearing calamitic and half-disc-like coligands.

mesophase transformations may occur through several intermediate mesophases between the nematic and smectic phases of the calamitic mesogens and the columnar phases of the discotic molecules, such as for instance 3D mesophases as in polycatenars systems [18], allowing for a better understanding of the intimate relationship between these different mesophases.

Thus, the first compound in the series, with a total number of four chains **(32ai)**, showed both SmC and SmA phases. Increasing the number of chains to five (**32aii**, 1 : 1 mixture of stereoisomers) led to destabilization of the mesophase and monotropic behavior, whereas with six chains the mesomorphism was either suppressed (**32aiii**; $R_1 = R_3 = OC_{10}H_{21}$; $R_2 = R_4 = H$) or a Col_h phase over small temperature-range as induced (**32aiv**, $R_1 = R_2 = OC_{10}H_{21}$; $R_2 = R_3 = H$, 1 : 1 mixture of stereoisomers). The temperature range of the Col_h phase was enhanced in the more disc-like mesogens bearing seven **(32av**, 1 : 1 mixture of stereoisomers) to eight chains **(32avi)**. Thus, the change from smectic to columnar mesophases is discontinuous. Melting temperatures decreased from 110–120 **(32ai)** to 60 °C **(32avi)** and rose to 80 °C **(32av–32avi)**. Similarly, clearing temperatures first decreased (130–100 down to 75 °C) then rose to 160 °C. The decrease in the mesophase stability of both the unsymmetrically compounds (six- and seven-chained compounds) relative to the corresponding symmetrical analogues (eight-chained compounds) was attributed to the fact that the former were obtained as a 1 : 1 cis/trans isomeric mixtures with respect to the pyrimidine ring. The reason for the absence of mesomorphism in the six-chained compound **32aiii** is not yet understood. Interestingly, a binary phase diagram between two unsymmetrical compounds **(32aii–32av)** revealed the induction of another birefringent mesophase at the contact region, with the destabilization of both mesophases of the pure compounds. The cycloplatinated complexes behaved almost identically to their palladium counterparts [150]. The structures of the compounds forming columnar phases facilitates face-to-face contact of the organometallic cores (alternated stacking), and the molten chain fills the space around these polar core regions. As for the smectic phases, the board-like molecules occupy the layers with no polar order of the cores.

The existence of the biaxial SmA phase (SmA_b) [151], also known as the McMillan phase, has been demonstrated by textural observations and X-ray investigations in some CT complexes formed with the palladium metallomesogens **32a (i–iii)** and TNF. Two, novel mesophases were induced systematically. At low TNF concentration, an ill-defined mesophase, probably columnar, is induced that remained stable at high temperature and up to high concentrations of TNF (ca 60 mol.%). At higher concentrations of TNF, that is upwards of 20 mol.%, the three CT complexes formed the SmA_b phase that is self-organized into layers, with the flat molecules arranged parallel to one other, and orthogonal to the layers, with a long-range, face-to-face organization, and short, side-by-side correlations. This face-to-face interaction hindered the molecular rotation around the long axis considerably, reducing the symmetry, and thus giving rise to the biaxiality.

The study of complexes **32a** was later followed by reports of the behavior of the related Pd^{II} and Pt^{II} complexes **32b,c** [149], in which the authors set out to investigate the transition from the behavior associated with calamitic materials to that associated with more disc-like materials. Thus, the mesomorphism of **32b** varies as that of **32a**, with slightly higher transition temperatures, except for the occurrence of an N_{Col} phase for one compound **(32biv)**. Strong stabilization of the mesomorphism, and particularly the lamellar phases, was generally observed with **32c** although, as the number of chains increased, the stability of the smectic phases decreases accordingly. Unlike previously, and due to the elongated rod-like part, the molecules may form face-to-face or edge-to-edge (the more dominant) dimers, depending on the number and spatial distribution of the chains. These results are summarized in Table 11.3 and show that it is both the number and distribution of chains on the β-diketonate that determine the transition to disc-like behavior. Interestingly, none of the complexes showed behavior characteristic of *both* rods and discs as found in polycatenar systems [18]. Moreover, the platinum compounds were luminescent in both the solid state and solution, and their optical properties depended strongly on the number, position and length of the chains.

Table 11.3 Summary of the mesomorphism of complexes **33**.[a]

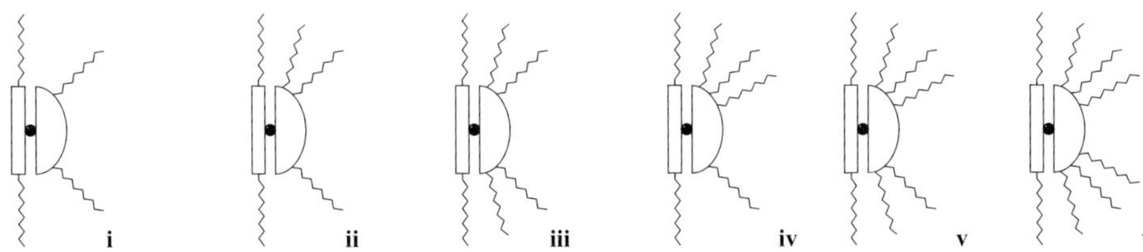

	i	ii	iii	iv	v	vi
32a	SmC and/or SmA	(SmA, N)	–	$Col_h(2)$	Col_h	Col_h
32b	SmA	(SmA, N)	–	$Col_h(2)$, N_{Col}	Col_h	Col_h
32c	SmA	SmC, SmA, N	(SmC), (SmA)	$Col_h(2)$	Col_X, $Col_h(2)$	Col_h

a $Col_h(2)$ represents a columnar hexagonal phase in which a repeat unit consists of two molecules of complex arranged edge-to-edge, Col_X an unidentified columnar phase; – complex not mesomorphic; monotropic phases are in parentheses.

Stronger emission occurred as the number of chains decreased or when the chains were shortened [150].

Hegmann *et al.* [152] modified this system further and synthesized novel, macrocyclic molecules combining two different molecular architectures namely rod-like *para*-cyclophanes with two half-disc-like 1,3-β-diketonato units, fused together by ortho-palladation (**33**: R^1, R^2 = $H/OC_{10}H_{21}$; *x*, *y* = 1–3). The complexes were mesomorphic, with a smectic-to-columnar phase cross-over observed on increasing the chain number, concomitantly with an important decrease in the transition temperatures. The paracyclophane units themselves showed only a monotropic nematic phase, below 150–160 °C.

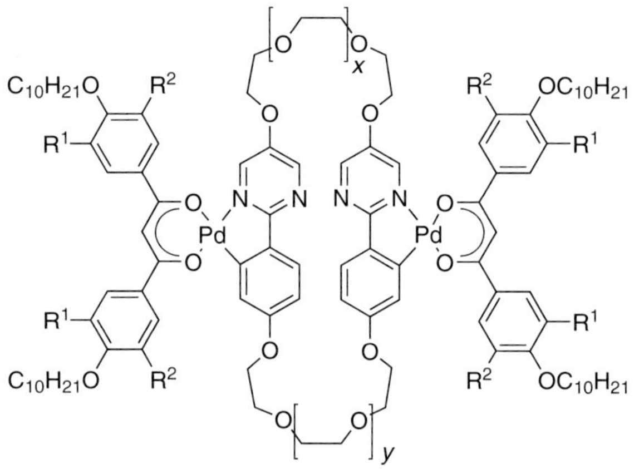

33

The four-chain compound exhibited a SmA phase near 200 °C (**33i**: R^1 = R^2 = H), whereas those with twelve chains (**33i–iii**: R^1 = R^2 = $OC_{10}H_{21}$) displayed an enantiotropic Col_h phase independently of the polyether spacers connecting the two central phenyl-pyrimidine moieties, between 110 and 180–210 °C; the compound with eight side-chains was not mesomorphic (**33i**: R^1 = H, R^2 = $OC_{10}H_{21}$). The two halves of the molecules were assumed planar, although, due to the flexible connectors, the complexes can adopt several conformations. The number of chains thus influenced the nature of the mesophase, while segregation between the central, polar cores and the aliphatic chains helped in stabilizing this arrangement (rather high transition temperatures), providing there were a required number of side chains.

11.4.6
Ortho-Metallated Pyridazine Complexes

Mesomorphic mono- (**34**: R = Me, Bu; *n* = 4–10) and di-cyclopalladated (**35**: R = Me, Bu, *n* = 4–10) pyridazine complexes have been reported [153]. The mononuclear complexes **34** (R = Me) exhibited a single SmA phase between 180 and ca 300 °C. For those with the bulkier diketonate (**34**: R = Bu), a significant depression in clearing temperatures of about 100 °C was observed, and an enantiotropic SmA phase was seen for all the homologs between 100 and 150 °C up to 190 °C. Among the dinuclear complexes, only the *cis*-dicyclopalladated acetylacetonato derivatives (**35**: R = Me) showed a SmA phase, well above 200 °C for the short chain-length homologs, with extensive decomposition. Note that these dimetallated complexes have a sterically induced twist in the molecule that renders them chiral. The related platinum species could not be isolated as their acetylacetonate derivatives [154].

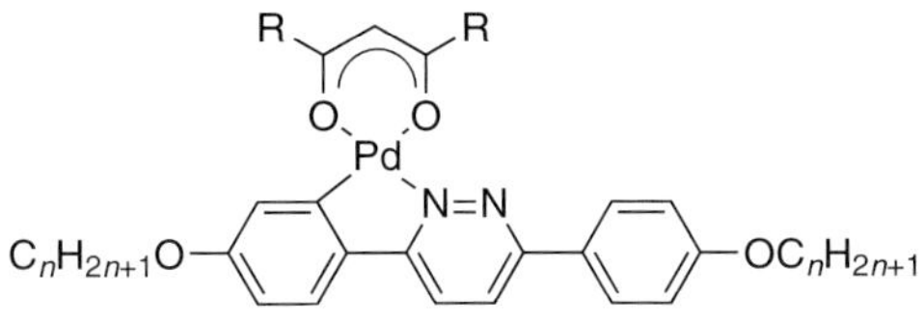

34

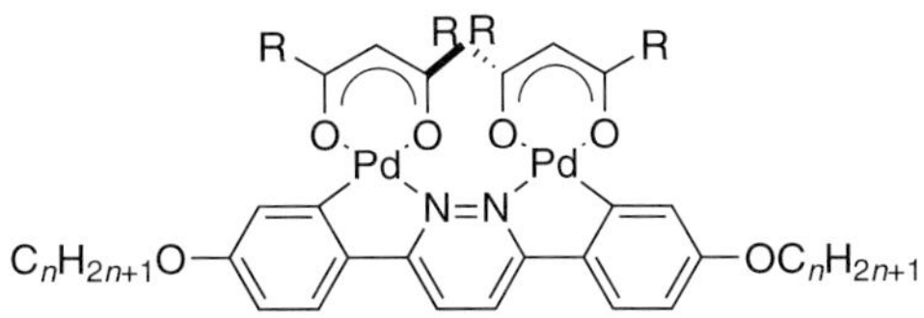

35

Another family of ortho-metallated pyridazine complexes (**36**) was reported by Guang *et al.* [145], where R was varied as Me, $ClCH_2$ and $BrCH_2$. For $n = m = 10$, the dichloro, chloro- and bromo-acetato precursor complexes showed a SmA phase between 160 and 180–210 °C; the simple μ-acetato complex was non-mesomorphic at all chain lengths reported. The lower symmetry complexes ($n = 6$; $m = 10$) were mesomorphic for the chloro- but not the bromo-acetate, and a more extensive series, where n was fixed as 10 and m was varied from 6 to 12, showed enantiotropic SmA phases for all except that with $m = 6$.

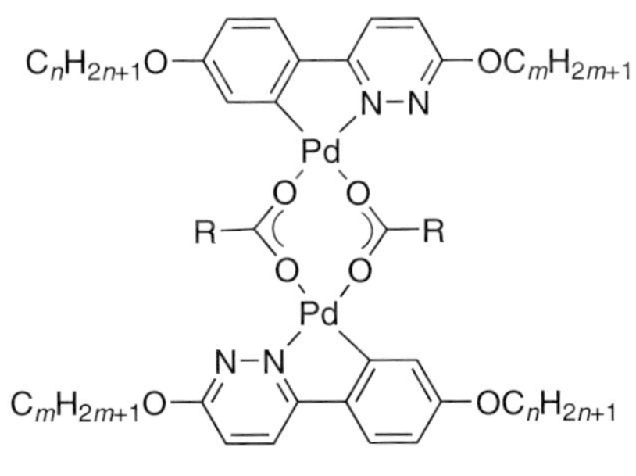

36

11.4.7
Other Ortho-Metallated Complexes

In contrast to **32b**, from which the molecular structure was inspired, ortho-metallated phenylpyridines (**37**: M = Pd, Pt; $R_1/R_2/R_3/R_4$ = $H/OC_{10}H_{21}$) revealed very poor mesomorphic properties. Indeed, only the eight-chain palladium complex exhibited a Col_h phase [149, 150].

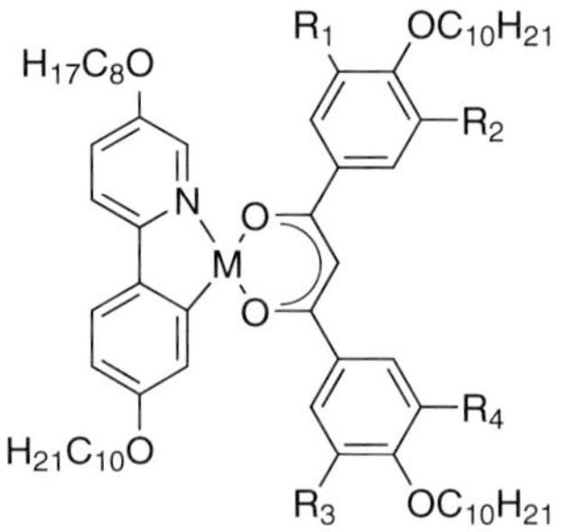

37

Ortho-palladated quinolines coordinated to non-mesomorphic 5,5′-disubstituted-2,2′-bipyridines (**38:** X = BF_4, $C_{12}H_{25}OSO_3$) [155] were found to exhibit liquid-crystallinity, but the thermal stability was much greater in the dodecyl sulfate (DOS) salt than the tetrafluoroborate salt, which degraded rapidly on heating at high temperature. One Col_r and one SmA phase were found for the DOS salt. The Col_r phase resulted from the stacking of the nearly flat aromatic cores on top of each other. However, the columns were not completely surrounded by aliphatic chains, and hence the polar centers of the columns were in lateral contact with one other, forming layers that were separated by layers of molten chains. Some local order of columns still remained in the SmA.

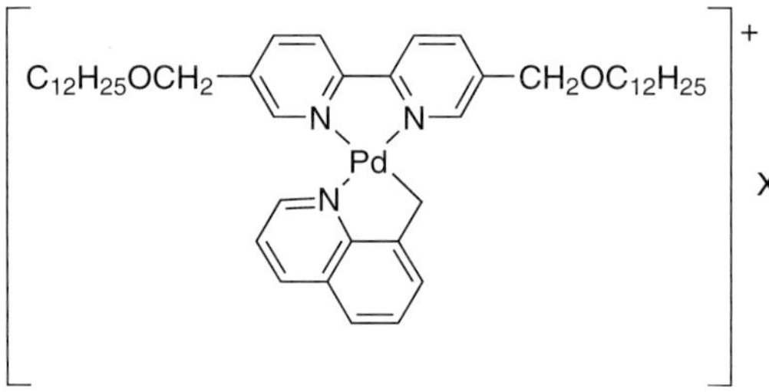

38

Some 6′-phenyl-2,2′-bipyridine ligands and the corresponding *C,N,N*-cyclometallated chloropalladium(II) complexes (**39**, n = 6, 8, 12) were found to be mesomorphic. The ligands were nematogenic between 140 and 170 °C for short chain-lengths, and between 130 and 145 °C for the dodecyloxy homolog. The complexes, however, exhibited very high temperature mesophases, all of which were monotropic, namely a nematic phase for the short chain-length complexes, replaced by a SmA phase for n = 12 [156, 157]. These high temperatures resulted from the rather elongated anisometric part, but shorter ligand anisotropy led to materials devoid of mesomorphism. Interest in such compounds also arises from their potentially interesting photophysical properties, and particularly their electroluminescent properties [158]. Related cationic orthometallated iridium(III) complexes were not mesomorphic [159].

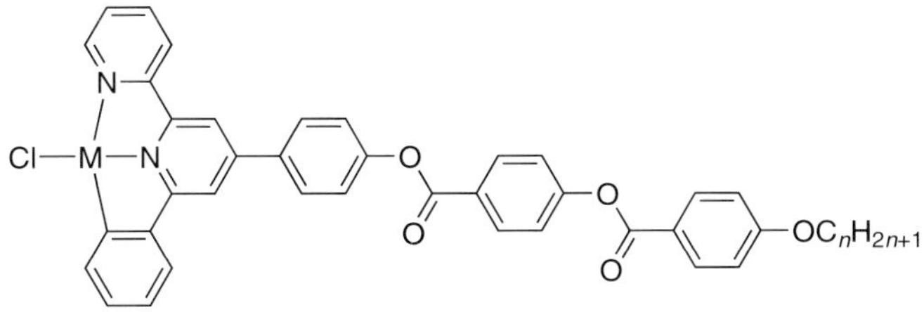

39

The dinuclear ortho-palladated complexes derived from (*S*)-2-(2-hydroxyaryl)oxazoline Schiff base ligands (**40:** X = Cl; x = 1, R = $CHMe_2$; x = 1, 2, R = $C^*HCH_3CH_2CH_3$) showed a broad SmA phase [160]. In the planar, chloro-bridged complexes (cis:trans mixture 1:2), the phase existed from 50 to 150–170 °C and up to 250 °C for the biphenyl derivative; as far as the non-planar, acetato-bridged complexes (only trans isomer) were concerned, only the biphenyl derivative showed a SmA, at elevated temperature (**40:** X = OAc, x = 2, R = $C^*HCH_3CH_2CH_3$). Used as chiral dopants (ca 10 mol.%), a chiral nematic phase (between room temperature and 85–115 °C) was induced systematically in both the chloro- and acetato-bridged series, with suppression of the SmA phase in the chloro-bridged systems (X = Cl, x = 1).

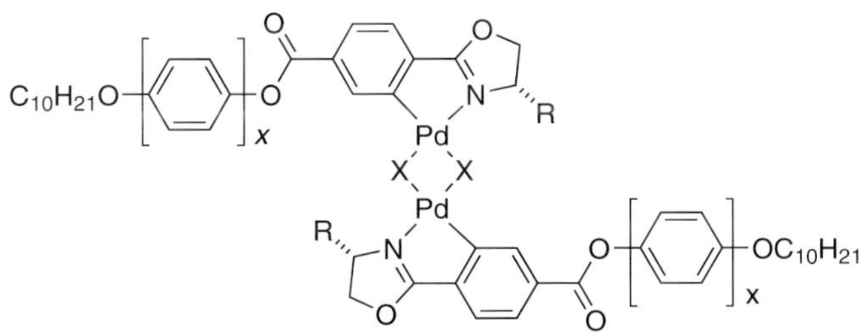

40

The related chiral, hexacatenar dinuclear metal complexes derived from lipophilic chiral oxazoline-based ligands were also synthesized [161]. None of the pure dinuclear compounds (**41:** X = OAc, Cl) was mesomorphic (room-temperature oils or glassy materials), likely due to the sterically demanding chiral central unit preventing molecular stacking and, hence, mesophase formation.

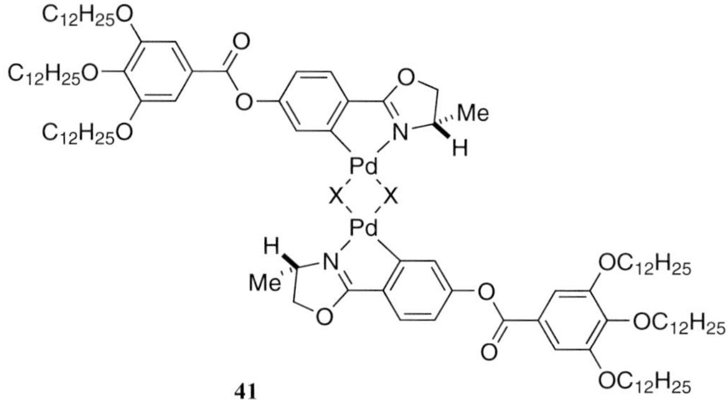

41

Finally, a new red-emitting chromophore resulting from cyclopalladation of 9-diethylamino-5H-benzo[a]phenoxazine-5-one palladium complex and curcumin equipped with terminal gallic substituents **(42)** was reported to show a Col_r phase from room temperature up to 173 °C [162].

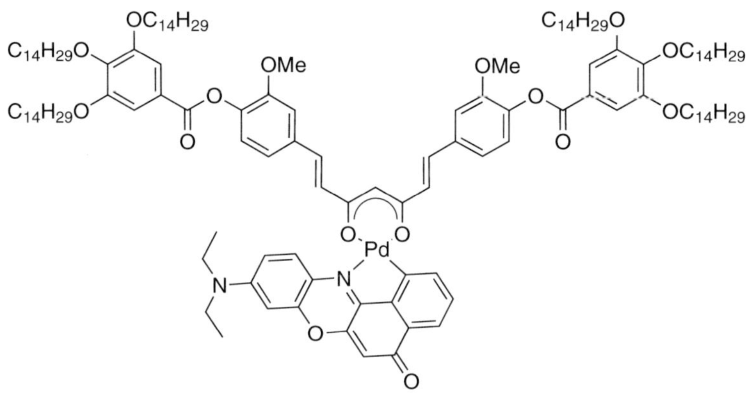

42

References

1 (a) Donnio, B., Guillon, D., Bruce, D.W. and Deschenaux, R. (2003) Metallomesogens, in Comprehensive Coordination Chemistry II: From Biology to Nanotechnology (eds J.A. McCleverty and T.J. Meyer), Elsevier, Oxford, UK. *Volume 7: From the Molecular to the Nanoscale: Synthesis, Structure, and Properties* (eds M. Fujita, A. Powell and C. Creutz), Chapter 7.9, pp. 357–627.
(b) Donnio, B., Guillon, D.W. Bruce and R. Deschenaux (2006) Metallomesogens, in Comprehensive Organometallic Chemistry III: From Fundamentals to Applications (eds R.H. Crabtree and D.M.P. Mingos), Elsevier, Oxford, UK. *Volume 12: Applications III: Functional Materials, Environmental and Biological Applications* (ed. D. O'Hare), Chapter 12.05, pp. 195–294.

2 Serrano, J.L. (1995) *Metallomesogens: Synthesis, Properties and Applications*, Wiley-VCH Verlag GmbH, Weinheim.

3 (a) Ringsdorf, H., Schlarb, B. and Venzmer, J. (1998) *Angewandte Chemie –*

International Edition, **27**, 113–58;(b) Tschierske, C. (1996) *Progress in Polymer Science*, **21**, 775–852.
4 Coates, D. and Gray, G.W. (1976) *The Microscope*, **24**, 117–50.
5 Schadt, S.M. and Helfrich, W. (1971) *Applied Physics Letters*, **18**, 127–8.
6 (a) Stewart, G.T. (2003) *Liquid Crystals*, **30**, 541–57;(b) Stewart, G.T. (2004) *Liquid Crystals*, **31**, 433–71.
7 Demus, D., Goodby, J.W., Gray, G.W., *et al.* (eds) (1998) *Handbook of Liquid Crystals*, Wiley-VCH Verlag GmbH, Weinheim.
8 Giroud-Godquin, A.M. and Maitlis, P.M. (1991) *Angewandte Chemie – International Edition in English*, **30**, 375–402.
9 Skoulios, A. and Guillon, D. (1988) *Molecular Crystals and Liquid Crystals*, **165**, 317–32.
10 Tschierske, C. (1998) *Journal of Materials Chemistry*, **8**, 1485–508.
11 Friedel, G. (1922) *Annales de Physique*, **18**, 273–505.
12 (a) Hendrikx, Y. and Levelut, A.M. (1988) *Molecular Crystals and Liquid Crystals*, **165**, 233–63.
(b) Demus, D. (1989) *Liquid Crystals*, **5**, 75–110.
13 Hird, M. and Toyne, K.J. (1998) *Molecular Crystals and Liquid Crystals*, **323**, 1–67.
14 Goodby, J.W., Mehl, G.H., Saez, I.M., *et al.* (1998) *Chemical Communications*, 2057–70.
15 (a) Peltz, G., Diele, S. and Weissflog, W. (1999) *Advanced Materials*, **11**, 707–24.
(b) Reddy, R.A. and Tschierske, C. (2006) *Journal of Materials Chemistry*, **16**, 907–61.
(c) Takezoe, H. and Takanishi, Y. (2006) *Japanese Journal of Applied Chemistry*, **45**, 597–625.
16 Tschierscke, C. (2001) *Annual Reports on the Progress of Chemistry. Section C*, **97**, 191–267.
17 Takezoe, H., Kishikawa, K. and Gorecka, E. (2006) *Journal of Materials Chemistry*, **16**, 2412–16.
18 (a) Malthête, J., Nguyen, H.-T. and Destrade, C. (1993) *Liquid Crystals*, **13**, 171–87;
(b) Nguyen, H.-T., Destrade, C. and Malthête, J. (1997) *Advanced Materials*, **9**, 375–88.
(c) Fazio, D., Mongin, C., Donnio, B., *et al.* (2001) *Journal of Materials Chemistry*, **11**, 2852–63.
19 (a) Tschierske, C. (2001) *Journal of Materials Chemistry*, **8**, 2647–71.
(b) Lee, M., Cho, B.K. and Zin, W.C. (2001) *Chemical Reviews*, **101**, 3869–92.
(c) Lee, M. and Yoo, Y.-S. (2002) *Journal of Materials Chemistry*, **12**, 2161–8.
20 (a) Imrie, C.T. (1999) *Structure and Bonding*, **95**, 149–92.
(b) Saez, I.M. and Goodby, J.W. (2005) *Journal of Materials Chemistry*, **15**, 26–40.
(c) Kato, T., Mizoshita, N. and Kishimoto, K. (2006) *Angewandte Chemie – International Edition*, **45**, 38–68.
21 (a) Ponomarenko, S.A., Boiko, N.I. and Shibaev, V.P. (2001) *Polymer Science*, **43**, 1–45.
(b) Donnio B. and Guillon, D. (2006) *Advances in Polymer Science*, **201**, 45–155.
22 Paleos, C.M. (1994) *Molecular Crystals and Liquid Crystals*, **243**, 159–83.
23 Jeffrey, G.A. and Wingert, L.M. (1992) *Liquid Crystals*, **12**, 179–202.
24 Paleos, C.M. and Tsiourvas, D. (2001) *Liquid Crystals*, **28**, 1127–61.
25 (a) Diele, S. (2002) *Current Opinion in Colloid and Interface Science*, **7**, 333–42.
(b) Kutsumizu, S. (2002) *Current Opinion in Colloid and Interface Science*, **7**, 537–543.
(c) Imperor-Clerc, M. (2005) *Current Opinion in Chemical Biology*, **9**, 370–6.
26 Kumar, S. (2006) *Chemical Society Reviews*, **35**, 83–109.
27 (a) Praefcke, K., Blunk, D., Singer, D., *et al.* (1998) *Molecular Crystals and Liquid Crystals*, **323**, 231–59.
(b) Bruce, D.W (2004) *The Chemical Record*, **4**, 10–22.
28 Luckhurst, G. (2001) *Thin Solid Films*, **393**, 40–52.
29 Gray, G.W. and Goodby, J.W. (1984) *Smectic Liquid Crystals; Textures and Structures*, Leonard Hill, Glasgow.
30 This classification was proposed in: Leadbetter, A.J. (1987) Structural classification of liquid crystals, in *Thermotropic Liquid Crystals* (ed. G.W. Gray), John Wiley & Sons, Ltd, Chichester, pp. 1–27.

31 (a) Levelut, A.M. (1983) *Journal de Chimie Physique*, **80**, 149–61.
(b) Guillon, D. (1999) *Structure and Bonding*, **95**, 41–82.
32 (a) Goodby, J.W. (1991) *Journal of Materials Chemistry*, **1**, 307–18.
(b) Goodby, J.W., Slaney, A.J., Booth, C.J. *et al.* (1994) *Molecular Crystals and Liquid Crystals*, **243**, 231–98.
33 Goodby, J.W. (1999) *Structure and Bonding*, **95**, 83–147.
34 Crooker, P.P. (1989) *Liquid Crystals*, **5**, 751–75.
35 Serrano, J.L. and Sierra, T. (2000) *Chemistry–A European Journal*, **6**, 759–66.
36 Ghedini, M., Armentano, S., Bartolino, R., *et al.* (1987) *Molecular Crystals and Liquid Crystals*, **151**, 75–91.
37 Ghedini, M., Longeri, M. and Bartolino, R. (1982) *Molecular Crystals and Liquid Crystals*, **84**, 207–11.
38 Ghedini, M., Licoccia, S., Armentano, S. and Bartolino, R. (1984) *Molecular Crystals and Liquid Crystals*, **108**, 269–75.
39 Ghedini, M., Armentano, S. and Neve, F. (1987) *Inorganica Chimica Acta*, **134**, 23–4.
40 Ghedini, M., Armentano, S. and Neve, F. (1988) *Journal of the Chemical Society–Dalton Transactions*, 1565–7.
41 Levelut, A.M., Veber, M., Francescangeli, O., *et al.* (1995) *Liquid Crystals*, **19**, 241–9.
42 Zhang, L., Huang, D., Xiong, N., Yang, J., *et al.* (1993) *Molecular Crystals and Liquid Crystals*, **237**, 285–97.
43 Crispini, A., Ghedini, M., Morrone, S., *et al.* (1996) *Liquid Crystals*, **20**, 67–76.
44 Ghedini, M., Pucci, D., Crispini, A., *et al.* (1999) *Appled Organometallic Chemistry*, **13**, 565–81.
45 Ghedini, M. and Crispini, A. (1999) *Comments on Inorganic Chemistry*, **21**, 53–68.
46 Cladis, P.E. (1998) *Handbook of Liquid Crystals*, Vol. **1**, Chapter VII (eds D. Demus, J. Goodby, G.W. Gray, *et al.*), Wiley-VCH Verlag GmbH, Weinheim, pp. 391–405.
47 Ghedini, M., Pucci, D., Cesarotti, E., *et al.* (1993) *Chemistry of Materials*, **5**, 883–90.
48 Ghedini, M., Morrone, S., Neve, F. and Pucci, D. (1996) *Gazzetta Chimica Italiana*, **126**, 511–15.
49 Ghedini, M., Pucci, D. and Neve, F. (1996) *Chemical Communications*, 137–8.
50 Ghedini, M., Pucci, D., Cesarotti, E., *et al.* (1994) *Liquid Crystals*, **16**, 373–80.
51 Huang, D.J., Xiong, N.Y., Yang, J., *et al.* (1993) *Molecular Crystals and Liquid Crystals*, **231**, 191–8.
52 Pucci, D., Francescangeli, O. and Ghedini, M. (2001) *Molecular Crystals and Liquid Crystals*, **372**, 51–68.
53 Talarico, M., Barberio, G., Pucci, D., *et al.* (2003) *Advanced Materials*, **15**, 1374–7.
54 Talarico, M., Termine, R., Berberio, G., *et al.* (2004) *Applied Physics Letters*, **84**, 1034–6.
55 An introduction and overview of molecular photoreactive materials: Ostroverkhova, O. and Moerner, W.E. (2004) *Chemical Reviews*, 104, 3267–314.
56 Ghedini, M., Neve, F. and Pucci, D. (1998) *European Journal of Inorganic Chemistry*, 501–4.
57 Ghedini, M., Pucci, D., Crispini, A. and Barberio, G. (1999) *Organometallics*, **18**, 2116–24.
58 Ghedini, M., Pucci, D. and Barberio, G. (2000) *Liquid Crystals*, **27**, 1277–83.
59 Hoshino, N., Hasegawa, H. and Matsunaga, Y. (1991) *Liquid Crystals*, **9**, 267–76.
60 Ghedini, M., Pucci, D., Armentano, S., *et al.* (1992) It. Pat., No VE92,000,003.
61 Ghedini, M., Morrone, S., Francescangeli, O. and Bartolino, R. (1994) *Chemistry of Materials*, **6**, 1971–7.
62 Ghedini, M., Morrone, S., Francescangeli, O. and Bartolino, R. (1992) *Chemistry of Materials*, **4**, 1119–23.
63 Ghedini, M., Pucci, D., Cesarotti, E., *et al.* (1993) *Liquid Crystals*, **15**, 331–44.
64 Ghedini, M., Morrone, S., De Munno, G. and Crispini, A. (1991) *Journal of Organometallic Chemistry*, **415**, 281–91.
65 Omenat, A. and Ghedini, M. (1994) *Journal of the Chemical Society–Chemical Communications*, 1309–10.
66 Espinet, P., Lalinde, E., Marcos, M., *et al.* (1990) *Organometallics*, **9**, 555–60.

67 Elder, R.C., Cruea, R.D.P. and Morrison, R.F. (1976) *Inorganic Chemistry*, **15**, 1623–6.

68 Espinet, P., Etxebarria, J., Marcos, M., *et al.* (1989) *Angewandte Chemie – International Edition*, **28**, 1065–6.

69 Zhang, L., Huang, D., Xiong, N. and Li, G. (1992) *Chinese Chemistry Letters*, **3**, 805–6.

70 Espinet, P., Perez, J., Marcos, M., Ros, *et al.* (1990) *Organometallics*, **9**, 2028–33.

71 Levelut, A.M. (1992) *Molecular Crystals and Liquid Crystals*, **215**, 31–46.

72 Huang, D.J., Xiong, N.Y., Yang, J., *et al.* (1993) *Molecular Crystals and Liquid Crystals*, **231**, 191–8.

73 Barberá, J., Espinet, P., Lalinde, E., *et al.* (1987) *Liquid Crystals*, **2**, 833–42.

74 Buey, J. and Espinet, P. (1996) *Journal of Organometallic Chemistry*, **507**, 137–45.

75 Baena, M.J., Espinet, P., Ros, M.B. and Serrano, J.L. (1996) *Journal of Materials Chemistry*, **6**, 1291–6.

76 Ciriano, M.A., Espinet, P., Lalinde, E., *et al.* (1989) *Journal of Molecular Structure*, **196**, 327–41.

77 Marcos, M., Ros, M.B. and Serrano, J.L. (1988) *Liquid Crystals*, **3**, 1129–36.

78 Ghedini, M., Armentano, S., De Munno, G., *et al.* (1990) *Liquid Crystals*, **8**, 739–44.

79 Ros, M.B., Ruiz, N., Serrano, J.L. and Espinet, P. (1991) *Liquid Crystals*, **9**, 77–86.

80 Baena, M.J., Barberá, J., Espinet, P., *et al.* (1994) *Journal of the American Chemical Society*, **116**, 1899–906.

81 Diez, L., Espinet, P., Miguel, J.A. and Ros, M.B. (2002) *Journal of Materials Chemistry*, **12**, 3694–8.

82 López de Murillas, D., Piñol, R., Ros, M.B., Serrano, J.L., *et al.* (2004) *Journal of Materials Chemistry*, **14**, 1117–27.

83 Baena, M.J., Buey, J., Espinet, P., *et al.* (1993) *Angewandte Chemie – International Edition*, **32**, 1201–3.

84 Buey, J., Diez, G.A., Espinet, P., *et al.* (1998) *European Journal of Inorganic Chemistry*, 1235–41.

85 Buey, J., Espinet, P., Kitzerow, H.S. and Strauss, J. (1999) *Chemical Communications*, 441–2.

86 Baena, M.J., Buey, J., Espinet, P. and García-Prieto, C.E. (2005) *Journal of Organometallic Chemistry*, **690**, 998–1010.

87 Arias, J., Bardají, M. and Espinet, P. (2006) *Journal of Organometallic Chemistry*, **691**, 4990–9.

88 Díez, L., Espinet, P. and Miguel, J.A. (2001) *Journal of the Chemical Society – Dalton Transactions*, 1189–95.

89 Díez, L., Espinet, P., Miguel, J.A. and Rodriguez-Medina, M.P. (2005) *Journal of Organometallic Chemistry*, **690**, 261–8.

90 Espinet, P., Hernández, C., Martin-Alvarez, J.M. and Miguel, J.A. (2004) *Inorganic Chemistry*, **43**, 843–5.

91 Baena, M.J., Espinet, P., Ros, M.B. and Serrano, J.L. (1991) *Angewandte Chemie – International Edition in English*, **30**, 711–12.

92 Omnès, L., Timimi, B.A., Gelbrich, T., *et al.* (2001) *Chemical Communications*, 2248–9.

93 Omnès, L., Cîrcu, V., Hutchins, P.T., *et al.* (2005) *Liquid Crystals*, **32**, 1437–47.

94 Cîrcu, V., Gibbs, T.J.K., Omnès, L., *et al.* (2006) *Journal of Materials Chemistry*, **16**, 4316–25.

95 Thompson, N.J., Iglesias, R., Serrano, J.L., *et al.* (1996) *Journal of Materials Chemistry*, **6**, 1741–4.

96 Baena, M.J., Espinet, P., Ros, M.B., *et al.* (1993) *Angewandte Chemie – International Edition*, **32**, 1203–5.

97 Thompson, N.J., Serrano, J.L., Baena, M.J. and Espinet, P. (1996) *Chemistry – A European Journal*, **2**, 214–20.

98 Espinet, P., Etxebarría, J., Folcia, C.L., *et al.* (1996) *Advanced Materials*, **8**, 745–8.

99 Buey, J., Díez, L., Espinet, P., *et al.* (1996) *Chemistry of Materials*, **8**, 2375–81.

100 Buey, J., Díez, L., Espinet, P., *et al.* (1998) *Applied Physics B*, **66**, 355–8.

101 Ortega, J., Folcia, C.L., Etxebarria, J., *et al.* (1997) *Liquid Crystals*, **23**, 285–91.

102 Díez, L., Espinet, P., Miguel, J.A. and Ros, M.B. (2002) *Journal of Materials Chemistry*, **12**, 3694–8.

103 Usol'tseva, N., Espinet, P., Buey, J. and Serrano, J.L. (1997) *Journal of Materials Chemistry*, **7**, 215–19.

104 Usol'tseva, N., Espinet, P., Buey, J., *et al.* (1997) *Molecular Crystals and Liquid Crystals*, **299**, 457–65.

105 Praefcke, K., Singer, D. and Gündogan, B. (1992) *Molecular Crystals and Liquid Crystals*, **223**, 181–95.

106 Praefcke, K., Bilgin, B., Pickardt, J. and Borowski, M. (1994) *Chemische Berichte*, **127**, 1543–5.

107 Praecke, K., Bilgin, B., Pickardt, J. and Borowski, M. (1999) *Journal of Organometallic Chemistry*, **592**, 155–61.

108 Bilgin-Eran, B., Singer, D., Pickardt, J. and Praefcke, K. (2001) *Journal of Organometallic Chemistry*, **620**, 249–55.

109 Singer, D., Liebmann, A., Praefcke, K. and Wendorff, J.H. (1993) *Liquid Crystals*, **14**, 785–94.

110 Praefcke, K. and Singer, D. (1994) *Molecular Materials*, **3**, 265–70.

111 Bilgin Eran, B., Singer, D. and Praefcke, K. (2001) *European Journal of Inorganic Chemistry*, 111–16.

112 Bilgin Eran, B., Tschierske, C., Diele, S. and Baumeister, U. (2006) *Journal of Materials Chemistry*, **16**, 1136–44.

113 Bilgin Eran, B., Tschierske, C., Diele, S. and Baumeister, U. (2006) *Journal of Materials Chemistry*, **16**, 1145–53.

114 Lydon, D.P., Cave, G.W.V. and Rourke, J.P. (1997) *Journal of Materials Chemistry*, **7**, 403–6.

115 Cave, G.W.V., Lydon, D.P. and Rourke, J.P. (1998) *Journal of Organometallic Chemistry*, **555**, 81–8.

116 Saccomando, D.J., Black, C., Cave, G.W.V., *et al.* (2000) *Journal of Organometallic Chemistry*, **601**, 305–10.

117 Kadkin, O., Galyametdinov, Y. and Rakhmatullin, A. (1999) *Molecular Crystals and Liquid Crystals*, **332**, 109–18.

118 Lydon, D.P. and Rourke, J.P. (1997) *Chemical Communications*, 1741–2.

119 Bruce, D.W. (1994) *Advanced Materials*, **6**, 699–701.

120 Bruce, D.W. and Liu, X.H. (1994) *Journal of the Chemical Society D–Chemical Communications*, 729–30.

121 Bruce, D.W. and Liu, X.H. (1995) *Liquid Crystals*, **18**, 165–6.

122 Liu, X.H., Heinrich, B., Manners, I., *et al.* (2000) *Journal of Materials Chemistry*, **10**, 637–44.

123 Liu, X.H., Manners, I. and Bruce, D.W. (1998) *Journal of Materials Chemistry*, **8**, 1555–60.

124 Liu, X.H., Abser, M.N. and Bruce, D.W. (1998) *Journal of Organometallic Chemistry*, **551**, 271–80. *Publisher's erratum* (1999), **577**, 150–2.

125 Guillevic, M.A. and Bruce, D.W. (2000) *Liquid Crystals*, **27**, 153–6.

126 Guillevic, M.A., Gelbrich, T., Hursthouse, M.B. and Bruce, D.W. (2001) *Molecular Crystals and Liquid Crystals*, **362**, 147–70.

127 Guillevic, M.A., Danks, M.J., Harries, S.K., *et al.* (2000) *Polyhedron*, **19**, 249–57.

128 Guillevic, M.A., Light, M.E., Coles, S.J., *et al.* (2000) *Journal of the Chemical Society–Dalton Transactions*, 1437–45.

129 Praefcke, K., Holbrey, J.D., Usol'tseva, N. and Blunk, D.N. (1996) *Molecular Crystals and Liquid Crystals*, **288**, 189–200.

130 Praefcke, K., Singer, D., Gündogan, B., *et al.* (1993) *Berichte der Bunsen-Gesellschaft Physical Chemistry Chemical Physics*, **97**, 1358–61. Corrigedum (1994) **98**, 118–121.

131 Praefcke, K., Bilgin, B., Usol'tseva, N., *et al.* (1995) *Journal of Materials Chemistry*, **5**, 2257–64.

132 Gündogan, B. and Praefcke, K. (1993) *Chemische Berichte*, **126**, 1253–5.

133 Heinrich, B., Praefcke, K. and Guillon, D. (1997) *Journal of Materials Chemistry*, **7**, 1363–72.

134 Praefcke, K., Diele, S., Pickardt, J., *et al.* (1995) *Liquid Crystals*, **18**, 857–65.

135 Usol'tseva, N., Praefcke, K., Singer, D. and Gündogan, B. (1994) *Molecular Materials*, **4**, 253–63.

136 Usol'tseva, N., Praefcke, K., Singer, D. and Gündogan, B. (1994) *Liquid Crystals*, **16**, 601–16.

137 Praefcke, K., Holbrey, J.D., Usol'tseva, N. and Blunk, D. (1997) *Molecular Crystals and Liquid Crystals*, **292**, 123–39.

138 Nesrullajev, A., Bilgin-Eran, B. and Kazanci, N. (2002) *Materials Chemistry and Physics*, **76**, 7–14.

139 Usol'tseva, N., Hauck, G., Koswig, H.D., *et al.* (1996) *Liquid Crystals*, **20**, 731–9.

140 Praefcke, K. and Holbrey, J.D. (1996) *Journal of Inclusion Phenomena and Macrocyclic Chemistry*, **24**, 19–41.

141 Usol'tseva, N., Praefcke, K., Singer, D. and Gündogan, B. (1994) *Liquid Crystals*, **16**, 617–23.

142 Ghedini, M. and Pucci, D. (1990) *Journal of Organometallic Chemistry*, **395**, 105–12.

143 Ghedini, M., Pucci, D., De Munno, G., *et al.* (1991) *Chemistry of Materials*, **3**, 65–72.

144 Ghedini, M., Pucci, D., Bartolino, R. and Francescangeli, O. (1993) *Liquid Crystals*, **13**, 255–63.

145 Guang, W., Han, J., Wan, W., *et al.* (2003) *Liquid Crystals*, **30**, 1259–65.

146 Pucci, D., Barberio, G., Bellusci, A., *et al.* (2006) *Journal of Organometallic Chemistry*, **691**, 1138–42.

147 Neumann, B., Hegmann, T., Wolf, R. and Tschierske, C. (1998) *Chemical Communications*, 105–6.

148 Hegmann, T., Peidis, F., Diele, S. and Tschierske, C. (2000) *Liquid Crystals*, **27**, 1261–5.

149 Hegmann, T., Kain, J., Diele, S., *et al.* (2003) *Journal of Materials Chemistry*, **13**, 991–1003.

150 Damm, C., Israel, G., Hegmann, T. and Tschierske, C. (2006) *Journal of Materials Chemistry*, **16**, 1808–16.

151 Hegmann, T., Kain, J., Diele, S., *et al.* (2001) *Angewandte Chemie – International Edition*, **40**, 887–90.

152 Hegmann, T., Neumann, B., Kain, J., *et al.* (2000) *Journal of Materials Chemistry*, **10**, 2244–8.

153 Slater, J.W., Lydon, D.P. and Rourke, J.P. (2002) *Journal of Organometallic Chemistry*, **645**, 246–55.

154 Slater, J.W., Lydon, D.P., Alcock, N.W. and Rourke, J.P. (2001) *Organometallics*, **20**, 4418–23.

155 El-ghayoury, A., Douce, L., Skoulios, A. and Ziessel, R. (1998) *Journal of Organometallic Chemistry*, **37**, 1255–8.

156 Neve, F., Ghedini, M. and Crispini, A. (1996) *Chemical Communications*, 2463–4.

157 Neve, F., Ghedini, M., Francescangeli, O. and Campagna, S. (1998) *Liquid Crystals*, **24**, 673–80.

158 Neve, F., Crispini, A. and Campagna, S. (1997) *Inorganic Chemistry*, **36**, 6150–6.

159 Neve, F. and Crespini, A. (2000) *European Journal of Inorganic Chemistry*, 1039–43.

160 Lehmann, M., Marcos, M., Serrano, J.L., *et al.* (2001) *Chemistry of Materials*, **13**, 4374–81.

161 Lehmann, M., Sierra, T., Barberá, J., *et al.* (2002) *Journal of Materials Chemistry*, **12**, 1342–50.

162 Ghedini, M., Pucci, D., Crispini, A., *et al.* (2007) *Inorganic Chemistry Communications*, **10**, 243–6.

12
Photophysical Properties of Cyclopalladated Compounds

Francesco Neve

12.1
Introduction

The last decade has seen a phenomenal surge of interest in electronic devices that rely on the photo- and electroluminescence of organic materials. Today, organic light-emitting diodes (OLEDs) [1], as well as their promising light-emitting electrochemical cells (LECs) [2] counterparts, are the most viable alternative to liquid crystals for the flat-panel displays of the future. Owing to the enormous potential of this technology, present research in this field is devoted to exploiting the luminescence properties of several classes of materials [3, 4], including transition metal organometallic complexes [5–8]. Among the latter, cyclometallated complexes of d^6 heavy metals are emerging as the most versatile candidates for the construction of multilayer OLEDs [6] or single-layer LECs [7, 8].

It is well known that the nature of the emitting excited states and other luminescence characteristics of the complexes can be tuned by chemical modification of the ligands. Variation of external conditions (solvent, medium, temperature and pressure) may be used in addition to this chemical tuning. Square-planar d^8 Pd(II) and Pt(II) complexes exhibit a wide range of luminescence properties that depend strongly on the individual characteristics of the coordinated ligands. Several classes of luminescent Pd(II) and Pt(II) complexes are known, including porphyrin [9], phthalocyanine [10] and diimine complexes [11]. Luminescence quenching (partial or complete) is always possible, especially for Pd, but it can be turned into a useful feature for applications (e.g. oxygen sensing [12, 13]).

Both Pd- and Pt-cyclometallated complexes represent a large class of emissive compounds that are synthetically accessible, chemically stable and easily afford luminophores with strong and tunable ligand fields. The latter characteristics are of great importance as they may enable the preparation of complexes wherein low-lying metal-centered excited states responsible for room-temperature radiationless processes are deactivated [14]. Quantum yields can be very high (even for Pd), and luminescence lifetimes are found in a wide range (10^{-4}–10^{2} μs) [14].

One of the major limitation of luminescent Pd complexes vs. Pt analogues (including cyclometallated ones) is the reduced heavy-atom effect of Pd, which

Palladacycles: Synthesis, Characterization and Applications. Edited by Jairton Dupont and Michel Pfeffer

ISBN: 978-3-527-31781-3

usually leads to weaker (or absent) room-temperature emissive behavior. Except for some notable exceptions, room-temperature luminescence of Pd complexes is often assigned as fluorescence as opposed to the more common phosphorescence of Pt species. By virtue of phosphorescent triplet excited states, cyclometallated Pt complexes have already been incorporated into electroluminescent devices [15, 16]. Indeed, the common perception is that the luminescence of cyclopalladated complexes represents a mere curiosity, and nothing new or interesting may emerge from further studies.

12.2
The Early Days

The first report on the photophysical properties of cyclometallated Pd(II) complexes appeared more than 20 years ago [17]. By coincidence, the species under study contained orthometallated azobenzene ligands, as in Cope and Siekman's seminal work [18] that started the cyclopalladation era in the mid-1960s. Soon after, several groups were involved in the extensive photophysical and electrochemical characterization of already existing or *ad hoc* prepared cyclopalladated complexes. Quite often, data on Pd(II) species were complemented by information on the analogous Pt(II) derivatives. Two main driving forces were behind the interest for the emissive and photochemical properties of cyclopalladated complexes. First, as stated by Wakatsuki and Kutal: "little information currently exists concerning the excited-state properties of orthometallated complexes" [17]. The second, less specific motivation was the great expectation that aromatic cyclometallated species could possibly replicate the success of polypyridine complexes.

At the beginning of 1990s, it was already time for assessment. Maestri *et al.* [19] reviewed the subject in such an influential way that it strongly affected later work. With reference to the emissive properties of cyclopalladated compounds, it was recognized that Pd(II) species (+2 being the only oxidation number of interest for Pd) posed several problems with respect to d^6 metal ions or even to other d^8 metals [especially Pt(II)]. The main problem noted was the almost general absence of room temperature luminescence, because of thermally activated radiationless processes. Further negative characteristics referred to the low efficiency and fast luminescence decay processes. Reduced heavy-atom effects and high oxidation/reduction potentials for Pd were recognized as major causes.

Some 15 years on, it is again time to monitor the progress of investigations on the luminescence of palladacycles, and to verify the degree of advance in terms of experimental results and theoretical interpretation of data. Our survey will therefore cover the literature that has appeared since 1991. Scattered and partial coverage of photophysical properties of cyclopalladated compounds has recently appeared, but mostly as part of more general reviews [14, 20, 21]. Figures 12.1–12.4 show the chemical structures and numbering of reported species.

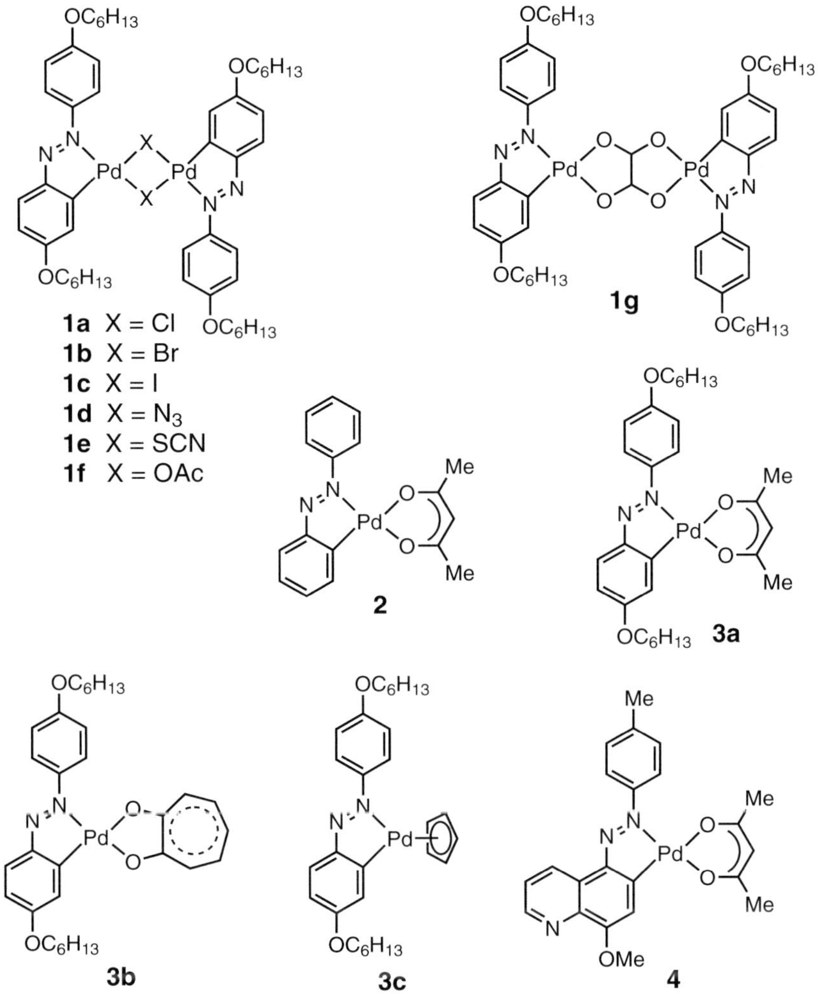

Figure 12.1 Structures of emissive azobenzene complexes.

12.3
Electronic Absorption Spectra of Cyclopalladated Complexes

Low-lying excited electronic states are responsible for the emissive behavior of complexes. A description of these electronic states (and the corresponding electronic transitions) has been reported following the so-called *localized* Molecular Orbitals (MO) model [19, 22]. Nevertheless, it is useful to be reminded that excited states for organometallic complexes can be largely delocalized and their description should always consider a possible mixed character for these states. Four

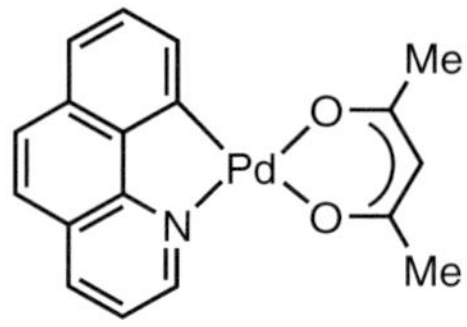

5

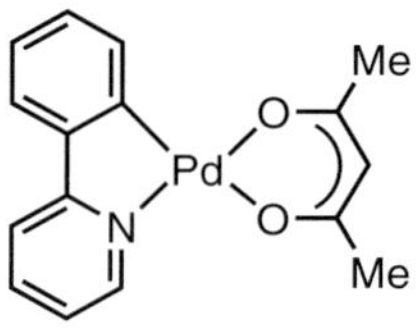

6

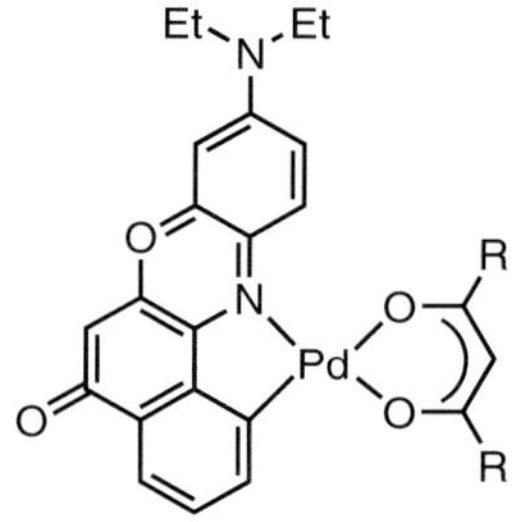

7a R = CH_3
7b R = CF_3
7c R = C_6H_5

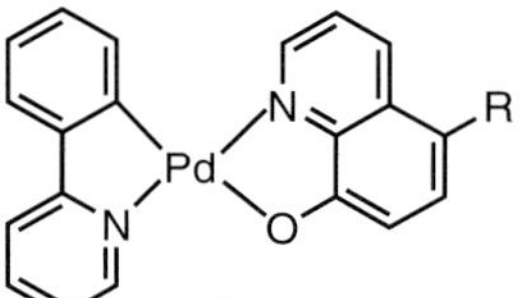

8

8a R = H
8b R = CHO
8c R = $CH{=}N{-}(CH_2)_{11}CH_3$
8d R = CH_2-NH-$(CH_2)_{11}CH_3$
8e R = NH-CO-$(CH_2)_{10}CH_3$

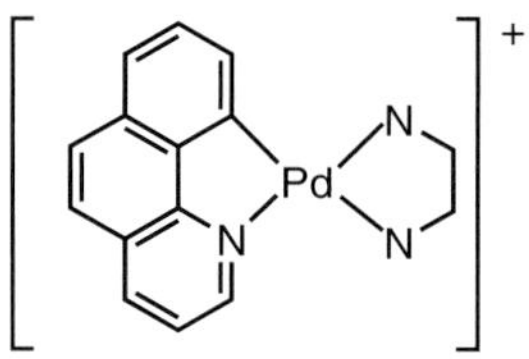

9

9a N-N = $NH_2(CH_2)_2NH_2$
9b N-N = $NMe_2(CH_2)_2NMe_2$
9c N-N = bpy
9d N-N = 5,5'-Me_2-bpy
9e N-N = 5,5'-$(nonyl)_2$-bpy
9f N-N = *o*-phen
9g N-N = 4,7-Me_2-*o*-phen

Figure 12.2 Structures of palladacycles with other cyclometallating bidentate ligands.

different electronic states (transitions) are listed: metal-centered states, dubbed MC (associated to d-d transitions of low intensity), ligand-centered states, LC or IL (corresponding to π-π* or n-π* excitations), and charge-transfer states with metal-to-ligand charge transfer (MLCT, or d-π*) or ligand-to-metal charge transfer (LMCT, or π-d) character. Recent studies on luminescent transition metal complexes enlarged the terminology of CT excited states to metal-metal-to-ligand charge transfer (MMLCT or dσ*-π*) and to ligand-to-ligand charge transfer (LLCT). For the latter, an alternative designation is that of sigma bond-to-ligand charge transfer (SBLCT).

UV/visible absorption spectra of cyclopalladated complexes exhibit absorption bands of medium–high intensity with the lowest-energy transitions rarely exceeding 500 nm (Table 12.1). Spin-allowed intraligand transitions involving the anionic cyclometallating ligand are easily recognized in the 270–350 nm range. Large red-shifts relative to the corresponding bands for the free protonated ligands are assigned to metal perturbation through an increased ligand rigidity and electronic delocalization upon metallation. Ligand-centered transitions with large charge-transfer character (ILCT, intraligand charge transfer) are of high intensity ($\varepsilon \approx 10^4\,M^{-1}\,cm^{-1}$) and can be easily recognized upon changing solvent polarity. Ligand-field transitions are not usually observed (or become buried under more intense features), while spin-allowed MLCT transitions give bands of medium intensity ($\varepsilon = 10^2$–$10^4\,M^{-1}cm^{-1}$) and often occur as low-energy tails of more intense bands.

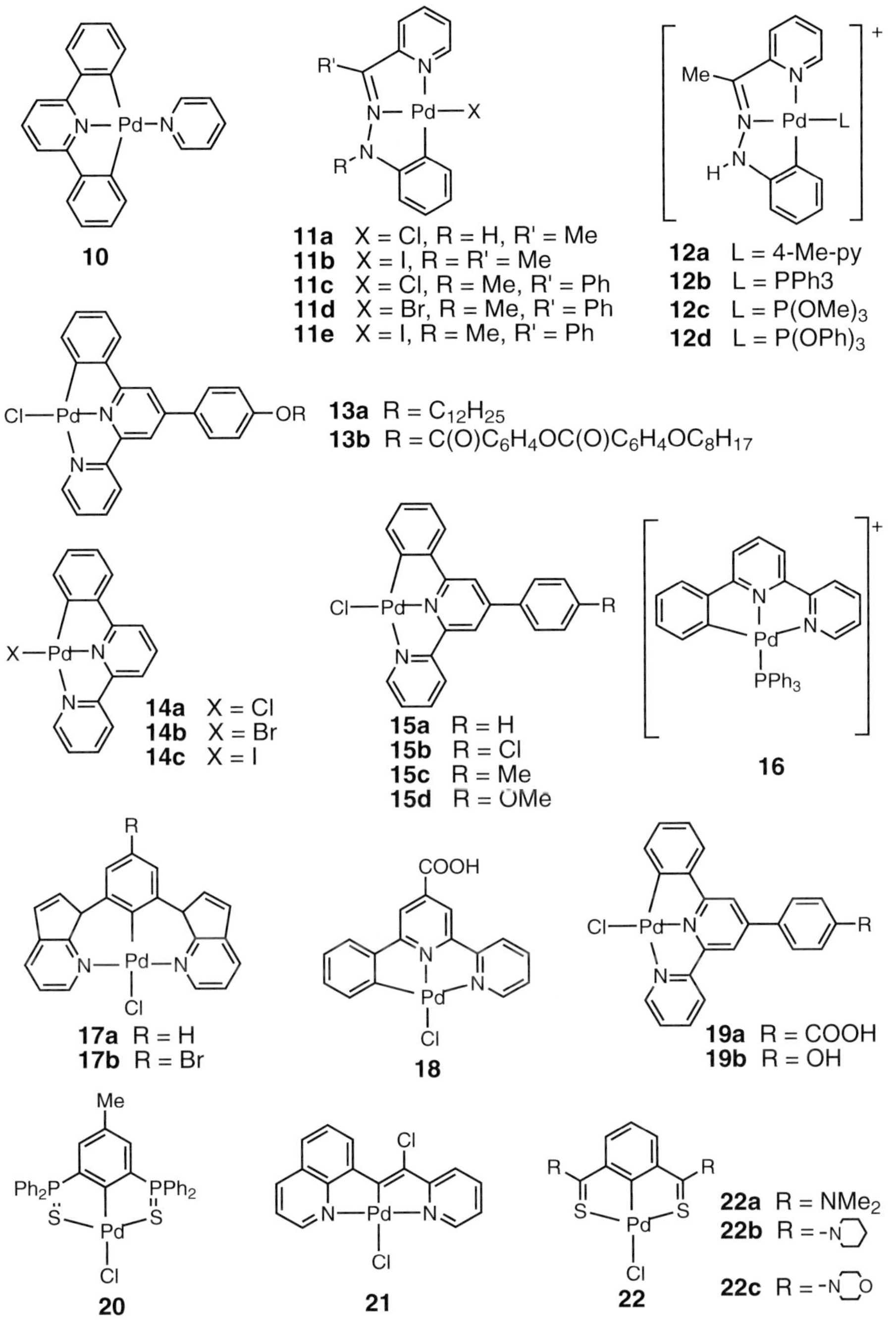

Figure 12.3 Structures of palladacycles with cyclometallating terdentate ligands.

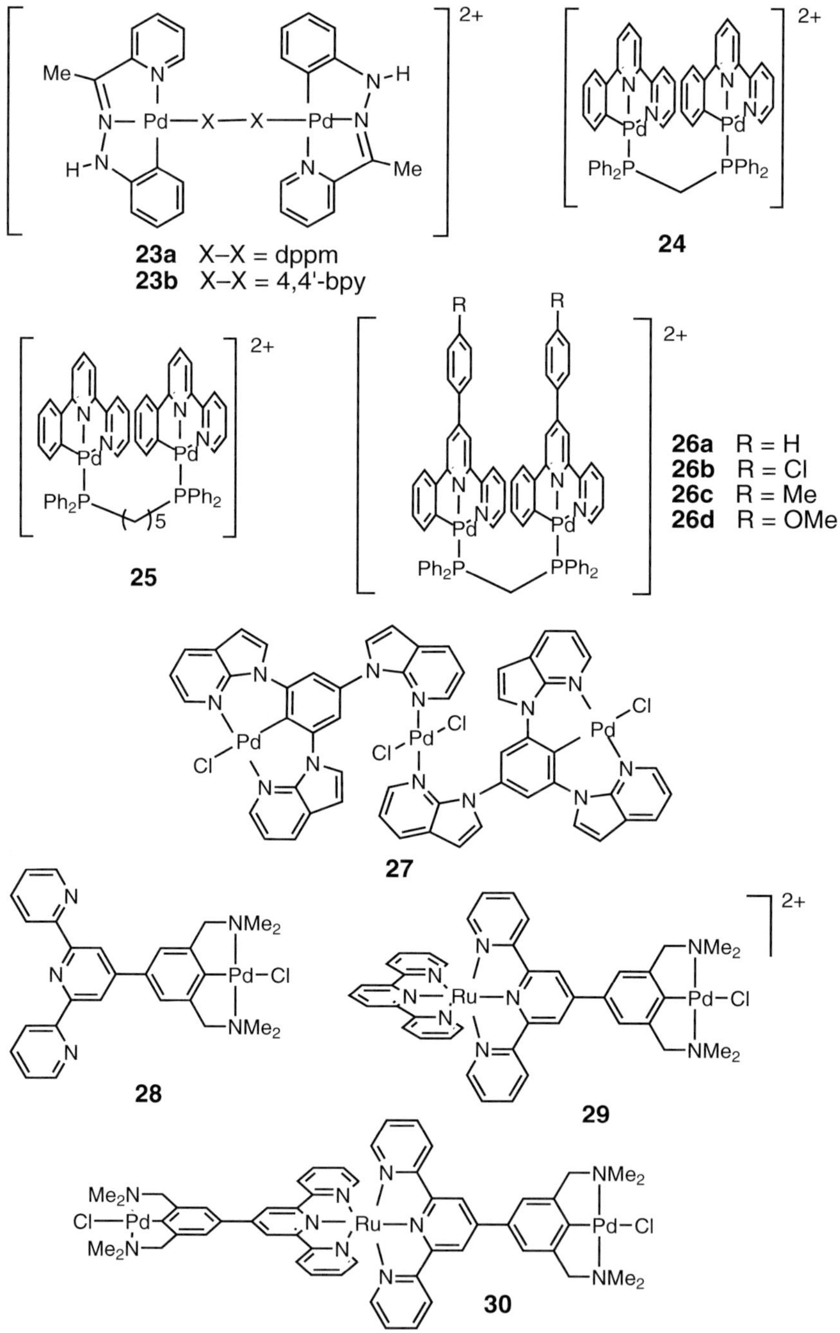

Figure 12.4 Structures of polynuclear luminescent palladacycles.

Table 12.1 Electronic absorption spectra of luminescent cyclopalladated complexes.

Complex	Medium	$\lambda_{max}{}^{a}$ (nm)	ε ($M^{-1} cm^{-1}$)	Ref.
Mononuclear				
3a	CH_2Cl_2	475	15 300	[23]
3b	CH_2Cl_2	478	16 600	[23]
3c	CH_2Cl_2	540	11 600	[23]
4	CH_2Cl_2	513	7 400	[24]
5	CH_2Cl_2	392	4 490	[24]
7a	CH_2Cl_2	608	30 000	[24]
7b	CH_2Cl_2	614	15 000	[25]
7c	CH_2Cl_2	610	50 100	[25]
8a	CH_2Cl_2	434	2 440	[26]
8b	CH_2Cl_2	431	4 180	[26]
8c	CH_2Cl_2	452	43 190	[26]
8d	CH_2Cl_2	443	4 350	[26]
8e	CH_2Cl_2	444	3 840	[26]
10	CH_2Cl_2	380	8 400	[27]
11a	CH_3CN	447	9 600	[28]
11b	CH_3CN	441	16 000	[28]
11c	CH_3CN	470	4 300	[29]
11d	CH_3CN	470	4 000	[29]
11c	CH_3CN	470	4 300	[29]
11e	CH_3CN	470	4 300	[29]
12a	CH_3CN	445	4 400	[28]
12b	CH_3CN	450	5 000	[28]
12c	CH_3CN	446	4 900	[28]
12d	CH_3CN	445	4 600	[28]
13a	CH_2Cl_2	320	15 690	[30, 31]
13b	CH_2Cl_2	332	18 000	[30, 32]
14a	CH_2Cl_2	326, 421[b]	15 900, 350	[33]
14b	CH_2Cl_2	326, 418[b]	14 700, 410	[33]
14c	CH_2Cl_2	328, 418[b]	15 000, 940	[33]
15a	CH_2Cl_2	328, 422[b]	18 600, 510	[33]
15b	CH_2Cl_2	329, 428[b]	18 600, 370	[33]
15c	CH_2Cl_2	329, 422[b]	18 800, 450	[33]
15d	CH_2Cl_2	312, 423[b]	34 500, 540	[33]
16	CH_3CN	306, 375[b]	13 000, 1 950	[33]
17a	CH_2Cl_2	330	nr	[34]
17b	CH_2Cl_2	314	nr	[34]
18	DMSO	326	12 100	[35]
19a	DMSO	332	17 600	[35]
19b	DMSO	368	13 600	[35]
20	CH_2Cl_2	323	5 500	[36]
21	CH_2Cl_2	431	13 000	[37]
22a	CH_2Cl_2	450	3 100	[38]
22b	CH_2Cl_2	460	3 400	[38]
22c	CH_2Cl_2	460	3 200	[38]

Table 12.1 *Continued*

Complex	Medium	$\lambda_{max}{}^{a}$ (nm)	ε ($M^{-1} cm^{-1}$)	Ref.
Polynuclear				
1a	CH_2Cl_2	472	22 000	[23, 39]
1b	CH_2Cl_2	472	27 000	[39]
1c	CH_2Cl_2	467	19 000	[39]
1d	CH_2Cl_2	475	30 000	[39]
1e	CH_2Cl_2	465	21 000	[39]
1f	CH_2Cl_2	467	27 000	[39]
1g	CH_2Cl_2	480	17 500	[39]
23a	CH_3CN	457	4 800	[28]
23b	CH_3CN	446	7 200	[28]
24	CH_2Cl_2	378	4 900	[33]
25	CH_2Cl_2	378	4 700	[33]
26a	CH_2Cl_2	386	4 900	[33]
26b	CH_2Cl_2	390	4 800	[33]
26c	CH_2Cl_2	387	4 700	[33]
26d	CH_2Cl_2	396	7 100	[33]
27	CH_2Cl_2	326	nr	[40]
29	CH_3CN	485	20 400	[41]
30	CH_3CN	496	18 000	[41]

a Lowest-energy feature unless otherwise stated.
b Shoulder peak.
"nr", not reported.

Recognition of MLCT bands is not trivial (solvent dependence cannot always be tested due to solubility limits), although it can be easier for certain bimetallic species wherein MMLCT can also occur [33]. When available, considerable help may come from a comparison with the electronic absorption features of the analogous cycloplatinated complexes, which are expected to reveal MCLT absorption bands at lower energy. The assignment of electronic transitions in the visible range to MLCT transitions is rare, and their presence probably goes unrecognized due to superposition of MLCT bands with more intense LC bands.

MO calculations are necessary to better describe the electronic excited states of cyclopalladated complexes and to establish orbital parentage. Calculations have been carried out at the extended Hückel level for Pd(Bab)Cl **(17a)** and Pd(Br-Bab)Cl **(17b)** [where BabH = 3,5-bis(7-azaindolyl)benzene and Br-BabH = 1-bromo-3,5-bis(7-azaindolyl)benzene] [34]. The LUMOs of **17a** and **17b** are π* orbitals centered on the ligands and with almost no contribution from the metal. The HOMO of **17a** receives large contributions from the d(z^2) orbital and the σ orbitals of the ligands. The lowest-energy electronic transition thus may be described mainly as MLCT with partial LC character. On the other hand, the HOMO of **17b** is a π orbital (with significant contribution from both the metal and the ligand) that also extends to a lone pair pπ orbital of the Br substituent on the central metallated phenyl ring. The electronic transition for **17b** may be described as an LC (π-π*) transition with strong MLCT mixing. This assignment seems to be confirmed by the nature of the HOMO-1 orbital of **17b**, which is Pd(d(z^2))-based and very close in energy to the HOMO [34].

More accurate excited-state energy calculations were reported for another pincer-type palladacycle (complex **21** in Figure 12.3). Time-dependent density functional calculations (at the B3LYP level [42]) led to LUMO orbitals (LUMO, LUMO+1, LUMO+2, and LUMO+3) with no metal character. The HOMO levels are either π (HOMO and HOMO-2) or σ orbitals [HOMO-4 has a strong d(z^2) contribution]. The strongest transitions of **21** in the 390–480 nm range were therefore reported as π-π*, receiving contributions from HOMO-2→LUMO (393 nm, oscillator strength 0.19), HOMO→LUMO+1 (409 nm, oscillator strength 0.02) and HOMO→LUMO (480 nm, oscillator strength 0.07).

An interesting DFT study was reported recently for the metalloligand Pd(TP)Cl (**28**, Figure 12.4), a pincer-type palladacycle obtained through insertion of Pd into the C—Br bond of the heteroditopic TPBr ligand (TPBr = 4′-{4-$BrC_6H_2(CH_2NMe_2)_2$-3,5}-2,2′:6′,2″-terpyridine) followed by bromide scrambling in the presence of chloride [41]. Although complex **28** has no emissive behavior, it is the precursor for heterodi- and trimetallic luminescent species $[Ru(tpy)(TP)PdCl]^{2+}$ **(29)** and $(Ru[Pd(TP)Cl]_2(^{2+}$ **(30)**. The electronic transitions in the UV/visible range calculated for **28** (coupled to similar calculations for the free TP-Br ligand) led to assignment of the more complicated electronic transitions for the polynuclear Ru-Pd species. The results are consistent with a relatively weak coupling between the {Pd(NCN-pincer)} and $\{Ru(tpy)_2\}$ chromophoric units in **29** and **30**.

12.4 Luminescence Studies

Discussion on the luminescence of palladacycles is organized according to the denticity of the cyclometallating ligands, most of which contain aromatic C donor atoms and nitrogen donors in bidentate C^N or tridentate C^N^N, N^C^N or C^N^C donor sets. However, luminescence literature data have been gathered according to the operating temperature of the experiments (Tables 12.2 and 12.3).

12.4.1 Azobenzene Palladacycles

The discovery of luminescence for cyclopalladated azobenzene complexes [17] was resumed by Ghedini *et al.* with comparative studies on mononuclear and dinuclear complexes obtained upon metallation of the symmetric 4,4′-bis(hexyloxy)azobenzene (HAzo-6) [23, 39]. As with Kutal's compounds [17], both the dinuclear $[Pd(Azo\text{-}6)(\mu\text{-}X)]_2$ **(1a–g)** and mononuclear [Pd(Azo-6)L] **(3a–c)** species are weakly luminescent at room temperature, with maxima in the range 520–580 nm. Luminescence lifetimes are in the nanoseconds regime with no notable exception (Table 12.2),

Table 12.2 Room-temperature luminescence data for cyclopalladated complexes.

Complex	Medium	λ_{max}[a] (nm)	Φ	τ (ns)	Ref.
1a	CH_2Cl_2	536	0.9×10^{-4}	3.4	[23, 39]
1b	CH_2Cl_2	522	0.9×10^{-4}	4.6	[39]
1c	CH_2Cl_2	522	0.8×10^{-4}	4.6	[39]
1d	CH_2Cl_2	578	4.7×10^{-3}	1.2	[39]
1e	CH_2Cl_2	574	2.5×10^{-4}	1.0	[39]
1f	CH_2Cl_2	552	1.4×10^{-4}	3.5	[39]
1g	CH_2Cl_2	582	2.0×10^{-4}	0.4	[39]
3a	CH_2Cl_2	560	4.7×10^{-4}	1.7	[23]
3b	CH_2Cl_2	556	0.8×10^{-4}	<1	[23]
3c	CH_2Cl_2	560	1.2×10^{-4}	1.1	[23]
4	CH_2Cl_2	600	0.001	<0.5	[24]
5	CH_2Cl_2	430	<0.004	4.0	[26]
7a	CH_2Cl_2	660	0.23	4.4	[24]
	MeOH	712	0.03	1.4	[25]
	C_6H_{12}	585, 638	0.12	0.9	[25]
7b	CH_2Cl_2	670	0.35	6.0	[25]
	MeOH	705	0.02	1.5	[25]
	C_6H_{12}	595, 650	0.50	4.5	[25]
7c	CH_2Cl_2	660	0.31	4.8	[25]
	MeOH	709	0.04	1.5	[25]
	C_6H_{12}	587, 636	0.14	4.5	[25]
8a	CH_2Cl_2	540	0.006	nr	[26]
8b	CH_2Cl_2	493	0.0025	nr	[26]
8c	CH_2Cl_2	485	0.004	nr	[26]
8d	CH_2Cl_2	570	0.008	nr	[26]
8e	CH_2Cl_2	570	0.0024	nr	[26]
11a	MeCN	608	5.0×10^{-4}	<1	[28]
11b	MeCN	612	nr	nr	[28]
11c	MeCN	610	nr	nr	[29]
	Solid[b]	~610	nr	nr	[29]
11d	Solid[b]	613	nr	nr	[29]
11e	Solid[b]	618	nr	nr	[29]
12a	MeCN	608	4.0×10^{-4}	<1	[28]
12b	MeCN	610	7.0×10^{-4}	<1	[28]
12c	MeCN	610	nr	nr	[28]
12d	MeCN	608	nr	nr	[28]
13a	Solid[b]	490	nr	nr	[31]
13b	Solid[b]	430	nr	nr	[30, 32]
18	Solid[b]	650	nr	1×10^3	[35]
21	CH_2Cl_2	442	nr	nr	[37]
	Toluene	447	nr	nr	[37]
	MeCN	457	0.0068	0.2×10^3 (2×10^3)[c]	[37]
	Solid[b]	501	nr	nr	[37]
23a	MeCN	608	nr	<1	[28]
23b	MeCN	606	7×10^{-4}	<1	[28]

a Lowest energy feature unless otherwise stated.
b Microcrystalline sample.
c Deoxygenated solution.

Table 12.3 Low-temperature (77 K) luminescence data for cyclopalladated complexes.

Complex	Medium	$\lambda_{max}{}^{a}$(nm)	Φ	τ (μs)	Ref.
10	MeOH/EtOH	460	nr	740	[27]
11a	C_3H_7CN	550	nr	$<1 \times 10^{-3}$	[27]
11b	C_3H_7CN	545	nr	1.4×10^{-3}	[27]
12a	C_3H_7CN	550	nr	$<1 \times 10^{-3}$	[28]
12b	C_3H_7CN	550	nr	2.8×10^{-3}	[28]
12c	C_3H_7CN	555	nr	$<1 \times 10^{-3}$	[28]
12d	C_3H_7CN	555	nr	2.9×10^{-3}	[28]
13a	C_3H_7CN	495	nr	74	[31]
13b	C_3H_7CN	476	nr	110	[30, 32]
16	MeOH/EtOH	467–576	nr	nr	[33]
17a	Solid^{b}	554	nr	nr	[34]
17b	Solid^{b}	573	nr	nr	[34]
18	DMF/toluene	510	nr	40	[35]
19a	DMF/toluene	610	nr	15	[35]
19b	DMF/toluene	495	nr	170	[35]
20	Glass^{c}	580	0.14	nr	[36]
22a	CH_2Cl_2/THF	570,^{d} 713	0.08	85	[38]
22b	CH_2Cl_2/THF	575,^{d} 713	0.12	89	[38]
22c	CH_2Cl_2/THF	585,^{d} 711	0.10	83	[38]
23a	C_3H_7CN	545	nr	2.2×10^{-3}	[28]
23b	C_3H_7CN	545	nr	$<1 \times 10^{-3}$	[28]
24	MeOH/EtOH	480, 510	nr	160	
		626		90	[33]
	Solid^{b}	526	nr	nr	[33]
25	MeOH/EtOH	474–586	nr	nr	[33]
	Solid^{b}	498	nr	nr	[33]
26a	MeOH/EtOH	480, 523	nr	64	
		633		88	[33]
	Solid^{b}	516	nr	nr	[33]
26b	MeOH/EtOH	480–658	nr	nr	[33]
	Solid^{b}	529	nr	nr	[33]
26c	MeOH/EtOH	478–632	nr	nr	[33]
	Solid^{b}	544	nr	nr	[33]
26d	MeOH/EtOH	479, 517	nr	74	
		634		16	[33]
	Solid^{b}	526	nr	nr	[33]
27	Solid^{b}	575	nr	nr	[40]
29	C_3H_7CN	624	nr	15	[41]
30	C_3H_7CN	630	nr	13	[41]

a Lowest energy feature unless otherwise stated.
b Microcrystalline sample.
c MeOH/EtOH/CH2Cl2 mixture.
d Shoulder peak.
"nr", not reported.

and the emission has very low efficiency ($\Phi \sim 0.001$). While the emission is reported as metal-perturbed intraligand fluorescence, and is assigned to a singlet n-π^* excited state for all compounds [23, 39], it is still possible for **1** and **3** to have an emitting state with $^1\pi$-π^* character [17]. Although Pd metallation is responsible for the appearance of room-temperature luminescence (the free HAzo-6 ligand is not emissive), no electronic role of the alkoxy substituents was envisaged. In this respect, it is worth noting the lack of luminescence (at least at room temperature) for the mononuclear [Pd(Azo)(acac)] **(2)** (where HAzo = azobenzene; acac = acetylacetonate), a species containing an unsubstituted azobenzene ligand, and which is isoelectronic with **3a,b** [24].

A low-quantum yield emission was observed at room temperature for the mononuclear species [Pd(QAzo)(acac)] (4) [HQAzo = 5-(4-methylphenylazo)-8-methoxyquinoline] [24]. The broad and unstructured emission band of 4 is redshifted relative to mononuclear 3, and is assigned the same n-π^* character. Once again, the orbital contribution of the palladacycle to the excited state is not well defined, and low-temperature data are not available.

12.4.2 Palladacycles with Other Orthometallating Bidentate Ligands

Substitution of azobenzene ligands with more rigid and more delocalized C^N ligands afforded emitters of composition [Pd(C^N)(acac)] with different characteristics [24]. Thus, [Pd(BhQ)(acac)] **(5)** (HBhQ = benzo[*h*]quinoline) has a short-lived (τ = 4 ns) room-temperature emission in the blue region (λ_{max} = 430 nm) with a broad emission spectral band tailing to lower energies. Based on the absence of vibronic progression in the emission spectrum, the redshift with respect to the free HBhQ ligand [43] and the short lifetime, the emission of **5** was assigned as fluorescence and a ^{1}MLCT state was suggested as the emitting excited state [24]. To support this assignment, an electrochemical study was carried out on **5** (and similar compounds) [44]. Analysis of the results suggested a LUMO level mainly located on the cyclometallated ligand and a HOMO orbital with contributions from the metal, the BhQ and the β-diketonate ligands. This picture is quite similar to that proposed by Thompson *et al.* [45] for the isoelectronic [Pt(C^N)(β-diketonate)] species, which revealed strong phosphorescence from an emitting state as a result of mixing of the ^{1}MLCT with ^{3}MLCT and $^3\pi$-π^* states.

A great improvement in emission efficiency for [Pd(C^N)(β-diketonate)] complexes – **5** has $\Phi < 0.004\%$ and the 2-phenylpyridinato (ppy) analogous [Pd(ppy)(acac)] **(6)** [24] has no detectable luminescence – was obtained through palladation of the fluorescent dye Nile Red, 9-diethylamino-5*H*-benzo[*a*]phenoxazine-5-one (HNR) [24, 26]. At room temperature, [Pd(NR)(acac)] **(7a)** gives an intense red emission (λ_{em} = 660 nm in CH_2Cl_2) with a remarkable quantum yield (Φ = 0.23), approaching that of the free laser dye HNR (Φ = 0.44). An even higher quantum yield (Φ = 0.35 in CH_2Cl_2 and 0.50 in cyclohexane) could be reached upon substitution of the acac ligand for hfacac (hfacac = 1,1,1,5,5,5-hexafluoroacetylacetonate). The red emission of [Pd(NR)(hfac)] **(7b)** (λ_{em} = 670 nm) is redshifted with respect to that of **7a**, and reveals a clear solvatochromism (Table 12.2). A third β-diketonate derivative [Pd(NR)(dppd)] (7c) (dppd = 1,3-diphenyl-1,3-propanedionate) reveals close similarities to **7a**. Its room-temperature emission is less intense although of comparable energy (λ_{em} = 660 nm in CH_2Cl_2) and efficiency (Φ = 0.31). Luminescence lifetimes of **7a–c** are in the 1–6 ns range, which likely account for fluorescence emission. A comparison of the free HNR luminescence features with those of the three homologous palladacycles **7a–c**, coupled to the solvatochromic behavior of their emission bands, led the authors to suggest a partially CT, ligand-localized π-π* singlet as the lowest-energy excited state. Given the characteristic chemical structure of the cyclometallating NR ligand, which can be assimilated to a donor–acceptor chromophore, a twisted intramolecular charge transfer (TICT) state [46] could be involved in the emissive behavior of **7a–c** [25].

A new class of mixed-ligand [Pd(C^N)(N^O)] neutral species exhibiting room-temperature photoluminescence was reported recently by Ghedini and coworkers [26]. Cyclopalladated ppy derivatives [Pd(ppy)(Q)] **(8a)** and [Pd(ppy)(Q-R)] **(8b–d)** were prepared that contained O-deprotonated 8-hydroxyquinoline (Q) or its 5-substituted Q-R derivatives as ancillary ligands (Figure 12.2). The emission energy of [Pd(ppy)(Q-R)] depended largely on the nature of R substituents on the quinolinate ligand. A green emission was observed for [Pd(ppy)(Q)] **(8a)** at 540 nm in dichloromethane solution, although with rather low efficiency (Φ = 0.006). Electron-acceptor substituents at the para position to the oxygen atom of the quinolinato ligand led to a blue-shift of the emission energy maximum, which was observed at 493 nm for **8b** (R = CHO) and at 485 nm for **8c** (R = –CH—NR′). On the other hand, R substituents such as $-CH_2N(H)R'$ **(8d)** or –NHC(O)R′ **(8e)** were responsible for a redshift of the emission band to 570 nm. Photoluminescence quantum yields for **8b–e** are also low, reaching a maximum of 0.008 for **8d** (Table 12.2). Thus, emission energy tuning was obtained for complexes **8a–e**, whose luminescent behavior probably originates from a metal-perturbed ILCT state receiving a large contribution from the quinolinate ligand [26]. The absence of a complete photophysical characterization prevents full appreciation of these interesting room-temperature emitters. More recent preliminary work from the same group [47] is devoted to the photophysical study of cationic palladacycles of composition $[Pd(C^{\wedge}N)(N^{\wedge}N)]^+$, where N^N ligands are either aliphatic diamines or α-diimines. Although two homologous classes were prepared (containing ppy or benzoquinolinate cyclometallating ligands, respectively), only $[Pd(BhQ)(N^{\wedge}N)]^+$ (**9a–g** in Figure 12.2) were reported to exhibit very weak room-temperature luminescence ($\Phi \sim 1 \times 10^{-4}$) in the blue region. While no more details on the emissive behavior can be found in that study (solubility problems were invoked for partial results), it is rather surprising that previous studies by Craig and Watts on some identical $[Pd(ppy)(N^{\wedge}N)]^+$ species [48] went overlooked.

12.4.3
Luminescent Palladacycles with Terdentate Ligands

The class of emitters described in this section is, photophysically, the most diverse. The cyclometallating ligands are terdentate, with different combinations of C and

N donors (C^N^C, C^N^N, N^C^N) but also with other donor sets (e.g. S^C^S) (Figure 12.3).

The first account of a luminescent palladacycle containing a terdentate cyclometallating ligand actually predates the review of Maestri *et al.* [19], and still remains the only report of a CNC pincer-type emissive palladacycle. [Pd(dppy)(py)] **(10)** (H_2dppy = 2,6-diphenylpyridine) has an intense luminescence only in a glass matrix at 77 K [27]. The energy maximum of the structured emission band is redshifted by circa 1000 cm^{-1} with respect to the emission of the free ligand under the same conditions. The emission lifetime is 740 μs. Both features are indicative of a very small MLCT character, while confirming the main LC nature of the emitting excited state. However, electrochemical results for complex **10** seem to point to a larger contribution from metal d orbitals to the excited state. In the authors' view, failure of the Koopmann's theorem in this case might be due to the very small energy difference between MC states and π and π* orbitals of the cyclometallating dppy ligand [27].

The next group of palladacycles contains N^N^C ligands in square-planar $[Pd(N^\wedge N^\wedge C)(L)]^{n+}$ (n = 0, 1) species or moieties. The first class to be considered is based on cyclometallated 2-acetylpyridine-*N*-phenylhydrazone (Haph) [28], 2-acetylpyridine-*N*-methyl-*N*-phenylhydrazone (Hamph) [28] or 2-benzoylpyridine-*N*-methyl-*N*-phenylhydrazone (Hbmph) [29]. Neutral [Pd(N^N^C)(L)] **(11)** or cationic $[Pd(N^\wedge N^\wedge C)(L)]^+$ **(12)** and $[Pd(N^\wedge N^\wedge C)(L)Pd(C^\wedge N^\wedge N)]^{2+}$ **(23)** were obtained with anionic or neutral (terminal or bridging) L co-ligands, respectively. In the most thorough study [28], the complexes showed very weak (Φ = 10^{-4}–10^{-5}) and short-lived (1–3 ns) luminescence both in solution at room temperature and at 77 K in a glass matrix. The luminescence of [Pd(bmph)(X)] [X = Cl **(11c)**, Br **(11d)**, I **(11e)**] was studied only at room temperature [29]. In all cases, the emission is ascribed to an ILCT excited state based mainly on the terdentate C^N^N ligand due to the invariance of the emission energy irrespective of the nature of the fourth ligand. For example, [Pd(aph)Cl] **(11a)**, $[Pd(aph)(PPh_3)]^+$ **(12b)**, and $[Pd(aph)(4,4'\text{-}bpy)Pd(aph)]^{2+}$ **(23b)** have the same emission energy (608 vs 610 vs 606 nm). The emitting state must also have strong CT character, as indicated by the large Stokes shift and the dependence from the temperature and the solvent polarity. Finally, intense room-temperature luminescence was observed for a microcrystalline sample of **11c**. In this case, a sort of excimer luminescence (due to π-π* interactions) was proposed, as suggested by the presence of π-π close contacts in the crystalline ground state [29]. However, the degree of redshift of the solid emission band with respect to the solution value is negligible, and the assignment of the emitting state is therefore rather questionable. No excimer luminescence in solution was reported.

Another large group of emissive palladacycles containing terdentate ligands based on a common 6′-phenyl-2,2′-bipyridine (pbpy) core has been investigated by Neve [30–32, 35] and Che [33]. Physical and photophysical properties were modulated either by chemical modification of the cyclometallating ligand or through substitution of the fourth ligand. Indeed, the simplest species [Pd(pbpy)X] [X = Cl **(14a)**, Br **(14b)**, I **(14c)**] were found non-emissive at room temperature and only weakly emissive in frozen dichloromethane at 77 K [33], as was also observed for compounds **15a–d**, which contain various 4′-(*p*-substituted-phenyl)-6′-phenyl-2,2′-bipyridine ligands (Figure 12.3). The latter observations were rather at odds with earlier reports by Neve *et al.* of strong luminescence from [Pd(N^N^C)Cl] species with similar ligands [30–32]. Compounds **13a** and **13b** showed a structured emission in a rigid matrix at 77 K with maxima in the range 475–510 nm. Luminescence decay is monoexponential with lifetimes of 110 **(13a)** and 74 μs **(13b)**. Based on the emission energy, lifetime values and the spectral shape, the luminescence was assigned to a ^{3}LC excited state partially mixed with a closely-lying ^{3}MLCT state [31]. More important, both complexes revealed a room temperature luminescence in the solid state. However, while **13a** exhibited a broad emission slightly blue-shifted from the 77 K value (λ_{max} = 490 nm) [31], the ipsochromic shift for **13b** was more pronounced and the emission band revealed a structured shape (λ_{max} = 430 and 450 nm) [32]. For both compounds the solid-state emission was assigned to the same ^{3}LC excited state responsible for the 77 K emission in glass matrix.

Che's work [33] was also important to assess the role of excimer emission in a series of binuclear $[Pd(N^\wedge N^\wedge C)(\mu\text{-dppm})Pd(C^\wedge N^\wedge N)]^{2+}$ species (**24** and **26a–d**, dppm = bis(diphenylphosphino)methane) (Figure 12.4). All binuclear complexes were found to be emissive at 77 K both in glass and the solid state. The corresponding excited states were highly dependent on the solvent and the medium. In glassy MeOH/EtOH solutions **24** and **26a–d** displayed dual emission with high energy structured bands and low energy structureless features (Table 12.3). The high-energy luminescence was assigned to an intraligand (^{3}IL) excited state involving the cyclometallated ligand, while the diffuse luminescence at low energy was tentatively explained as resulting from weak excimeric contributions involving intramolecular π-π interactions within the Pd_2 units [enabled by the face-to-face arrangement of the Pd(N^N^C) moieties]. Confirmation of the assignment came from the absence of the low-energy emission component for the mononuclear $[Pd(pbpy)(PPh_3)]^+$ **(16)** and the dinuclear $[(pbpy)Pd(\mu\text{-dppp})Pd(pbpy)]^{2+}$ (**25**, dppp = 1,5-bis(diphenylphosphino)pentane), both of which do not allow close π-π contacts [33]. All the cationic complexes **24–26** are luminescent in the solid state at 77 K. The emission bands can be either diffuse or structured, and are always assigned to metal-perturbed ^{3}LC excited states. No apparent excimeric emission was observed in the solid state, which is confirmed by the absence of ideal aromatic stacking in the crystal lattice. In all cases, and contrary to the Pt(II) congeners [49], a MMLCT(dσ*→π*) component of the emission was dismissed on the grounds of minimal metal–metal interaction [33].

A room-temperature excimeric emission is the distinctive feature of complex **18** [35]. A broad, unstructured emission band was observed for a microcrystalline sample at 650 nm, with a lifetime of 1 μs. Emission band shape and energy are consistent with an excimeric excited state. Two types of excimer emission from square planar d^8 complexes should be considered: (i) a canonical π-π* luminescence resulting from stacking interactions of aromatic portions of the ligands; (ii) a metal-to-ligand charge transfer assisted by substantial metal–metal interaction (MMLCT), which is usually quite effective for certain Pt(II) terpyridine species

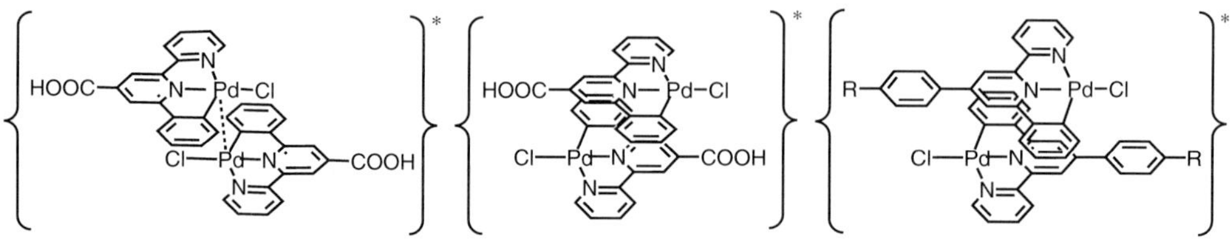

Figure 12.5 Proposed excimeric dimers for palladacycles **18 (A, B)** and **19 (C)**.

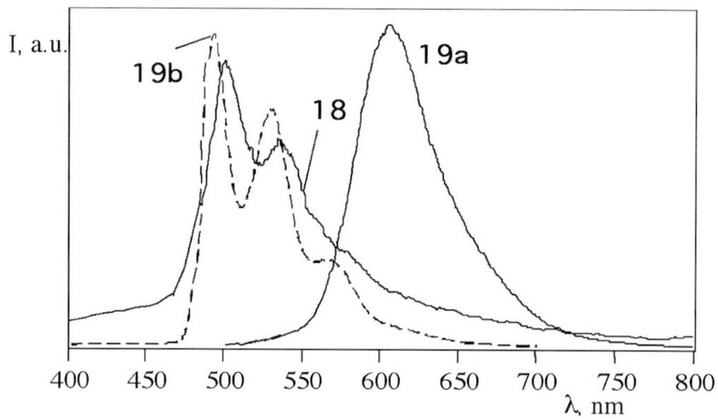

Figure 12.6 Emission spectra of palladacycles **18, 19a** and **19b** in a DMF/toluene rigid matrix at 77 K. (Reprinted with permission from Reference [35]. Copyright 2002 American Chemical Society.)

[49, 50]. For **18**, both excimers **A** and **B** (Figure 12.5) are strongly supported by the solid-state features of the ground-state molecule. However, it is difficult to assign this luminescence to a specific excimeric excited state or to both components.

Unlike **18**, the related palladacycles **19a,b** do not emit in the solid state at room temperature. The reasons for this may reside in the different electronic nature of the ligands, but also in a substantially different arrangement of the molecules in the crystal lattice. A completely different picture holds at 77 K. In glassy DMF/toluene solutions all three species **(18, 19a,b)** were found to luminesce with monoexponential lifetimes (Table 12.3). Interestingly, the emission of **19a** is unstructured, strongly redshifted (Figure 12.6), and short-lived with respect to both **18** and **19b**. The 77 K luminescence of **18** and **19b** can be safely assigned to a metal-perturbed ^{3}LC state. However, evaluation of the lifetime values led the authors to suggest in the case of complex **18** a further contribution to the emitting state from an upper MLCT state. The same contribution was considered less effective for **19b** due to the expected higher energy of the MLCT state [35]. In this framework, the longer lifetime for **19b** can be taken as a confirmation of this assumption.

The 77 K luminescence of **19a** was attributed to an excimeric excited state [35] (excimer **C** in Figure 12.5 is a possibility). Thus, stacking interactions leading to excimer formation become more facilitated in frozen solutions than in the solid state for **19a**, a molecule containing a cyclometallated ligand with a larger aromatic extension than **18**. The absence of excimer emission for **18** at low temperature is thus reasonable, while the same argument does not explain the luminescence characteristics of **19b**. In frozen solution, it is possible that strong H-bonding interactions sustain excimer formation for **19a** but are inefficient (or absent) for **19b**.

Solid-state luminescence seems to be a common characteristic of palladacycles based on terdentate ligands. Ligands BabH, Br-BabH and 1,3,5-tris(7-azaindolyl)benzene (TabH) are easily metallated, affording Pd(Bab)Cl **(17a)**, Pd(Br-Bab)Cl **(17b)**, as well as their Pt(II) analogues [34], and $Pd_3(Tab)_2Cl_4$ (**27**, Figure 12.4) [34, 40]. The blue luminescence of the free ligands is drastically affected by metallation. In contrast to the uncoordinated ligands, and to the Pt(II) derivatives, **17a, 17b** and **27** reveal no room-temperature luminescence. The yellow-orange luminescence of all Pd(II) species could be observed only at 77 K in the solid state. The emission is not particularly bright (no emission lifetimes measured) and emission bands are unstructured. MMLCT contribution to the emission was ruled out on the grounds of the X-ray crystal structures. The possible role of MLCT and LC(π-π*) excited states was explored by means of EHMO calculations (Section 12.3). The results for Pd(Bab)Cl **(17a)** were consistent with a prevalent MLCT character of the emission, whereas the solid-state emission of Pd(Br-Bab)Cl **(17b)** could be equally due to MLCT or to LC (π-π*) components (or to both).

Although weak, room-temperature solid-state emission is featured by the pincer palladacycle **21** [37]. Reaction of 2-pyridinyl-8-quinolinyl-acetylene with Li_2PdCl_4 led to the flat palladacycle **21** containing an asymmetric N^C^N ligand metallated trough a C-vinyl atom. Complex **21** emits in solution at room temperature in the blue region. The emission energy is solvent-dependent, shifting to lower energy on increasing solvent polarity (Figure 12.7). Laser-flash photolysis experiments allowed the authors to assign the luminescence of **21** as fluorescence, while quantum yields measurements showed an emission efficiency for the complex (6.8×10^{-3}) lower than that (3.0×10^{-2}) of the uncomplexed ligand. TD-DFT calculations were also used to suggest an assignment of the room-temperature luminescence as a metal-perturbed ^{3}LC(π-π*) fluorescence. As mentioned above, palladacycle **21** is a room temperature solid-state emitter. A single broad emission band was observed at 501 nm, redshifted with respect to solution luminescence. As with **18**, the solid-state luminescence of **21** is assigned to an excimeric emitting state, with a π-π* excited state as the most likely candidate [37].

In addition to terdentate nitrogen donors, mixed donor sets incorporating sulfur atoms were considered for the formation of luminescent palladacycles. Kambara *et al.* reported on the cyclometallation of a phosphine sulfide-based SCS-pincer ligand, 3,5-bis(diphenylphosphinothioyl)toluene (Hdptt), leading to isoelectronic and isostructural M(dptt)Cl species [M = Pt, Pd **(20)**] [36]. Both the Pt(II) and Pd(II) dptt species are non-emissive in solution at room temperature, whereas strong

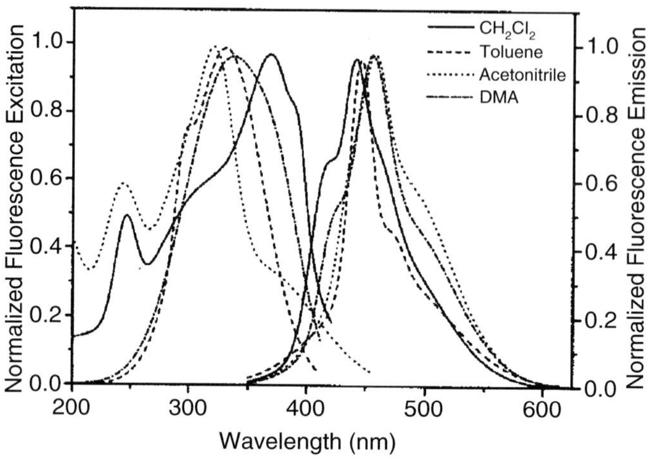

Figure 12.7 Fluorescence excitation and emission spectra of palladacycle **21** in various solvents at room temperature (Reprinted with permission from Reference [37]. Copyright 2004 American Chemical Society.)

luminescence was observed in a glass matrix at 77 K. While the emitting state for the Pt(II) species was tentatively assigned to a ^{3}MLCT state, no discussion on the photoluminescence of **20** was reported. It is also unclear whether **20** could exhibit luminescent behavior in the solid state at room temperature, as actually observed for Pt(dptt)Cl [36]. A more accurate study on luminescent (SCS)PdCl species was recently reported by the same group [38]. Palladacycles **22a–c** contain some 1,3-benzenedicarbothioamides in their deprotonated form. The complexes are structurally similar to **20**, and all of them exhibit strong emission in a glassy frozen state as well as in the solid state at room temperature. The 77 K emission spectra revealed two-component bands with a major contribution in the red region (Table 12.3). Considerations on the relatively long (80–90 μs) emission lifetimes in the glassy state led to the assignment of the luminescence of **22a–c** as phosphorescence, and to a guess of a ^{3}MLCT state as the lowest energy excited state. At room temperature, microcrystalline samples of **22a–c** revealed a strong red emission featuring broad and unstructured bands in all cases. By analogy with the 77 K emission, the solid-state phosphorescence is suggested to originate from a ^{3}MLCT state, ruling out the possibility of excimeric excited states.

Finally, we may consider some luminescent species obtained through reaction of the Pd(II) NCN-pincer **28**, which contains a free terpyridine fragment available for further coordination (Figure 12.4). Although palladacycle **28** is not luminescent by itself, dinuclear **(29)** and trinuclear **(30)** Ru/Pd species exhibit weak luminescence both at room temperature in acetonitrile and in a frozen glass at 77 K [41]. However, the presence of the Pd(II) ion within the polynuclear species does not correspond to a relevant modification of the emissive properties of the {Ru(tpy)$_2$} chromophoric units, which indicates a very weak intercomponent coupling [41].

12.5
Conclusions and Prospects

Over the last decade, revived interest in the applications of palladacycles has stimulated the search for better performances of known physical properties and for new ones. Among the latter, photoconductivity and photorefractivity have been reported for cyclopalladated complexes [51], making them interesting candidates for future applications [21]. On the other hand, luminescence is a more fundamental characteristic of many palladium complexes that may eventually find a role in photonic [14] and biophotonic [52] devices and techniques.

Photoluminescent palladacycles are becoming increasingly studied, yet palladium species suffer a sort of "inferiority complex" with respect to other organometallic luminophores. This is reflected in incomplete studies or no study at all. In general, low-temperature luminescence is not sought when room-temperature emission in solution is observed. Solid-state emission is also poorly studied, making it difficult to evaluate the potential use of this feature.

What would be needed to stimulate the search for novel luminescent palladacycles and for advancing the knowledge of their emission features? First of all, photophysical studies should be more systematic (including variation of both the cyclometallated ligands and co-ligands), and luminescence should be examined almost routinely. Luminescence quantum yields and lifetimes should always be measured – a condition not met by many past studies. To assign excited states, theoretical calculations could complement experimental studies and, as a further aid, electrochemical properties should be investigated.

It is true that triplet phosphorescent emitters – the most valuable luminophores for electroluminescent devices – are rare among palladacycles. Still, OLED technology is not the only one that could benefit from a wider scrutiny of the emissive properties of such a large and stable family of organometallic compounds. Rather, work needs to be done to shine light on luminescent palladacycles and to boost a brighter future.

References

1 Tang, C.W. and Van Slyke, S.A. (1987) *Applied Physics Letters*, **51**, 913.

2 Pei, Q., Yu, G., Zhang, C., *et al.* (1995) *Science*, **269**, 1086.

3 Burroughes, J.H., Bradley, D.D.C., Brown, A.R., *et al.* (1990) *Nature*, **347**, 539.

4 Friend, R.H., Gymer, R.W., Holmes, A.B., *et al.* (1999) *Nature*, **397**, 121.

5 Baldo, M.A., O'Brien, D. F., You, Y., *et al.* (1998) *Nature*, **395**, 15.

6 Baldo, M.A., Thompson, M.E. and Forrest, S.E. (2000) *Nature*, **403**, 750.

7 Slinker, J.D., Bernards, D., Houston, P.L., *et al.* (2003) *Chemical Communications*, 2392.

8 Slinker, J.D., Gorodetsky, A.A., Lowry, M.S., *et al.* (2004) *Journal of the American Chemical Society*, **126**, 2763.

9 Spellane, P.J., Gouterman, M., Antipas, A., *et al.* (1980) *Inorganic Chemistry*, **19**, 386.

10 Lawrence, D.S. and Whitten, D.G. (1996) *Photochemistry and Photobiology*, **64**, 923.

11 Hissler, M., McGarrah, J.E., Connick, W.B., *et al.* (2000) *Coordination Chemistry Reviews*, **208**, 115.

12 Papkovsky, D.B. and O'Riordan, T.C. (2005) *Journal of Fluorescence*, **15**, 569.

13 Borisov, S.M., Vasylevska, A.S., Krause, C. and Wolfbeis, O.S. (2006) *Advance Functional Materials*, **16**, 1536.

14 Evans, R.C., Douglas, P. and Winscom, C.J. (2006) *Coordination Chemistry Reviews*, **250**, 2093.

15 (a) D'Andrade, B.W., Brooks, J., Adamovich, V., *et al.* (2002) *Advanced Materials*, **14**, 1032;(b) Adamovich, V., Brooks, J., Tamayo, A., *et al.* (2002) *New Journal of Chemistry*, **26**, 1171.

16 Lu, W., Mi, B.-X., Chan, M.C.W., *et al.* (2002) *Chemical Communications*, 206.

17 Wakatsuki, Y., Yamazaki, H., Grutsch, P.A., *et al.* (1985) *Journal of the American Chemical Society*, **107**, 8153.

18 Cope, A.C. and Siekman, R.W. (1965) *Journal of the American Chemical Society*, **87**, 3272.

19 Maestri, M., Balzani, V., Deuschel-Cornioley, C. and von Zelewsky, A. (1992) *Advances in Photochemistry*, **17**, 1.

20 Dupont, J., Consorti, C.S. and Spencer, J. (2005) *Chemical Reviews*, **105**, 2527.

21 Ghedini, M., Aiello, I., Crispini, A., *et al.* (2006) *Coordination Chemistry Reviews*, **250**, 1373.

22 Crosby, G.A. (1975) *Accounts of Chemical Research*, **8**, 231.

23 Ghedini, M., Pucci, D., Calogero, G. and Barigelletti, F. (1997) *Chemical Physics Letters*, **267**, 341.

24. Aiello, I., Ghedini, M. and La Deda, M. (2002) *Journal of Luminescence*, **96**, 249.

25 La Deda, M., Ghedini, M., Aiello, I., *et al.* (2005) *Journal of Organometallic Chemistry*, **690**, 857.

26 Ghedini, M., Aiello, I., La Deda, M. and Grisolia, A. (2003) *Chemical Communications*, 2198.

27 Maestri, M., Deuschel-Cornioley, C. and von Zelewsky, A.(1991) *Coordination Chemistry Reviews*, **111**, 117.

28 Garcia-Herbosa, G., Munoz, A. and Maestri, M. (1994) *Journal of Photochemistry and Photobiology A: Chemistry*, **83**, 165.

29 Ghedini, M., Aiello, I., Crispini, A. and La Deda, M. (2004) *Dalton Transactions*, 1386.

30 Neve, F., Ghedini, M. and Crispini, A. (1996) *Chemical Communications*, 2463.

31 Neve, F., Crispini, A. and Campagna, S. (1997) *Inorganic Chemistry*, **36**, 6150.

32 Neve, F., Ghedini, M., Francescangeli, O. and Campagna, S. (1998) *Liquid Crystals*, **24**, 673.

33 Lai, S.-W., Cheung, T.-C., Chan, M.C.W., *et al.* (2000) *Inorganic Chemistry*, **39**, 255.

34 Song, D., Wu, Q., Hook, A., *et al.* (2001) *Organometallics*, **20**, 4683.

35 Neve, F., Crispini, A., Di Pietro, C. and Campagna, S. (2002) *Organometallics*, **21**, 3511.

36 Kambara, T. and Yamamoto, T. (2003) *Journal of Organometallic Chemistry*, **688**, 15.

37 Consorti, C.S., Ebeling, G., Rodembusch, F., *et al.* (2004) *Inorganic Chemistry*, **43**, 530.

38 Akaiwa, M., Kambara, T., Fukumoto, H. and Yamamoto, T. (2005) *Journal of Organometallic Chemistry*, **690**, 4192.

39 Ghedini, M., Pucci, D., Crispini, A., *et al.* (1999) *Applied Organometallic Chemistry*, **13**, 565.

40 Wu,Q., Hook, A. and Wang, S. (2000) *Angewandte Chemie – International Edition*, **39**, 3933.

41 Gagliardo, M., Rodriguez, G., Dam, H.H., *et al.* (2006) *Inorganic Chemistry*, **45**, 2143.

42 (a) Lee, C.T., Tang, W.T. and Parr, R.G. (1988) *Physical Review B*, **37**, 785.(b) Becke, A.D. (1988) *Physical Review A*, **38**, 3098.

43 Maestri, M., Sandrini, D., Balzani, V., *et al.* (1987) Literature data for HBhQ report the phosphorescence emission features at 77 K. *Chemical Physics Letters*, **122**, 375. Privately communicated room-temperature data in CH_2Cl_2 give a structured fluorescence emission at 348 nm (highest energy feature).

44 Pugliese, T., Godbert, N., La Deda, M., *et al.* (2005) *Chemical Physics Letters*, **410**, 201.

45 Brooks, J., Babayan, Y., Lamansky, S., *et al.* (2002) *Inorganic Chemistry*, **41**, 3055.

46 Rettig, W. (1986) *Angewandte Chemie – International Edition in English*, **25**, 971.

47 Pugliese, T., Godbert, N., Aiello, I., *et al.* (2006) *Inorganic Chemistry Communications*, **9**, 93.

48 Craig, C.A. and Watts, R.J. (1989) *Inorganic Chemistry*, **28**, 309.

49 Lai, S.-W., Chan, M.C.W., Cheung, T.-C., *et al.* (1999) *Inorganic Chemistry*, **38**, 4046.

50 Bailey, J.A., Hill, M.G., Marsh, R.E., *et al.* (1995) *Inorganic Chemistry*, **34**, 4591.

51 (a) Aiello, I., Dattilo, D., Ghedini, M. and Golemme, A. (2001) *Journal of the American Chemical Society*, **123**, 5598; (b) Aiello, I., Dattilo, D., Ghedini, M., *et al.* (2002) *Advanced Materials*, **14**, 1233.

52 Gross, S., Gilead, A., Scherz, A., *et al.* (2003) *Nature Medicine*, **9**, 1327.

13
Cyclopalladated Compounds as Enzyme Prototypes and Anticancer Drugs

Alexander D. Ryabov

13.1
Introduction

Nature does not use palladium. Nature ignores the square planar coordination of biometals. Indeed, Nature uses organometallic compounds only in a few exceptional cases [1]. Cyclopalladated compounds exemplify square planar organometallic derivatives of palladium(II). Consequently, biomedical applications of cyclopalladated compounds, at first sight, may seem to be in jeopardy. Fortunately, this is not true and chemists are capable of finding biological applications in areas where a search for such does not seem to be justified or make much sense. There is always room for bio-inspired applications and cyclometallated compounds have created a noticeable niche. The present chapter demonstrates that cyclopalladated complexes are used as *functioning* mimics of hydrolytic enzymes (hydrolases), oxidoreductases and anticancer drugs. Palladium complexes comprise the core of this chapter but relevant examples of platinum complexes are selectively included. This will help to draw a more general picture of bio-applications of the complexes, highlight main achievements and visualize new trends in designing novel applications. Cyclometallated derivatives of amino acids, their derivatives and other biomolecules are not considered in this chapter.

13.2
Cyclopalladated Compounds as Mimetics of Hydrolases

13.2.1
Hydrolysis of Activated Esters

The catalytic role of metal ions in the mononuclear and dinuclear active sites of metalloenzymes such as carboxypeptidase, thermolysin, methionine aminopeptidase, leucine aminopeptidase, urease and so on is fairly well understood [2–5]. In particular, there are three accepted mechanism of the catalysis by metals in the

Palladacycles: Synthesis, Characterization and Applications. Edited by Jairton Dupont and Michel Pfeffer

ISBN: 978-3-527-31781-3

Figure 13.1 Biochemical mechanisms of catalysis of ester and amide hydrolysis by metal ions. (a) A hypothetical example of coordination of amino acid derivative to a metal center with an intramolecular attack of a coordinated nucleophile at the sp^2 carbon. (b) Metal ion effects in the active sites of enzymes (see text for details).

hydrolysis of bio-relevant esters and amides [6, 7]. A metal center can be responsible for:

1. Increasing the acidity of the coordinated water (type "1" action in Figure 13.1b). As a result, a powerful nucleophile (coordinated hydroxide) is generated under neutral conditions.
2. The metal acts as a Lewis acid and coordinates the ester or amide oxygen; this increases an effective positive charge at the carbonyl carbon and thus favors an internal or external nucleophilic attack (type "2" action).
3. A combination of mechanisms "1" and "2".

The mechanistic principles stated above have successfully been realized within a structural framework of cyclometallated compounds. Moreover, structural advantages of metallacycles, which have been numerously emphasized in other chapters of this book, allow us to verify and strengthen the principles and introduce novel features related to stereospecific hydrolysis with chiral recognition. Figure 13.2 presents the "bio"-advantageous features square planar metallacycles. The metallacycle itself (M—C—X) with a donor center X is exceptionally stable in aqueous solutions over a very broad pH range [8]. Two neighboring sites, A and B, provide binding opportunities, respectively, for a substrate and an internal nucleophile, usually water or hydroxide; the properties of metallacycles with aqueous/hydroxo ligands have been thoroughly reviewed recently by Vicente and Arcas [9]. An extra hanging or external nucleophile (C) is easily incorporated into the organic core of a metallacycle. A chiral unit (D) somewhere within a metallacycle is crucial for biomimetic stereospecific transformations. The possibilities here are exceptionally broad because the introduced chirality can be of three different types at least: (i) central carbon, (ii) planar [10] and (iii) N-centered (namely, when X = N) [11].

Features A–D, independently or combined, give rise to a remarkable set of functioning catalysts that are described below. Prior to that, a negative feature of catalytic metallacycles – their limited solubility in water – should be mentioned. Cyclometallated complexes with hydrophobic metallacycles are usually neutral. Earlier attempts to increase the solubility in water by incorporating charged func-

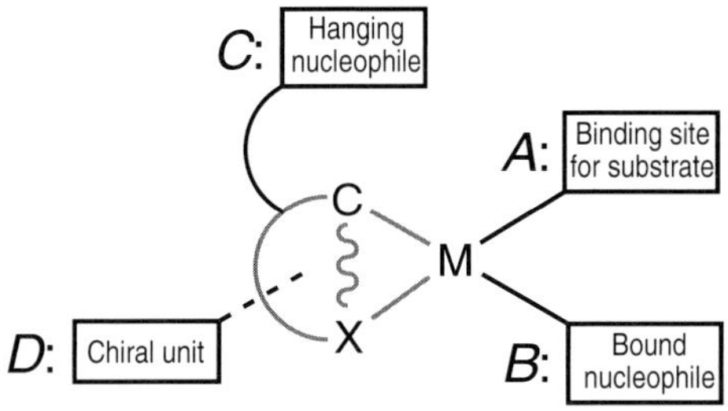

Figure 13.2 Structural features of square planar metallacycles essential for biomimetic catalysis (see text for explanations).

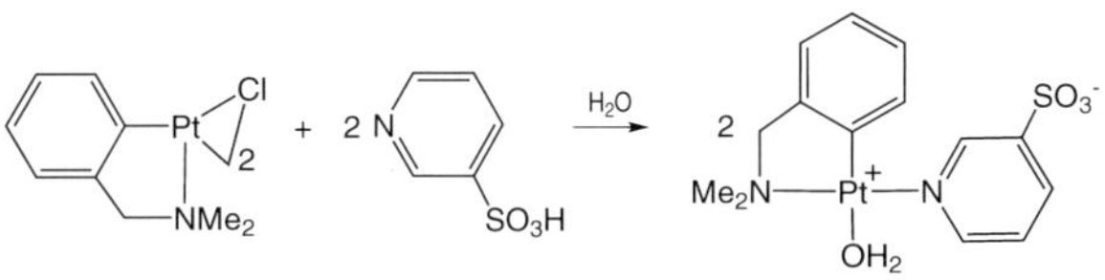

1

Scheme 13.1

tional groups were insufficiently successful [12]. For example, the reaction with pyridine-3-sulfonic acid in water afforded poor water-soluble zwitterionic species **1** (Scheme 13.1).

Attaching a 5-crown-15 motif to the orthopalladated aryl oxime to form compound **2** significantly increased the solubility of the metallacycle in water. Its aqueous solubility is more than ten times higher than that of the related complex without the crown ether fragment and increases further in the presence of magnesium(II) salts [13]. An application of palladacycle **2** in the biomimetic hydrolysis of 4-nitrophenyl-2,3-dihydroxybenzoate is described below.

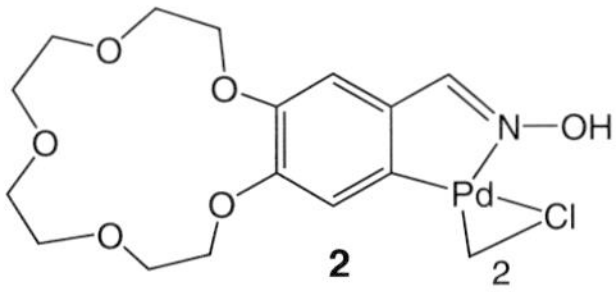

2

The first example of using palladacycles in the biomimetic hydrolysis of 2,4-dinitrophenyl acetate (DNPA) was described by Yatsimirsky *et al.* [14]. Hydrolysis of DNPA into 2,4-dinitrophenolate and acetate is easy [15] and catalysis by ortho-palladated aryl oximes **3a,b** was anticipated, as well as the fact that more nucleophilic **3b** was more active than **3a** (Scheme 13.2). The reaction is first order both in **3** and DNPA; the observed second-order rate constant k_{cat} is pH dependent (Equation 13.1):

$$k_{cat} = \frac{k_{3a}[H^+] + k_{3b}K_a}{[H^+] + K_a} \tag{13.1}$$

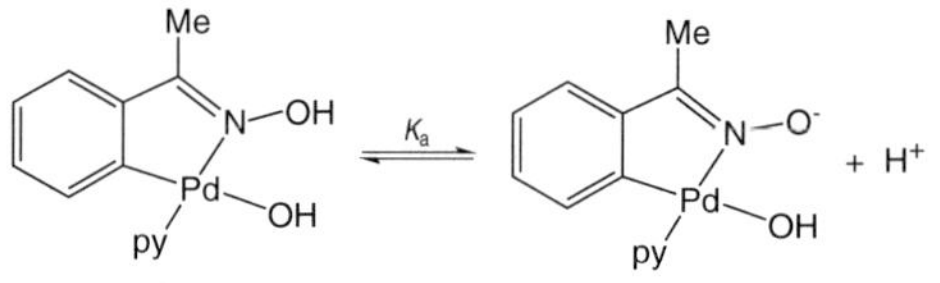

3a 3b

Scheme 13.2

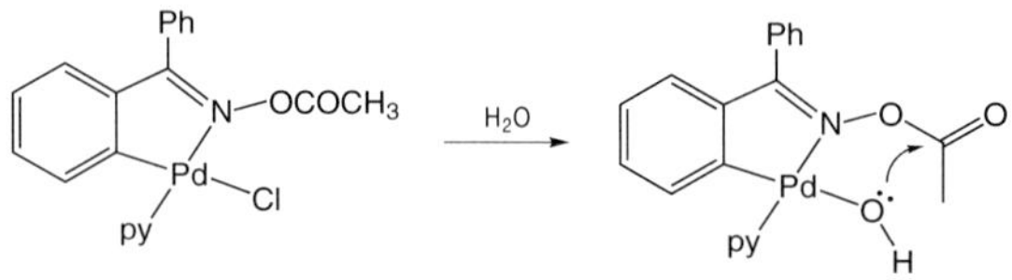

4 5

Figure 13.3 Isolated *O*-acetyl oxime complex **4** and postulated product of its rapid hydrolysis **(5)**.

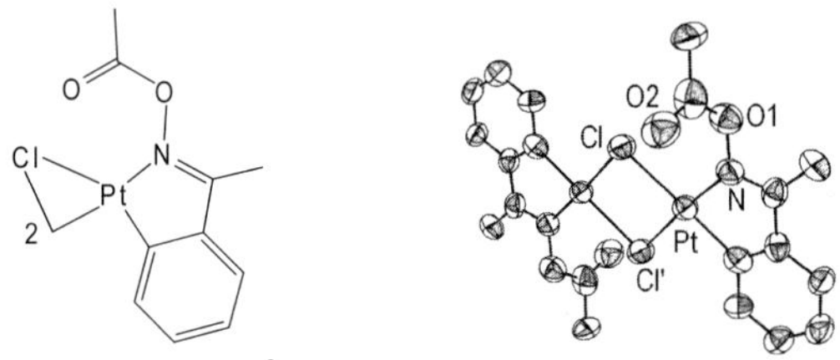

6

Figure 13.4 Complex **6** and its ORTEP diagram. Thermal ellipsoids are drawn at the 50% probability level. From Reference [16].

Cyclopalladated aryl oxime **3** has two advantages over the use of free aryl oxime. The activity of **3b** was a factor of 1000 higher than that of the free oximate and the hydrolysis turned out to be truly catalytic. Therefore, palladium(II) plays an important role in catalysis. In fact, deacylation of complex **4**, a plausible intermediate formed as a result of nucleophilic attack of **3b** at DNPA, occurs very rapidly and the pH profile of deacylation gives evidence that the coordinated hydroxide of intermediate **5** rapidly derived from **4** (Figure 13.3) in water is involved in the intramolecular nucleophilic attack [14].

The intimate mechanistic features of this acceleration have been analyzed using complex **6**, the platinum counterpart of complexes **4** and **5** [16]. The X-ray structural data for **6** (Figure 13.4) show that the axially oriented carbonyl oxygen is located almost above the platinum plane. The Pt···O2 separation is 3.65 Å and may be compared with the distance of the closest approach (3.48 Å) or approximate sum of platinum and oxygen van-der-Waals radii of ca 3.3 Å estimated according to Bondi [18]. However, *ab initio* quantum chemical calculations indicate that the

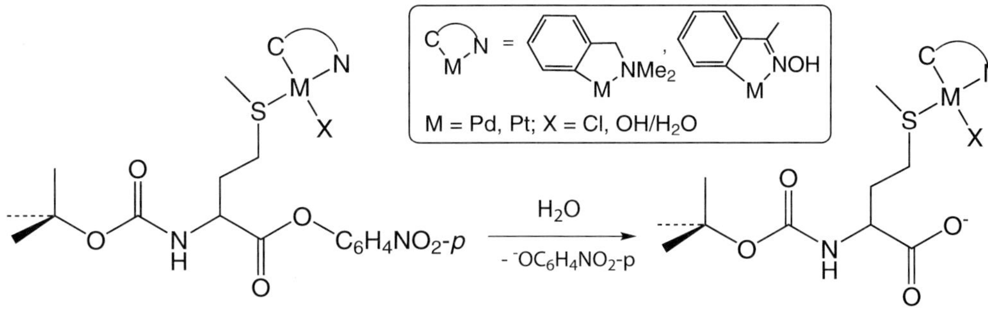

Figure 13.5 Complexes of *N*-t-BOC-L-methionine *p*-nitrophenyl ester with pallada- and platinacycles (**7**, X = Cl) undergo rapid substitution of chloride by water in aqueous solution, followed by hydrolysis of *p*-nitrophenolate, the rate of which is dictated by the nature of metal center.

location of carbonyl oxygen above the metal square plane does not increase the effective positive charge at the carbonyl carbon but may favor the intramolecular nucleophilic attack by the coordinated hydroxide or aqua ligand at the carbonyl group due to stereochemical control, which arises from the close proximity of the coordinated OH (or H_2O) ligand and the carbonyl sp^2 carbon (the estimated O···C separation is 3.47 Å) in the reactive complex **5** generated in aqueous solution. Coordinated nucleophiles other than hydroxide in complexes such as **5** have also been shown to be active in the intramolecular deacylation [17].

Donor ligands convert readily dimeric cyclopalladated complexes such as **2** into the corresponding monomeric species. The monomerization reaction is a tool for targeted delivery of pallada- and platinacycles to the corresponding parts of biomolecules. For example, chloro-bridged cyclometallated dimers [M(*o*-$C_6H_4CH_2NMe_2$)(μ-Cl)]$_2$ (M = Pd^{II}, Pt^{II}) and [Pd(*o*-C_6H_4CMe=NOH)(μ-Cl)]$_2$, as well as the monomer [Pt(*o*-C_6H_4CMe=NOH)Cl(C_6H_5CMe=NOH)] react with *N*-*t*-BOC-L-methionine *p*-nitrophenyl ester (BOC = butyloxycarbonyl) in benzene to give monomeric compounds **7** (X = Cl) (Figure 13.5) [19]. Aquation of **7** in aqueous solution results in the substitution of aqua/hydroxo ligand for chloride, the coordinated nucleophile being near the ester moiety. The hydrolysis of *p*-nitrophenolate ion within **7** was followed by UV/vis spectrophotometry at pH 8.0 (0.01 M phosphate), 23 °C. The highest, 535-fold, rate increase compared to free *N*-*t*-BOC-L-methionine *p*-nitrophenyl ester was found for the platinum(II) complex of acetophenone oxime, the first-order rate constant for which equals $6.0 \pm 0.6 \times 10^{-2}\,s^{-1}$.

The proximity of donor center and a bond to be cleaved plays an important role. This has been probed in a study of the hydrolysis of *N*-acetyl protected *p*-nitrophenyl esters of *S*-methylcysteine and methionine catalyzed by the dimeric palladacycle [Pd(*o*-$C_6H_4CH_2NMe_2$)(μ-Cl)]$_2$ [20]. The formation of *p*-nitrophenolate

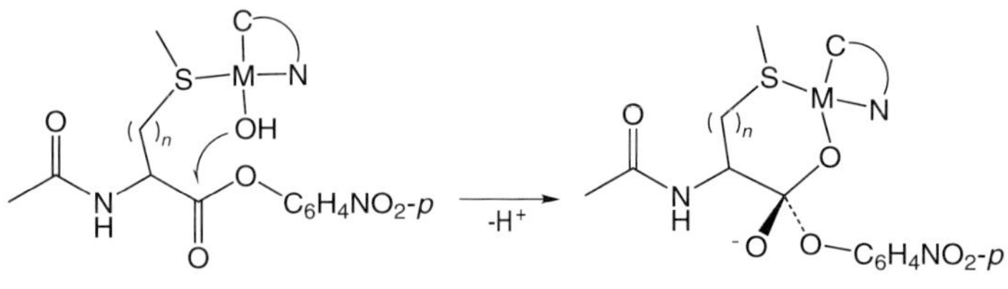

8

Figure 13.6 Postulated formation of the six-membered cyclic tetrahedral intermediate **8** in the hydrolysis of *N*-acetyl-*S*-methylcysteine *p*-nitrophenyl ester (n = 1) catalyzed by dimeric palladacycle $[Pd(o\text{-}C_6H_4CH_2NMe_2)(\mu\text{-}Cl)]_2$.

in the *S*-methylcysteine and methionine cases follows the rate law shown in Equation 13.2:

$$k_{obs} = k_0 + k_{cat}[Pd^{II}] \quad (13.2)$$

The catalytic rate constant k_{cat} equals 880 ± 90 and 45 ± 3 $M^{-1}s^{-1}$ for the *S*-methylcysteine and methionine derivatives, respectively, at pH 8.0 and 25 °C. Interestingly, the equilibrium constants K_1 for the monomerization of $[Pd(o\text{-}C_6H_4CH_2NMe_2)(\mu\text{-}Cl)]_2$ by *N*-acetyl protected *p*-nitrophenyl esters of *S*-methylcysteine and methionine to form monomers such as **7** are $2.9 \pm 0.3 \times 10^3$ and $3.0 \pm 0.6 \times 10^4\,M^{-1}$, respectively, in chloroform at 25 °C. Thus, the *S*-methylcysteine derivative is substantially more reactive, though it forms a weaker complex with the palladacycle. The higher reactivity of *N*-acetyl-*S*-methylcysteine *p*-nitrophenyl ester is rationalized in terms of the intramolecular mechanism of hydrolysis via the formation of the six-membered cyclic tetrahedral intermediate **8** (n = 1) (Figure 13.6). The less thermodynamically favorable seven-membered intermediate (n = 2) should be formed in the case of the methionine counterpart.

The catalytic activity of the cyclopalladated complexes of tertiary and primary benzylamines $[Pd(o\text{-}C_6H_4CHR'NR_2)Cl(py)]$ (**9**: R = Me, H; R′ = H, *S*-Me and *R*-Me) in the hydrolysis of *N*-*t*-BOC-L-methionine *p*-nitrophenyl ester appeared to be rather unexpected [21]. The pseudo-first-order rate constants k_{obs} for release of *p*-nitrophenolate followed Equation 13.2. Truly remarkable was the dependence of k_{cat} on the nature of cyclopalladated organic ligand (Figure 13.7).

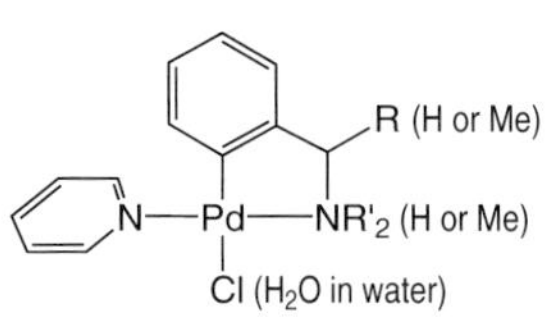

9

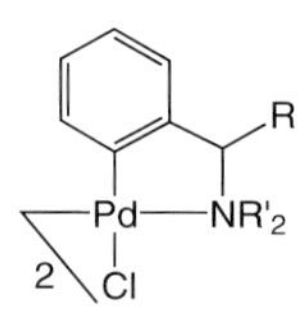

10

The lowest catalytic activity ($k_{cat} \sim 2.5\,M^{-1}s^{-1}$) was observed for all *N,N*-dimethylbenzylamine complexes tested. The transition to the primary benzylamine complex resulted in 10 times better catalysis ($k_{cat} = 32\,M^{-1}s^{-1}$). The highest activity

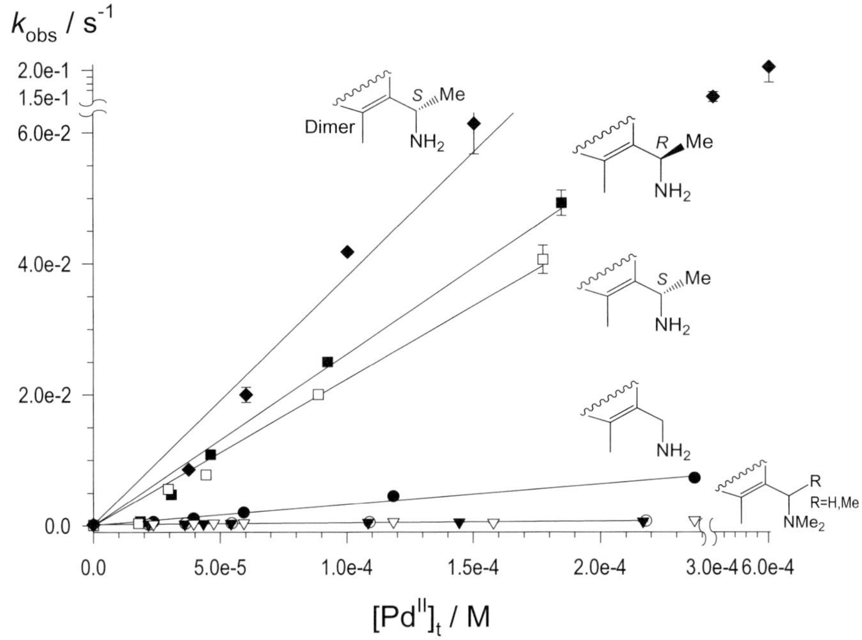

Figure 13.7 Pseudo-first-order rate constants for the hydrolysis of *N-t*-BOC-L-methionine *p*-nitrophenyl ester catalyzed by monomeric palladacycles **9** derived from primary and tertiary benzylamines and the dimer *(SS)*-**10** (R = Me, R′ = H) as a function of total Pd^{II} concentration (calculated for the monomeric form for the dimer). Conditions: pH 8.0 (0.01 M phosphate), 10% MeCN, 25 °C. From Reference [21].

was registered for the *(S)*- and *(R)*-enantiomers of α-methylbenzylamine, that is 220 and 260 $M^{-1}s^{-1}$, respectively. Discrimination is observed when the activated ester has a donor center capable of binding with Pd^{II}. In accordance with this is the fact that all complexes studied displayed similar activity ($k_{cat} \approx 0.8\,M^{-1}s^{-1}$) in the hydrolysis of 2,4-dinitrophenyl acetate. Even the most active complexes in Figure 13.7 did not catalyze the hydrolysis of *N-t*-BOC-leucine *p*-nitrophenyl ester. These facts are consistent with a hypothesis that the discrimination occurs when a target ester has a donor center for binding Pd^{II}. The catalytic hydrolysis of methionine derivative should occur intramolecularly (Figure 13.6). Deprotonation of the primary NH_2 group plays no role in catalysis and, therefore, a ten-fold rate increase on going from tertiary to primary amine complexes may have a steric origin. The effect of α-methylation is more puzzling. If the tetrahedral intermediate/transition state is formed, the α-methyl group sits right above the phenyl ring of the leaving 4-nitrophenolate (Figure 13.8). Thus, hydrophobic stabilization of the intermediate is possible in aqueous solution. The chiral center of the optically active palladacycle is directed oppositely to the asymmetric carbon of the amino acid and this explains the low enantioselectivity.

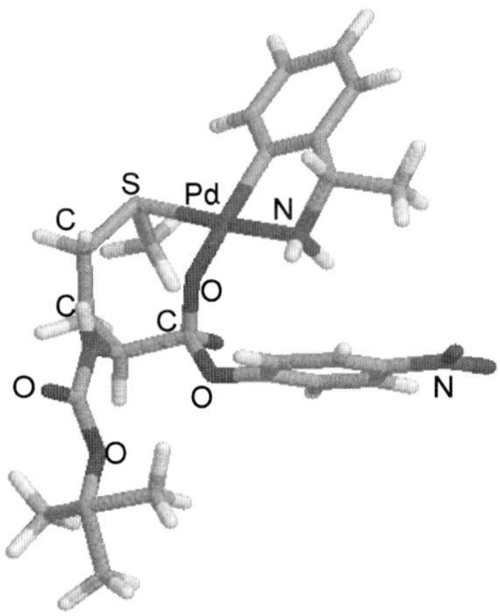

Figure 13.8 Model of the tetrahedral reactive intermediate formed in the hydrolysis of *N-t*-BOC-L-methionine *p*-nitrophenyl ester catalyzed by cyclopalladated α-methylbenzylamine (key atoms are labeled, see text for details). From Reference [21].

13.2.2
Enantioselective Hydrolysis of Activated Esters

Ester cleavage of the (*R*)- and (*S*)-isomers *N*-CBZ-leucine *p*-nitrophenyl ester (**11**, CBZ = benzyloxycarbonyl), which is intermolecularly catalyzed by the (*R*)- and (*S*)-stereoisomers of Pd^{II} metallacycle [Pd(*o*-$C_6H_4C^*HMeNMe_2$)Cl(py)] (**9**), follows Equation 13.2 [22].

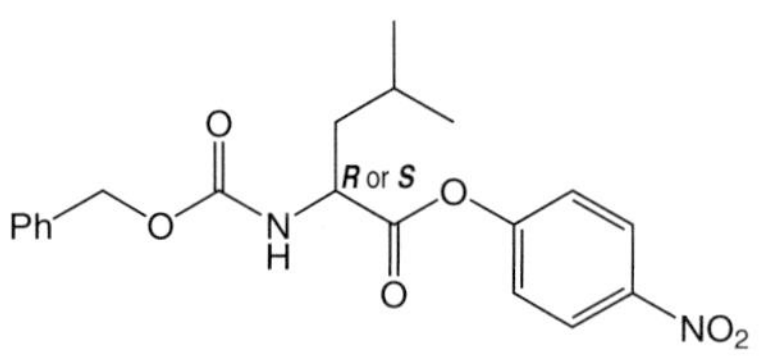

11

The best catalysis occurs when the asymmetric carbons of the leucine ester and palladium(II) complex are configured *R* and *S* or *S* and *R*, respectively. The enantioselectivity factor, defined as a ratio of the rate constants k_{cat} for *R* and *S* enantiomers of the ester, equals 4.5 (*R*/*S*) and 3.5 (*S*/*R*), respectively (Table 13.1).

The highest catalytic activity is observed when the absolute configuration of complex **9** is *R* and of the ester is *S*, or vice versa, whereas the rate constants k_{cat} are much lower for the *R*-*R* or *S*-*S* pairs, suggesting a transition state for the catalytic process. Modeling has been performed using the conventional molecular models and supported by calculations using the Alchemy III package. The results shown in Figure 13.9 demonstrate a possible approach of the (*S*)-isomer of **9** to the (*R*)-isomer of the leucine ester. The chiral recognition seems to be due to a

Table 13.1 Rate constants k_{cat} ($dm^3 mol^{-1} s^{-1}$) and k_o (s^{-1}) for the hydrolysis of *(R)*- and *(S)*-enantiomers of **11** in the presence of *(R)*- and *(S)*-enantiomers **9** (R = R′ = Me), pH 6.23, 25 °C.

Complex configuration	(*S*)-Ester		(*R*)-Ester	
	$10^3 \times k_o$	k_{cat}	$10^3 \times k_o$	k_{cat}
R	2.4 ± 0.1	25.8 ± 0.4	2.4 ± 0.1	7.6 ± 0.5
S	2.4 ± 0.1	5.7 ± 0.6	2.4 ± 0.1	26.7 ± 0.5

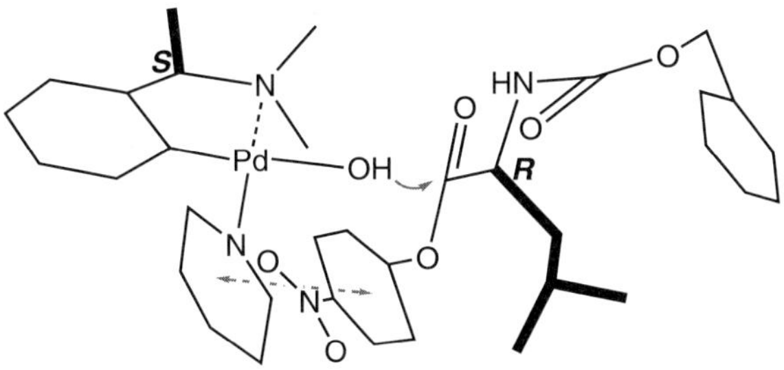

Figure 13.9 Possible mode of interaction between complex *(S)*-**9** and *N*-CBZ-*(R)*-leucine *p*-nitrophenyl ester **11**, accounting for the origin of the enantioselectivity in the ester hydrolysis (for details, see text).

mutual orientation of alkyl radicals at the asymmetric carbons of the complex and the ester, namely of the methyl and isopropyl groups, respectively. If the approach of the catalyst to the carbonyl carbon of the ester is such as shown in Figure 13.9, less steric repulsion is expected between the alkyl groups when ester **11** and catalyst **9** have either *S* and *R*, as in Figure 13.9, or *R* and *S* absolute configurations, respectively. The transition state geometry suggested is also advantageous in view of a possible hydrophobic/stacking interaction between the leaving 4-nitrophenolate and coordinated pyridine. This could minimize repulsion between the alkyl groups. Thus, there are three types of interaction that provide stereoselectivity, namely, between (i) the coordinated hydroxide and the carbonyl carbon, (ii) the nitro-substituted aromatic ring and coordinated pyridine and (iii) the alkyl groups at the ligand and catalyst stereogenic centers. Combined these bring about a sound kinetic enantioselectivity [23, 24]. A disadvantage of the system is that the enantioselectivity is observed along with spontaneous, catalyst-independent, hydrolysis of the activated ester. This minimizes the applicability of chiral palladacycles for kinetic resolution of racemic esters. Lower intramolecular enantioselectivity was found in the hydrolysis of *N-t*-BOC-(*S*)-methionine *p*-nitrophenyl ester with coordinated to sulfur *(R)*- and *(S)*-enantiomers of [Pd(*o*-$C_6H_4C^*HMeNMe_2$)Cl] [22].

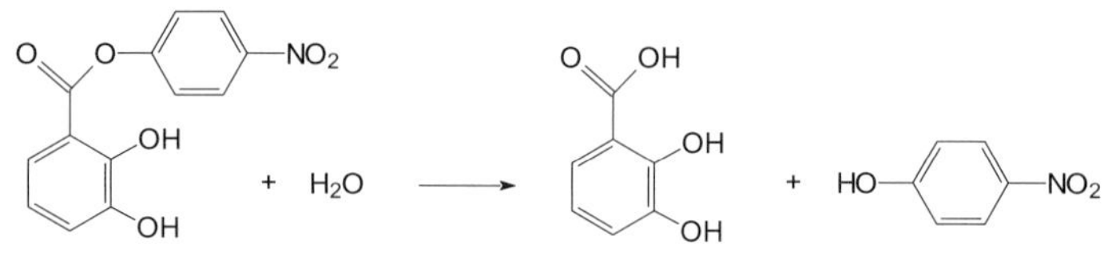

12

Scheme 13.3

Hydrolysis of 4-nitrophenyl 2,3-dihydroxybenzoate is used for modeling the catalysis by serine proteases [25]. To illustrate the advantages of using complex **2** as a catalyst, palladacycles Pd(*o*-C_6H_4CMe=NOH)(μ-Cl)]$_2$ and [Pd(*o*-$C_6H_4CH_2NMe_2$)(μ-Cl)]$_2$ were also investigated as effectors of biomimetic hydrolysis of 4-nitrophenyl 2,3-dihydroxybenzoate **(12)**. Its vicinal hydroxy groups may serve as a trap for certain molecules, which, on coordination, can both increase or decrease the reaction rate and alter the intramolecular general base mechanism of the hydrolysis of **12** [26, 27]. Such regulation of the reactivity by binding extra molecules makes **12** a functioning mimetic for the simulation of fine regulatory and/or allosteric effects encountered in the enzymatic systems [28]. *C,N*-chelates have a proper geometry for complexation with 1,2-diols. They exist in a monomeric form in MeCN and water and, hence, two cis coordinative sites are available for the complexation with the hydroxy groups of **12**. The binding of Pd^{II} species with aromatic 1,2-diols is of precedent [29]. Therefore, the rate of hydrolysis of **12** (Scheme 13.3) has been investigated as a function of the concentration of Pd^{II}.

Figure 13.10 shows the dependencies of k_{obs} for Scheme 13.3 against concentrations of different palladacycles. Oxime complexes **2** and [Pd(*o*-C_6H_4CMe=NOH)(μ-Cl)]$_2$ accelerate the hydrolysis of **12**, whereas the tertiary amine palladacycle [Pd(*o*-$C_6H_4CH_2NMe_2$)(μ-Cl)]$_2$ retards the rate. The rate constants k_{obs} display Michaelis-type dependence and level off at higher concentrations of Pd^{II}, indicative of the formation of reactive (**2** and Pd(*o*-C_6H_4CMe=NOH)(μ-Cl)]$_2$) and inactive ([Pd(*o*-$C_6H_4CH_2NMe_2$)(μ-Cl)]$_2$) intermediates. The kinetic data are interpreted in terms of a general mechanism in Figure 13.11; **12** is shown as a semi-anion (pK_a 8.8) [30]. There are two types of oxime-diol Pd^{II} complexes and only **B** is reactive, since the ester and oxime functionalities in complex **A** are, crucially, separated.

If fast reversible formation of **A** and **B** is characterized by similar equilibrium constants K (Figure 13.11), one arrives at Equation 13.3 for k_{obs} with n = 1 for **2** and 2 for [Pd(*o*-$C_6H_4CH_2NMe_2$)(μ-Cl)]$_2$:

$$k_{obs} = \frac{k_0 + nk_1K[\mathrm{Pd^{II}}]}{1 + 2K[\mathrm{Pd^{II}}]} \tag{13.3}$$

The ratio of the rate constants k_1/k_0 indicates that the coordinated palladated oxime **2** increases the rate 116-fold, whereas the inhibiting effect of [Pd(*o*-$C_6H_4CH_2NMe_2$)(μ-Cl)]$_2$ is a factor of 14. The hydrolysis of free **12** occurs via the

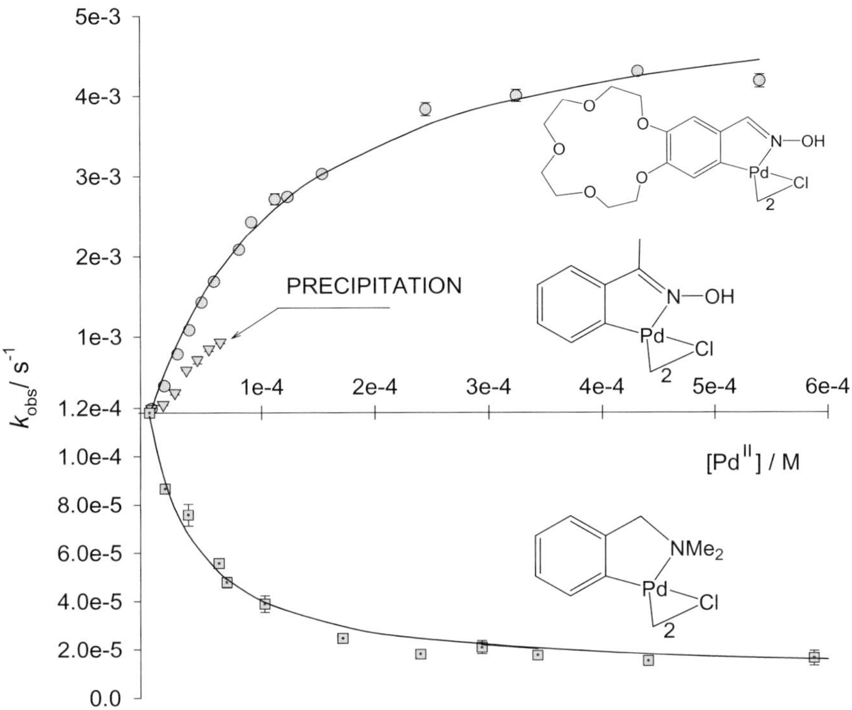

Figure 13.10 Dependence of the observed pseudo first order rate constants for the hydrolysis of **12** (2×10^{-5} M) against concentrations of three cyclometallated complexes at pH 8.0 (phosphate buffer, 25 °C). From Reference [13].

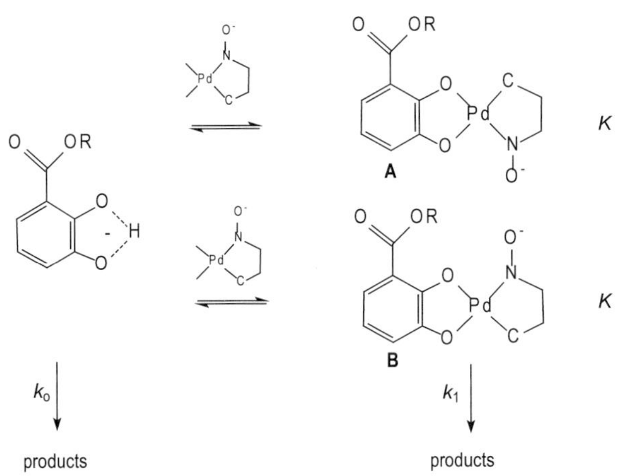

Figure 13.11 Mechanism of hydrolysis of **12** catalyzed by cyclopalladated complexes.

intramolecular general base mechanism [27, 30, 31]. Complex formation between **12** and [Pd(*o*-$C_6H_4CH_2NMe_2$)(μ-Cl)]$_2$, namely the palladacycle without nucleophile within the C,N-chelate, lowers the nucleophilicity of the semi-deprotonated "intramolecular" base. The activation of water is no longer feasible and [Pd(*o*-$C_6H_4CH_2NMe_2$)(μ-Cl)]$_2$ retards hydrolysis. Complexes **2** and [Pd(*o*-C_6H_4CMe=NOH)(μ-Cl)]$_2$ accelerate hydrolysis of **12** because oximate rather than the hydroxo group functions as a general base [the measured solvent isotope effect $k(H_2O)/k(D_2O)$ equals 2.3 ± 0.2 at pD 9.0], and, since the nucleophilicity of the metallated aryl oximate is higher than that of the phenolic hydroxide, significant catalysis by cyclopalladated aryl oximes is observed.

The mechanism proposed involving the O,O′-bonded intermediate is supported by the fact that hydrolysis of 4-nitrophenyl salicilate and benzoate is not catalyzed by **2** under the same conditions. Independent spectrophotometric measurements of the binding between catechol as an unreactive analog of **12** and **2** support the mechanism shown in Figure 13.11. The mechanism agrees with observations that the platinum(II) complex [Pt(*o*-C_6H_4C(Me)=NOH)Cl(py)] is catalytically inactive and does not show evidence for binding with catechol.

13.2.3
Hydrolysis of Phosphoric Acid Esters

The catalytic rate constants for the hydrolysis of amino acid esters described above are marginal though the catalysis has features typical of metallohydrolases. Pd^{II} and Pt^{II} are the "soft" acids and therefore "softer" bases could be better substrates for good catalysis. Such are thiophosphoric acid esters **13, 14, 16** and **17** (Figure 13.12) and, therefore, metallacycles **18** and **19** have been tested as catalysts for the hydrolysis of thiophosphate pesticides and neurotoxins **13–17**. The metallacycles performed superbly [32].

Spontaneous hydrolysis of parathion **13** is immeasurably slow at pH 8.5. Addition of a catalytic amount of Pd^{II} complex **18** (R′ = H, Z = py) enhances the rate of hydrolysis of parathion (Table 13.2), which is complete and follows first-order kinetics. Platinum(II) metallacycles **19** display even higher catalytic activity. Methyl parathion **(14)** and coumaphos **(16)** behave similarly. No intercept has been

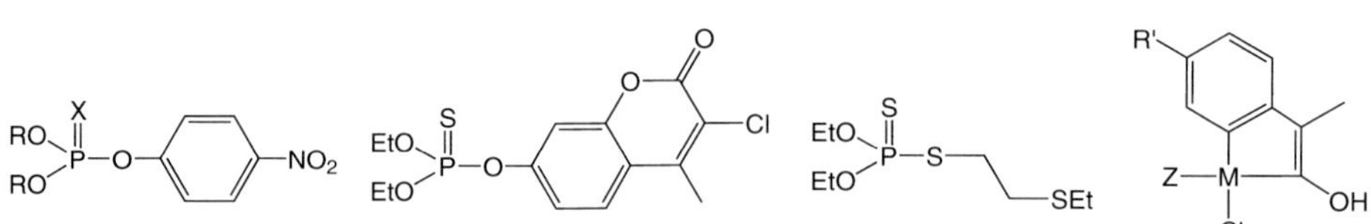

13 R = Et; Z = S: parathion
14 R = Me; Z = S: methyl parathion
15 R = Et; Z = O: paraoxon

16 coumaphos

17 demeton-S

M=Pd (**18**), Pt (**19**)
R'= H,Me,MeO,F Cl
Z = dmso, py

Figure 13.12 Some phosphoric acid esters, the hydrolysis of which was studied in the presence of cyclometallated complexes **18** and **19**.

Table 13.2 Observed second-order rate constants (k_4) for the hydrolysis of phosphoric acid triester parathion **(13)** catalyzed by **18** and **19** at pH 8.5, 25 °C, 0.01 M $NaClO_4$.

Complexes (Z/R′)	k_4 ($M^{-1}s^{-1}$)
18a (py/H)	54
19a (dmso/H)	310
19b (dmso/MeO)	914
19c (dmso/Me)	773
19d (dmso/F)	429
19e (dmso/Cl)	452
19f (py/H)	230
OH^-	9.5×10^{-5}

Table 13.3 Observed second-order rate constants (k_4) for the hydrolysis of **13–17** catalyzed by **19** (Z = dmso, R′ = H) at pH 8.5, 25 °C, 0.01 M $NaClO_4$.

Triester	k_4 ($M^{-1}s^{-1}$)
13 (parathion)	310
14 (methyl parathion)	175
15 (paraoxon)	10.2
16 (coumaphos)	141
17 (demeton-S)	27.8

observed on the plots of k_{obs} versus concentrations of **18** or **19**, indicating good catalysis according to the rate law $k_{obs} = k_4$[catalyst].

$$RR'P(=X)YZ + H_2O \rightarrow RR'P(=X)OH + HYZ(X/Y = O \text{ and/or } S) \qquad (13.4)$$

The kinetic data obtained are summarized in Tables 13.2 and 13.3. The catalytic activity of **18** and **19** in hydrolysis of **13** is 10^6–10^7 times higher than that of hydroxide. The catalyzed hydrolysis is a factor of 10^9 more efficient than the spontaneous one at concentrations of **19** as low as 10^{-4} M! This is a significant number compared to other systems where the hydrolysis of nitrophenyl phosphates catalyzed by biomimetically relevant transition metal catalysts was investigated [33–38]. The catalytic effects are usually in the range 10^4–10^6. However, these estimates should be treated with care, since, first, they are often referred to the promoted hydrolysis rather than the catalyzed one and, second, the kinetic data are sometimes not used carefully enough. For example, the estimate of 10^{11} reported by Williams *et al.* corresponds to the *promoted* cleavage of 4-nitrophenyl methyl phosphate coordinated to a dinuclear Co^{III} complex [39].

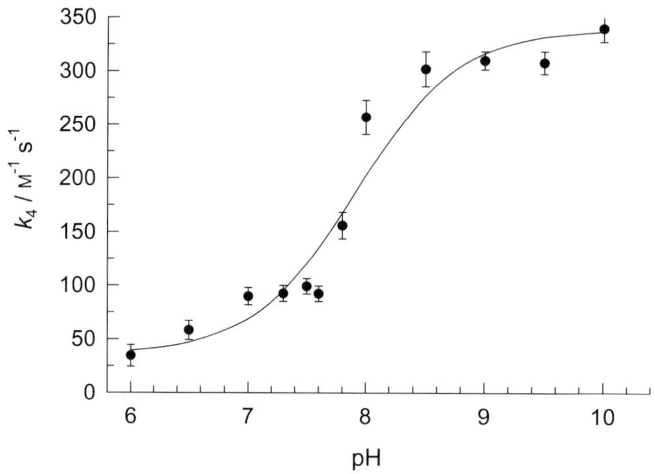

Figure 13.13 pH profile for **19a**-catalyzed hydrolysis of parathion **(13)**. Conditions: **[19a]** 1×10^{-4} M, 0.01 M $NaClO_4$, 25 °C. Buffer components: 0.005 M $Na_2B_4O_7$ (pH 8–11), 5,5′-diethylbarbituric acid (pH 6–7.8) and NaOAc (pH 4.5–5.5). From Reference [32].

If one compares the catalytic activity of **19** and organophosphate hydrolases (OPH) – a class of metal-dependent enzymes, a vital virtue of which is in detoxification of organophosphoric species via Scheme 13.4 – in the hydrolysis of **13**, the low molecular weight catalyst is fairly competitive although OPH is a very reactive enzyme (k_{cat} 600 s^{-1} and K_M 2.8×10^{-4} M) [40]. Recalculated to per gram of catalyst, k_4 and k_{cat}/K_M equal ca 1 and 60 $g^{-1} s^{-1}$, respectively, and this shows nicely the effect of metallacycles, taking into account that the rate of the enzymatic reaction levels off at high concentrations of **13**.

The data in Tables 13.2 and 13.3 are indicative of several features. The catalysis of paraoxon **(15)** hydrolysis by **19** is 30-times less efficient than that of parathion **(13)**. Thus, a soft sulfur donor center is in fact more favorable than harder oxygen. Interestingly, the base hydrolysis of paraoxon is ca. 1000 times faster than of parathion. The catalytic activity of **19** can further be increased three-fold by introducing substituents at the benzene ring of the cyclometallated ligand. Both electron-donating and -withdrawing groups favor the catalysis. Comparison of the Pt^{II} **(19)** and Pd^{II} **(18)** complexes both with Z = py shows that the former is ca five-times catalytically more active. The hard–soft principle [41] seems to hold here as well.

The **19**-catalyzed hydrolysis of **13** is pH-dependent in the range 6–10 (Figure 13.13). The profile corresponds to the rate law shown in Equation 13.5. There are two catalytically active species, which are characterized by the rate constants k_{AH} and k_A of 40 ± 20 and 340 ± 20 $M^{-1} s^{-1}$, respectively:

$$k_4 = \frac{k_{AH}[H^+] + k_A K_a}{[H^+] + K_a} \tag{13.5}$$

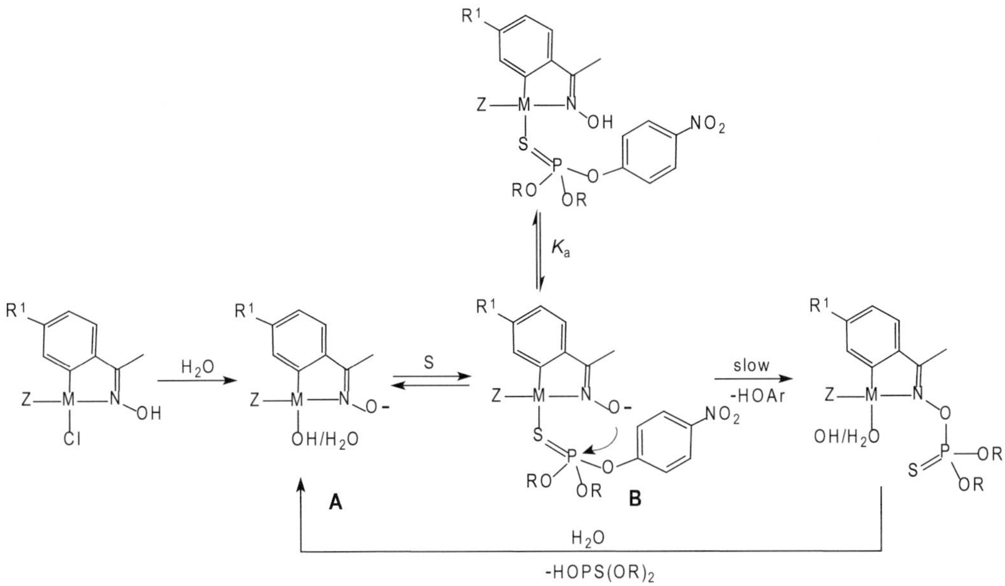

Figure 13.14 Postulated mechanism of hydrolysis of thiophosphoric acid esters (S) catalyzed by cyclopalladated and cycloplatinated complexes.

The value of K_a is $1.2 \pm 0.5 \times 10^{-8}\,M^{-1}$ (pK_a 7.9). The acid–base equilibrium is ascribed to deprotonation of the NOH group of the orthometallated oxime, a key nucleophilic center in the reaction. The coordinated halides trans to the σ-bound phenyl carbon in **18** or **19** are rapidly and completely hydrolyzed in water and, therefore, a plausible reaction mechanism is that envisioned in Figure 13.14. Esters **13–16** form sulfur-bonded intermediates **B** with aqua/hydroxo species **A** followed by intramolecular nucleophilic attack of the oximate oxygen at phosphorus(V). The solvent isotope effect of 1.2 ± 0.1 determined in the reaction of **13** and **19c** at pD 10.05 (borate buffer) is consistent with the nucleophilic mechanism [42]. Thus, Pd^{II} or Pt^{II} play a dual role in the catalysis. The metal centers are binding sites for the substrate in a close proximity to the coordinated oxime and its pK_a is substantially lowered due to the coordination with the metal. As a result, a strong nucleophile is generated at neutral pH.

Coordination of the esters and the metals should also increase an effective positive charge at P^V and thus facilitate the intramolecular nucleophilic attack by the oximate. The replacement of dmso or pyridine ligands at Pt^{II} in **18** and **19** is unlikely since no characteristic bands from free pyridine were observed in the UV/vis spectrum of a mixture of **17** and **19f**. The mechanism in Figure 13.14 is supported by a UV/vis and ^{31}P NMR study of the **19a**-promoted hydrolysis of neurotoxin demeton-S **(17)**. The resonance from intact **17** is at δ 30.005 in aqueous 1×10^{-4} M $NaClO_4$. Addition of **19a** generates a new signal at δ 0.655 from the reaction product $(EtO)_2P(=O)OH$ [43]. Although the rate of hydrolysis of **17**

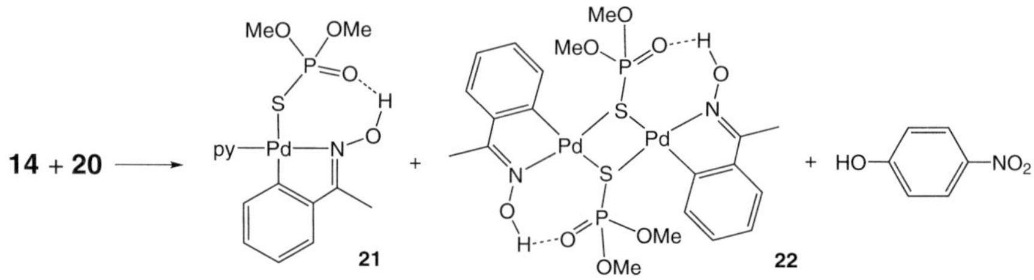

Figure 13.15 Stoichiometric study of the reaction between methyl parathion **(14)** and complex **20**, the acetato counterpart of **18a**.

increases tremendously in the presence of **19a**, there is no true catalysis and a stoichiometric amount of **19a** is necessary. The cleavage of **17** liberates a potentially good bidentate ligand $HSCH_2CH_2SEt$, which on interaction with the *C,N*-metallacycle blocks the substrate binding site, making the intramolecular nucleophilic attack inaccessible [44].

This initial work stimulated a search for similar metallacycle-based catalytic systems for the hydrolysis of thiophosphoric acid esters and related studies. Kim and Gabbai have investigated a stoichiometric reaction between methyl parathion **(14)** and complex **20**, the acetato counterpart of **18a**, in wet acetone [45]. After 24 h the reaction affords two products (**21** and **22**) in a molar ratio of 9 : 1 (Figure 13.15), both of which have been characterized by X-ray crystallography. The P—S bond distances is 1.982(1) and 2.061(1) Å in **21** and **22**, respectively. These should be compared with the double (1.94 Å) and single (2.09 Å) phosphorus–sulfur bonds.

A compound structurally similar to **22** was isolated under similar conditions when cyclopalladated *(S)*-4-carbomethoxy-2-phenyl-2-oxazoline **23** was treated with methyl parathion [46]. Interestingly, this reaction is accompanied by unexplained racemization of the oxazoline ligand. At pH 9–10, complex **23** serves as a precatalyst for the hydrolysis of methyl parathion. The reported second-order rate constant is $726\,M^{-1}\,s^{-1}$ at pH 9.0 and, probably, ambient temperature. Cyclopalladated ferrocenylketimines **24** also catalyze the hydrolysis of methyl parathion **(14)** and other thiophosphoric acid ester pesticides. The oxime complex **24** (R = OH) is more active and exhibits an increased selectivity towards sulfur-containing pesticides [47].

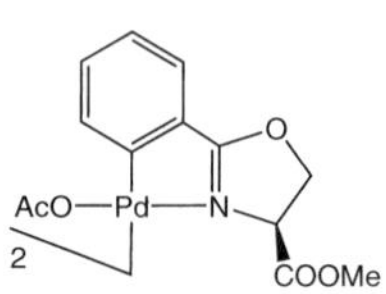

23

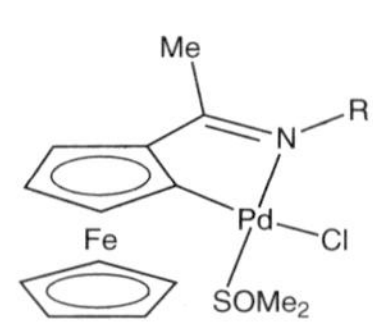

24 (R=OH, $(CH_2)_2OH$, C_6H_4OMe-o/p)

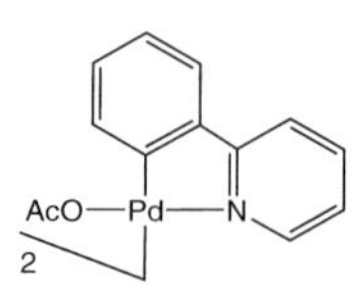

25

High catalytic activity in the hydrolysis of methyl parathion is reported for palladacycle **25** (derived from 2-phenylpyridine) [48]. The reaction product in wet THF is structurally identical to complex **22**. It reacts with PPh_3 in THF to afford the

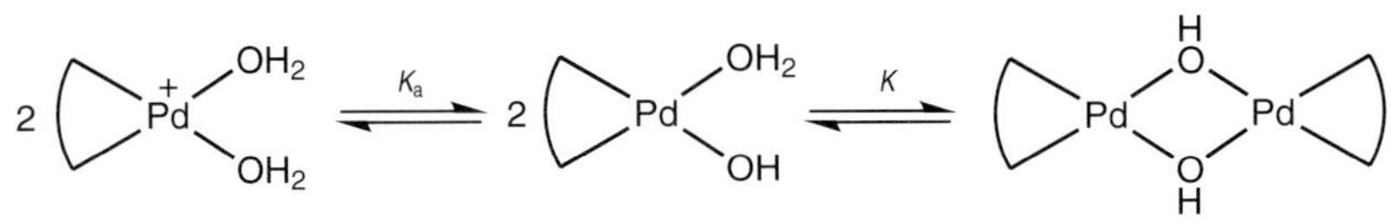

25a **25b** **25c**

Scheme 13.4

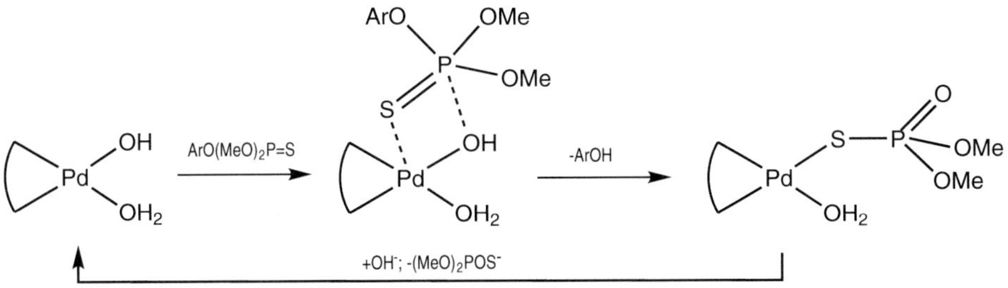

Figure 13.16 An associative mechanism of hydrolysis of methyl parathion **(14)** catalyzed by **25b** at basic pH.

corresponding mononuclear complex in which the S-bound anion $SP(=O)(OMe)_2^-$ is in a trans position relative to the σ-bound carbon. Kinetic and spectroscopic studies of **25** dissolved in water suggest that it exists in aqueous solution as a cationic diaqua complex (**25a**, Scheme 13.4). At basic pH, it undergoes deprotonation to form **25b** and dimerization to afford a dinuclear complex, presumably **25c**. Accordingly, the catalyzed hydrolysis of methyl parathion at pH 7 is both first order in substrate and palladium catalyst (the second-order rate constant is $146\,M^{-1}\,s^{-1}$ at 303 K). The reaction rate increases significantly at pH 8–9 but the reaction order in the catalyst switches to 1/2, indicating that dimer **25c** is catalytically inactive. Analysis of the data obtained at pH 9 and 298 K yields the equilibrium constant $K = (\mathbf{25c})/(\mathbf{25b})^2 = 6.6 \times 10^6\,M^{-1}$ and the second-order rate constant $k = 8.6 \times 10^3\,M^{-1}\,s^{-1}$ that characterizes the intrinsic reactivity of **25b**. An associative mechanism is suggested. The rate-limiting step at basic pH involves nucleophilic attack of the phosphorus center of methyl parathion by a hydroxide ligand of **25b** (Figure 13.16) [48].

Ortho-palladated compounds catalyze the solvolysis of fenitrothion **(26)** and P=S pesticides **27–29** in methanol as solvent [49]. Running the reactions in MeOH introduces some new features as compared with using water as the medium. The catalyzed reaction itself turns into transesterification, namely the leaving aryl oxide group is replaced by methoxide. Amine buffer components and chloride both inhibit the methanolysis due stronger binding to Pd^{II} than in water. The catalytic activity of chloro complex **9** (R = H, R′ = Me) was minor and therefore compound **30**, with triflate as a better leaving group, was synthesized.

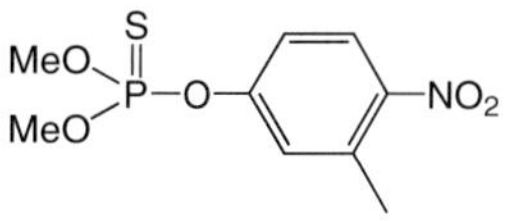

26

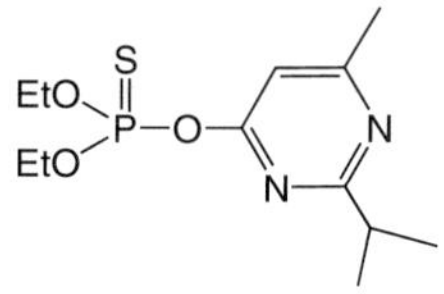

27

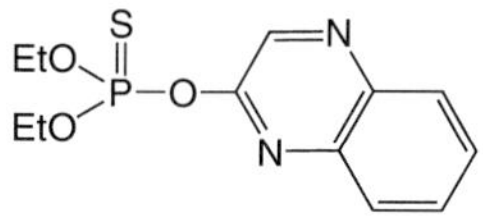

28

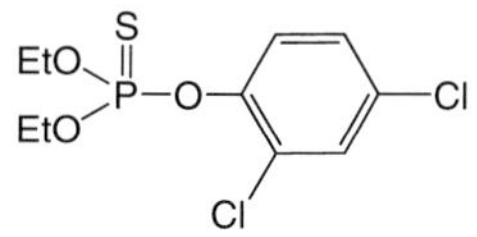

29

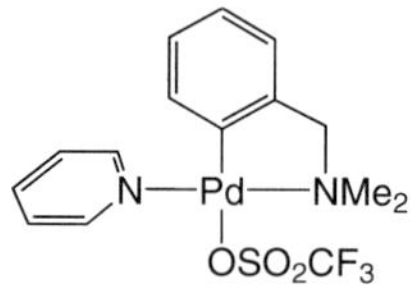

30

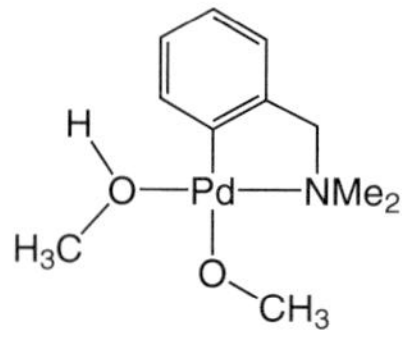

31

In neutral methanolic solutions the triflate is replaced by MeOH. In acidic MeOH the pyridine ligand dissociated to give $[Pd(o\text{-}C_6H_4CH_2NMe_2)(HOMe)_2]^+$. Ionization of ligated methanol under basic conditions gave $[Pd(o\text{-}C_6H_4CH_2NMe_2)py(OMe)]$, from which the pyridine can dissociate. Kinetic studies under buffered conditions reveal that **30** is an effective catalyst for the methanolysis of **26–29**. The active form of the catalyst is a basic one, having one associated methoxide such as **31** generated with an apparent ${}^s_s pK_a$ of 10.8 (here, the pK_a and pH values refer to electrode calibration and measurement in the same solvent, methanol). This catalytic system promotes the methanolysis of **26**, diazinon **(27)**, quinalphos **(28)**, coumaphos **(16)** and dichlofenthion **(29)** at 0.05 M Et_3N buffer, pH 10.8, 25 °C, the second-order turnover rate constants being 36.9, 0.45, 0.12, >146.7 and 44.3 $M^{-1}\,s^{-1}$, respectively. An associative mechanism for the catalyzed methanolysis is proposed, in which a transiently coordinated substrate is intramolecularly attacked by the Pd^{II}-coordinated methoxide [49].

The applications of pallada- and platinacycles described in this section are unexpected, unique and very significant. Organophosphorus triesters are the largest group of crop protectants [50]. Their widespread use in agriculture is linked to environmental concerns associated with cholinergic toxicity and, in some cases, delayed neuropathy [51]. Some thiophosphate pesticides are associated with endocrine disruption [52]. Disposal of obsolete pesticides and remediation of associated contaminated sites are of worldwide concern [53, 54]. Phosphoric acid esters are warfare agents such as of the VX type [55], and the catalytic activity of pallada- and platinacycles in the hydrolysis of organophosphorus triesters and less toxic analogs of VX is an essential achievement in creation of new catalytic systems. The catalytic activity of pallada- and platinacycles exceeds many man-made artificial hydrolases and is only lower than that of the organophosphate hydrolase enzyme, the specific substrates of which are thiophosphoric acid esters [55].

13.3
Biologically Relevant Deoxygenation of Dimethyl Sulfoxide by Orthoplatinated Oximes: Oxidoreductase Mimetics

Orthometallated aryl oxime species **32**, likewise other transition-metal complexes [56], undergo deoxygenation of dimethyl sulfoxide (DMSO) in methanol in the presence of HCl to afford the Pt^{IV} dimethyl sulfide platinacycles **33** (Scheme 13.5) [57, 58].

This system could be viewed as a functioning mimetic of the Mo-dependent enzyme DMSO reductase [59, 60] because of several common features observed in catalysis. The stoichiometry of the enzymatic process (Scheme 13.6) parallels that of Scheme 13.5.

The enzymatic reduction is a multistep process involving two one-electron steps. At pH 8.5 two oxidations for *Rhodobacter sphaeroides* DMSO reductase observed at +37 and +83 mV (versus NHE) are associated with the $Mo^{IV/V}$ and $Mo^{V/VI}$ redox transitions, respectively [60]. DMSO is coordinated with Mo via oxygen in the key enzymatic intermediate. The proton plays an important role in the enzymatic system [59]. There are a few mechanistic similarities in Schemes 13.5 and 13.6. In particular, the acid catalysis is a key feature in the platinum(II)-promoted Scheme 13.5, the mechanism of which was investigated using the cyclic voltammetry, UV/vis and ^1H NMR spectrometry at 40–60 °C [58]. The conversion of **32** into **33** does not occur intramolecularly and involves two time-resolved phases. The first is the substitution of chloride for DMSO to afford the anionic reactive complexes **34**, which are involved in the acid-promoted interaction with free dimethyl sulfoxide in the second phase (Figure 13.17). The formation of **34** follows the usual two-term rate law $k_{obs1} = k_s + k_{Cl}[LiCl]$. Complexes **34**, in contrast to their precursors **32**, are more susceptible to oxidation and the irreversible peak E(p1) is ca 300 mV more cathodically than that of **32**. The second phase is acid-catalyzed and at low LiCl concentrations follows the rate expression $k_{obs2} = (k' + k[LiCl])[H^+]$. Complexes with the electron-withdrawing substituents R react faster and there is

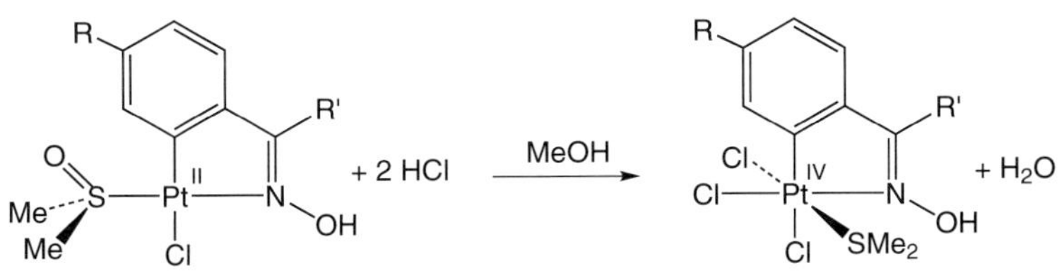

32 **33**

R = H (a), MeO, Me, F, and Cl

Scheme 13.5

Enzyme-Mo(IV) + $Me_2S{=}O$ + $2H^+$ → Enzyme-Mo(VI) + Me_2S + H_2O

Scheme 13.6

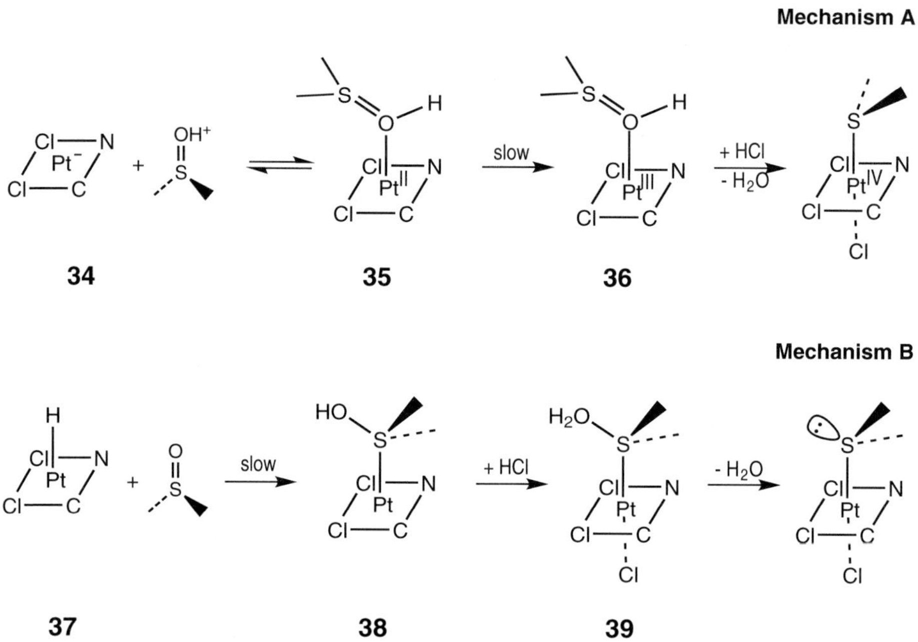

Figure 13.17 Two kinetically indistinguishable mechanisms for the reaction shown in Scheme 13.5.

a linear correlation between log*k* and *E*(p1). The first-order in the acid could be rationalized in terms of two kinetically indistinguishable mechanisms. Similarly to the enzymatic reaction, there are two one-electron oxidations in mechanism A (Figure 13.17). There is the rate-limiting electron transfer from Pt^{II} to protonated DMSO within intermediate **35**. The second electron transfer within **36** leads to S–O bond cleavage and oxygen dissociation as water in the presence of HCl. Mechanism B (Figure 13.17) involves insertion of the S=O bond of free DMSO into the Pt–H bond of the reactive hydride complex **37** of Pt^{IV} to afford **38**. The protonation of **38** by HCl and subsequent dissociation of water gives the final product. The ^{1}H NMR evidence supports the formation of the hydride species **37**, whereas a density functional study of the two mechanisms indicates that mechanism B is less energy demanding.

Scheme 13.5 represents an interesting mechanistic example in which the intuitively *intramolecular* conversion of **32** into **33** does not occur as such because the participating fragments of the starting complex **32** are insufficiently tuned to undergo deoxygenation of DMSO coupled with the oxidation of Pt^{II} into Pt^{IV}. The system adopts, at a first glance, a less advantageous *intermolecular* pathway, which allows independent adjustment of the reactivity of the two participating species. The deoxygenation mechanism involves insertion of the sulfoxide S=O bond into the Pt–H bond formed on protonation of the anionic intermediate **34**.

13.4 Labeling of Biological Molecules

Labeling of biomolecules by organotransition metal complexes has several objectives [61–64]. The most advanced area is diagnostics. Metal centers are involved either in electron transfer relays [62, 64] or serve as analytical markers [63]. Though octahedral cyclometallated Ru^{II} [65–68] and Os^{II} complexes [69] have proved their potential in biosystems, applications of square-planar pallada- and platinacycles are significantly less developed. Examples include water- and acid-resistant organoplatinum(II) "pincer" complexes covalently bonded to the N-terminus of L-valine **40** [70]. Owing to the ^{195}Pt nucleus (I = 1/2), these building blocks are potentially versatile biomarkers (e.g. MRI). They display *in vitro* biosensor characteristics since they detect SO_2 gas selectively and fully reversibly by an instantaneous change of the spectroscopic properties, including a diagnostic ^{195}Pt NMR signal [71, 72].

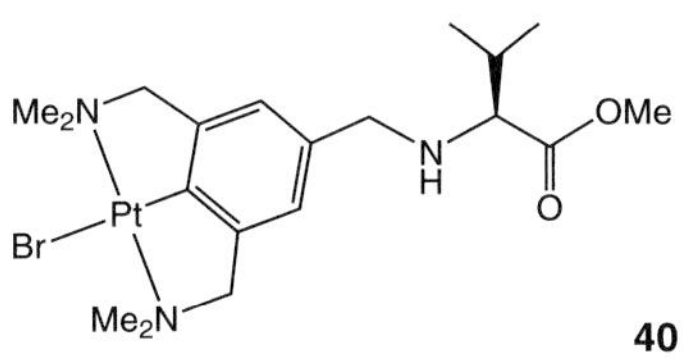

40

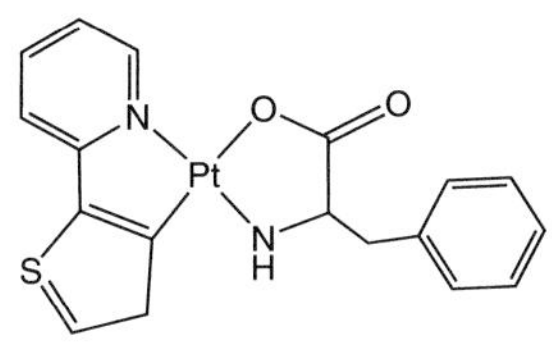

41a

There have been reports on binding of pallada- and platinacycles to proteins. Luminescent cyclometallated Pt^{II} 2-(2-thienyl)pyridine complex **41a** showed a high binding affinity (ca $10^6\,M^{-1}$) and selectivity towards human serum albumin (HSA). The binding is accompanied by an enhancement of photoluminescence at 562 nm [73].

Pincer-based nitrophenol phosphonate esters **42–44**, known for their lipase inhibitory activity, have been hydrolyzed in the presence of the lipase cutinase. After dissociation of 4-nitrophenolate the residues of **42–44** stay bound to the enzyme to form metal-complex/protein hybrids containing exactly one organometallic unit per protein [74].

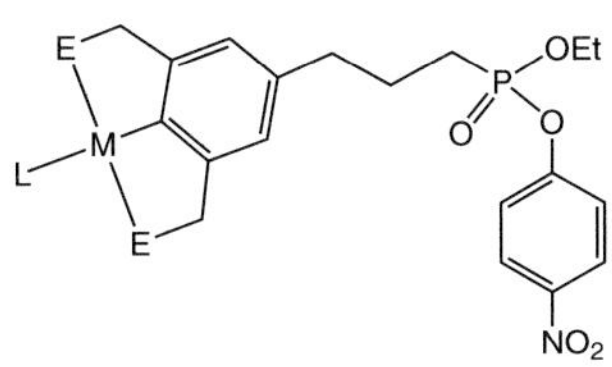

42 (E/M/L = NMe_2/Pt/Cl)
43 (SPh/Pd/Br)
44 (SMe/Pd/Br)

13.5 Inhibitors of Enzymatic Activity

Organometallic compounds are used as inhibitory probes for mechanistic investigations of enzymatic activity and topology of enzyme active sites [75]. Recently,

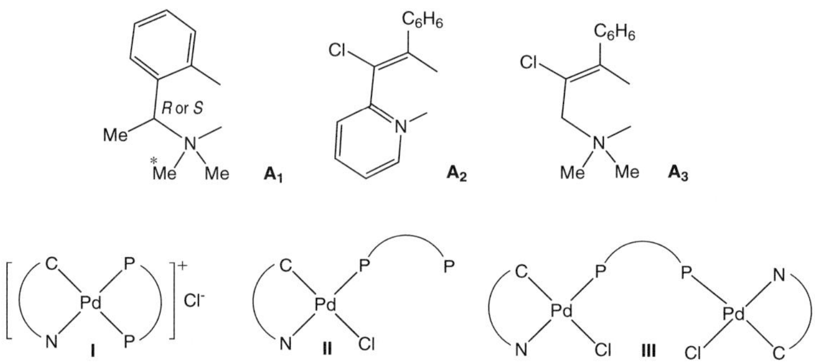

Figure 13.18 Dimeric and monomeric palladacycles with a dppe ligand studied against a murine melanoma cell line.

cyclopalladated compounds **45** were used as inhibitors of the endopeptidase cathepsin B [76], which is structurally related to the papain superfamily [77]. These enzymes contain a thiolate-imidazole pair in the active site. Studies on tumor tissues and cell lines have shown changes in expression, activity and distribution of cysteine cathepsins in numerous human cancers and, therefore, the activity of cathepsin B toward its fluorogenic substrate benzyloxycarbonyl-L-phenylalanyl-L-arginine-7-(4-methyl)coumarylamide (*Z*-Phe-Arg-MCA) was studied in the presence of palladacycles of type **46** (*S*-A_1-**III** in Figure 13.18 below), which possess antitumor activity (Section 13.6).

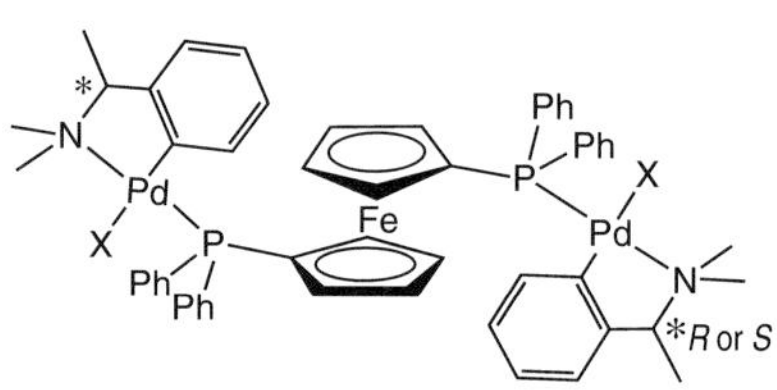

45 (**a** (R), **b** (S))

Enantiomers **45a,b** are inhibitors of cathepsin B with IC_{50} of ca. 4 μM. Detailed kinetic studies performed with **45b** have indicated that the cyclopalladated complex decreases both the apparent maximal rate $V_{M,app}$ and the apparent Michaelis constant $K_{M,app}$, the ratio $k_{cat}/K_{M,app}$ being approximately constant ($4.5 \times 10^5\ M^{-1}s^{-1}$ at pH 6.4 and 37 °C). These observations are consistent with a mixed type inhibition as depicted in Scheme 13.7.

The experimental data were analyzed in terms of Equation 13.6, where $V_{max} = k_{cat}[E]$:

$$v = \frac{v_{max}[S]\left(1+\frac{\beta[I]}{\alpha K_1}\right)}{K_s\left(1+\frac{[I]}{K_1}\right)+[S]\left(1+\frac{[I]}{\alpha K_1}\right)} \tag{13.6}$$

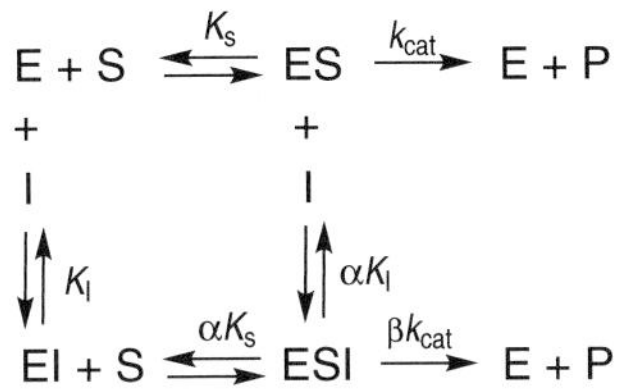

Scheme 13.7 E = cathepsin B, S = *(Z)*-Phe-Arg-MCA, and I = **45.**

The results show that the palladacycle binds to free cathepsin B (E) with a dissociation constant K_S of 12 μM, and also binds to the enzyme–substrate complex (ES) with a dissociation constant αK_S of 2.4 μM. The complex also induced a 5.3-fold increase in the affinity of cathepsin B for the substrate *Z*-Phe-Arg-MCA as shown by the reduction in the K_S value from 110 μM in the presence of **45b**, corresponding to $\alpha = 0.19$. On the other hand, the k_{cat} value in the presence of **45b** was also decreased 5.6-fold corresponding to $\beta = 0.18$. Interestingly, the cyclopalladated complex decreased the rate constant for product formation to the same extent as it increased the affinity of cathepsin B for the substrate implying that $\alpha \approx \beta$. This accounts for the fact that the overall activity of cathepsin B is unaffected by **45b** at low substrate concentrations when the second term of denominator of Equation 13.6 is negligible as compared to the first term.

This inhibitory effect was also observed in other cysteine proteinases of the papain family. The IC_{50} values for papain, cathepsin B, and cathepsin L equal 1.3, 1.5, and 1.6 μM, respectively. The cyclopalladated compounds with the bridging biphosphine–ferrocene ligand are effective inhibitors of papain-like cysteine proteinases. This result can be involved in the tumor growth delay observed in Walker tumor-bearing rats treated with cyclopalladated complexes (see below), since several studies have demonstrated that cathepsin B is involved in metastatic tumor development.

Cyclopalladated complexes produced by substitution of the bridging ferrocene ligand by 1,2-bis(diphenylphosphino)ethane are unable to inhibit the cathepsin B activity. The substitution of chloride ion by azide in **45** produces potently cytotoxic compounds, which do not inhibit the enzyme. Nevertheless, they can intercalate into DNA molecules promoting apoptosis in tumor cells. It is suggested that the inhibitory features of the chloride-containing palladacycles in aqueous solutions arise due to facile generation of the aqua/hydroxo ligand. Their origin is similar to that that makes cyclometallated complexes of Pd^{II} and Pt^{II} functioning mimetics of metallopeptidases (Section 13.2).

13.6 Medical Applications

The entire present book is a convincing illustration of the vast applications of cyclopalladated compounds. Total number of compounds synthesized is overwhelming. There is little doubt that the creators of new palladacycles are all perfectly aware of cisplatin [78, 79] and related platinum anticancer drugs. A challenge of adding palladacycles to the drug list was taken but overall the results could be classified as cautiously optimistic. In fact, cyclopalladated compounds are not found in the subject index in the recent books edited by Sessler *et al.* [80] or by Sigel and Sigel [81]. However, the antitumor features of some Pd and Pt cyclometallated thiosemicarbazones were reviewed in 2004 [82]. This section describes DNA and cancer therapy related properties of cyclopalladated compounds and their platinum counterparts.

A pioneering role here belongs to Newkome *et al.* who studied the ability of cis and trans complexes **47** and **48**, respectively, to produce single-stranded nicks (single-strand breaks) in supercoiled phage PM2 DNA [83, 84]. Though the reported property of **47** to produce nicks was subsequently abandoned [85], the report ignited further activities. In particular, it was demonstrated that complexes **47** bind to PM2 DNA. Structurally related complexes without or with single Pd–C bond showed weaker binding, the highest being registered for **47a**. The exact binding mode was unknown though an intercalation mechanism seemed probable.

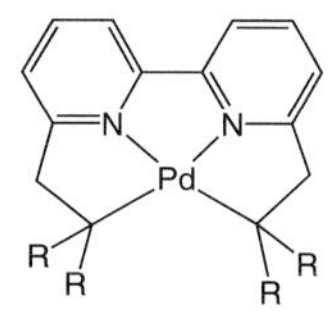

47 R = CO_2Me (**a**), CO_2Et (**b**)

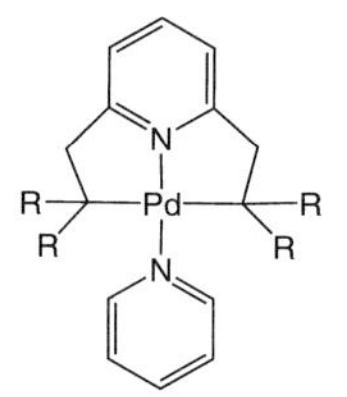

48

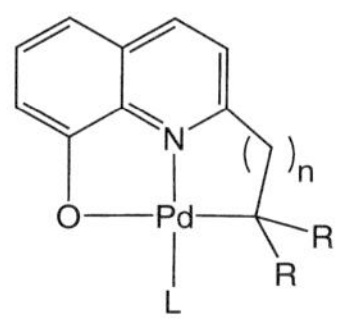

49 (n = 2, R = CO_2Et, L = py)
50 (n = 1, R = CO_2Me, L = 1-(2-hydroxyethyl)-2-methyl-5-nitroimidazole(metronidazole))

The antitumor activity of **49** and **50** against P-388 leukemia in mice was investigated [86]. The authors concluded that these cyclometallated complexes are considerably positive in lower doses.

The first systematic study of cytotoxicity of cyclometallated palladium(II) complexes was undertaken by Higgins III *et al.* [87]. The complexes were screened for cytotoxicity against a panel of seven human tumor cell lines. It was assumed that the results could allow useful extrapolations from *in vitro* cytotoxicity activity to actual *in vivo* antitumor activity. The strategy was as follows. Compounds that display different cytotoxicity, as opposed to uniform cytotoxicity to all cell lines, should have the most potential of possessing antitumor activity, since compounds that are uniformly cytotoxic to a tumor cell panel are likely to be toxic to normal cells as well when tested *in vivo*. Table 13.4 summarizes the evaluated complexes and IC_{50}.

Most of the complexes are quite cytotoxic towards the tumor panel, having IC_{50} values in the $10\,mg\,mL^{-1}$ range. The only complexes that displayed differential cytotoxicity were *N,N*-dimethylbenzylamine derivatives **51** and **52**. A 3–5-fold differential response was observed, for example, between the HT1376 and SW6020

Table 13.4 Cytotoxicity of selected palladacycles against a seven human tumor cell lines [87].

Complex		Cell line/IC_{50} ($mg\,mL^{-1}$)						
		SW6020	SW1116	SW403	ZR75-1	HT29/219	HT1376	SK-OV-3
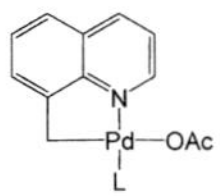	L = py	7	8	6	10	6	6	12
	$H_2N(CH_2)_3OH$	6	6	6	10	6	5	6
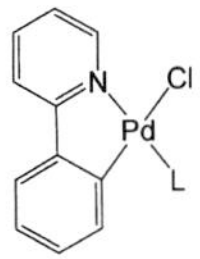	L = $H_2N(CH_2)_3OH$	7	8	7	16	5	5	11
	Me_2CHNH_2	12	12		7	8	7	8
	py	7	7		7	6	5	6
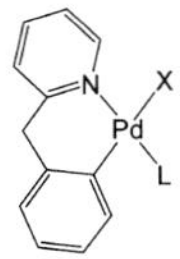	L = py, X = Cl	10	10	9	17	7	6	8
	L = Me_2CHNH_2, X = OAc	1	5	5	7	6	4	4
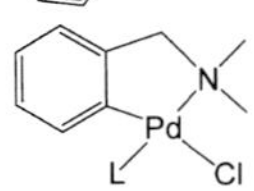	L = py **(51)**	44	44		50	33	8	20
	Me_2CHNH_2**(52)**	30	26		9	18	9	10
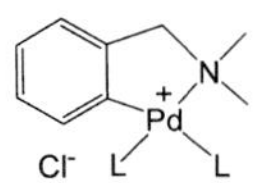	L = Me_2CHNH_2	51	59	61	49	36	35	46
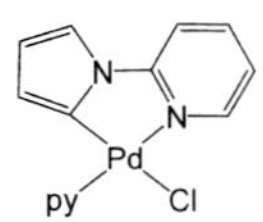		10	7		7	6	4	5
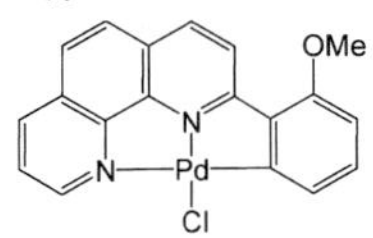		5	7	6	7		8	7

cell lines. The cyclometallated portion of the complexes may have some effect on the overall cytotoxicity. *N,N*-Dimethylbenzylamine palladacycles with a tertiary amine aliphatic chelate arm are less cytotoxic than the pyridine and quinoline complexes. The nature of auxiliary non-metallated ligands has a minor effect on cytotoxicity. The differential cytotoxicity observed for **51** and **52** suggests that this structural type might be the most promising.

Navarro-Ranniger *et al.* were interested in the antiproliferative activity of benzoylbenzylideneamine Pd^{II} and Pt^{II} complexes **53**, which were screened against leukemic HL-60 cells, mammary cancer MDA-MB 468 cells and MDA-MB 468 (breast carcinoma) tumor cells [88–90]. The platinum complexes show higher activity than the corresponding palladium compounds, since their IC_{50}s are 2–9-fold lower. Compound **53** also modifies the DNA structure of the open circular (oc) and covalently closed circular (ccc) forms of plasmid DNA. The acetate-bridged platinum complex **53** (R = OMe) showed the highest antiproliferative activity which is even higher than that of cisplatin (*cis*-$PtCl_2(NH_3)_2$) [88].

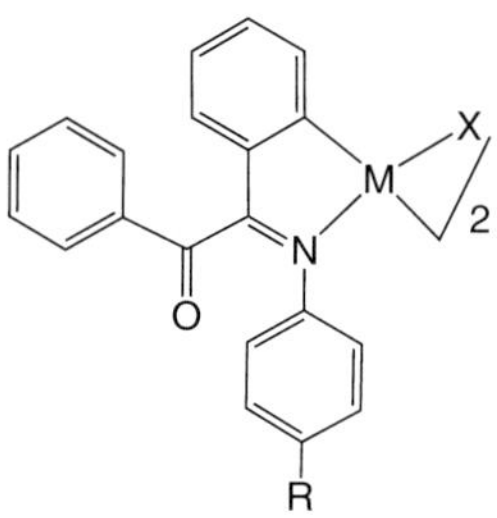

53 (M = Pd, Pt; X = OAc, Cl, R = OMe, Cl)

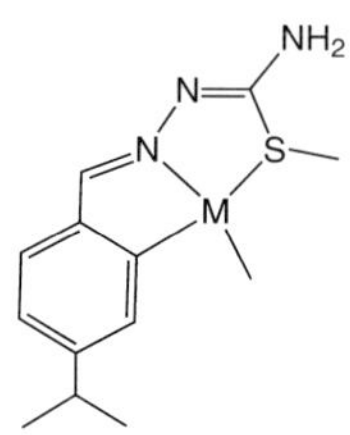

54 (M = Pd, Pt)

A targeted search for antitumor drugs includes cyclometallation of molecules that have, for example, antibacterial, antiviral and antitumor activities in a free state. This explained the interest in the cyclometallation of thiosemicarbazones [91]. Cyclometallation of *p*-isopropylbenzaldehyde thiosemicarbazone furnished tetranuclear Pd^{II} and Pt^{II} complexes with a cyclometallated structural unit **(54)**. Tests of the cytotoxic activity of these compounds against several human cell lines sensitive and resistant to cisplatin suggest that compounds **54** may be endowed with important anticancer properties since they elicit IC_{50} values in the mM range, as does the clinically used drug cisplatin, and, moreover, they display cytotoxic activity in tumor lines resistant to cisplatin. Analysis of the interaction of the tetrameric cyclometallated compounds with DNA suggests that they form DNA interhelical crosslinks. When used at IC_{90} doses, complexes **54** produce characteristic features of apoptosis – the main types of programmed cell death that involves a series of biochemical events leading to a characteristic cell morphology and death [92]. Typical of apoptosis, there is a drastic decrease in levels of H-*ras* protein (enzyme that converts GTP into GDP) in Pam-*ras* cells (cisplatin resistant cells that overexpress the H-*ras* oncogene). These effects are not observed when the cells are treated with the IC_{90} of cisplatin drug or free isopropylbenzaldehyde thiosemicarbazone ligand. Altogether, these results suggest that the platinum compound **54** might have potential as antitumor agents in view of their specific induction of apoptosis in cisplatin resistant cells.

There is a positive example of when a cyclometallated motif brings about obvious advantages over structurally related complexes without the M—C bond [93, 94]. S-180cisR is mouse sarcoma 180 cell with a 25-fold higher cisplatin resistance. The S-180cisR cells grow quite slowly in the presence of high concentration of cisplatin. This may show that S-180cisR cells modulate the cell cycle to acquire cisplatin resistance. P-Glycoprotein is selectively expressed on the surface of S-180cisR, which is not on cisplatin-sensitive S-180 parent cells. In an experiment using the verapamil inhibitor of P-glycoprotein, the cytotoxicity of cisplatin against S-180cisR increases significantly, as well as accumulation of CDDP in S-180cisR cells. These results indicate that enhanced pumping-out of cisplatin by P-glycoprotein should be one of the reasons for the cisplatin resistance of S-180cisR. Cycloplatinated compound **55**, incorporating two differently bound 2-phenylpyridine ligands, has a cytotoxicity against S-180cisR (IC_{50} 8.6 mM) higher than that of cisplatin (59 mM) and related complexes **56** with 2- or 3-substituted pyridines. Complex **55** is incorporated in S-180cisR to an enormously greater extent than cisplatin; the ratio of accumulated platinum after 3 h is 61.9. In S-180 parent cells, on the other hand, the ratio remains 8.1. This high accumulation of complex **55** into S-180cisR must account for its higher activity of against S-180cisR compared to cisplatin. Hydrophobicity seems to be one of the reasons for this efficient accumulation because transportation of drugs through the cell membrane is generally influenced by their hydrophobicity. Complex **55** is more hydrophobic than cisplatin in view of the fact that the retention time of **55** is longer than that of cisplatin in HPLC using an octadecylsilane column [93]. Cyclopalladated phenylpyridine with 4-hydroxyacridine as an *N,O*-bidentate ligand is active against human ovarian cancer cells A2780, OVCAR 5 and OVCAR 8 [95].

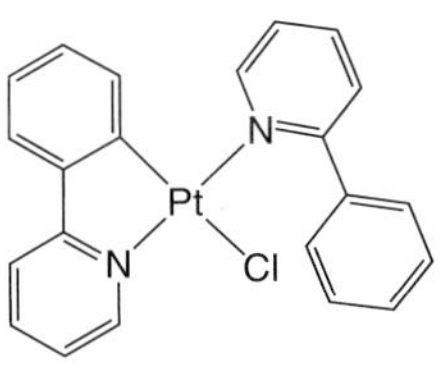

55

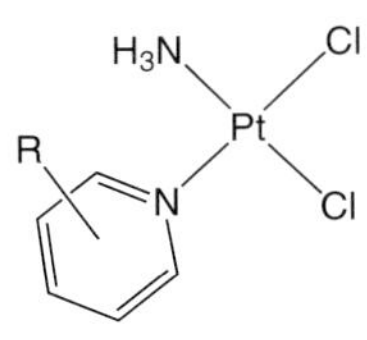

56 (R = 2-Me, 3-Me, 2-Et, 3-Et, 2-CH_2Ph)

Structural type **55** has common features with related 2-(2-thienyl)pyridine complex **41a**, which is cytotoxic to five human carcinoma cell lines, HeLa, HepG2, SF-268, NCI-H460 and MCF-7 and normal CCD-19Lu [73]. The cytotoxicity of **41a** and its tryptophan **(41b)** and glycine **(41c)** analogs was measured by a 3-(4,5-dimethylthiazol-2-yl)-2,5-tetrazolium bromide assay, and the IC_{50} values were determined from the dose-dependence of surviving cells after exposure to the Pt^{II} complexes for 48 h. The IC_{50} values under the experimental conditions (final concentration ≤4% DMSO) are in the order: **41a** > cisplatin ≅ **41b** > **41c** > free ligands

[2-(2-thienyl)pyridine, Phe, Trp or Gly]. Importantly, the decrease in IC_{50} values from **41a** to **41b** and **41c** parallels the corresponding decrease in the HSA binding affinity.

Evidence was presented that incorporating phosphine ligands may increase the antitumor activity of palladacycles in certain cases [96, 97]. Figure 13.18 shows the complexes studied [97]. There were combinations of three cyclometallated ligands, namely *N,N*-dimethyl-1-phenethylamine (A_1), 1-chloro-1-(2-pyridinyl)-2-phenylethene (A_2) and 1-phenyl-2-chloro-3-dimethylaminepropene (A_3) with the bidentate ligand 1,2-bis(diphenylphosphino)ethane (dppe) (Figure 13.18).

These cyclopalladated complexes were tested *in vitro* and *in vivo* against syngeneic BIX16F10-Nex2 murine melanoma cells of low immunogenicity implanted subcutaneously in mice. The complexes were not toxic to mice injected three times with as much as 60 mM per animal per week. Of three palladacycles (A_1-**I**, *R*-A_1-**III** and *S*-A_1-**III**) that were inhibitory *in vitro* at low concentrations (<1.25 mM), complex **46** (*S*-A_1-**III**) was the most active *in vivo*, delaying tumor growth and prolonging animal survival. The enantiomer of **46**, that is, similar complex of *R*-A_1, showed no activity *in vivo*. *In vitro*, dinuclear complex **46** caused a collapse of the respiratory activity with an abrupt decrease of extracellular acidification on short incubation (up to 100 min), followed by DNA degradation after 24 h. The apoptosis-like reaction induced by **46** was not accompanied by an increased level of the caspase-1 and caspase-3 enzymatic activity. Complex **46** binds to a bacterial plasmid DNA and after 24 h causes late conformation changes as suggested by circular dichroism measurements. The very fact that complexes A_2-**I** and A_3-**II** were less active *in vitro* but had pronounced antitumor properties *in vivo* indicates that the drug search among cyclometallated compounds is still significantly random. As a continuation of studies of phosphine ligands holding cyclopalladated moieties, the authors looked at dinuclear complexes of type **45**, which contain a ferrocene unit [76, 98]. This is an interesting approach as ferrocenes themselves show antitumor activity [81]. In addition, as the authors believe that though in all other cases the cytotoxicity of cyclopalladated compounds is mostly due to intercalation into DNA, the action of **45** could also be due to inhibition of proteolytic enzymes, such as the serine and cysteine proteinases, especially cathepsin B [99], which is involved in tumor development (Section 13.5).

The antitumor activity of complex **45b** against Walker tumor-bearing rats was analyzed in detail [76]. It was verified that in 90% of the animals studied the tumor growth was totally inhibited. Figure 13.19, which is representative of ten animals studied, demonstrates results obtained from a rat that received 106 Walker-256 mammary carcinoma tumor cells in the right thigh and the same implant of 106 tumor cells plus 2.0 mg kg^{-1} of complex **45b** in the left thigh. In the non-treated thigh (control) the solid tumor growth reached a maximal mass of 4.0 ± 1.0 g 12 days after the Walker tumor cell inoculation. However, in the left leg, which received **45b** during inoculation, the tumor mass was reduced to 0.3 ± 0.1 g. The efficacy of the palladacycle was confirmed when it was verified that it reverted installed Walker tumors in nine out of the ten cases studied.

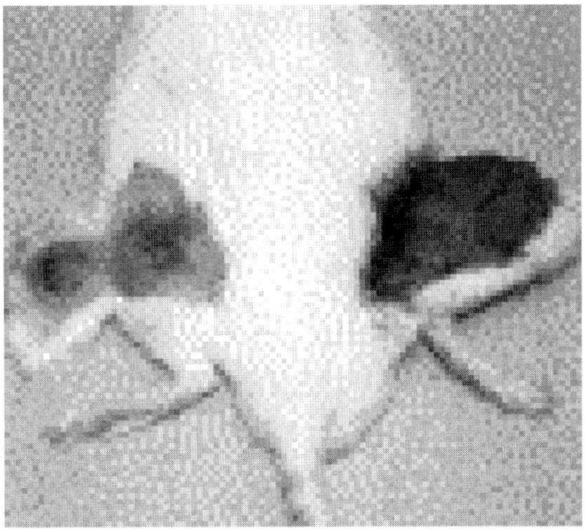

Figure 13.19 A rat showing two implants of tumor cells (Walker-256 mammary carcinoma). The left side implant (treated with 45b at a single dose of $2.0\,mg\,kg^{-1}$) shows a tumor growth of 0.3 g in 12 days. The right-side implant (untreated) shows tumor growth of 4.0 g in 12 days. From Reference [76]. Published with permission from publishers.

These results suggest that the palladacycle **45b** might be a promising antimetastatic drug. Notably, the authors did not observe any collateral effects in the treated rats. There were no lesions in the kidney, liver and spleen of mice after 14 days of complex **45b** treatment with one dose of $100\,mg\,kg^{-1}$. The specificity of **45b** was verified by analyzing its antitumor effects using the Ehrlich ascite tumor (EAT), a non-metastatic tumor model, which grows rapidly in almost any mouse strain, killing its host even when given in extremely small doses. The administration of **45b** for four days, beginning 72 h after tumor inoculation, did not present any protection to EAT-treated mice. These results could be explained, at least in part, by the fact that EAT is a non-metastatic tumor. In favor of this hypothesis is the fact that palladacycle, especially complex **45b**, inhibited cathepsin B activity *in vitro* at low concentrations because this enzyme is involved in the metastatic process. It is suggested that the palladacycles are more selective for malignant tumors such as the Walker carcinoma and compound **45b** should be considered for introducing into cancer chemotherapy after appropriate pre-clinical and clinical investigation [76]. It is worth mentioning that palladacycle **45b** was chosen among a series of related complexes, including its enantiomer **45a**, after detailed kinetic study of the inhibition of the enzymatic activity of endopeptidase cathepsin B described in Section 13.5. The palladacycle **45a** also inhibits the cathepsin B activity but it is more cytotoxic *in vivo* than **45b.**

The examples of using cyclopalladated compounds in tumor therapy summarized in this section reveal that though their screening for antitumor activity is worth trying there is still no straightforward and clear guidance for a targeted search. So far none of the compounds seem to have promising clinical trials. The future of using palladacycles as anticancer drugs is unclear. The virtue of these studies is that they have launched investigations of other metallacycles in cancer therapy. If the properties of recently reported anticancer cyclometallated derivatives of gold [100] or ruthenium [101] appear to be superior to those of palladium and platinum species described here, all previous efforts are worth while.

References

1 Dolphin, D. (ed.) (1982) Vol. **1** and **2**, John Wiley & Sons, Inc., New York.
2 Bertini, I., Gray, H.B., Lippard, S.J. and Valentine, J.S. (1994) *Bioinorganic Chemistry*, Univ. Sci. Books, Mill Valley, CA.
3 Lippard, S.J. and Berg, J.M. (eds) (1995) *Bioinorganic Chemistry*, Akad Verlag, Berlin, Germany.
4 Bertini, I., Sigel, A. and Sigel, H. (eds) (2001) *Handbook on Metalloproteins*, Marcel Dekker, New York.
5 Messerschmidt, A., Cygler, M. and Bode, W. (eds) (2001) *Handbook on Metalloproteins*, Vol. **1–3**, John Wiley & Sons, Inc., Chichester, New York.
6 Chin, J. (1991) *Accounts of Chemical Research*, **24**, 145–52.
7 Murphy, N.N. and Karlin, K.D. (1995) *Adv. Chem. Ser., 246 (Mechanistic Bioinorganic Chemistry)* (eds H.H. Thorp and V.L. Pecoraro), ACS, Washington, DC, pp. 165–93.
8 Ryabov, A.D., Kazankov, G.M., Yatsimirsky, A.K., *et al.* (1992) *Inorganic Chemistry*, **31**, 3083–90.
9 Vicente, J. and Arcas, A. (2005) *Coordination Chemistry Reviews*, **249**, 1135–54.
10 Sokolov, V.I. (1995) *Journal of Organometallic Chemistry*, **500**, 299–306.
11 Dunina, V.V., Zalevskaya, O.A. and Potapov, V.M. (1988) *Uspekhi Khimii*, **57**, 434–73.
12 Schmülling, M., Ryabov, A.D. and van Eldik, R. (1992) *Journal of the Chemical Society D–Chemical Communications*, 1609–11.
13 Bezsoudnova, E.Y. and Ryabov, A.D. (2001) *Journal of Organometallic Chemistry*, **622**, 38–42.
14 Yatsimirsky, A.K., Kazankov, G.M. and Ryabov, A.D. (1992) *Journal of the Chemical Society–Perkin Transactions 2*, 1295–300.
15 Levashov, A.V. and Ryabov, A.D. (1986) *Biochemical Education*, **14**, 34–6.
16 Kazankov, G.M., Dyachenko, O.G., Nemukhin, A.V. and Ryabov, A.D. (1997) *Mendeleev Communications*, 159–62.
17 Kazankov, G.M. and Ryabov, A.D. (1998) *Russian Journal of Organic Chemistry* (Translation of *Zh. Org. Khim*), **34**, 205–9.
18 Bondi, A. (1964) *Journal of Physical Chemistry*, **68**, 441–51.
19 Krooglyak, E.V., Kazankov, G.M., Kurzeev, S.A., *et al.* (1996) *Inorganic Chemistry*, **35**, 4804–6.
20 Kurzeev, S.A., Kazankov, G.M. and Ryabov, A.D. (2000) *Zhurnal Organicheskoi Khimii*, **136**, 1371–4.
21 Kurzeev, S.A., Kazankov, G.M. and Ryabov, A.D. (2000) *Inorganica Chimica Acta*, **305**, 1–6.
22 Ryabov, A.D., Kazankov, G.M., Kurzeev, S.A., *et al.* (1998) *Inorganica Chimica Acta*, **280**, 57–61.
23 Kagan, H.B. and Fiaud, J.C. (1988) *Topics in Stereochemistty*, **18**, 249–330.
24 Ward, R.S. (1995) *Tetrahedron Asymmetry*, **6**, 1475–90.
25 Jencks, W.P. (1969) *Catalysis in Chemistry and Enzymology*, McGraw-Hill Book Company, New York.
26 Bender, M.L., Kezdy, F.J. and Zerner, B. (1963) *Journal of the American Chemical Society*, **85**, 3017–24.
27 Capon, B. and Ghosh, B.C. (1966) *Journal of the Chemical Society B*, 472–8.
28 Fersht, A. (1999) *Structure and Mechanism in Protein Science: A Guide to Enzyme*

Catalysis and Protein Folding, Freeman, New York.

29 Kapteijn, G.M., Dervis, A., Grove, D.M., *et al.* (1995) *Journal of the American Chemical Society*, **117**, 10939–49.

30 Yatsimirsky, A.K., Bezsoudnova, K.Y. and Sakodinskaya, I.K. (1993) *Bioorganic and Medicinal Chemistry Letters*, **3**, 635–8.

31 Khan, M.N. (1987) *Journal of Molecular Catalysis*, **40**, 195–210.

32 Kazankov, G.M., Sergeeva, V.S., Efremenko, E.N., *et al.* (2000) *Angewandte Chemie – International Edition in English*, **39**, 3117–19.

33 Hendry, P. and Sargeson, A.M. (1989) *Journal of the American Chemical Society*, **111**, 2521–7.

34 Chung, Y., Akkaya, E.U., Venkatachalam, T.K. and Gzarnik, A.W. (1990) *Tetrahedron Letters*, **31**, 5413–16.

35 Kuo, L.Y., Kuhn, S. and Ly, D. (1995) *Inorganic Chemistry*, **34**, 5341–5.

36 Chapman, W.H. and Breslow, R. (1995) *Journal of the American Chemical Society*, **117**, 5462–9.

37 Kady, I.O., Tan, B., Ho, Z. and Scarborough, T. (1995) *Journal of the Chemical Society D – Chemical Communications*, 1137–8.

38 Wall, M., Linkletter, B., Williams, D., *et al.* (1999) *Journal of the American Chemical Society*, **121**, 4710–11.

39 Williams, N.H., Cheung, W. and Chin, J. (1998) *Journal of the American Chemical Society*, **120**, 8079–87.

40 Dumas, D.P., Caldwell, S.R., Wild, J.R. and Raushel, F.M. (1989) *Journal of Biological Chemistry*, **33**, 19659–65.

41 Pearson, R.G. (1966) *Science*, **151**, 1721–7.

42 Bender, M.L., Bergeron, R.J. and Komiyama, M. (1984) *The Bioorganic Chemistry of Enzymatic Catalysis*, John Wiley & Sons, Inc., New York.

43 Lai, K., Stolowich, N.J. and Wild, J.R. (1995) *Archives of Biochemistry and Biophysics*, **318**, 59–64.

44 Kazankov, G.M., Sergeeva, V.S., Borisenko, A.A., *et al.* (2001) *Russian Chemical Bulletin* (translation of *Izvestiia – Akademiia Nauk Seriia Khimicheskaia*), **50**, 1844–8.

45 Kim, M. and Gabbai, F.P. (2004) *Journal of the Chemical Society – Dalton Transactions*, 3403–7.

46 Kim, M., Liu, Q. and Gabbaie, F.P. (2004) *Organometallics*, **23**, 5560–4.

47 Tao, J.C., Jia, J., Wang, X.W. and Zang, S.T. (2002) *Chinese Chemistry Letters*, **13**, 1170–3.

48 Kim, M., Picot, A. and Gabbaie, F.P. (2006) *Inorganic Chemistry*, **45**, 5600–6.

49 Lu, Z.-L., Neverov, A.A. and Brown, R.S. (2005) *Organic and Biomolecular Chemistry*, **3**, 3379–87.

50 Barr, D.B., Bravo, R., Weerasekera, G., *et al.* (2004) *Environmental Health Perspectives*, **112**, 186–200.

51 Casida, J.E. and Quistad, G.B. (2004) *Chemical Research in Toxicology*, **17**, 983–98.

52 Okubo, T., Yokoyama, Y., Kano, K., *et al.* (2004) *Archives of Environmental Contamination and Toxicology*, **46**, 445–53.

53 Food and Agriculture Organization. (2001) The Ticking Time Bomb: Toxic Pesticide Waste Dumps; News Highlights, http://www.fao.org/News/2001/010502-e.htm (accessed 9 May 2001).

54 Food and Agriculture Organization. (2001) Prevention and Disposal of Obsolete Pesticides, http://www.fao.org/WAICENT/FAOINFO/AGRICULT/AGP/AGPP/Pesticid/Disposal/index_en.htm

55 Morales-Rojas, H. and Moss, R.A. (2002) *Chemical Reviews*, **102**, 2497–521.

56 Kukushkin, V.Y. (1995) *Coordination Chemistry Reviews*, **139**, 375–407.

57 Ryabov, A.D., Kazankov, G.M., Panyashkina, I.M., *et al.* (1997) *Journal of the Chemical Society – Dalton Transactions*, 4385–91.

58 Alexandrova, L., D'yachenko, O.G., Kazankov, G.M., *et al.* (2000) *Journal of the American Chemical Society*, **122**, 5189–200.

59 Baugh, P.E., Garner, C.D., Charnock, J.M., *et al.* (1997) *Journal of Biological Inorganic Chemistry*, **2**, 634–43.

60 George, G.N., Hilton, J., Temple, C., *et al.* (1999) *Journal of the American Chemical Society*, **121**, 1256–66.

61 Ryabov, A.D. (1991) *Angewandte Chemie – International Edition in English*, **30**, 931–41.

62 Heller, A. (1990) *Accounts of Chemical Research*, **23**, 128–34.

63 Jaouen, G., Vessières, A. and Butler, I.S. (1993) *Accounts of Chemical Research*, **26**, 361–9.

64 Ryabov, A.D. (2004) *Advances in Inorganic Chemistry*, **55**, 201–70.

65 Ryabov, A.D., Sukharev, V.S., Alexandrova, L., *et al.* (2001) *Inorganic Chemistry*, **40**, 6529–32.

66 Le Lagadec, R., Rubio, L., Alexandrova, L., *et al.* (2004) *Journal of Organometallic Chemistry*, **689**, 4820–32.

67 Ryabov, A.D., Kurova, V.S., Ivanova, E.V., *et al.* (2005) *Analytical Chemistry*, **77**, 1132–9.

68 Le Lagadec, R., Alexandrova, L., Estevez, H., *et al.* (2006) *European Journal of Inorganic Chemistry*, **2006**, 2735–8.

69 Ryabov, A.D., Soukharev, V.S., Alexandrova, L., *et al.* (2003) *Inorganic Chemistry*, **42**, 6598–600.

70 Albrecht, M., Rodriguez, G., Schoenmaker, J. and van Koten, G. (2000) *Organic Letters*, **2**, 3461–4.

71 Albrecht, M., Lutz, M., Spek, A.L. and van Koten, G. (2000) *Nature*, **406**, 970–4.

72 Albrecht, M., Gossage, R.A., Lutz, M., *et al.* (2000) *Chemistry – A European Journal*, **6**, 1431–45.

73 Siu Kit-Man, P., Ma, C.-M. and Che, D.-L. (2005) *Chemical Communications*, 1025–7.

74 Kruithof, C.A., Casado, M.A., Guillena, G., *et al.* (2005) *Chemistry – A European Journal*, **11**, 6869–77.

75 Shevel'kova, A.N., Ryabov, A.D. and Sinitsyn, A.P. (1993) *Biokhimiya (Moscow)*, **58**, 928–37.

76 Bincoletto, C., Tersariol, I.L.S., Oliveira, C.R., *et al.* (2005) *Bioorganic & Medicinal Chemistry*, **13**, 3047–55.

77 Jedeszko, C. and Sloane, B.F. (2004) *Biological Chemistry*, **385**, 1017–27.

78 Rosenberg, B., Vancamp, L. and Krigas, T. (1965) *Nature*, **205**, 698–9.

79 Rosenberg, B., VanCamp, L., Trosko, J.E. and Mansour, V.H. (1969) *Nature*, **222**, 385–6.

80 Sessler, J.L., Doctrow, S.R., McMurry, T. and Lippard, S.J. (2005) *Medicinal Inorganic Chemistry [ACS Symp. Ser.; 2005, 903]*, American Chemical Society, Washington, DC, p. 453.

81 Sigel, A. and Sigel, H. (2004) *Metal Complexes in Tumor Diagnosis and as Anticancer Agents. [In: Met. Ions Biol. Syst.; 2004, 42]*, Marcel Dekker, Inc., New York, NY, p. 534.

82 Quiroga, A.G. and Navarro Ranninger, C. (2004) *Coordination Chemistry Reviews*, **248**, 119–33.

83 Newkome, G.R., Onishi, M., Puckett, W.E. and Deutsch, W.A. (1980) *Journal of the American Chemical Society*, **102**, 4551–2.

84 Newkome, G.R., Kawato, T., Kohli, D.K., *et al.* (1981) *Journal of the American Chemical Society*, **103**, 3423–9.

85 Newkome, G.R., Puckett, W.E., Kiefer, G.E., *et al.* (1985) *Inorganic Chemistry*, **24**, 811–26.

86 Yoneda, A., Ouchi, M., Hakushi, T., *et al.* (1993) *Chemistry Letters*, **1993**, 709–12.

87 Higgins, J.D. III, Neely, L. and Fricker, S. (1993) *Journal of Inorganic Biochemistry*, **49**, 149–56.

88 Navarro-Ranniger, C., Lopez-Solera, I., Perez, J.M., *et al.* (1993) *Journal of Medicinal Chemistry*, **36**, 3795–801.

89 Navarro-Ranniger, C., Lopez-Solera, I., Perez, J.M., *et al.* (1993) *Applied Organometallic Chemistry*, **7**, 57–61.

90 Navarro-Ranninger, C., Lopez-Solera, I., Gonzalez, V.M., *et al.* (1996) *Inorganic Chemistry*, **35**, 5181–7.

91 Quiroga, A.G., Perez, J.M., Lopez-Solera, I., *et al.* (1998) *Journal of Medicinal Chemistry*, **41**, 1399–408.

92 Perez, J.M., Quiroga, A.G., Montero, E.I., *et al.* (1999) *Journal of Inorganic Biochemistry*, **73**, 235–43.

93 Okada, T., El-Mehasseb, I.M., Kodaka, M., *et al.* (2001) *Journal of Medicinal Chemistry*, **44**, 4661–7.

94 El-Mehasseb, I.M., Kodaka, M., Okada, T., *et al.* (2001) *Journal of Inorganic Biochemistry*, **84**, 157–8.

95 Pucci, D., Albertini, V., Bloise, R., *et al.* (2006) *Journal of Inorganic Biochemistry*, **100**, 1575–8.

96 Caires, A.C.F., Almeida, E.T., Mauro, A.E., *et al.* (1999) *Quimica Nova*, **22**, 329–34.

97 Rodrigues, E.G., Silva, L.S., Fausto, D.M., *et al.* (2003) *International Journal of Cancer*, **107**, 498–504.

98 Caires, A.C.F., Trindade, C.B. and Tersariol, I.L.d.S. (2004) Patent 2004019924, Brazil, p. 69.

99 Nomura, T. and Katunuma, N. (2005) *The Journal of Medical Investigation*, **52**, 1–9.

100 Li, C.K.-L., Sun, R.W.-Y., Kui, S.C.-F., *et al.* (2006) *Chemistry–A European Journal*, **12**, 5253–66.

101 Pfeffer, M., Sirlin, C., Loeffler, J.P., *et al.* (2006) Patent 2873037, France, p. 55.

14
Thermomorphic Fluorous Palladacycles

John A. Gladysz

14.1
Introduction

As detailed in the other chapters in this book, palladacycles have been designed with many target properties in mind [1, 2]. One recurring theme has been recyclability, for which numerous strategies are available. The present chapter reviews palladacycles that have highly temperature-dependent solubilities. These can be called thermomorphic, a broad term denoting some type of temperature-dependent property. Such palladacycles are amenable to one of the conceptually simplest approaches to recycling: precipitation by cooling, followed by solid/liquid phase separation.

Much recent research involving catalysts and reagents with temperature-regulable solubilities has focused on fluorous substances [3, 4]. These commonly feature "ponytails" of the formula $(CH_2)_m(CF_2)_{n-1}CF_3$, which can be abbreviated $(CH_2)_mR_{fn}$. However, branched and/or oxygen-containing ponytails see occasional use [5]. In the original Horváth/Rábai protocol for recovering fluorous catalysts [6], the ponytails were used to enhance solubilities in fluorous liquid phases. These include perfluoroalkanes and perfluoroethers but not fluorinated arenes [7]. Fluorous solvents are normally immiscible with organic solvents at room temperature, but often a single phase is obtained upon moderate warming. Fluorous catalysts and reagents containing sufficient percentages of fluorine by weight exhibit extremely high fluorous/organic liquid/liquid partition coefficients – measures of relative solubilities [8].

As fluorous chemistry developed, it was further realized that the absolute solubilities of fluorous compounds could be tailored by varying the lengths and/or numbers of $(CF_2)_{n-1}CF_3$ segments. It was also noticed that many exhibited immense solubility increases upon warming, both in fluorous and organic solvents. This phenomenon is conceptually related to the highly temperature-dependent miscibilities of fluorous and organic solvents. These two trends were documented in early work by Hughes [9] involving fluorous ferrocenes. However, their generality was not appreciated until some time later.

Scheme 14.1 illustrates how suitably designed fluorous catalysts might be employed under homogeneous conditions at elevated temperatures, and recovered

Palladacycles: Synthesis, Characterization and Applications. Edited by Jairton Dupont and Michel Pfeffer

ISBN: 978-3-527-31781-3

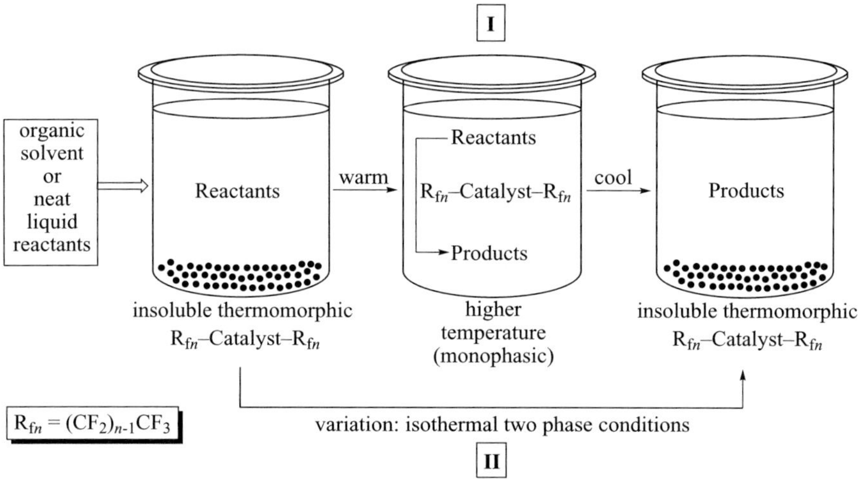

Scheme 14.1 Recovery of fluorous catalysts via solid/liquid phase separation.

at lower temperatures by solid/liquid phase separations. This avoids the use of fluorous solvents, which are commonly expensive and environmentally persistent [7]. To the author's knowledge, the first example of this protocol was reported by Sheldon, who recycled perfluorinated di(*n*-octyl)ketone, an epoxidation catalyst used in conjunction with H_2O_2 [10].

Since many types of palladacycle-catalyzed reactions are conducted at elevated temperatures, the author's coworkers set out to prepare fluorous palladacycles that might be applied as in Scheme 14.1. Since numerous palladacycles are active at very low loadings, often only minuscule concentrations would need to be realized. However, as sketched in variation II in Scheme 14.1, there is no a priori requirement that any catalyst dissolve. If rates are sufficiently fast under heterogeneous conditions, solubilization is not a concern.

Any catalyst to be recovered as in Scheme 14.1 should be designed to have a solubility as low as possible under the separation conditions. This will determine the upper bound for leaching. Various refinements have now been reported that utilize insoluble fluorous supports [4]. These allow attractive forces between the supports and the catalysts to be bought into play. Non-thermal means for regulating the solubilities of fluorous compounds have also been developed [11].

However, the reader should be alert to an overlooked design element. With any recoverable catalyst, what is actually being recycled is the rest state. This often can be much different than the catalyst precursor, and can greatly depend upon conditions (e.g. which of two reactants is used in excess). Fortunately, in many cases the solubility properties of the fluorous catalyst precursor appear to be shared by the rest state [12].

Finally, another way of viewing thermomorphic fluorous catalysts is as follows. When molecular catalysts are covalently bound to polymer supports, they assume the solubility properties of the host polymer. In the case of fluorous catalysts, a host molecule is derivatized with short segments of Teflon, which is insoluble in common solvents. As the segments are lengthened, they impart progressively more of the solubility characteristics of Teflon. Hence, solubilities can be tailored to virtually any target value with respect to the low- and high-temperature limits in Scheme 14.1.

14.2
Palladacycles Derived from Aromatic Imines and Thioethers [13]

First-generation fluorous catalysts are usually derived from established non-fluorous analogs. Accordingly, the initial efforts of the author's coworkers were directed at fluorous versions of aromatic imine-based palladacycles developed by Milstein [14]. To ensure high fluorous phase affinities for parallel efforts involving liquid/liquid biphase catalysis, systems with three R_{f8} ponytails were targeted.

As shown in Scheme 14.2, a synthesis was developed starting from commercial *p*-iodobenzaldehyde (**1**). A Wittig reaction with a fluorous phosphonium salt [15] introduced the first ponytail. This very reliable reaction has been employed with several other aromatic aldehydes [15, 16]. It was not possible to hydrogenate the alkene moiety in **2** or the analogous bromide without partial hydrogenolysis of the aryl halide. Accordingly, *i*-PrMgCl was used to effect iodine/magnesium exchange,

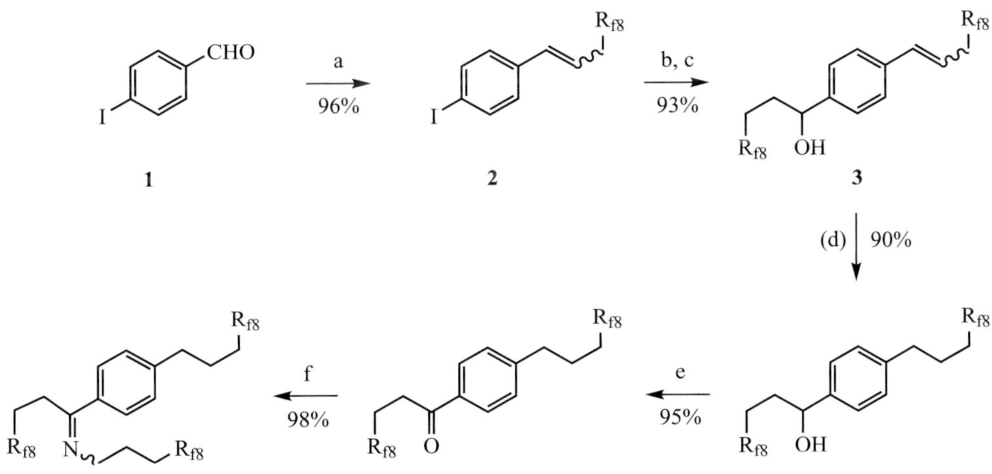

Scheme 14.2 Synthesis of fluorous imine **6**. Conditions: (a) $R_{f8}CH_2CH_2PPh_3{}^+I^-$, K_2CO_3, *p*-dioxane–H_2O, 95 °C; (b) *i*-PrMgCl, THF; (c) $R_{f8}CH_2CH_2CHO$; (d) $(Ph_3P)_3RhCl$, H_2 (75 psi), EtOH / $CF_3C_6H_5$, 40 °C; (e) Dess-Martin periodinane, $CF_3C_6H_5$; (f) $NH_2CH_2CH_2CH_2R_{f8}$, $SnCl_2(H_2O)_2$, toluene, reflux, Dean Stark trap.

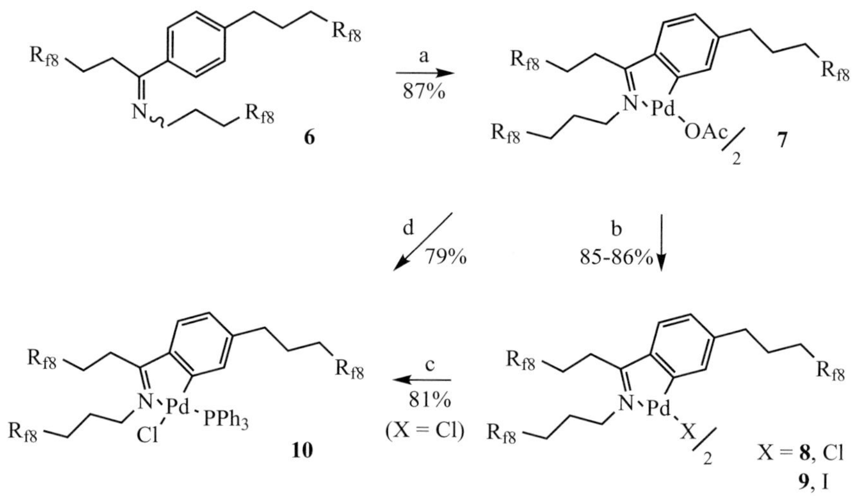

Scheme 14.3 Syntheses of fluorous N-donor palladacycles. Conditions: (a) $Pd(OAc)_2$, AcOH, 95 °C; (b) LiX (X=Cl, I), $CF_3C_6H_5$–MeOH; (c) PPh_3, CH_2Cl_2; (d) LiCl, PPh_3, THF.

and subsequent addition of the readily available fluorous aldehyde $R_{f8}CH_2CH_2CHO$ [17] gave the doubly ponytailed benzylic alcohol **3** in 93% yield.

The alkene moiety in **3** could be hydrogenated without competing carbon–oxygen bond hydrogenolysis, affording **4** in 90% yield. Oxidation of **4** with the Dess-Martin reagent gave the aryl ketone **5** in 95% yield. Subsequent condensation with the readily available fluorous amine $H_2NCH_2CH_2CH_2R_{f8}$ [17] generated the triply ponytailed imine **6**, which was isolated in 98% yield (*Z*/*E* mixture). Although the synthesis of **6** required five steps, product purifications – often the most important considerations in fluorous syntheses – were easy and the overall yield was good.

As shown in Scheme 14.3, **6** and $Pd(OAc)_2$ reacted under standard conditions [1] to give the acetate-bridged dimeric N-donor palladacycle **7** in 87% yield. Subsequent additions of LiCl or LiI afforded the corresponding chloride- and iodide-bridged dimeric palladacycles **8** and **9** in 85–86% yields. Reactions of the palladacycle chloride **8** with PPh_3, or of the palladacycle acetate **7** with LiCl/PPh_3, gave the monomeric phosphine-substituted palladacycle **10** in 79–81% yields.

A crystallized sample of **7** melted at 78–80 °C and was thermally stable to 225 °C. When DMF or $CF_3C_6H_5$ solutions of **7** were kept at 140 °C (2 h) or 100 °C (16 h), no decomposition was detected visually or by NMR. Palladacycles **8–10** exhibited similar stabilities, but melted at higher temperatures (176, 182, 135 °C, respectively). Note that above the melting point of the catalyst the protocol in Scheme 14.1 involves liquid/liquid miscibility.

At room temperature, **7** was soluble in the fluorous solvents $CF_3C_6F_{11}$ (perfluoromethylcyclohexane) and $C_8F_{17}Br$, and the non-fluorous aromatic solvents $CF_3C_6F_5$ (perfluorotoluene) and $CF_3C_6H_5$ [(trifluoromethyl)benzene]. The last is often

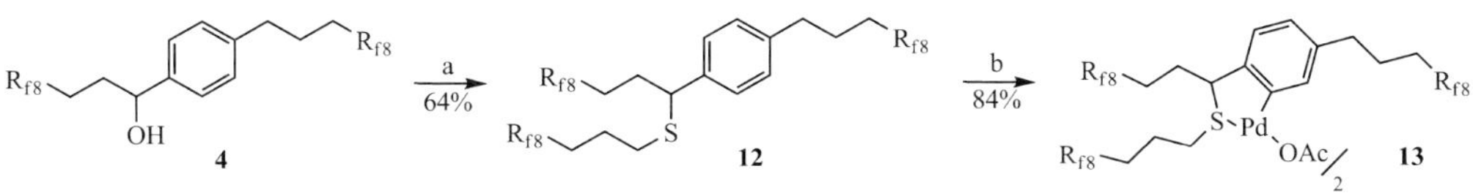

Scheme 14.4 Synthesis of a fluorous S-donor palladacycle. Conditions: (a) $R_{f8}CH_2CH_2CH_2SH$ (**11**), ZnI_2, $CF_3C_6H_5$ (**14**), 60 °C; (b) $Pd(OAc)_2$, AcOH, 95 °C.

termed a hybrid solvent, as it is broadly able to solubilize both fluorous and organic substances [7, 18]. In contrast, **7** was poorly soluble in common organic solvents such as CH_2Cl_2, $CHCl_3$, acetone and THF, and insoluble in DMF. The higher-melting palladacycles **8** and **9** exhibited no detectable solubility in organic solvents, and were poorly soluble in the preceding fluorinated solvents. However, solubilities in $CF_3C_6F_5$ were much higher above 50 °C, allowing NMR spectra to be recorded. The monomeric palladacycle **10** was much more soluble than **7** in CH_2Cl_2, $CHCl_3$, acetone and THF, and was also very soluble in $CF_3C_6F_{11}$, $CF_3C_6F_5$ and $CF_3C_6H_5$.

Attention was next turned to fluorous thioethers that could serve as precursors to similar palladacycles. These would model non-fluorous complexes developed by Dupont [19]. Benzylic alcohols and thiols readily condense in the presence of Lewis acids [20]. Thus, reaction of the fluorous benzylic alcohol **4** from Scheme 14.2 and the readily available fluorous thiol $R_{f8}CH_2CH_2CH_2SH$ (**11**) [13, 21] gave the triply ponytailed fluorous thioether **12** in 64% yield (Scheme 14.4). Treatment with $Pd(OAc)_2$ as in Scheme 14.3 afforded the dimeric S-donor palladacycle **13** in 84% yield. Unlike the imine **6**, the thioether **12** is chiral. Therefore, **13**, which features two carbon stereocenters, can exist as a mixture of diastereomers. NMR spectra were more complex than those of **7**, but only a single HPLC peak was observed. The solubility properties of **13** were very similar to those of **7**.

Table 14.1 summarizes the $CF_3C_6F_{11}$/toluene partition coefficients of selected compounds. That of the palladacycle precursor **6** was very high (98.7 : 1.3) but that of **5** was – as expected from the number of pony tails – lower (84.6 : 15.4). The partition coefficient of palladacycle **7** (95.5 : 4.5) was slightly lower than that of **6**, which is consistent with the diminished fluorine weight%. The value in $C_8F_{17}Br$/DMF, a solvent system employed in catalysis experiments, was similar (95.9 : 4.1). The partition coefficient of palladacycle **13**, which contains a larger and more polarizable heteroatom than **7**, was slightly lower (90.7 : 9.3); those of **8** and **9** could not be determined due to their miniscule solubilities in organic solvents.

14.3
Pincer Palladacycles: PC(sp²)P [22]

Numerous diphosphines of the general formula $1,3\text{-}C_6H_4(CH_2PX_2)_2$ (**III**, Scheme 14.5) have been elaborated to pincer-type palladacycles [2b]. Thus, related fluorous diphosphines were sought. Many free-radical chain additions of $R_{3-n}PH_n$ species

Table 14.1 $CF_3C_6F_{11}$/toluene partition coefficients for fluorous ligands, palladacycles and related compounds (23–25 °C).

Compound	Partition coefficient
5	84.6 : 15.4
6	98.7 : 1.3
7	95.5 : 4.5[a]
12	99.5 : 0.5
13	90.7 : 9.3[a]
23-R_{f8}	96.4 : 3.6
24-R_{f8}	98.0 : 2.0
26-R_{f6}	95.3 : 4.7
26-R_{f8}	98.8 : 1.2
26-R_{f10}	99.3 : 0.7
28-R_{f6}	97.4 : 2.6
28-R_{f8}	99.3 : 0.7
28-R_{f10}	99.7 : 0.3
33-R_{f10}	83.6 : 16.4
34-R_{f8}	70.9 : 29.1
34-R_{f10}	78.9 : 21.1
35-R_{f10}	76.0 : 24.0

a Data for $C_8F_{17}Br$/DMF: 95.9 : 4.1 (7); 91.4 : 8.6 (13).

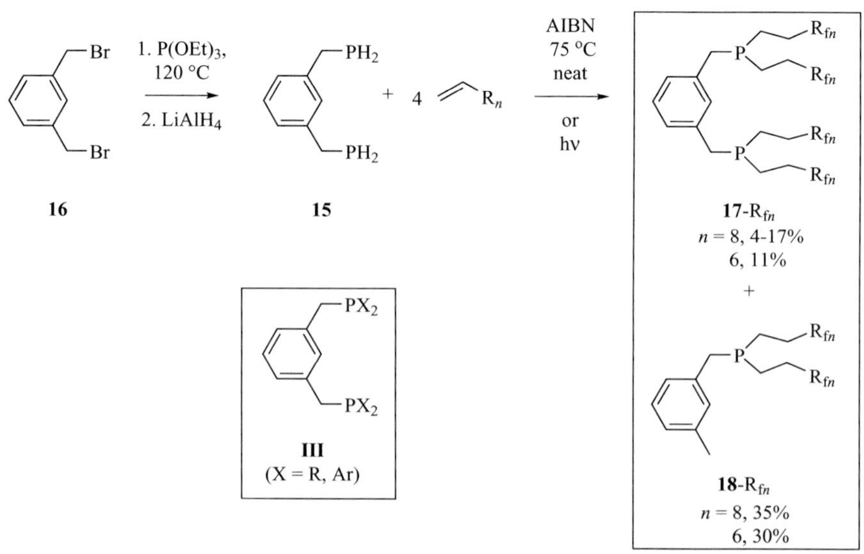

Scheme 14.5 Syntheses of fluorous pincer ligands **17**-R_{fn} via free-radical chain additions.

Table 14.2 Qualitative solubility profiles of PC(sp^2)P pincer ligands and complexes at room temperature.

	$C_6F_{11}CF_3$	$C_6H_5CF_3$	Hexane	Ether	THF	CH_2Cl_2	Acetone
17-R_{f8}	High	High	None	Low	High	Med	Med
18-R_{f8}	High	High	Low	Med	High	Med	Med
17-R_{f6}	High	High	Low	Low	High	Med	High
18-R_{f6}	High	High	Low	Med	High	Med	High
23-R_{f8}	Med	High	None	Low	Med	Med	Med
23-R_{f6}	High	High	None	Low	Med	Med	Med
24-R_{f8}	High	High	Low	Med	Med	Med	Med

to fluorous terminal alkenes $H_2C{=}CHR_{fn}$ have been described [6, 23, 24]. Hence, the author's attention was drawn to the diprimary diphosphine 1,3-$C_6H_4(CH_2PH_2)_2$ (**15**) as a building block. As shown in Scheme 14.5, this compound is readily available from α,α′-dibromo-*m*-xylene (**16**) via an Arbuzov/reduction sequence [25].

The reaction of **15** and excess $H_2C{=}CHR_{f8}$ at 75 °C in the presence of AIBN (radical initiator) was monitored by ^{31}P NMR. There was a sequence of intermediates as the phosphorus–hydrogen bonds were consumed, followed by a signal appropriate for the target molecule **17**-R_{f8} (Scheme 14.5). However, another species (**18**-R_{f8}) also formed, and was always the major product when addition was complete. On a good day, **17**-R_{f8} could be isolated in 15–17% yields. Isolated yields of **18**-R_{f8} were typically 35%. Spectroscopic data established the structure shown in Scheme 14.5, which requires the cleavage of a benzylic carbon–phosphorus bond. Both **17**-R_{f8} and **18**-R_{f8} were air-sensitive solids. As summarized in Table 14.2, they were soluble in fluorous and non-fluorous solvents, with the exception of hexane.

All efforts to increase the **17**-R_{f8}/**18**-R_{f8} ratio were unsuccessful. These included higher AIBN loadings, and other temperatures, initiators and solvents. Photochemical reactions of **15** and $H_2C{=}CHR_{f8}$ (0–20 °C) gave similar yields of **17**-R_{f8} and **18**-R_{f8}. Analogous thermal and photochemical reactions of **15** and $H_2C{=}CHR_{f6}$, which has a shorter ponytail, gave comparable yields of the lower homologs **17**-R_{f6} and **18**-R_{f6}. Thus, the formation of **18**-R_{fn} under free-radical chain conditions is general, but the exact mechanistic origin remains obscure.

Other approaches to **17**-R_{fn} were investigated. Following a literature precedent for non-fluorous systems [26], **16** and the fluorous secondary phosphine $HP(CH_2CH_2R_{f8})_2$ (**19**-R_{f8}) [24] were combined. No reaction occurred under homogeneous conditions in refluxing acetone, toluene and THF. This is consistent with the diminished reactivities of fluorous nucleophiles, especially when only two methylene groups separate the perfluoroalkyl segments and the Lewis basic site [5]. Hence, as shown in Scheme 14.6, **16** (mp 75–77 °C) and **19**-R_{f8} (mp 80 °C) were heated to 80 °C in the absence of solvent (1 : 2 mol ratio). The sample melted and resolidified, and could not be redissolved in any common solvent. Mass spectrom-

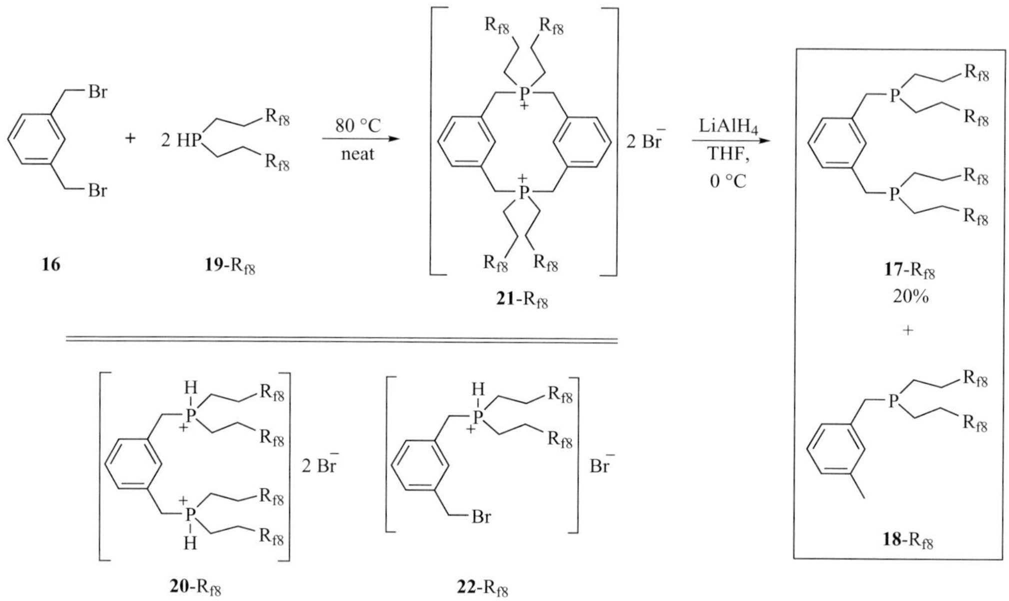

Scheme 14.6 Synthesis of fluorous pincer ligand **17**-R$_{f8}$ via nucleophilic substitutions.

etry did not show an ion consistent with the dication of the desired primary product, the bis(phosphonium salt) **20**-R$_{f8}$ (Scheme 14.6). However, a strong ion appropriate for the dication of the metacyclophane **21**-R$_{f8}$ was apparent.

Efforts were made to exploit the unanticipated formation of **21**-R$_{f8}$. In accord with an established procedure for the reductive cleavage of carbon–phosphorus bonds of phosphonium salts [27], the crude product was treated with $LiAlH_4$. Chromatography gave **17**-R$_{f8}$ in 20% overall yield from **16**. A ^{31}P NMR spectrum of the reaction mixture showed some **18**-R$_{f8}$ as well as other byproducts. Many attempts were made to refine this synthetic approach. For example, the first-formed species in Scheme 14.6 must be the monophosphonium salt **22**-R$_{f8}$. A higher concentration of **19**-R$_{f8}$ should increase the rate of displacement of the remaining benzylic bromide to give **20**-R$_{f8}$. However, reactions using 1:4 mol ratios of **16** and **19**-R$_{f8}$ gave nearly identical results. Further analyses and investigations, which included other types of phosphorus nucleophiles, are detailed elsewhere [22]. Unfortunately, all were unsuccessful.

Despite the modest yields of the fluorous pincer ligands **17**-R$_{fn}$ in Schemes 14.5 and 14.6, the masses commonly isolated were greater than the masses of the educts **15** or **16**. Thus, some coordination chemistry could be developed. As shown in Scheme 14.7, **17**-R$_{fn}$ and $Pd(O_2CCF_3)_2$ were combined in THF. In accord with much precedent [2], workups gave the air-stable palladium pincer complexes **23**-R$_{fn}$ in 80–90% yields. Interestingly, **17**-R$_{fn}$ could be used as precursors to other classes of metallacycles. Accordingly, the reaction of **17**-R$_{f8}$ and $[IrCl(COE)_2]_2$ in THF at

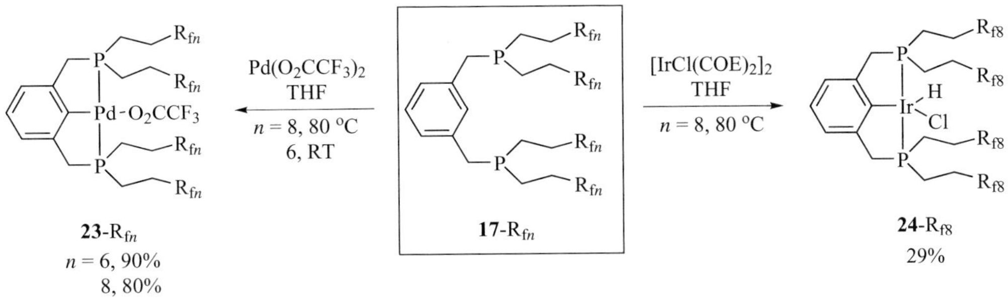

Scheme 14.7 Metal complexes of fluorous pincer ligands **17**-R_{fn} (COE=Cyclooctene).

80 °C gave the iridium(III) adduct **24**-R_{f8} in 29% yield. Several closely related iridations have been reported [26d, 28].

As summarized in Table 14.2, **23**-R_{fn} were soluble in most fluorous and non-fluorous solvents at room temperature. Although hexane was an exception, solubilities in hot hexane were appreciable. When such solutions of **23**-R_{f8} cooled, single crystals were obtained. The crystal structure was determined, and has been thoroughly analyzed elsewhere [22]. Most relevant to the theme of this chapter is the packing diagram in Figure 14.1a. The crystal lattice is conspicuously divided into fluorous and non-fluorous domains, with the ponytails of neighboring molecules in van-der-Waals contact. Other square planar metal complexes with fluorous phosphines crystallize similarly [29].

As illustrated in Table 14.1, the $CF_3C_6F_{11}$/toluene partition coefficients of the quadruply ponytailed pincer palladacycles **23**-R_{f8} and **24**-R_{f8} (96.4–98.0 : 3.6–2.0) were somewhat greater than those of the triply ponytailed species **7** and **13**. Owing to the air sensitivity of the pincer ligands **17**-R_{fn}, reliable partition coefficients could not be determined.

14.4
Pincer Palladacycles: PC(sp³)P [30]

Pincer complexes based upon diphosphines with aliphatic (sp³) carbon backbones have also received much attention [31]. Thus, fluorous analogs were sought. Towards this end, the known diprimary 1,5-diphosphine $H_2P(CH_2)_5PH_2$ (**25**) was synthesized from $Br(CH_2)_5Br$ by an easily scalable Arbuzov/reduction sequence analogous to **16** → **15** in Scheme 14.5 [32]. As shown in Scheme 14.8 (top), **25** and excess $H_2C{=}CHR_{fn}$ ($n = 6, 8, 10$) were reacted at 60 °C in the presence of a radical initiator. Workups gave the fluorous ditertiary diphosphines $(R_{fn}CH_2CH_2)_2P(CH_2)_5P(CH_2CH_2R_{fn})_2$ (**26**-R_{fn}) as air sensitive liquids (**26**-R_{f6}) or solids (**26**-R_{f8}, **26**-R_{f10}) in 68–74% yields on 3–5-gram scales.

Reactions of non-fluorous 1,5-diphosphines $R_2P(CH_2)_5PR_2$ with L_2PdCl_2 or L_2PtCl_2 species often give bimetallic complexes of the type **27** (Scheme 14.8) [31a,b] that can be converted at elevated temperatures into monometallic PC(sp³)P pincer

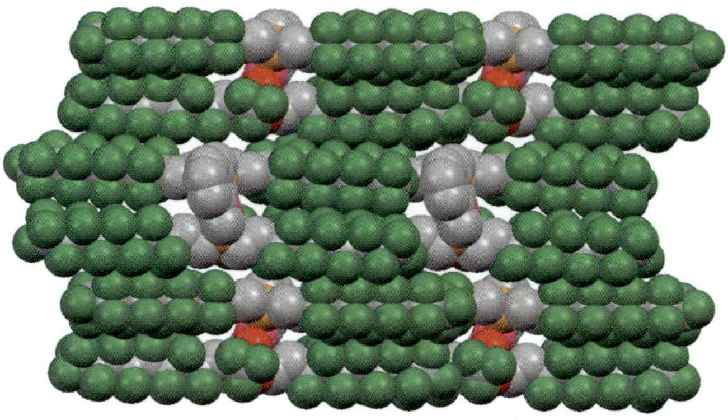

(a)

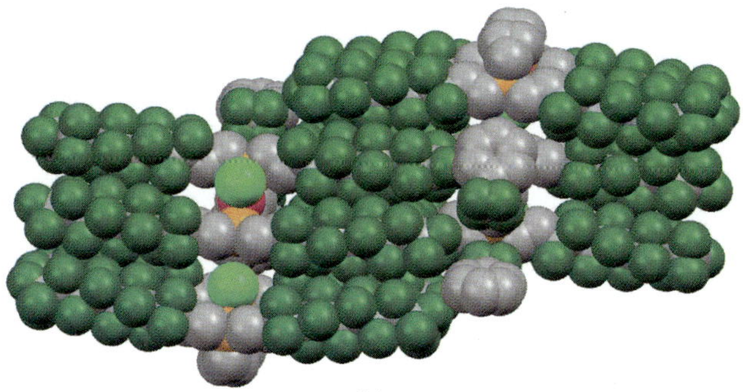

(b)

Figure 14.1 Crystal structures of **23**-R_{f8} (a) and **28**-R_{f8}·$CF_3C_6H_5$ (b): packing diagrams with atoms at van-der-Waals radii.

complexes. Reactions of **26**-R_{f8} with $(PhCN)_2PdCl_2$ and $(COD)_2PtCl_2$ appeared to give similar adducts (96–97%). However, when they were (a) heated as solids or (b) refluxed in trifluoroacetic acid for extended periods – conditions that typically give palladium and platinum pincer complexes [31a,b,d] – either no reaction (platinum) or decomposition (palladium) occurred.

As shown in Scheme 14.8 (top), alternative strategies were investigated. In accord with much precedent [2, 31e], reaction of $Pd(O_2CCF_3)_2$ and **26**-R_{fn} in $CF_3C_6F_5$ at 80 °C gave target palladium pincer complexes **28**-R_{fn} in 18–51% yields. Although these systems are achiral, the palladacycle rings are not planar, resulting in a variety of diastereotopic groups. Thus, their NMR spectra are more complex than those of the PC(sp^2)P pincer complexes.

As summarized in Table 14.3, the diphosphine **26**-R_{f6} was very soluble in the $CF_3C_6F_{11}$, $CF_3C_6H_5$ and $CF_3C_6F_5$, and moderately soluble in THF and CH_2Cl_2. The higher homologs **26**-R_{f8} and **26**-R_{f10} exhibited progressively lower solubilities. The pincer complexes **28**-R_{fn} showed analogous trends, and exhibited enhanced solubilities at 60 °C. The solvate **28**-R_{f8} · $CF_3C_6H_5$ could be crystallized, and the

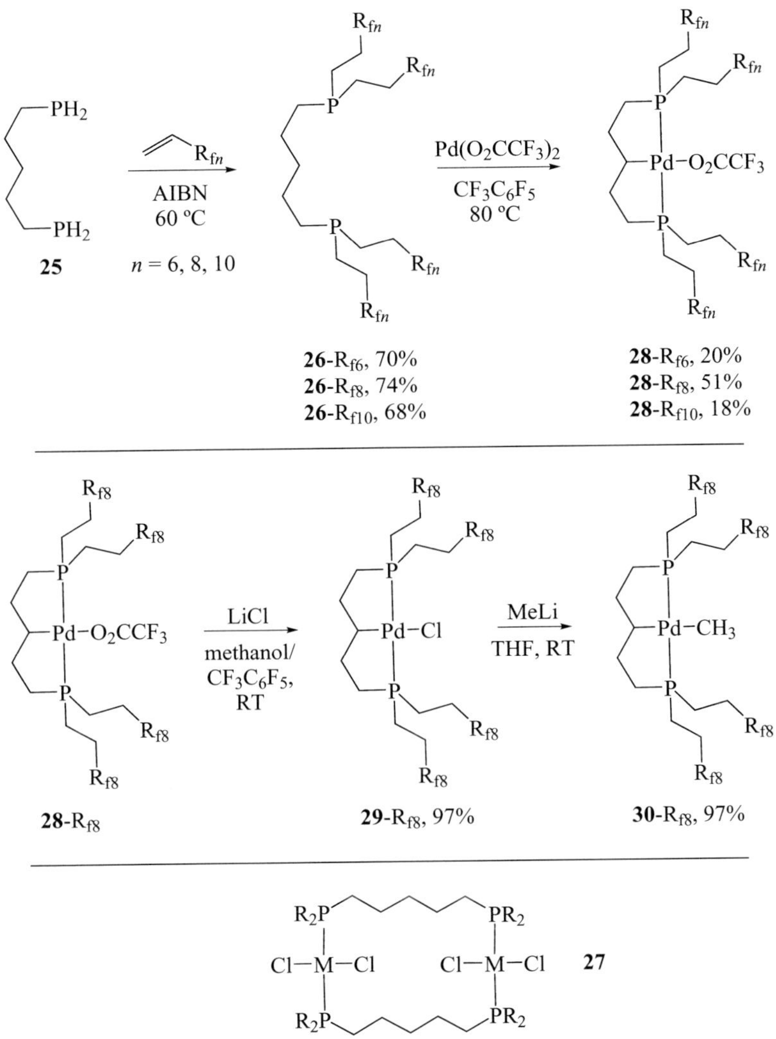

Scheme 14.8 Synthesis of fluorous PC(sp³)P pincer ligands and palladium complexes.

structure was determined. As shown in Figure 14.1b, distinct fluorous domains were evident. Additional features are analyzed elsewhere [30].

As summarized in Table 14.1, The $CF_3C_6F_{11}$/toluene partition coefficients of the diphosphines ranged from 95.3 : 4.7 (**26**-R_{f6}) to 99.3 : 0.7 (**26**-R_{f10}). Those of the pincer complexes were somewhat higher, indicating enhanced fluorophilicity despite the diminished wt% of fluorine (e.g. 67.3% for **26**-R_{f8} versus 63.1% for **28**-R_{f8}). Similar phenomena have been documented with other metal complexes of fluorous ligands [8, 33]. As would be expected [8], the partition coefficient of **28**-R_{f8} was greater than that of its aromatic counterpart **23**-R_{f8} (99.3 : 0.7 versus 96.4 : 3.6).

Table 14.3 Qualitative solubility profiles of PC(sp^3)P pincer ligands and complexes.

	Room temperature					60 °C				
	$CF_3C_6F_{11}$	$CF_3C_6H_5$	$CF_3C_6F_5$	THF	CH_2Cl_2	$CF_3C_6H_5$	$CF_3C_6F_5$	Hexane	THF	CH_2Cl_2
26-R_{f6}	High	High	High	Moderate	Moderate	High	High	Moderate	Moderate	Moderate
26-R_{f8}	High	High	High	Low	Low	High	High	Low	Low	Low
26-R_{f10}	High	High	High	V. low	V. low	High	High	V. low	V. low	V. low
28-R_{f6}	High	High	High	High	High	High	High	High	High	High
28-R_{f8}	Moderate	High	High	Moderate	V. low	High	High	Moderate	Moderate	Moderate
28-R_{f10}	Moderate	Moderate	Moderate	V. low	V. low	Moderate	High	V. low	V. low	V. low

Some further chemistry was developed. As shown in Scheme 14.8 (bottom), **28**-R_{f8} and excess LiCl reacted to give the chloride complex **29**-R_{f8} in 97% yield. No trace of **29**-R_{f8} was detected during the pyrolysis of the corresponding adduct **27**. Certain methyl derivatives of palladium PC(sp^3)P pincer complexes have been reported to be effective catalyst precursors [31d]. Hence, **29**-R_{f8} and MeLi were combined in THF. Workup gave the crude methyl complex **30**-R_{f8} in 97% yield as an air stable white powder that was quite labile in solution. Also, reaction of the diphosphine **26**-R_{f8} and $[IrCl(COE)_2]_2$ (THF, 80 °C) gave an iridacycle analogous to **24**-R_{f8} (Scheme 14.7).

14.5
Pincer Palladacycles: SC(sp^2)S [34, 35]

Many bis(thioethers) of the formula 1,3-$C_6H_4(CH_2SR)_2$ have been elaborated into SC(sp^2)S pincer systems and applied in catalysis [19a, 36]. Thus, in efforts that predate those in the previous two sections, fluorous analogs were sought. Since sulfur does not support as many substituents as phosphorus, these would not be as fluorophilic as the PC(sp^2)P homologs above, unless additional ponytails were introduced elsewhere.

The precursor ligands **33**-R_{fn} were synthesized by the two complementary base-promoted routes shown in Scheme 14.9 (top). One involves the dibenzylic dibromide **16** and the known fluorous thiols $R_{fn}CH_2CH_2SH$ (**31**) [21a, 37], which are easily synthesized from the corresponding fluorous iodides [34, 35]. The other

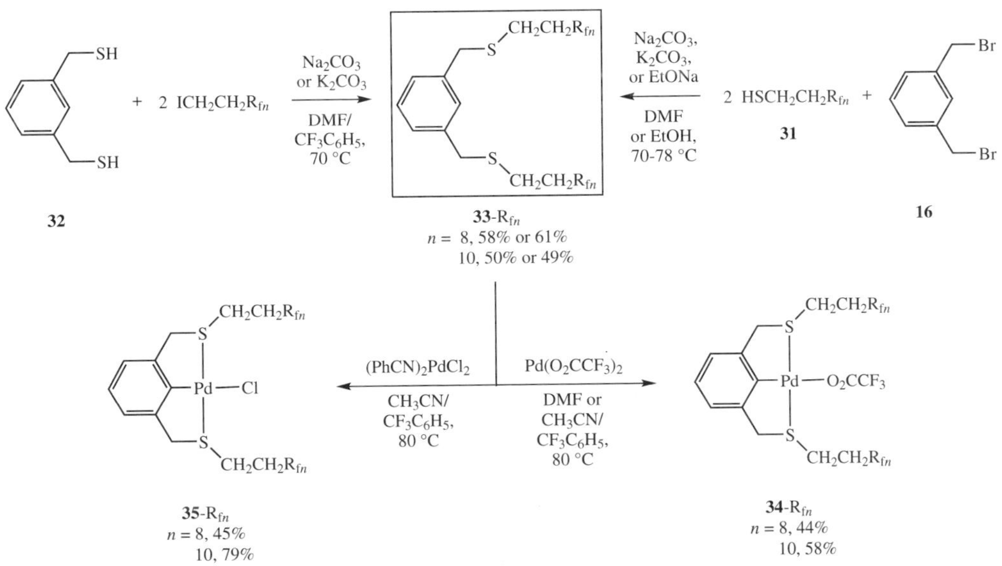

Scheme 14.9 Syntheses of fluorous SC(sp^2)S pincer ligands and palladium complexes.

Table 14.4 Qualitative solubility profiles of SC(sp^2)S pincer ligands and complexes at room temperature.

	$CF_3C_6F_{11}$	$CF_3C_6H_5$	THF	DMF
33-R_{f8}	High	High	High	Moderate
33-R_{f10}	High	High	Moderate	Low
34-R_{f8}	High	High	High	Moderate
34-R_{f10}	Moderate	High	Moderate	Low
35-R_{f8}	High	High	High	Moderate
35-R_{f10}	Moderate	High	Moderate	Low

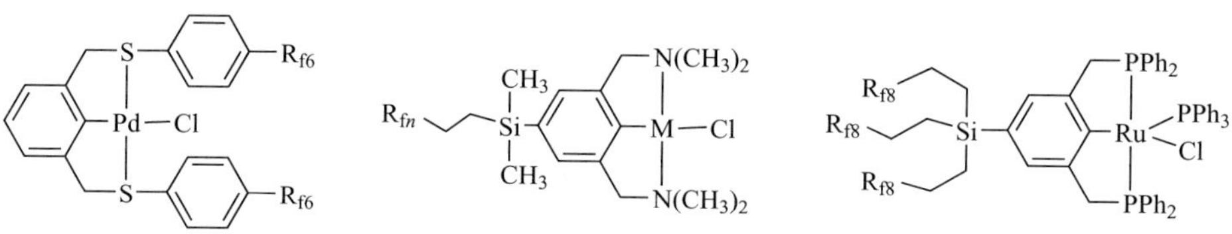

36 **37** **38**

$n = 6, 10$; M = Pt, Ni

Figure 14.2 Fluorous metallacycles prepared by other researchers.

involves the dibenzylic dithiol **32**, which is commercially available but more expensive than **16**, and fluorous iodides. As summarized in Scheme 14.9 (bottom), cyclopalladations could be effected with $Pd(O_2CCF_3)_2$ or $(PhCN)_2PdCl_2$, giving the target pincer complexes **34**-R_{fn} and **35**-R_{fn}, respectively, in 44–79% yields.

The doubly ponytailed compounds **33–35**-R_{fn} exhibited the unexceptional solubility profiles shown in Table 14.4. As summarized in Table 14.1, the partition coefficients of the palladium complexes (70.9–78.9 : 29.1–21.1) were lower than that of quadruply ponytailed PC(sp^2)P species **23**-R_{f8} (96.4 : 3.6).

14.6
Related Complexes from Other Groups [38–40]

As summarized in Figure 14.2, other researchers have also prepared fluorous palladacycles or closely related species. Among these, the SC(sp^2)S complex **36** is the most relevant [38]. This adduct, reported by Curran, is by virtue of the shorter ponytails and additional aryl rings less fluorous than the related systems in Scheme 14.9. Accordingly, it was not designed to be recovered by the protocol in Scheme 14.1, but rather by a solid-phase extraction involving fluorous silica gel.

A palladium analog of the nickelacycle and platinacycle **37** [39] could presumably be readily prepared. These complexes illustrate an alternative method of ponytail attachment, employing a trialkylsilyl group para to the metal. Although metallacycles **37** feature only a single ponytail, analogs with up to three ponytails should be easily accessed, as exemplified by the ruthenacycle **38** [40]. Such palladacycles would be good candidates for highly thermomorphic systems. A promising direction for future research would involve the introduction of *p*-$(R_{fn}CH_2CH_2)_3Si$ substituents on the $SC(sp^2)S$ and $PC(sp^2)P$ complexes described in previous sections. This would give extraordinarily fluorous species with highly temperature-dependent solubilities.

14.7 Catalysis

As detailed in other chapters, palladacycles can be employed in various catalytic reactions [1, 2, 41]. Most of the fluorous systems described above were modeled after complexes that are efficient catalyst precursors for Heck and/or Suzuki coupling reactions. Often, only very low loadings were necessary.

Indeed, the fluorous palladacycles **7, 13, 23**-R_{f8}, **34**-R_{f8}, **35**-R_{f8} and **36** are highly effective catalyst precursors for Heck and Suzuki reactions [13, 34, 35, 38]. Results with the acetate-bridged dimers **7** and **13,** optimized with respect to turnover numbers, are summarized in Table 14.5 [13]. Although both complexes are insoluble in DMF at room temperature, they readily dissolve under the reaction conditions (100–140 °C). Upon cooling, the palladacycles precipitate as halide-bridged dimers, with the halide ligand derived from the aromatic coupling partner (e.g. iodide-bridged **9** when using C_6H_5I). In view of the very small catalyst quantities involved, recycling experiments used somewhat higher loadings (e.g. 0.2 mol.%).

However, numerous control experiments establish that these palladacycles serve as steady-state sources of non-fluorous colloidal palladium nanoparticles, formed anew with each cycle until the palladacycles are consumed [13, 34, 35]. For example, every cycle exhibits an induction period, and the catalytically active species can be extracted into non-fluorous solvents. Transmission electron micrographs show a distribution of nanoparticle sizes. The addition of mercury quenches all activity [34, 35]. Either the nanoparticles or low-valent Pd(0) species generated therefrom are the active catalysts. Parallel results, including many other types of mechanistic tests, were reported for closely related non-fluorous palladacycles [42, 43] during the course of the above investigations.

However, one should not lose sight of the numerous other reactions that can be catalyzed by palladacycles [1, 2, 41]. Furthermore, highly effective enantioselective catalysts for Heck reactions are known, and these are undoubtedly molecular in nature. Hence, the thermomorphic fluorous palladacycles described above, as well as new generation systems, retain a promising future.

Table 14.5 Representative data for Heck reactions under high turnover conditions.[a,b]

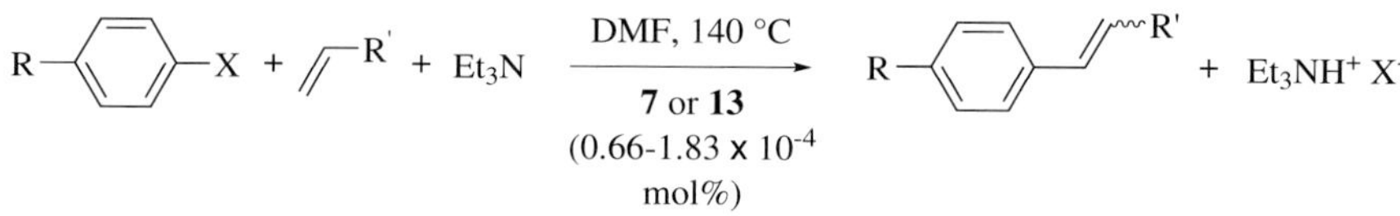

X	R	R′	Catalyst, mmol	*t* (h)	Conversion (%)	Yield (%)	TON[c]
I	H	CO_2CH_3	**7**, 0.00000344	14	100	100	1 460 000
			13, 0.0000415	18	100	100	1 510 000
I	H	C_6H_5	**7**, 0.00000344	24	94	88	1 270 000
			13, 0.0000415	48	100	100	1 510 000
Br	CH_3CO	CO_2CH_3	**7**, 0.00000917	48	77	49	266 000
			13, 0.0000831	48	66	50	301 000

a Conditions for **7**: ArX (ca 5.000 mmol), alkene (ca 1.25 equiv), Et_3N (ca 2 equiv), DMF (6.00 mL).
b Conditions for **13**: ArX (ca 5.000–6.279 mmol), alkene (ca 2 equiv), Et_3N (ca 2 equiv), DMF (8.00 mL).
c In accord with most literature data, not normalized to the number of palladium atoms in the catalyst.

14.8
Summary and Outlook

This chapter has summarized the syntheses and key physical properties of all presently known fluorous palladacycles. In general, their solubilities dramatically increase with temperature. This constitutes a type of thermomorphism, for which quantitative data are available for other families of fluorous catalysts [4, 44]. The solubilities can be fine-tuned by varying the lengths of the perfluoroalkyl groups, allowing efficient recycling by solid/liquid phase separation as sketched in Scheme 14.1. Second-generation protocols involving supports and other fluorous catalysts are reviewed elsewhere [4]. However, all fluorous palladacycles examined to date yield small quantities of non-fluorous nanoparticles under typical conditions for Heck reactions. These or subsequently-formed species are the active catalysts. Nonetheless, such thermomorphic systems hold much promise for the many other types of palladacycle-catalyzed processes [1, 2, 41].

Acknowledgment

The Deutsche Forschungsgemeinschaft (DFG, GL 300/3-3), the Bundes ministerium für Bildung und Forschung (BMBF), the Welch Foundation and Johnson Matthey (precious metal loans) are thanked for their generous support.

References

1 Dupont, J., Consorti, C.S. and Spencer, J. (2005) *Chemical Reviews*, **105**, 2527.

2 (a) Albrecht, M. and van Koten, G. (2001) *Angewandte Chemie – International Edition*, **40**, 3750.
(b) van der Boom, M.E. and Milstein, D. (2003) *Chemical Reviews*, **103**, 1759.
(c) Albrecht, M. and van Koten, G. (2001) *Angewandte Chemie*, **113**, 3866.

3 Gladysz, J.A., Curran, D.P. and Horváth, I.T. (eds) (2004) *Handbook of Fluorous Chemistry*, Wiley-VCH Verlag GmbH, Weinheim.

4 Gladysz, J.A. and Tesevic, V. (2008) *Topics in Organometallic Chemistry*, **23**, 62.

5 Gladysz, J.A. (2004) Pony tails: structural and electronic considerations, in *Handbook of Fluorous Chemistry* (eds J.A. Gladysz, D.P. Curran and I.T. Horváth), Wiley-VCH Verlag GmbH, Weinheim, pp. 41–55.

6 Horváth, I.T. and Rábai, J. (1994) *Science*, **266**, 72.

7 Gladysz, J.A. and Emnet, C. (2004) Fluorous solvents and related media, in *Handbook of Fluorous Chemistry* (eds J.A. Gladysz, D.P. Curran and I.T. Horváth), Wiley-VCH Verlag GmbH, Weinheim, pp. 11–23.

8 Gladysz, J.A., Emnet, C. and Rábai, J. (2004) Partition coefficients involving fluorous solvents, in *Handbook of Fluorous Chemistry* (eds J.A. Gladysz, D. P. Curran and I.T. Horváth), Wiley-VCH Verlag GmbH, Weinheim, pp. 56–100.

9 Hughes, R.P. and Trujillo, H.A. (1996) *Organometallics*, **15**, 286.

10 van Vliet, M.C.A., Arends, I.W.C.E. and Sheldon, R.A. (1999) *Chemical Communications*, 263.

11 (a) Jessop, P.G., Olmstead, M.M., Ablan, C., *et al.* (2002) *Inorganic Chemistry*, **41**, 3463.
(b) Jessop, P.G. and Subramaniam, B. (2007) *Chemical Reviews*, **107**, 2666.

12 (a) Dinh, L.V. and Gladysz, J.A. (2005) *Angewandte Chemie – International Edition*, **44**, 4095.
(b) Dinh, L.V. and Gladysz, J.A. (2005) *Angewandte Chemie*, **117**, 4164.

13 Rocaboy, C. and Gladysz, J.A. (2003) *New Journal of Chemistry*, **27**, 39.

14 Ohff, M., Ohff, A. and Milstein, D. (1999) *Chemical Communications*, 357.

15 Rocaboy, C., Rutherford, D., Bennett, B.L. and Gladysz, J.A. (2000) *Journal of Physical Organic Chemistry*, **13**, 596.

16 (a) Soós, T., Bennett, B.L., Rutherford, D., *et al.* (2001) *Organometallics*, **20**, 3079.
(b) Rocaboy, C., Hampel, F., Gladysz, J.A. (2002) *Journal of Organic Chemistry*, **67**, 6863.

17 Rocaboy, C., Bauer, W. and Gladysz, J.A. (2000) *European Journal of Organic Chemistry*, 2621.

18 Maul, J.J., Ostrowski, P.J., Ublacker, G.A., *et al.* (1999) *Topics in Current Chemistry*, **206**, 79.

19 (a) Gruber, A.S., Zim, D., Ebeling, G., *et al.* (2000) *Organic Letters*, **2**, 1287.
(b) Zim, D., Gruber, A.S., Ebeling, G., *et al.* (2000) *Organic Letters*, **2**, 2881.
(c) Dupont, J., Gruber, A.S., Fonseca, G.S., *et al.* (2001) *Organometallics*, **20**, 171.

20 Guindon, Y., Frenette, R., Fortin, R. and Rokach, J. (1983) *Journal of Organic Chemistry*, **48**, 1357.

21 (a) Naud, C., Calas, P., Blancou, H., Commeyras, A. (2000) *Journal of Fluorine Chemistry*, **104**, 173.
(b) Mureau, N., Guittard, F. and Géribaldi, S. (2000) *Tetrahedron Letters*, **41**, 2885.

22 Tuba, R., Tesevic, V., Dinh, L.V., *et al.* (2005) *Dalton Transactions*, 2275.

23 (a) Alvey, L.J., Rutherford, D., Juliette, J.J.J. and Gladysz, J.A. (1998) *Journal of Organic Chemistry*, **63**, 6302.
(b) Klose, A. and Gladysz, J.A. (1999) *Tetrahedron Asymmetry*, **10**, 2665.
(c) Alvey, L.J., Meier, R., Soós, T., *et al.* (2000) *European Journal of Organic Chemistry*, 1975.

24 (a) Emnet, C. and Gladysz, J.A. (2005) *Synthesis*, 1012.
(b) Emnet, C., Tuba, R. and Gladysz, J.A. (2005) *Advanced Synthesis Catalysis*, **347**, 1819.

25 (a) Musina, E.I., Nikonov, G.N., Balueva, A.S., *et al.* (1999) *Russian Journal of General Chemistry*, **69**, 407.

(b) Caplan, N.A., Pogson, C.I., Hayes, D. J. and Blackburn, G.M. (2000) *Journal of the Chemical Society–Perkin Transactions 1*, 421.

26 (a) Gusev, D.G., Madott, M., Dolgushin, F.M., *et al.* (2000) *Organometallics*, **19**, 1734.
(b) Hollink, E., Stewart, J.C., Wei, P. and Stephan, D.W. (2003) *Journal of the Chemical Society–Dalton Transactions*, 3968.
(c) Grimm, J.C., Nachtigal, C., Mack, H.-G., *et al.* (2000) *Inorganic Chemistry Communications*, **3**, 511.
(d) Mohammad, H.A.Y., Grimm, J.C., Eichele, K., *et al.* (2002) *Organometallics*, **21**, 5775.

27 (a) Bailey, W.J. and Buckler, S.A. (1957) *Journal of the American Chemical Society*, **79**, 3567.
(b) Gough, S.T.D. and Trippett, S. (1961) *Journal of the Chemical Society*, 4263.
(c) Sasse, K. (1963) *Houben-Weyl Methoden der Organischen Chemie*, Vol. **XII**, Part 1. Georg Thieme Verlag, Stuttgart, Germany, pp. 52–3.

28 (a) Göttker-Schnetmann, I. and Brookhart, M. (2004) *Journal of the American Chemical Society*, **126**, 9330.
(b) Morales-Morales, D., Rédon, R., Yung, C. and Jensen, C.M. (2004) *Inorganica Chimica Acta*, **357**, 2953.

29 (a) Guillevic, M.-A., Rocaboy, C., Arif, A.M., *et al.* (1998) *Organometallics*, **17**, 707.
(b) Fawcett, J., Hope, E.G., Kemmitt, R.D.W., *et al.* (1997) *Journal of the Chemical Society, Chemical Communications*, 1127.
(c) Haar, C.M., Huang, J. and Nolan, S.P. (1998) *Organometallics*, **17**, 5018.
(d) Smith, D.C. Jr, Stevens, E.D. and Nolan, S.P. (1999) *Inorganic Chemistry*, **38**, 5277.
(e) Stibrany, R.T. and Gorun, S.M. (1999) *Journal of Organometallic Chemistry*, **579**, 217.

30 Consorti, C.S., Hampel, F. and Gladysz, J.A. (2006) *Inorganica Chimica Acta*, **359**, 4874.

31 Group 10 PC(sp3)P complexes: (a) Al-Salem, N.A., Empsall, H.D., Markham, R., *et al.* (1979) *Journal of the Chemical Society–Dalton Transactions*, 1972.
(b) Al-Salem, N.A., McDonald, W.S., Markham, R., *et al.* (1980) *Journal of the Chemical Society–Dalton Transactions*, 59.
(c) Briggs, J.R., Constable, A.G., McDonald, W.S. and Shaw, B.L. (1982) *Journal of the Chemical Society–Dalton Transactions*, 1225.
(d) Seligson, A.L. and Trogler, W.C. (1993) *Organometallics*, **12**, 738 and 744.
(e) Sjövall, S., Wendt, O.F. and Anderson, C. (2002) *Journal of the Chemical Society–Dalton Transactions*, 1396.
(f) Neo, K.E., Neo, Y.C., Chien, S.W., *et al.* (2004) *Dalton Transactions*, 2281 and subsequent correction of NMR data (DOI: 10.1039/b403051f).

32 Alder, R.W., Ganter, C., Gil, M., *et al.* (1998) *Journal of the Chemical Society–Perkin Transactions 1*, 1643.

33 (a) Richter, B., Spek, A.L., van Koten, G. and Deelman, B.-J. (2000) *Journal of the American Chemical Society*, **122**, 3945.
(b) Dinh, L.V. and Gladysz, J.A. (2005) *Chemistry–A European Journal*, **12**, 7211.

34 da Costa, R.C. (2006) Doctoral Dissertation, Universität Erlangen-Nürnberg.

35 da Costa, R.C., Jurisch, M. and Gladysz, J.A. (2008) *Inorganica Chimica Acta*, in press. DOI: 10.1016/j.ica.2007.11.016.

36 (a) Bergbreiter, D.E., Osburn, P.L. and Liu, Y.-S. (1999) *Journal of the American Chemical Society*, **121**, 9531.
(b) Giménez, R. and Swager, T.M. (2001) *Journal of Molecular Catalysis B*, **166**, 265.
(c) Nakai, H., Ogo, S. and Watanabe, Y. (2002) *Organometallics*, **21**, 1674.
(d) Kjellgren, J., Sundén, H. and Szabo, K. J. (2005) *Journal of the American Chemical Society*, **127**, 1787.
(e) Cervantes, R., Castillejos, S., Loeb, S.J., *et al.* (2006) *European Journal of Organic Chemistry*, 1076.

37 Szonyi, F. and Cambon, A. (1989) *Journal of Fluorine Chemistry*, **42**, 59.

38 Curran, D.P., Fischer, K. and Moura-Letts, G. (2004) *Synlett*, 1379.

39 Kleijn, H., Jastrzebski, J.T.B.H., Gossage, R.A., *et al.* (1998) *Tetrahedron*, **54**, 1145.

40 Dani, P., Richter, B., van Klink, G.P.M. and van Koten, G. (2001) *European Journal of Organic Chemistry*, 125.

41 Szabó, K.J. (2006) *Synlett*, 811 (account).

42 (a) Yu, K., Sommer, W., Richardson, J.M., *et al.* (2005) *Advanced Synthesis Catalysis*, **347**, 161.
(b) Bergbreiter, D.E., Osburn, P.L. and Frels, J.D. (2005) *Advanced Synthesis Catalysis*, **347**, 172.

43 Lead references to a now-extensive literature: (a) Eberhard, M.R. (2004) *Organic Letters*, **6**, 2125.
(b) Consorti, C.S., Flores, F.R. and Dupont, J. (2005) *Journal of the American Chemical Society*, **127**, 12054.
(c) Phan, N.T.S., van der Sluys, M. and Jones, C.W. (2006) *Advanced Synthesis Catalysis*, **348**, 609.

44 Wende, M. and Gladysz, J.A. (2003) *Journal of the American Chemical Society*, **125**, 5861.

15
Palladacycles on Dendrimers and Star-Shaped Molecules

Niels J. M. Pijnenburg, Ties J. Korstanje, Gerard van Koten and Robertus J. M. Klein Gebbink

15.1
Introduction

15.1.1
Development and Synthesis of Dendrimers

Dendrimers are highly branched, three-dimensional macromolecules with a branching point at each monomer unit. This class of macromolecules was pioneered by Vögtle *et al.* in 1978 by publishing a procedure called cascade synthesis, using a repetition of similar synthetic steps, leading to highly symmetrical monodisperse macromolecules [1]. Following this publication, the field of dendrimers and dendritic structures was further investigated by the groups of Tomalia [2–4], Newkome [5, 6] and Fréchet [7–10] and has been a blossoming field of research ever since. Since major developments in NMR spectroscopy, mass spectrometry and size-exclusion chromatography made detailed characterization of dendrimers possible, these well-defined macromolecular compounds have been used for many applications: host–guest chemistry, drug delivery, self-assembly, usage as macromolecular templates or as sensor, and as catalytic materials [11].

15.1.2
Dendrimers in Catalysis

An interesting application of dendrimers is their use in catalysis. When considering the fields of homogeneous catalysis, where the catalyst is in the same phase as the reactants and the products, and heterogeneous catalysts, where the catalyst is in another (most likely solid) phase, several clear distinctions are obvious (Table 15.1). Quite notably, these distinctions seem to be complementary.

The idea of developing catalysts that combine the advantages of both homogeneous and heterogeneous catalysts therefore seems very attractive. One objective could be to arrive at recyclable homogeneous catalysts. Immobilization of homo-

Palladacycles: Synthesis, Characterization and Applications. Edited by Jairton Dupont and Michel Pfeffer

ISBN: 978-3-527-31781-3

Table 15.1 Homogeneous catalysis compared with heterogeneous catalysis

	Homogeneous	Heterogeneous
Activity	+++	–
Selectivity	+++	–
Catalytic description	+++	–
Reaction conditions	+++	+
Catalyst recycling	–	+++
Quantity of catalyst	++	+++
Total turnover number	+	+++

geneous catalysts on dendrimers is an attractive strategy for this objective [11], because of the macromolecular dimensions and excellent solubility properties of dendrimers.

Generally, dendritic catalysts combine the kinetic behavior, activity and selectivity of a homogeneous catalyst with the easy separation of the heterogeneous catalyst, namely, by applying precipitation or membrane separation techniques of the dendritic, nanosized catalyst. Other favorable characteristics of homogeneous catalysis are also retained, like the possibility of performing mechanistic studies (due to the well-defined structure of dendrimers) and the possibility to fine-tuning the catalytic centers.

Several excellent reviews on dendrimers in catalysis have been published [13–20].

15.1.3
Metallodendrimers

In the area of dendritic catalysis, metallodendrimers are most widely developed [21–24]. Metallodendrimers, or dendritic catalysts in general, can be roughly divided in five classes, depending on the position of the metal centers, or active centers, in the dendrimer (Figure 15.1).

Periphery bound metal complexes (Figure 15.1a) are the type of metallodendrimers used most frequently in catalysis. Knapen's carbosilane dendrimer [25] with twelve organonickel functional groups (Figure 15.2) is an example of this type. Herein the transition metals are in general readily accessible for substrates. In most cases, this results in reaction rates per metal site that are comparable to the parent homogeneous system with a monometallic species. In other cases, the peripheral active-site amplification may lead to a decrease in catalytic activity per metal site, for example due to an increased steric bulk and a decreased accessibility of the catalytic fragments for the substrates. This is called a negative dendritic effect. In contrast, the presence of catalytic sites in close proximity may lead to increased catalytic activity due to cooperation between the sites. Several examples of such positive dendritic effects in catalysis can be found in the literature [26–33].

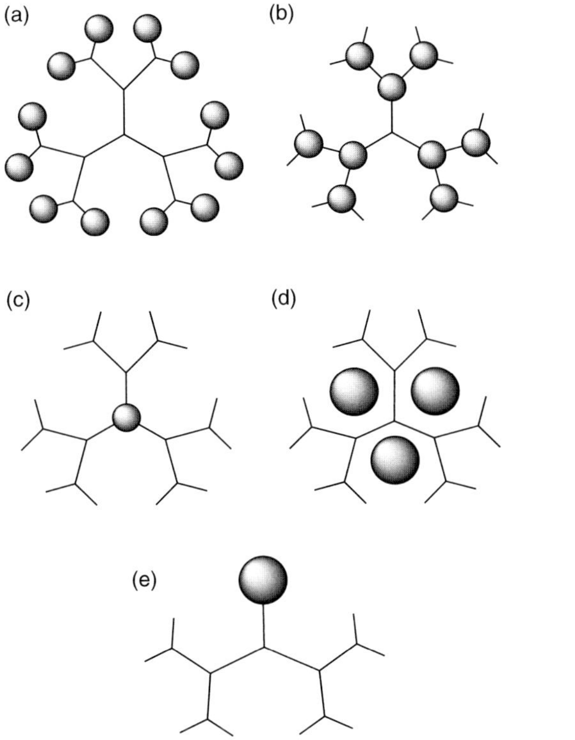

Figure 15.1 Different types of metallodendrimers. Transition metal sites can be located at the periphery (a), at or near each branching point (b), at the core (c) or as metal particles formed inside the voids of the dendrimer (d). A transition metal site acting as the focal point of a dendritic wedge is visualized in (e). The figure was adopted from Oosterom *et al.* [11].

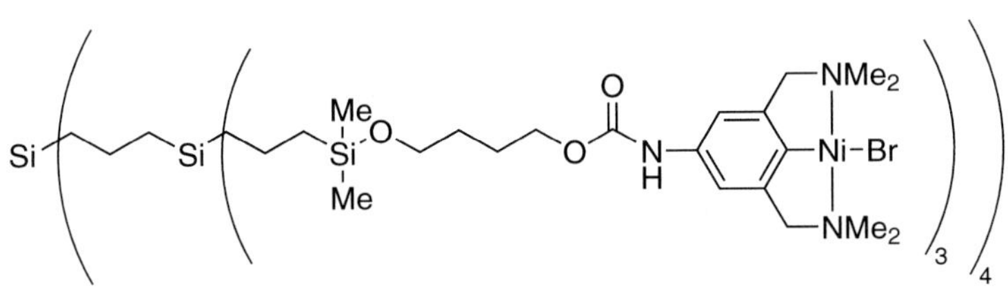

Figure 15.2 The first catalytically active metallodendrimer, G1-$(SiMe_2\text{-}ArNiBr)_2$ by Knapen *et al.* [25].

Core-functionalized dendrimers (Figure 15.1c) are mostly synthesized with another goal in mind. Here, the metal centers could benefit from a site-isolation effect, caused by the dendritic environment. This dendritic shielding provides an encapsulation of the catalytic core and creates a microenvironment, which affects the properties of the catalyst. This effect may be the result of both the intrinsic chemistry of the separate building blocks and of the decreased accessibility of the catalytic site [34]. Another important difference between periphery- and core-functionalized dendrimers lies in their molecular weight per catalytic site. The very high molecular weight per catalytic site of core-functionalized dendrimers may hamper their activity due to the difficult mass transport within the increasing steric bulk of the dendritic wedges [35].

Catalytic fragments or nanoclusters located at branching points or in the voids of a dendrimer (Figure 15.1b and d), and dendronized catalysts (Figure 15.1e), which are single catalytic fragments bearing one or more dendritic substituents, complete the topological picture of dendritic (metallo)catalysts.

Various palladium catalysts have been attached or immobilized to different types of dendrimers. Few of these "palladium dendrimers" contain palladacyclic structures in the true sense of the metallacycle definition, that is structures in which a covalent Pd–C bond is complemented by a Pd–donor interaction to form a palladacycle. Section 15.3 contains a concise description of these "palladadendrimers" in terms of structure, synthesis and application.

On the other hand, most "palladium dendrimers" do not exhibit palladium sites in the form of a palladacycle. To provide the reader with a conceptual overview of the field, selected examples of these "palladium dendrimers" with a focus on catalysis are discussed in Section 15.2 – prior to the discussion on "palladadendrimers" (Section 15.3).

15.2
Palladium Catalysts on Dendrimers: An Overview

15.2.1
Periphery-Bound Palladium Catalysts

In many examples palladium has been introduced on the dendrimer periphery (Figure 15.1a) where the palladium centers have been bound to the dendrimer by means of coordinative bonds. Here, palladium is mostly introduced by treatment of a presynthesized dendritic poly-ligand with an appropriate palladium salt. For more insight, the recent review of De Jesús and coworkers about catalysts based on palladium dendrimers is strongly recommended [23].

15.2.1.1 Dendritic Bis-Diphenylphosphino Palladium Complexes

Probably the most frequently used ligand to introduce palladium onto the periphery of a dendrimer is the bis-diphenylphosphino ligand. In 1997, Reetz and coworkers developed peripheral diphosphines on DAB-based (DAB = 1,4-diaminobutane) poly(propylene)imine (PPI) dendrimers, which were coordinated to several different metals, including palladium (Figure 15.3) [28]. Initial catalytic studies on the palladium dendrimers in the Heck reaction of bromobenzene and styrene were quite successful. Owing to the higher thermal stability of the dendritic compound compared to the parent monomeric one, a positive dendritic

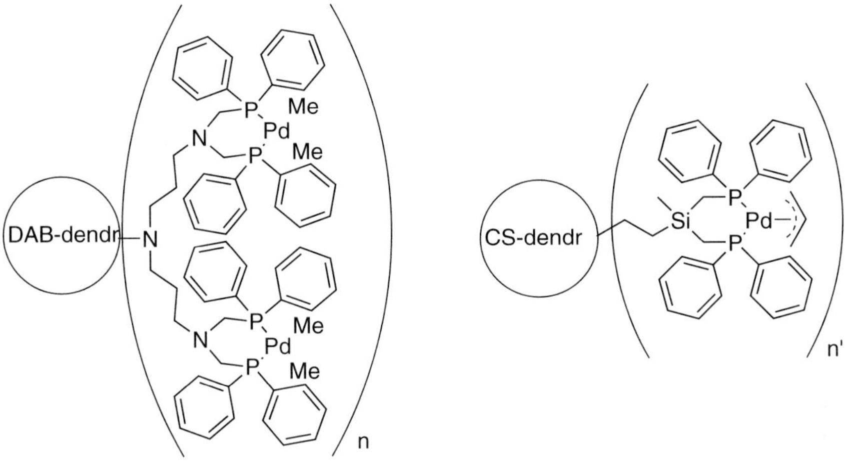

Figure 15.3 Palladium coordinated diphenylphosphino-functionalized DAB-based- and carbosilane dendrimer by Reetz [28] and de Groot [36], respectively; $n = 16$; $n' = 4$ or 12.

effect was even observed; the turnover number (TON) of the G2 dendrimeric compound is 50 mol product per mol of catalyst, whereas the TON of the monomeric compound is 16 mol product per mol of catalyst.

De Groot *et al.* [36] reported palladium complexes of diphenylphosphine-functionalized carbosilane dendrimers. These complexes have been used as catalysts in the allylic alkylation reaction of crotyl acetate or cinnamyl acetate with sodium diethyl 2-methylmalonate [37]. As it turned out, the catalytic activity per palladium atom was almost unaffected by the size of the dendrimer in batch reactions. Unfortunately, the yields for these allylic substitution reactions dropped rapidly in a continuous flow membrane reactor, presumably due to leaching of palladium from the dendrimer.

Several other research groups have also used bis-phosphino palladium complexes on dendrimers. Heuzé *et al.* investigated a series of four different generations of bis(di-*tert*-butylphosphino)- and bis(dicyclohexylphosphino)-functionalized palladium(II) poly(propylene imine) dendrimers for carbon–carbon coupling reactions [38]. A negative dendritic effect was found for the Sonogashira coupling reaction of iodobenzene with phenylacetylene. The largest tested dendrimers appeared to be the least active ones, which is explained by the authors as being due to a decreased accessibility of the catalytic centers. Mizugaki *et al.* synthesized PPI dendrimers containing peripheral bis-diphenylphosphino palladium complexes for allylic substitution reactions of allylic acetates with amines [39]. The selectivity for the cis-product increased dramatically with increasing dendrimer generation in the reaction of *cis*-3-acetoxy-5-carbomethoxycyclohex-1-ene with morpholine.

Alper and coworkers have performed extensive research on silica-supported polyamido amine (PAMAM) palladium complexes (Figure 15.4). These palladium

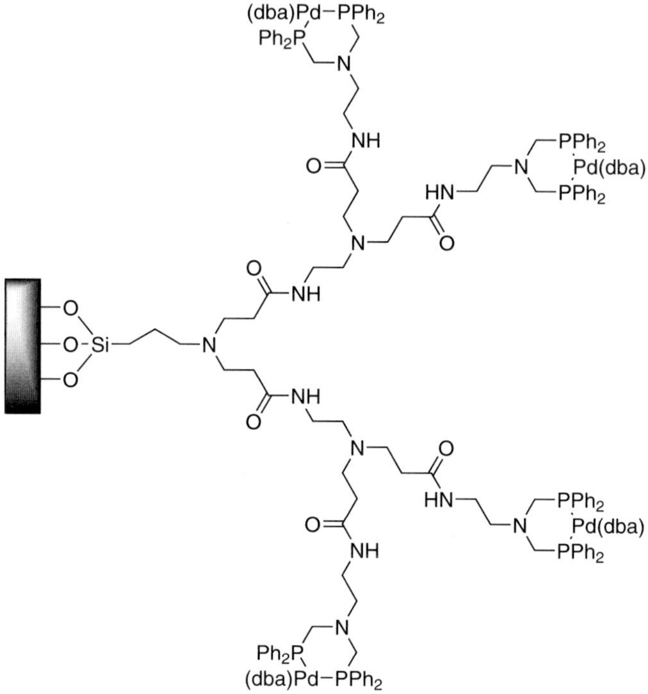

Figure 15.4 Recyclable palladium dendrimers on silica as used by Alper [40].

catalysts are used in various organic transformations, for example Heck [41], hydrogenation [42], carbonylation [40, 43], hydroesterification [44] and oxidation reactions [45]. These dendritic complexes display some interesting recycling properties, whereby the activity is largely retained for up to eight cycles. Palladium loading in these materials is achieved by means of ligand complexation with $Pd_2(dba)_3$ (dba = dibenzylideneacetone), but no full loading could be obtained.

Intramolecular carbonylation reactions were also performed with these palladium dendrimers on silica in the synthesis of numerous oxygen-, nitrogen- and sulfur-containing medium-sized fused heterocycles by using complex substrates [40]. A remarkable array of different heterocycles show excellent yields in these reactions. Again, the dendritic catalysts have been recovered by an easy filtration in air and were reused for up to eight cycles with only a slight loss in activity.

15.2.1.2 Other Periphery-Bound Palladium Complexes

Here, two representative examples of bi- and tridentate nitrogen-palladium complexes are shown. G0 and G1 PPI pyridylimine palladium dendrimers were synthesized by Mapolie *et al.* and were used as catalyst precursors for the polymerization of ethylene [46] and for Heck reactions (Figure 15.5) [47]. For the latter reaction type, a positive dendritic effect was observed in that the conversion of the arylations of styrene and 1-octene, respectively, dramatically increased compared to reactions

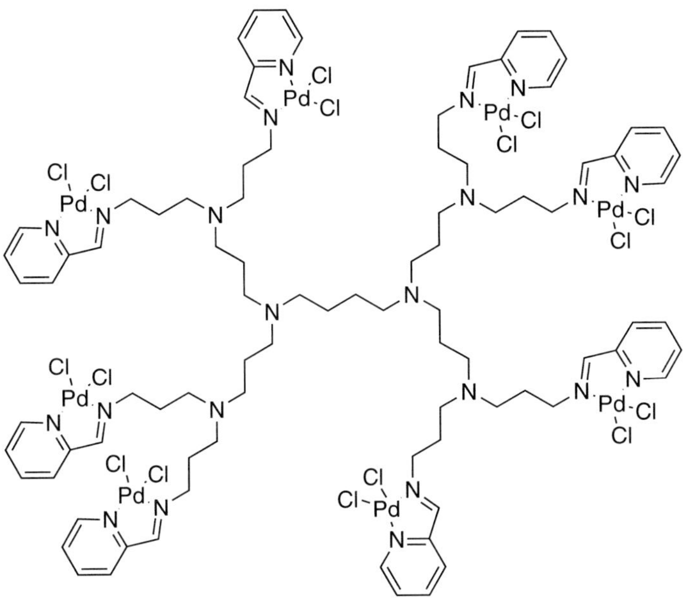

Figure 15.5 A first generation pyridylimine palladium dendrimer by Mapolie [46].

using an analogous monomeric palladium complex or palladium dichloride, which apparently agglomerate more readily.

A tridentate *(N,N,N)* palladium complex is formed by metallating a (2-pyridylimino)isoindolato (BPI) ligand. Gade *et al.* connected these ligands to the periphery of a carbosilane dendrimer [48] via an alkynyl linker unit. Subsequent palladation leads to the dendritic complex shown in Figure 15.6.

The same group also anchored polyether dendritic wedges to monomeric BPI ligands, forming a dendronized ligand [49]. After palladation, the resulting dendritic complexes were investigated as hydrogenation catalysts. Styrene and 1-octene were hydrogenated without decomposition with a TOF of $5\,h^{-1}$ [$T = 295$ K, $p(H_2) = 1$ bar, 2 mol.% cat.]. The G2 Fréchet-type dendron-functionalized catalyst was isolated and succcssfully reused several times.

15.2.1.3 Dendrimers and Star-Shaped Molecules Containing Covalent Pd–C Bonds

The above examples nicely illustrate the potential of palladium dendrimers in catalysis. Yet, due to the intrinsic properties of the palladium sites and the specific catalytic transformations, palladium leaching may be a serious limitation for the reuse or continuous use of some of these examples. One way to prevent Pd leaching is to include a covalent palladium–carbon bond. A few examples of these covalently bound palladium complexes bound to dendrimers or star-shaped molecules are described here. Obviously, palladacycle-containing dendrimers also include covalent Pd–C bonds (Section 15.3).

Chen and coworkers synthesized a series of metal acetylide star-shaped molecules in which the metal center is covalently connected to the dendrimer (Figure

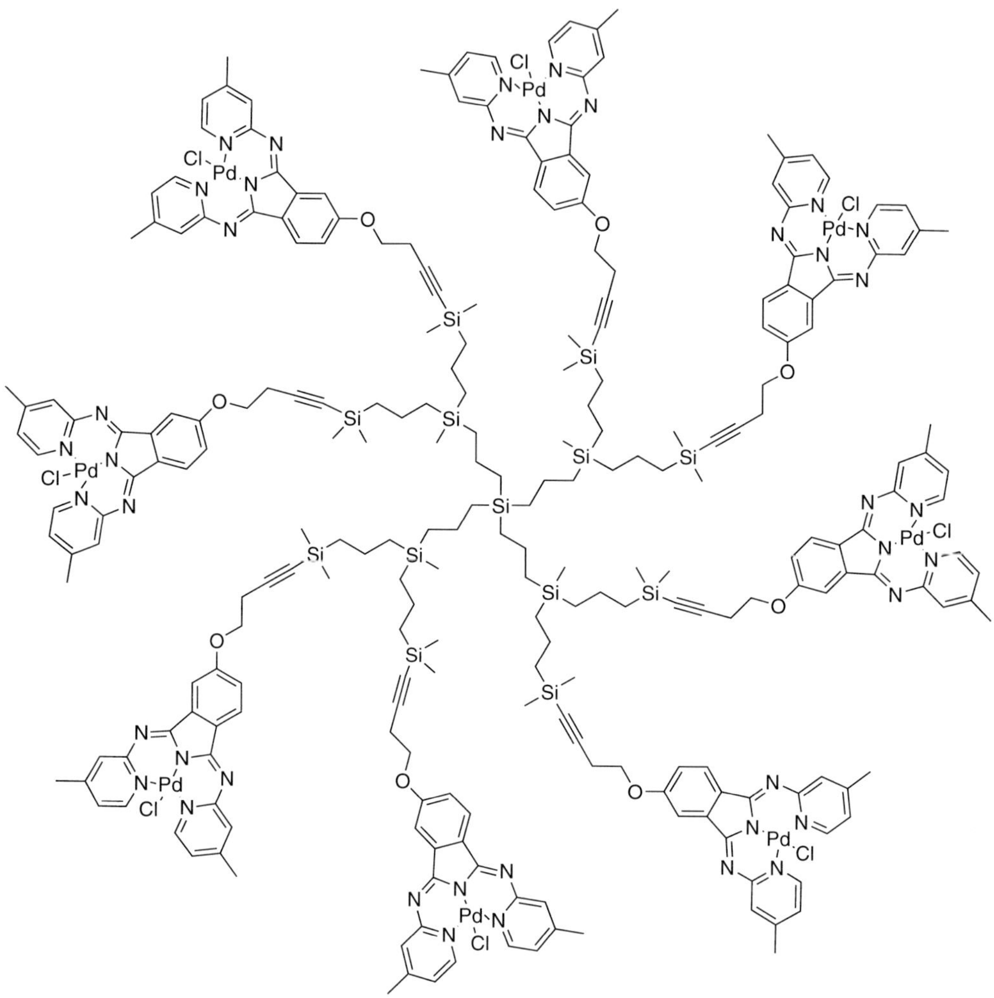

Figure 15.6 G1 carbosilane dendrimer functionalized with eight Pd(BPI) ligands by Gade [48].

15.7) [50]. Palladium metal moieties were also added to terminal carbon atoms of the poly-yne. These compounds were tested for their fluorescent behavior. Photophysical data of these metal complexes showed that the absorption peaks shift to longer wavelengths and showed higher extinction coefficients than non-metallated equivalents. In contrast, the luminescence quantum yield dramatically decreased for the palladated star-shaped molecule.

Another example of a periphery-palladated dendrimer was synthesized in the group of Van Koten [51]. $Pd_2(dba)_3$ and TMEDA (*N,N,N′,N′*-tetramethylethylenediamine) were reacted with a carbosilane dendrimer carrying 12 iodoarene groups and subsequent reaction with methyllithium and bipyridine (bpy) led to the air-stable first generation palladadendrimer depicted in Figure 15.8. These dendritic

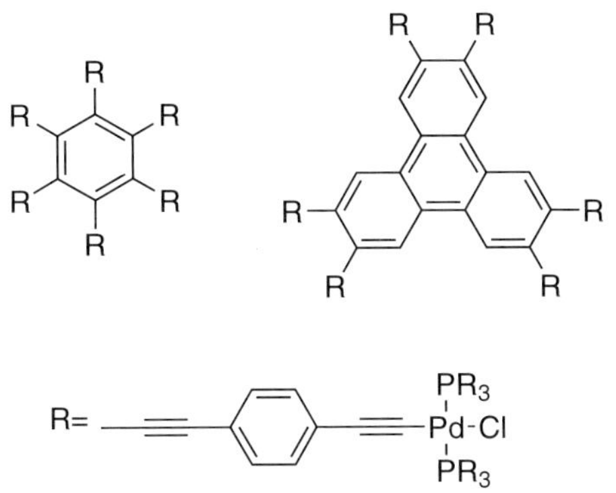

Figure 15.7 Star-shaped molecules containing a covalent carbon–palladium bond by Chen [50].

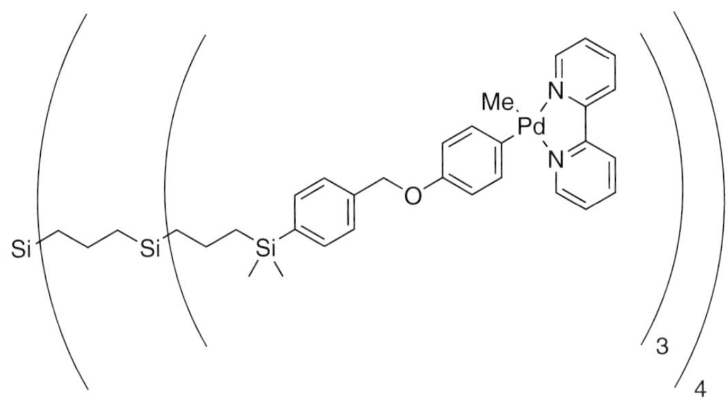

Figure 15.8 Van Koten's palladacarbosilane dendrimer containing covalent palladium bonds [51].

organopalladium(II) complexes were able to undergo oxidative addition with benzyl bromides, yielding palladium(IV) compounds [52].

15.2.2 Dendrimer-Encapsulated Palladium Nanoparticles

A completely different type of dendrimer-based palladium catalysis has been performed with dendrimer-encapsulated nanoparticles (DENs). These metallodendrimers are examples of category d in Figure 15.1. DENs combine the desirable physical and chemical features of metal nanoparticles with the tunable properties of dendrimers such as solubility, size, composition, structure and surface reactivity [53, 54].

Herein, the backbone of PPI or PAMAM dendrimers is used for coordinating palladium ions: it involves complexation of the palladium centers with interior tertiary amine functionalities. Once coordinated, the palladium ions can be reduced

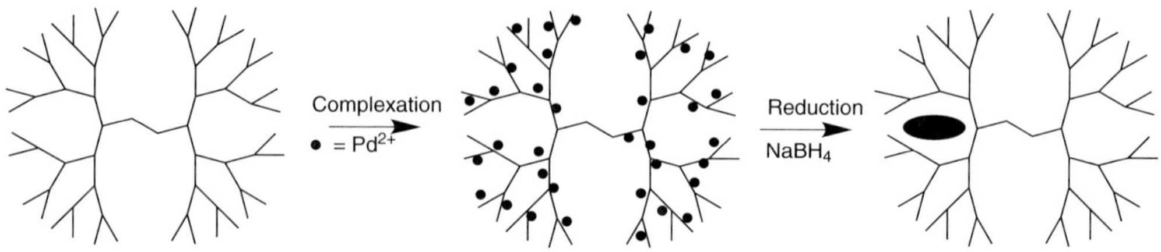

Scheme 15.1 Strategy for the synthesis of nanoparticles encapsulated in PAMAM or PPI dendrimers.

with an excess of sodium borohydride to obtain Pd(0), which aggregates into nanoparticles inside the voids of the dendrimer (Scheme 15.1). When the periphery of the dendrimer is functionalized with non-complexating entities, or in case the terminal amino groups are protonated prior to metal complexation, the metal binding can take place exclusively in the interior of the dendrimer [54]. Higher dendrimer generations (starting from G3) stabilize metal nanoparticles by a combination of electrostatic and complexation interactions. Crooks and coworkers were able to synthesize DENs with various metals like copper, gold, platinum, iron, ruthenium and palladium [54].

Because of the ability of dendrimers to act as a molecular box by encapsulating guest molecules [55], it is possible to filter out substrates of a certain size. This can be an interesting feature in catalysis. DENs have been used in this respect as size-selective hydrogenation catalysts in which the catalytic performance of Pd DENs is dependent on the substrate size. Larger substrates have more difficulty reaching the encapsulated nanoparticle, because of the steric bulk on the outside of the dendrimer. The same effect is observed when the generation of the dendrimer is increased: bulkier substrates are less likely reach the DEN's reactive site.

Niu *et al.* [56] reported such an effect in the hydrogenation of allylic alcohols. Comparing palladium nanoparticles in G4-OH PAMAM dendrimers with its G8-OH analog, it was found that the turnover frequency for prop-2-en-1-ol dramatically decreases from 480 to 120 mol-H_2 (mol Pd)$^{-1}$ h^{-1}. This effect was seen for all tested substrates. Therefore, the dendrimer acts as a molecular filter. For G4-OH(Pd_{40}) the TOF changed from 480 mol-H_2 (mol Pd)$^{-1}$ h^{-1} for prop-2-en-1-ol to just 100 mol-H_2 (mol Pd)$^{-1}$ h^{-1} for the more bulky enol 3-methyl-1-penten-3-ol.

Kaneda's group reported a way to hydrogenate 3-cyclohexene-1-methanol selectively out of a mixture with 3-cyclohexene [57]. Traditional Pd/C catalysts give incomplete conversions for both substrates, while PPI G5-TEBA(Pd) (TEBA = triethoxybenzoic acid) nanoparticles give a quantitative reduction of 3-cyclohexene-1-methanol, while cyclohexene stays unaffected. In addition, bimetallic (palladium in combination with platinum or gold) dendrimer encapsulated nanoparticles have been successfully used in this way [58]. Chandler and Gilbertson have reviewed bimetallic DENs [59].

Heck reactions have been performed by using Pd DENs as catalysts as well. DENs based on PPI dendrimers decorated with perfluorinated tails catalyze the

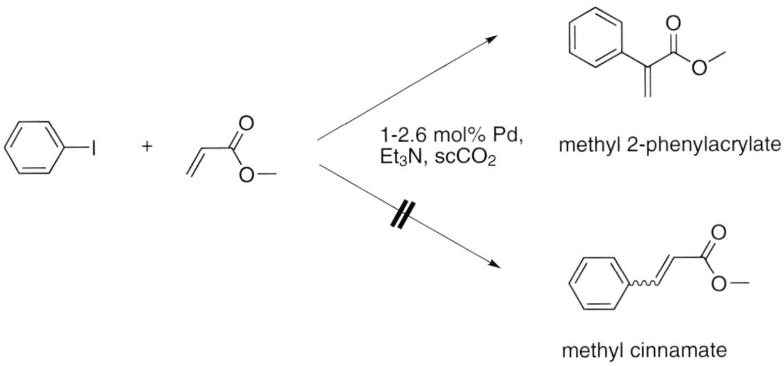

Scheme 15.2 Remarkable selectivity in the Heck reaction of palladium DEN's derived from PPI dendrimers in $scCO_2$, yielding only methyl 2-phenylacrylate; Yeung *et al.* [61].

Heck coupling of butyl acrylate with a series of aryl halides [60]. Milder conditions can be used with respect to more traditional colloidal catalysts (90 compared to 120 °C) and above all 100% isomer selectivity has been achieved in fluorous solvents, where colloidal catalysts usually produce selectivities of 80%.

Furthermore, Crooks *et al.* showed that tuning of reaction conditions, including solvent, dendrimer structure, and dendrimer generation, for reactions taking place in the voids of DENs may yield isomers that are unfavorable under "normal" conditions. A nice example concerning this is the use of palladium-DENs in catalysis in supercritical CO_2 [61]. Iodobenzene can be coupled with methyl acrylate via a Heck coupling reaction to yield exclusively methyl 2-phenylacrylate (Scheme 15.2), while in hydrocarbon/fluorocarbon solvents the same DENs yield only *trans*-cinnamate derivatives [60]. This selectivity is remarkable, since catalysis with standard palladium complexes or colloidal nanoparticles results in *cis*- and/or *trans*-methyl cinnamate in supercritical CO_2.

15.2.3
Miscellaneous

A bimetallic platinum–palladium dendritic catalyst (a representative of class b in Figure 15.1) was published in 1996 by Liu and Puddephatt (Figure 15.9) [62]. A divergent route to organo-platinum or -palladium dendrimers was described. Here, mono- and tris(2,2′-bipyridine)-containing molecules underwent metallation after treatment with a platinum precursor, and subsequent oxidative addition of the platinum complexes with a focal benzylic bromide dendron decorated with two 2,2′-bipyridine end groups led to a dendrimer with platinum(IV) branching points. After dendrimer growth, the free bipyridine ligands in the periphery were platinated or palladated (Figure 15.9 shows a bimetallic complex). The latter step was performed with the precursor $[Pd_2Me_4(m\text{-pyridazine})_2]$. Higher dendrimer generations could not be synthesized due to solubility problems.

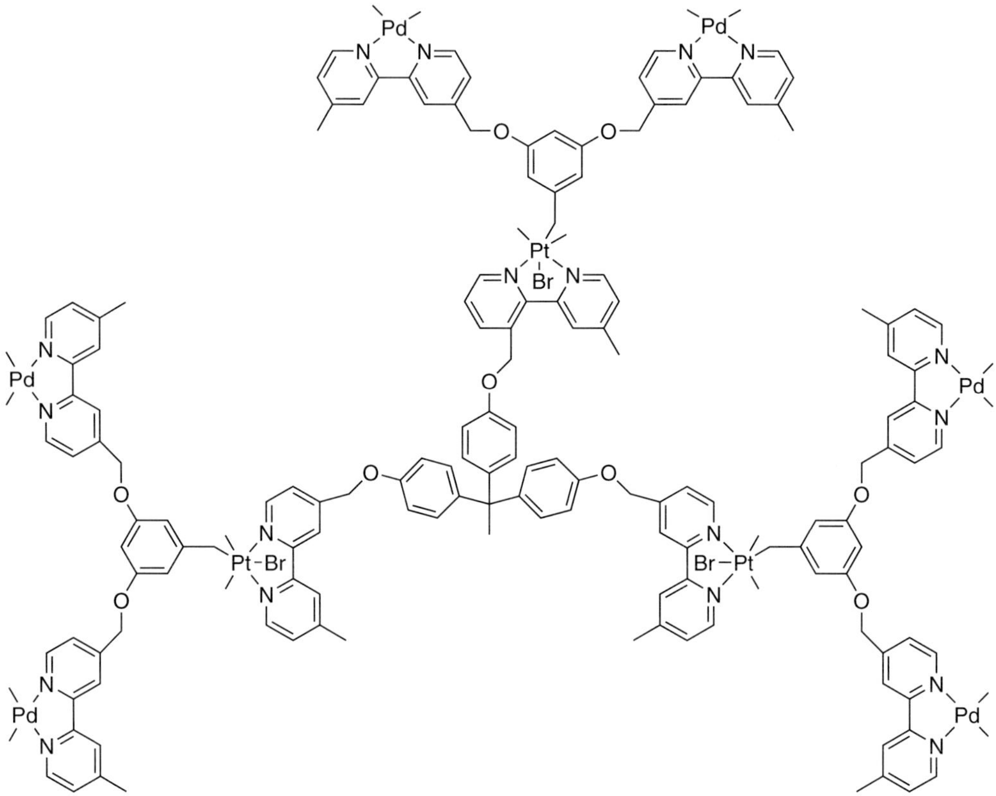

Figure 15.9 Puddephatt's double-metallated platinum–palladium dendrimer [62].

Dendrimers with repeating SNS-pincer (see below) units have been synthesized by Chessa *et al.* [63] A Mitsunobu reaction was used to couple unsymmetrical pyridylthioether monomers to dendritic structures by a convergent route. In the last step, complexation of each binding site with $Pd_2(dba)_3 \cdot CHCl_3$ led to pallada-dendrimers as depicted in Figure 15.10.

Another strategy for the synthesis of bimetallic dendrimers was developed by Angurell [64]. Ru(*p*-cymene)Cl_2 was attached to the branches of a diphenylphosphino-terminated dendrimer. The key step in this synthesis involves the selective binding of the bifunctional 4-(diphenylphosphine)pyridinyl ligand via the N-donor atom to the ruthenium atoms (Scheme 15.3). Metallating the resulting species with $[PdCl(\eta^3\text{-}2\text{-}MeC_3H_4)]_2$ renders Ru/Pd dendrimers. Besides palladium, also gold and rhodium were incorporated in this way.

Palladium complexes have been used as core functionality in a few cases (Figure 15.1c). Vinogradov synthesized a series of dendritic polyglutamic palladium-porphyrins via a divergent growth approach [65]. A widespread application of palladium(II)-porphyrin complexes is as phosphorescent indicators for oxygen measurements in various systems, including medical devices. These indicator

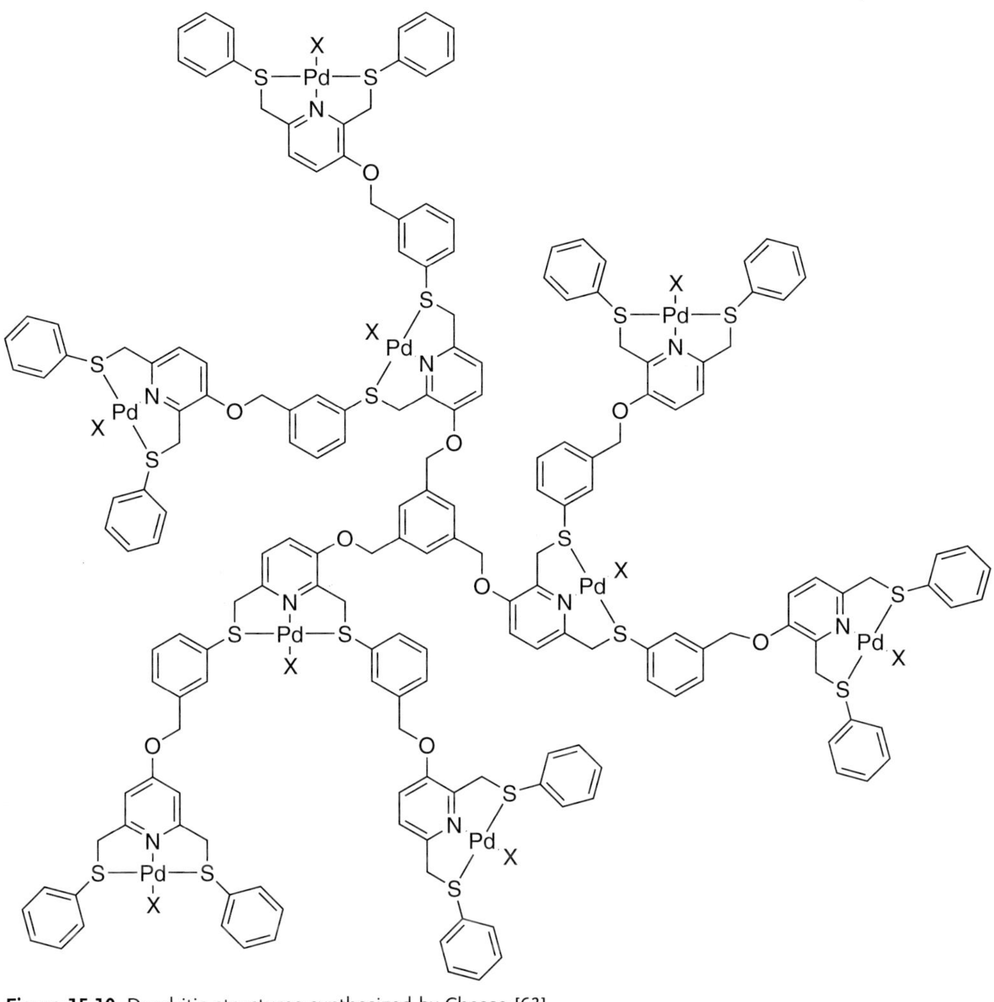

Figure 15.10 Dendritic structures synthesized by Chessa [63] with X = NC—CH=CH—CN (idealized structure).

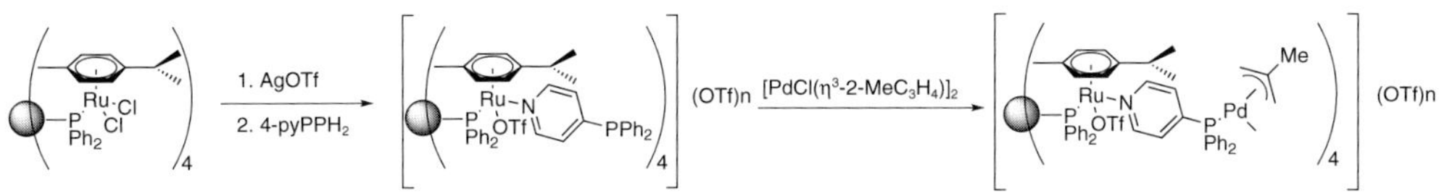

Scheme 15.3 Synthesis of Ru/Pd-containing dendrimers by Angurell [64].

molecules generally consist of the actual phosphor (here the metalloporphyrin) and a protective layer designed to provide biological compatibility of the molecule. Several properties of dendrimers (e.g. the globular shape and narrow molecular weight distribution) make this class of molecules interesting candidates to act as

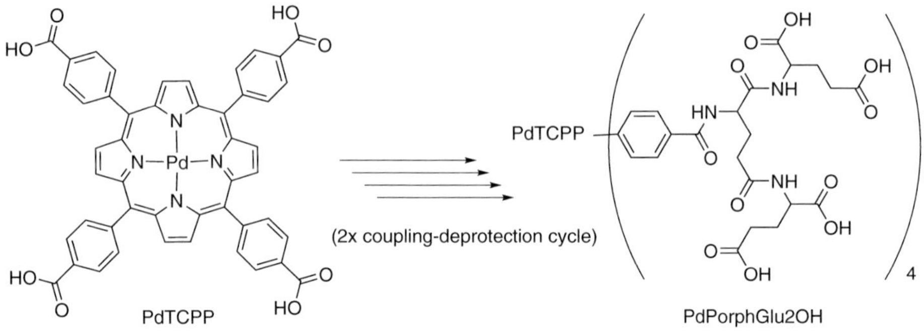

Scheme 15.4 Dendrimer growth out of PdTCPP in a multistep synthesis.

protective layer. Variation of the dendrimer size allows the fine-tuning of the oxygen quenching properties of the phosphorescent indicator.

PdTCPP (Pd-*meso*-tetra-*para*-carboxyphenylporphyrin) has been used as the core element in the dendrimer synthesis. Through an amide bond formation reaction with protected glutamic acid and subsequent basic deprotection, new carboxylic acid peripheral groups are formed. In this way the dendrimer is grown in a stepwise manner (Scheme 15.4).

All these compounds phosphoresce strongly with $\lambda_{max} = 690\,nm$, and this phosphorescence can be quenched by addition of O_2. The quenching constants in DMF are almost unaffected by the dendrimer generation, while in water the K_q values decrease significantly with increasing dendrimer generation. Possibly, in water the conformation of the branches has changed, leading to either open or compact systems. Thereby, the oxygen diffusion barrier will be affected, and so will the kinetics for oxygen diffusion towards the phosphor.

In 2002 Vinogradov's group reported a comparative study involving variation of the peripheral dendrimers [66]. Fréchet- and Newkome-type dendrimers were synthesized and compared to the earlier mentioned polyglutamate dendrimers. This investigation shows that the composition of the dendritic matrix has a major influence on the encapsulation properties of dendrimers.

15.3
Palladacyclic Pincers on Dendrimers and Star-Shaped Molecules

15.3.1
The ECE-Pincer Complex: An Introduction

All palladacycles attached to dendrimers and star-shaped molecules reported in the literature so far are so-called ECE-pincer palladium complexes, or highly related EC-half-pincer palladium compounds. The monoanionic pincer ligand, with general formula $[2,6\text{-}(ECH_2)_2C_6H_3]^-$ (ECE, where E is a group like NMe_2, PPh_2 or SPh, containing a neutral two-electron donor atom) and its metal complexes

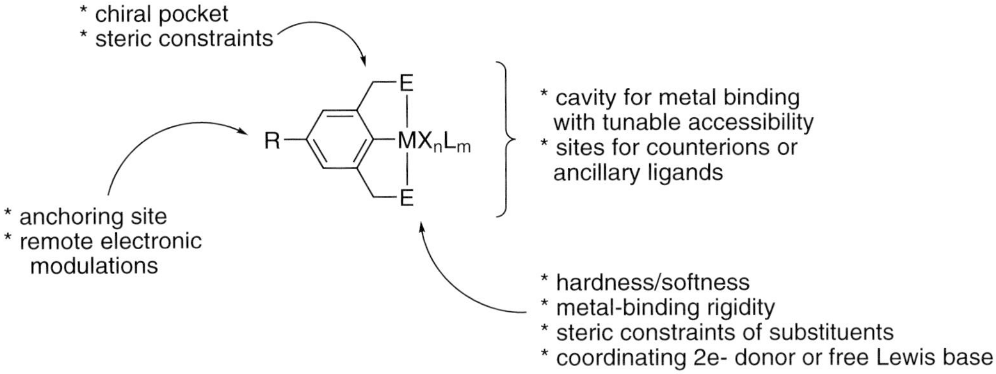

Figure 15.11 Common structure of ECE-pincer metal complexes, with potential modification sites [67]. X: halogen, L: ligand, E: neutral two-electron donor atom containing moiety, like NMe_2, PPh_2, SPh, R: substituents, usually an aliphatic, aromatic or dendritic structure.

with the common structure as shown in Figure 15.11 have been the subject of much research ever since their discovery in the late 1970s [68–71]. Probably the most interesting feature of pincer complexes is the presence of a stable σ carbon–metal bond combined with a high tunability of the coordination sphere around the metal. By varying the two-electron donor atoms E, the coordination sphere as well as the electronic properties of the metal site can be changed, while variation of the substituents of E and the para-substituent R modifies the steric and electronic properties of the metal site. Besides this, R is often used as an anchoring site for many types of (macro)molecules, including dendrimers. Furthermore, chirality can be introduced into the system by adapting the benzylic position of the pincer arms and by the use of stereogenic E groups [67].

To immobilize pincer complexes on dendrimers and star-shaped molecules the substituent R is used. For such dendritic pincer constructs, NCN [72], PCP [73] and SCS [74] pincer metal moieties have exclusively been used so far. In most of these constructs the palladium centers are introduced after connecting the ECE-pincer ligand to the dendrimer, because the resulting palladium metal complexes often do not withstand the reaction conditions used in previous steps. A disadvantage of this method is that a full dendrimer loading cannot always be achieved. Depending on the donor atom, different palladation routes have been developed.

In many cases, ECE-pincer arene ligand palladation can be achieved via an oxidative addition reaction on the pre-ligand, using either an activated C–halogen bond or otherwise a direct C–H bond activation step. In using these palladation procedures, selectivity may be an issue. Direct palladation with $[Pd(MeCN)_4](BF_4)_2$ at elevated temperatures in acetonitrile is the most often used route for synthesizing SCS- [74, 75] and PCP-pincer [76, 77] complexes from the corresponding arene ligands. Subsequent addition of a salt like LiCl or NaBr furnishes the halogenated complexes. In addition, $Pd(TFA)_2$ (TFA = trifluoroacetic acid) is widely used to synthesize PCP-pincer palladium complexes [78]. For NCN-pincer arene ligands, direct palladation cannot be performed selectively, so oxidative addition on NC(X)N (X = halogen) arene ligands with a $Pd(dba)_2$ or $Pd_2(dba)_3 \cdot CHCl_3$ precursor complex in benzene is favored [79, 80]. Starting from a non-halogenated NC(H)N-pincer arene ligand, the selective lithiation with *n*-BuLi at C_{ipso}, followed by a transmetallation of the corresponding NCN-lithium intermediate with a palladium complex like $PdCl_2(cod)$ (cod = 1,5-cyclooctadiene) also leads to the NCN-palladium complexes [81].

15.3.2 Pincer-Palladium Complexes on Star-Shaped Molecules

The group of Van Koten has reported star-shaped molecules containing multiple NCN-palladium pincers for which the name cartwheel-complexes was adopted. These structures consist of a central benzene ring, with either three or six pincer moieties attached (Figure 15.12). The aromatic backbone of these compounds ensures a high rigidity, which is expected to be important for the use of the complexes as catalysts in a membrane reactor, since in general a high rigidity enhances the retention of compounds [79].

These compounds have been tested as Lewis acidic catalysts in the double Michael reaction between ethyl α-cyanoacetate and methyl vinyl ketone[1] and were

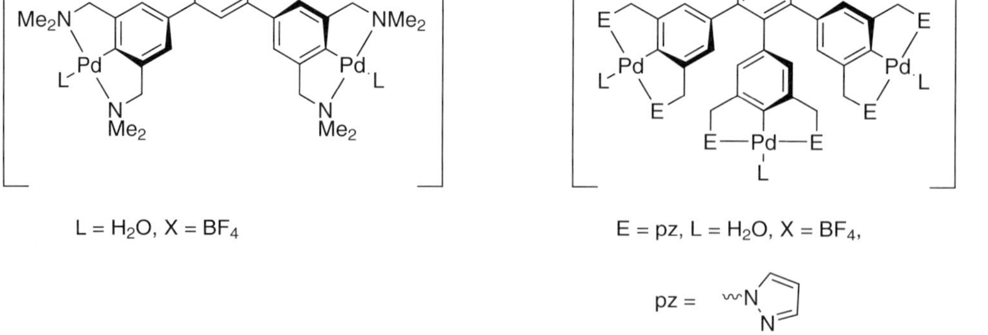

Figure 15.12 Star-shaped tris- and hexakis (NCN—Pd) "cartwheel" complexes synthesized by Dijkstra *et al.* [79].

1) Reaction conditions: 0.5 mol.% Pd, 10 mol.% *i*-Pr_2NEt, CH_2Cl_2, rt.

compared with a series of monomeric pincer complexes. In this monomeric series, the influence of the donor substituent E on the catalytic activity was investigated. For this particular reaction, NCN-pincer complexes were superior to PCP- and SCS-pincer complexes. Within the NCN-pincer complexes, further research has been performed by using dimethylamine, pyrazol-1-yl, and 3,5-dimethylpyrazol-1-yl donor substituents. All three complexes showed good catalytic performance.

Pyrazole, SPh and PPh_2 donor substituent-containing cartwheel palladium complexes have been synthesized successfully in a seven-step synthesis starting from 3,5-dimethylaniline [75, 79]. Unfortunately, direct palladation of the hexakis pincer-ligand with dimethylamine donor groups did not lead to the completely metallated product. Some of these star-shaped complexes have been used for catalysis.

Although the hexakis(aqua-pincer) complex with E = pyrazole (pz), L = H_2O and X = BF_4 (Figure 15.12) does not dissolve in dichloromethane it does show catalytic activity in this solvent (70% conversion after 22 h). To improve its solubility and thus the catalytic performance, the tris(pincer) complexes have also been synthesized. The tris(aqua-pincer) complex is soluble under the reaction conditions used, and shows a catalytic activity per palladium atom that is almost equal to the corresponding monopincer analog ($k_{obs} = 2.34 \times 10^{-4}\,s^{-1}$ as compared to $2.79 \times 10^{-4}\,s^{-1}$) [79].

These shape-persistent, multimetallic materials were also used in olefin metathesis (RCM) for the template-directed synthesis of macrocycles [82]. Diolefin-substituted pyridines were coordinated to each of the six metal sites in both Pd(II) and Pt(II) cartwheel compounds and were subsequently subjected to standard olefin metathesis conditions with use of Grubbs-type Ru-catalysts, aiming for a large macrocycle that would be formed after six olefin metathesis reactions of the twelve olefin arms. It was found that, in particular, disubstituted pyridines bound to a NCN-Pt(II) center gave fast metathesis reactions, yielding the desired large macrocycles. The same reaction with the palladium analog was less fast and less selective, since preliminary tests did not result in the formation of large macrocycles. Instead, isomerized products and a minor amount of intrapyridyl metathesis products were observed.

Further research on this type of compounds has provided a dodecakis(pincer) compound. This compound comprises an aromatic core to connect twelve (dimethylamine) NCN-pincer palladium complexes via ether bonds (Figure 15.13). Metallated monomers bearing a TBDMS protecting group on the para-position were deprotected and *in situ* coupled to a star-shaped dodecakis-bromide core molecule. Through this convergent approach, a complete palladium loading is ensured. This 12-fold coupling reaction was performed under mild conditions using $(NBu_4)F$, K_2CO_3 and 18-crown-6 as reagents, which do not damage the palladium centers at all [83].

A dendrimer, consisting of a tetraarylsilane core and eight NCN-pincer palladium complexes attached via ether bonds in the same way as in the dodecakis(pincer) was synthesized in the same manner [32, 72, 83].

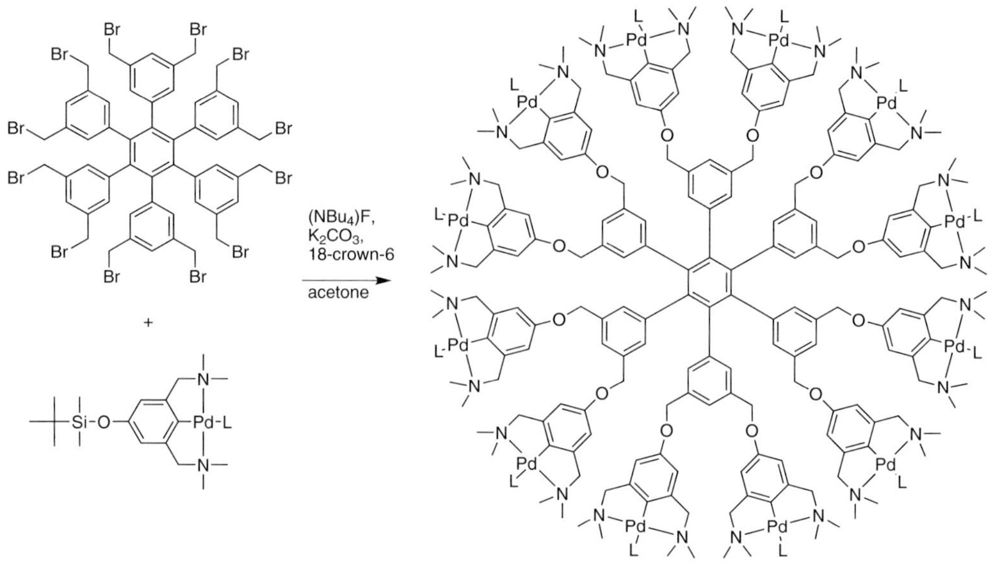

Figure 15.13 Synthetic route to a star-shaped dodecakis(NCN—Pd) cartwheel complex as synthesized by Dijkstra *et al.* [72].

These compounds were also tested in the double Michael reaction. The tetraarylsilane dendrimer shows a comparable activity per palladium center with the earlier mentioned monomeric and trimeric compounds ($k_{obs} = 2.1 \times 10^{-4}$ versus $2.3 \times 10^{-4}\ s^{-1}$). The dodecakis(pincer) compound, however, shows an almost threefold increase in catalytic activity per palladium(II) center ($k_{obs} = 8.1 \times 10^{-4}\ s^{-1}$): clearly a positive dendritic effect is observed. Probably, the closer proximity of the palladium-sites and thus possible involvement of cooperative effects between different palladium(II) centers is the reason for this increase. Another possible explanation is that aggregate formation could create highly polar micro-environments, which enhance the catalytic performance [32].

The high rigidity of these star-shaped molecules makes them excellent candidates for performing nanofiltration experiments. The retention of these compounds was tested by using the deep orange color of the corresponding neutral NCN-pincer Pt(II)-SO_2 complexes to monitor them with UV/Vis spectroscopy [67]. These five-coordinate adducts were obtained by reversible binding of SO_2 to NCN-platinum complexes and were used as an extremely sensitive gas sensor that is not disturbed by the presence of other gases [67]. By using the commercially available MPF-60 membrane with a molecular weight cut-off (MWCO) of 400 Da, a G1-carbosilane dendrimer bearing twelve Pt(NCN)(SO_2) pincer moieties shows a retention of 0.995. The dodecakis(pincer) cartwheel compound, however, shows a retention >0.999 when using the MPF-60 membrane and a retention of 0.999 is determined when an MPF-50 membrane (MWCO = 700 Da) is used [83].

The dodecakis(pincer) palladium complex was then tested for the double Michael addition in a continuous flow membrane reactor using an MPF-50 membrane.[2] The catalyst was found to be stable under the continuous reaction conditions as a constant activity was obtained over prolonged reaction times (26 h, 65 exchanged reactor volumes). The total turnover number of the catalyst increased by almost a factor 40 compared to the batch experiment (80 versus >3000 mol mol-Pd^{-1} for the continuous process). Finally the retention was determined at 0.995, thereby making these compounds suitable for the use in a continuous flow membrane reactor [72].

The group of Van Koten in collaboration with Beletskaya's group also synthesized rigid star-shaped molecules functionalized with PCP-pincer palladium complexes. These are very similar to the earlier mentioned hexakis(NCN-pincer Pd) compounds, yet in these compounds an acetylene bridge is used to connect the core and the peripheral arene groups, which further improves the rigidity of the compounds. Metallation of the tris-ligand proceeded quantitatively to yield the corresponding trinuclear palladium complexes, whereas metallation of the hexa-ligand did not fully proceed. The tris(pincer) complex was tested as pre-catalyst in the Heck reaction of iodobenzene and ethyl acrylate and compared to the monomeric species. In this reaction the tris(pincer) compound shows a lower activity and TON than the monomeric species, which is presumably due to steric factors [73, 77].

SCS-pincer-porphyrin hybrids (Figure 15.14) represent another type of star-shaped molecules [84]. These multi-ligand site compounds contain a porphyrin core and four pincer arms, in which each of the two different sites can be selectively metallated without metallating the other site.

A series of various *meso*-tetrakis(SCS-pincer PdCl)-(metallo)-porphyrin hybrids [with as central metal M = 2H, Mg(II), Mn(III)Cl, Ni(II)] were used as pre-catalyst for the Heck reaction of styrene with iodobenzene in DMF. Although none of these complexes reach the initial catalytic activity of the monomeric SCS-pincer palladium chloride complex, the catalytic performance of the reaction mixtures is strongly dependent on the metal present in the porphyrin center. The magnesium chelate turned out to have the highest initial activity of the bimetallic complexes, followed by the nickel-, the free base and the manganese complexes. The order of activity follows the order of the electron-donating ability of the (metallo)porphyrin, that is the more electron-donating the (metallo)porphyrin, the greater the initial catalytic activity. This research is a proof of principle that demonstrates that catalytic rate can be influenced by remote tuning of the catalyst by changing a distant modulating substituent in one step (in this case metallation of another "remote" metal complex) [85].

In particular, the fact that only one reaction step performed under mild conditions can lead to remote control makes this discovery even more intriguing.

2) Reaction conditions: 4.2 μmol Pd, 2 M methyl vinyl ketone, 0.064 M i-Pr_2NEt, 0.64 M ethyl α-cyanoacetate, 0.4 M n-decane, CH_2Cl_2, flow rate = 30 mL h^{-1}, residence time τ = 0.4 h, 23 °C, 20 bar.

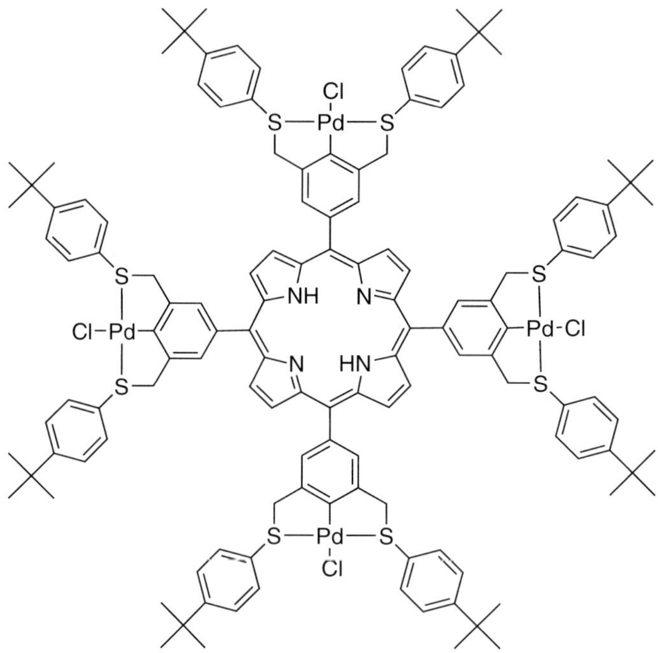

Figure 15.14 SCS-pincer palladium porphyrin hybrids synthesized by Suijkerbuijk *et al.* [84].

15.3.3
Non-covalently Bound Dendrimer–Pincer Palladium Complexes: Dendritic Catalysts

Non-covalent metallodendritic assemblies containing NCN-pincer palladium complexes have been reported by Van de Coevering *et al.* [86]. In this study, ionic core–shell dendrimers were used as a dendritic host molecule. This core–shell molecule consists of a tetraarylsilane core, eight quaternary ammonium groups and, accordingly, eight bromide counterions, and a dendritic shell of different generations of Fréchet-type dendrons. Three generations dendritic hosts have been used. Incorporation of (functional) guest molecules in these dendrimers can be achieved by means of anion exchange. Exchange of the bromide anions with eight anionic, sulfate-terminated pincer complexes yields the octakis-NCN-pincer Pd dendrimer shown in Figure 15.15 (as a typical example of class a in Figure 15.1). This exchange reaction can be performed by simply adding eight equivalents of the tetrabutylammonium salt of the pincer anionic monomer to the octacationic dendrimer with bromide counterions in dichloromethane. Subsequently, the metallodendritic assembly was isolated by aqueous washing steps to remove tetrabutylammonium bromide, followed by passive dialysis.

The formation and stoichiometry of this assembly and the strength of the ionic bonds were investigated by various analytical techniques. NOESY NMR measurements showed that the sulfato groups of the catalysts are located close to the octacationic core, and nano-ESI mass spectroscopy proved the stoichiometry of the

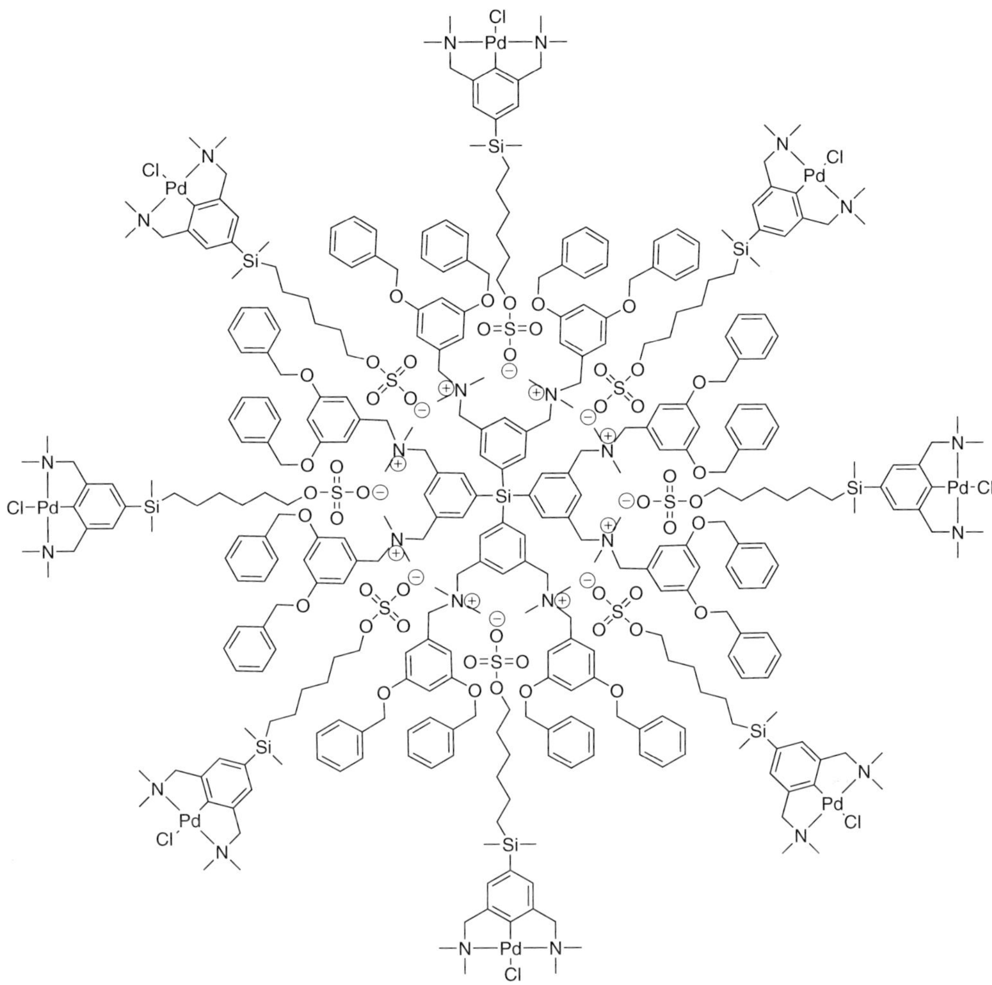

Figure 15.15 G1 Non-covalent metallodendritic assembly by Van de Coevering *et al.* [86].

non-covalent metallodendritic assemblies. Pulse gradient spin-echo (PGSE) NMR diffusion measurements were also performed on these dendritic assemblies to determine the diffusion coefficients for the octacation and the anions. Octacationic dendrimers of generation 1–3 were compared to gain insight into the freedom of movement of the anion in the dendritic shell of the assembly. For the first and second-generation dendritic assemblies, a small, but reproducible, difference (1.5% for G1, 3.2% for G2) in the diffusion coefficients between anion and cation was found, with the anion diffusing slightly faster than the cation. This might be due to an equilibrium in which small quantities of the separated cations and anions contribute. For G3 this difference in diffusion coefficients is not observed. Apparently, the anion is more tightly trapped in the dendrimer when the dendritic shell is larger.

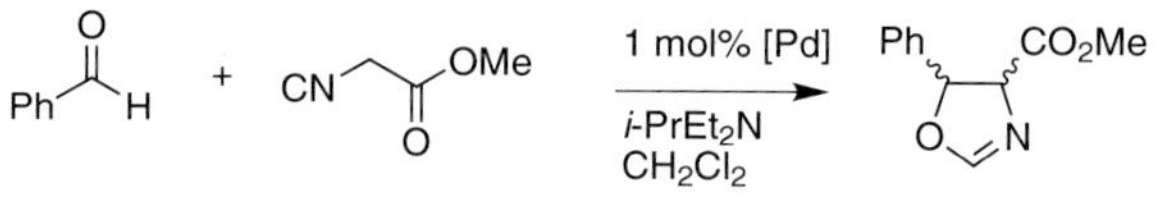

Scheme 15.5 Aldol condensation reaction between benzaldehyde and methyl isocyanoacetate.

The catalytic performance of the dendrimer-supported and unsupported NCN-pincer Pd-complexes was investigated by performing the aldol condensation reaction between methyl isocyanoacetate and benzaldehyde (Scheme 15.5) [81, 87]. It was found that the metallodendritic NCN-pincer palladium chloride assemblies show turnover frequencies (TOF), conversion and product distributions per Pd center that are comparable to their monomeric anionic analog with tetrabutylammonium.

In this study, two strategies were used to vary the structure of the dendrimer from exposed catalytic sites at the periphery towards encapsulated catalytic sites at the inside of the dendrimer. The first strategy was to alter the thickness of the dendritic shell, which was achieved by changing the generation of the dendrimer or by the attachment of dodecyl groups at the periphery. The latter method also heavily influences the polarity of the dendrimers. The other strategy was to vary the length of the tether between the pincer and the sulfato group. Using these strategies, only small differences in catalytic performance were found. Apparently, the structure of the dendritic shell in the assemblies is relatively open, which keeps the catalytic Pd(II)-sites reasonably well accessible for the reactants. The dendrimer decorated with dodecyl groups at the periphery, however, displays a significantly lower activity, pointing to a more dense structure, in which the palladium center is less accessible for reactants.

These non-covalent, metallodendritic assemblies offer an interesting addition to the palette of dendritic catalysts. Remarkably, they do not show a decrease of the catalytic activity upon incorporation of the homogeneous catalyst on the dendritic hosts, which is possibly due to the flexibility of the resulting metallodendritic assemblies. The modularity in the synthesis of the assemblies, furthermore, opens the way for the incorporation of many other homogeneous catalysts. The non-covalent nature of the assemblies also allows the recycling of either or both the dendritic support and the anion-tethered catalyst as well [88].

Non-covalent catalyst anchoring to functionalized dendrimers and other polymeric supports has been reviewed by Ribaudo *et al.* [89] and Van de Coevering *et al.* [90]

15.3.4
Non-covalently Bound Dendrimer–Pincer Palladium Complexes: Self-Assembled Dendrimers

The group of Reinhoudt has performed extensive research on the incorporation of SCS-pincer palladium complexes in dendrimers and has published a range of

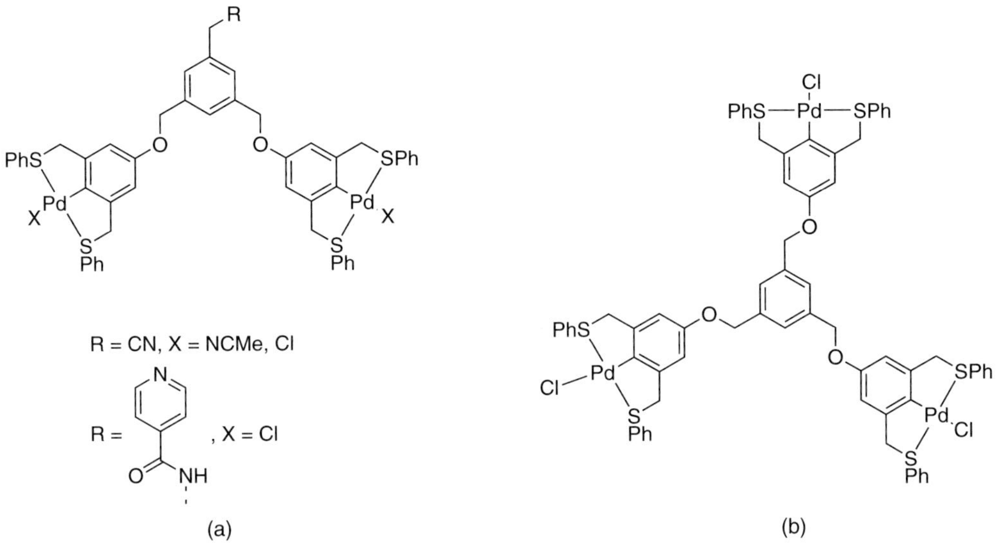

Figure 15.16 A tweezer SCS-pincer palladium dendron [91] (a) and a tris-SCS-pincer palladium complex dendrimer core moiety (b) [92].

articles starting from 1995. Unlike the non-covalently bound dendrimers shown in the previous section, these dendrimers have metal sites at every branching point (class b in Figure 15.1). Numerous self-assembled dendritic structures have been synthesized, all of which are based on coordinative interactions and hydrogen bonding between different building blocks.

The first report by Huck *et al.* [74] used a G0 tweezer dendron (Figure 15.16 with R = CN and X = NCMe), which upon evaporation of the acetonitrile solvent formed self-assembled spheres by coordination of the cyano group of one building block to the palladium center of another. In this manner non-covalent, hyperbranched polymers were obtained. Different size measurement techniques such as Quasi-Elastic Light Scattering (QELS), Atomic Force Microscopy (AFM) and Grazing-Angle FT-IR spectroscopy have been performed on these spheres. All techniques showed objects with a diameter of about 200 nm with a relative narrow distribution: 95% of the diameters were found within $2\sigma_d$ of the mean value.

This protocol was adapted by using the dendron depicted in Figure 15.16 with R = CN and X = Cl in combination with a tris(pincer) core (Figure 15.16) [92]. In this way, G0 dendrimers can be formed in a controlled manner: the chloride ions of the core molecule are replaced by non-coordinating BF_4^- ions by reaction with silver tetrafluoroborate. Addition of three equivalents of the dendron shown in Figure 15.16 yields the G0 dendrimer. Halogen removal and subsequent coupling enables further growth of the self-assembled organometallic dendrimer. This process can be repeated several times to obtain non-covalent dendrimers up to the fifth generation and proceeds in a very controlled manner due to the strongly

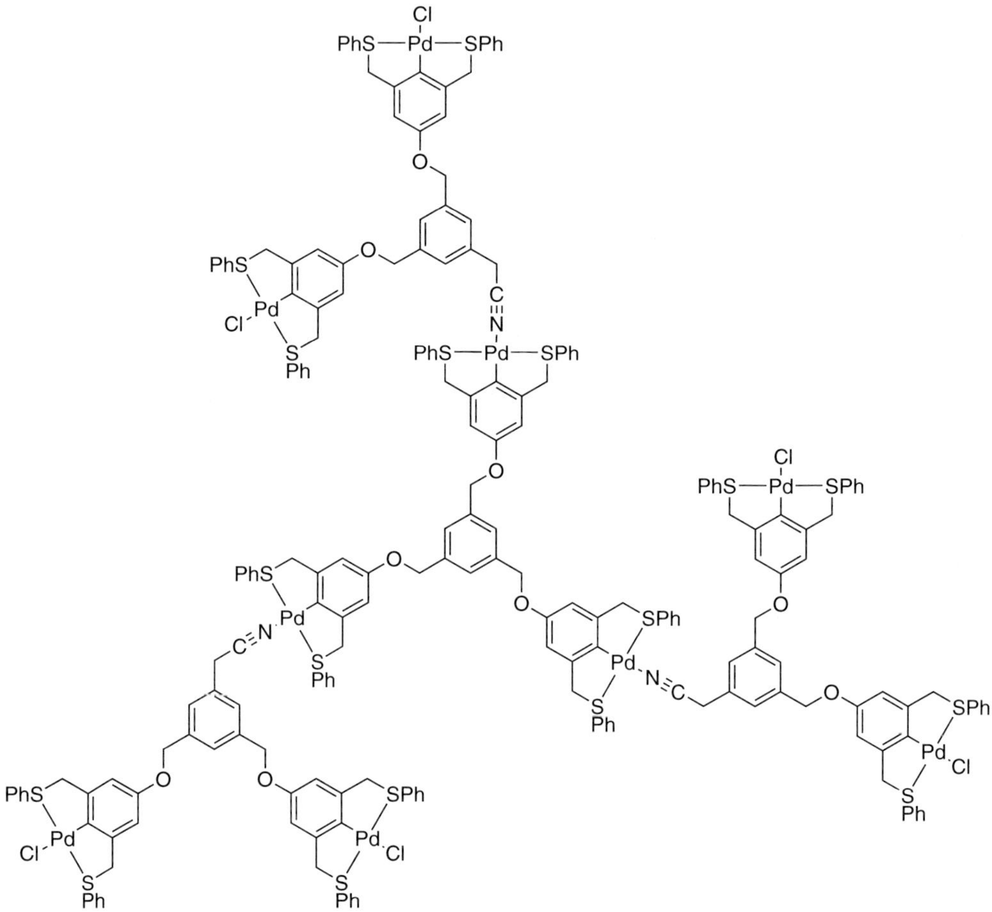

Figure 15.17 Non-covalently bound G1-pincer dendrimer by Huck *et al.* [92].

bound chloride ion, which temporarily protects the palladium center. The G1 resulting dendrimer is shown in Figure 15.17.

To gain insight into the diameter of these nanoscale molecules, Tapping Mode Atom Force Microscopy (TM-AFM) measurements have been performed. These measurements illustrated that the G5 dendrimers have diameters of approximately 15 nm. Similar structures containing star-shaped dendrimer, consisting of a benzyl ether core and three PCP-pincer palladium complexes, were later published by Huck as well [76].

For the controlled assembly of metallodendrimers in either a convergent or a divergent way, besides a cyano-based building block, also a pyridine-based building block was synthesized [Figure 15.16, R = isonicotinamide (=$NH_2C(O)$=C_5H_4N)] [93]. With a combination of pyridine- and cyano-based building blocks, dendrons and metallodendrimer assemblies up to the third generation were obtained. Stable dendrons are formed in a divergent way by using the relatively strongly coordinating pyridine-based ligands; however, for convergent assemblies to the trivalent core molecule (Figure 15.16b) dendrons bearing a cyano-based ligand have to be used. By doing this (in a convergent strategy), the coordinative strength of the ligands bound to the palladium(II) pincers from the periphery to the core decreases. In this way, scrambling of various building blocks is avoided.

Owing to the presence of SCS-pincer palladium complexes at the periphery of these polycationic metallodendrimers, their solubility in organic solvents is moderate. With the introduction of a hydrophobic dendritic layer this solubility increases [91, 94]. Again, ligands that are stronger than the interior ligands are required for a convergent synthetic route. Dendritic phosphine ligands (Fréchet-type dendrons were used) met these requirements and were successfully coordinated to the peripheral SCS-pincer palladium(II)-complexes. In this case, in every new generation, another class of functional group (i.e. cyano-, pyridine- and phosphine ligands) has been used as the coordinating assembly motif. For these assemblies, indeed, the solubility in apolar solvents increases accordingly.

This self-assembly strategy also allowed the introduction of redox or photoactive fragments in both the core and periphery, provided that those moieties can be functionalized with palladium pincer complexes or suitable ligands [95]. For example, Huck used porphyrins in the non-covalent assembly of dendrimers, functioning both as core molecules and as peripheral groups [95]. Starting from a porphyrin core moiety that contains four SCS-pincer palladium complexes at its meso-positions, dendrimers of generation 0 and 1 were grown by BF_4^- anion exchange and subsequent addition of the dendritic SCS pincer palladium-based building blocks of Figure 15.16a. Then porphyrin end groups that coordinate via a pyridine nitrogen to the palladium in the pincer (as the peripheral porphyrin groups in Figure 15.18) were introduced. This process has yielded dendrimers with a porphyrin core and either four ($G0Por_4$, Figure 15.18) or eight ($G1Por_8$) porphyrin end groups. These dendrimers may be interesting for the construction of donor–acceptor systems in which energy transfer from the core porphyrin to the periphery porphyrins and vice versa can take place.

In a similar way, the tris(pincer) core (Figure 15.16b) has been used as core molecule and was grown to G1 and G2 dendrimers with the building blocks depicted in Figure 15.16. These dendrimers were then functionalized with porphyrins in the same manner as described above to yield $G0Por_3$, $G1Por_6$ and $G2Por_{12}$.

For visualization and, ultimately, manipulation of individual nanosized dendritic molecules on a surface, G0, G1 and G2 pyridine-coordinated dendrimers have been attached to either a sulfide monolayer coated gold surface [96, 97] or gold nanoclusters [98]. Owing to their regular structure, these self-assembled monolayers (SAMs) are useful starting materials for the development of nanometer-scale devices, which require controllable positioning of functional nanosized molecules.

In a first attempt [96], it has been shown that metallodendrimers functionalized with long sulfide side chains can be inserted as individual particles in an

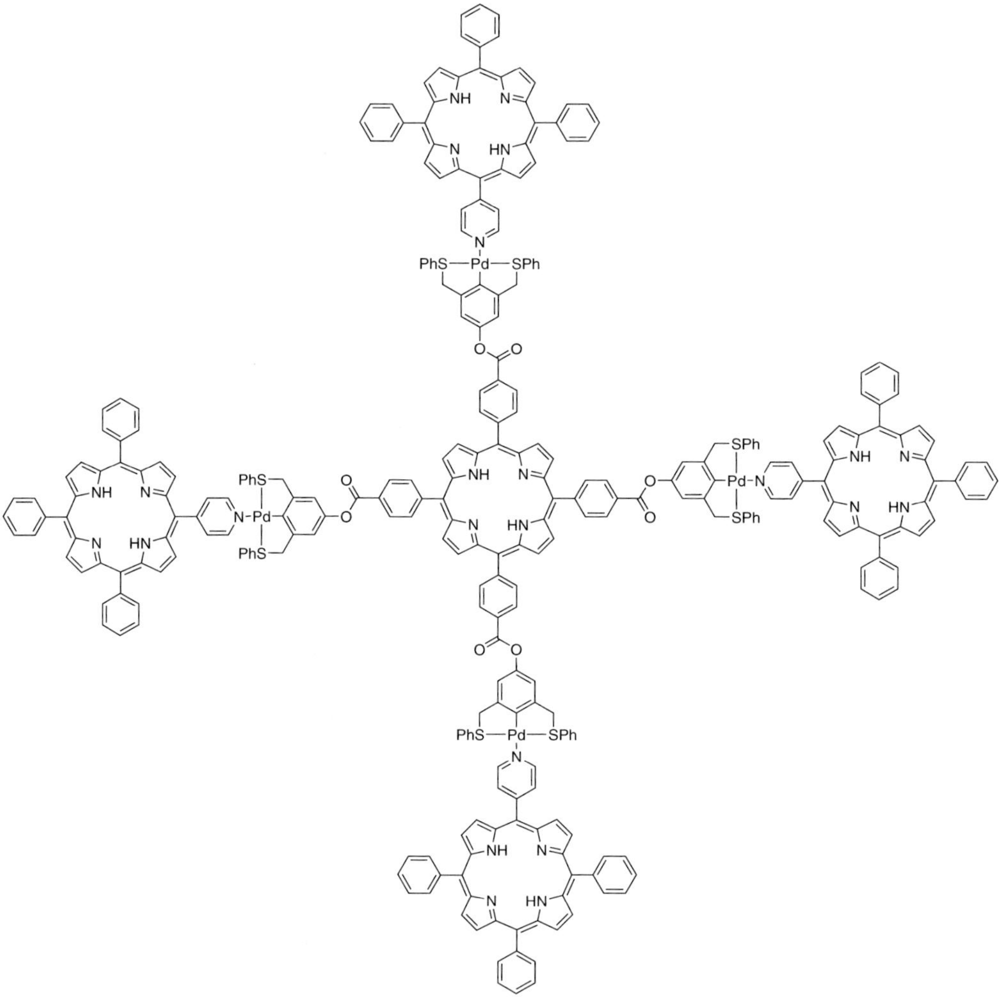

Figure 15.18 The multiporphyrin dendrimer G0Por$_4$ by Huck *et al.* [95].

alkanethiol monolayer. Moreover, the number of isolated dendrimers could be controlled, since the amount of desorption of the alkanethiol chains from the SAM and the subsequent adsorption of the derived metallodendrimers at these deficiencies in the surface are time-dependent processes.

Van Manen *et al.* synthesized the same sort of nanometer-scale dendritic structures [97]. A dendritic wedge containing peripheral pyridines and a focal dialkyl sulfide chain was synthesized. The peripheral pyridines were coordinated to a second-generation Fréchet-type dendron functionalized with a focal SCS-pincer moiety. The focal dialkyl sulfide tail has been used to anchor these nanomolecules to the gold surface. Highly complex assemblies have been made in this way (Figure 15.19). TM AFM height images clearly showed an increase in size when the second

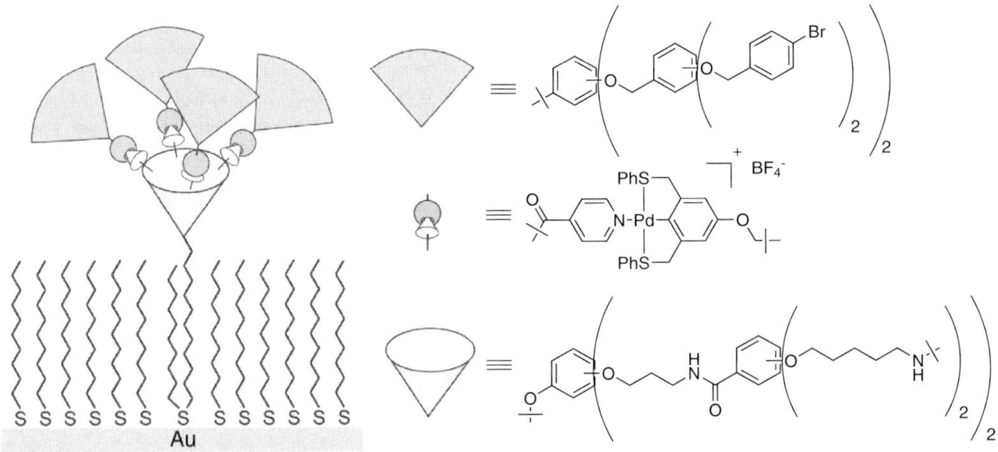

Figure 15.19 Schematic representation of the non-covalent surface-confined dendrimers synthesized by Van Manen *et al.* [97].

dendritic moiety with the focal SCS-pincer complex (gray cone in Figure 15.19) was coordinated to the dendron-containing SAM.

Friggeri monitored the growth of surface-confined, nanometer-sized, coordinative SCS-pincer dendrimers on gold nanoparticles [98]. The embedding of the isolated dendritic pincer palladium complexes bearing a long sulfide chain adsorbed onto the decanethiol SAMs on gold was carried out in dichloromethane. Dendrimer growth was performed in a similar way as in the examples shown above. TM-AFM measurements showed that individual molecules were obtained with a height and diameter of 4.3 ± 0.2 nm and 15.3 ± 4 nm, respectively.

A very interesting combination of coordination chemistry and hydrogen bonding was reported by Huck *et al.*, using a barbituric acid residue (Figure 15.20a) as a core moiety [99]. The tweezer-like dendrons of Figure 15.16 were used to build up the dendritic wedges via repetitive deprotection–coupling steps (see above). Upon acquiring the desired dendrimer generation, the melamine derivative is added at low temperature (−30 °C). Three wedges self-assemble by numerous, selective hydrogen bonds around three melamine derivatives. The core moiety of the obtained rosette assembly is shown in Figure 15.20b; its structure was confirmed by the use of low-temperature ^{1}H NMR (−30 to −60 °C) in combination with 2D NOESY and TOCSY NMR.

Furthermore, the group of Reinhoudt has published the synthesis of a non-covalent water-soluble dendrimer containing ligands based on either linear sugars or tetra(ethylene glycol) (Figure 15.21) [100]. These ligands were coordinated via the pyridine or phosphine group to either a tris- or hexakis(pincer) palladium compound with a benzyl ether as core molecule. Figure 15.22 depicts a peripheral fragment of a water-soluble dendrimer based on tetra(ethylene glycol). Three molecules, as shown in this figure, have been coordinated via the focal pyridine moiety to a tris(SCS-pincer palladium complex) (Figure 15.16b) leading to a large,

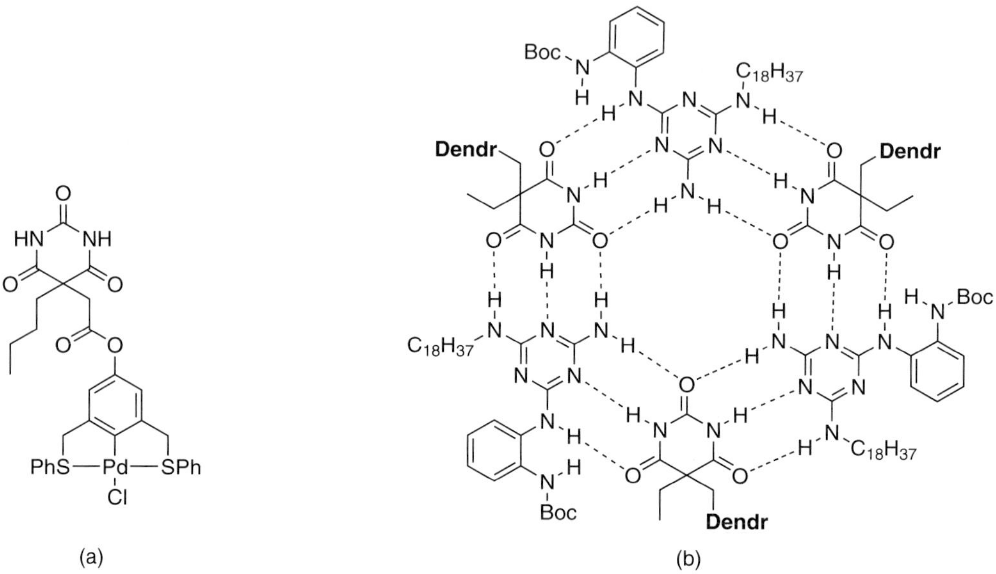

Figure 15.20 Barbituric acid functionalized SCS-pincer palladium complex used for hydrogen bonding-assisted self-assembly (a) and the hydrogen bonding based self-assembled dendrimers (b) by Huck *et al.* [99].

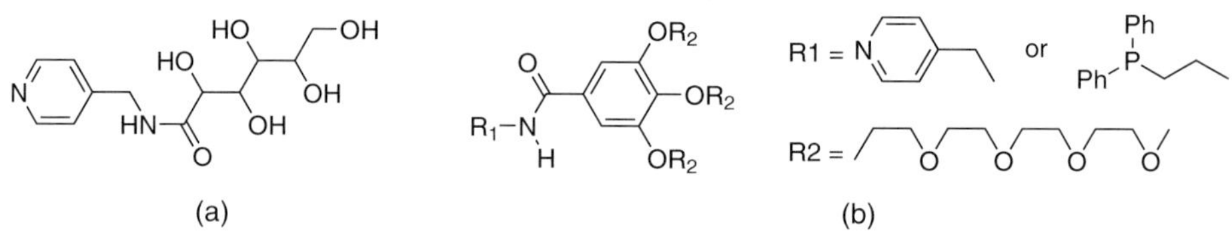

Figure 15.21 Water-soluble ligands based on a linear sugar (a) or tetra(ethylene glycol) (b) [100].

nanoscale-sized metallodendrimer having 18 peripheral tetra(ethylene glycol) groups.

The compounds based on the linear sugar, unexpectedly, turned out to be poorly soluble in water, resulting in an aqueous gel. The solubility of the pyridine-coordinated compounds that were decorated with tetra(ethylene glycol) chains in water was also moderate. However, the compounds containing phosphine-bound tetra(ethylene glycol) moieties were highly water-soluble. Therefore, this ligand was used to synthesize water-soluble G1 metallodendrimers bearing a hydrophilic periphery.

A general, comprehensive review on non-covalent metallodendrimers was published by Van Manen *et al.* in 2001 [101].

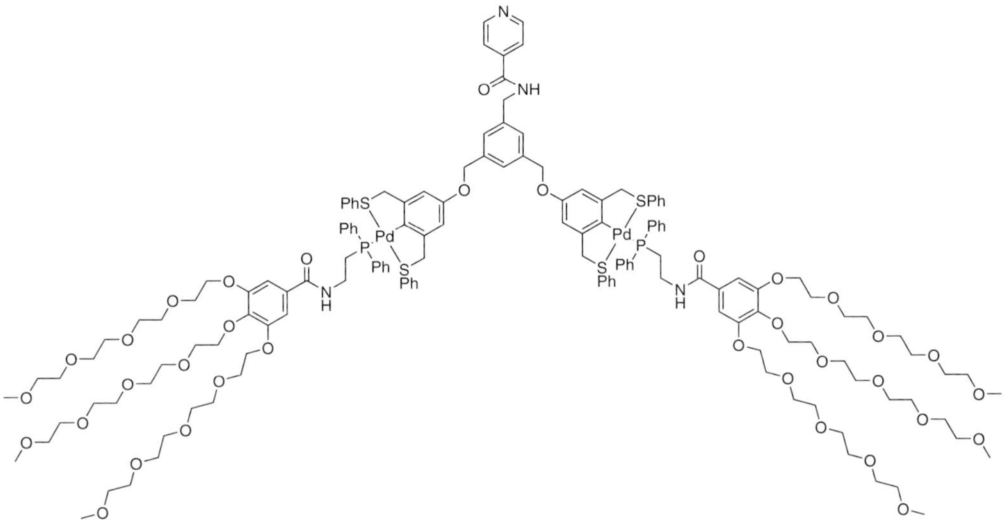

Figure 15.22 A peripheral fragment of a water-soluble wedge based on tetra(ethylene glycol) by Van Manen [100].

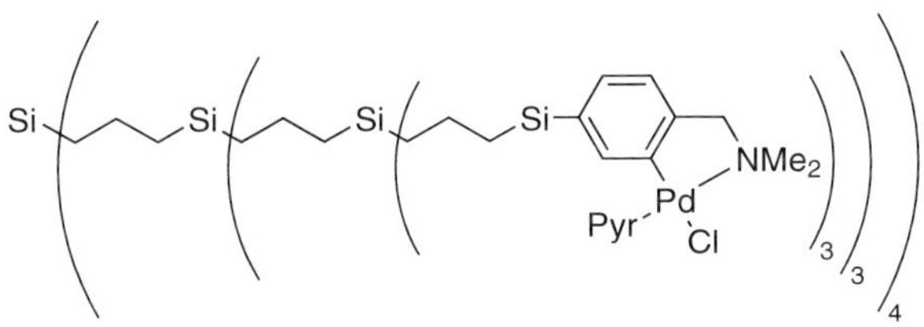

Figure 15.23 NC-half-pincer palladium-functionalized G2-dendrimers by Kleij *et al.* [102].

15.3.5
EC-Half-Pincer Palladium Complexes on Dendrimers

One of the simplest ways of obtaining a palladacyclic compound is the reaction of benzylamines with $Pd(OAc)_2$ to yield so-called "half-pincer" complexes. These complexes have, for example, been used in catalysis [102] and as resolving agents [103]. In a report by Kleij *et al.* NC-half-pincer palladium compounds were coupled to a second-generation carbosilane dendrimer, resulting in a dendritic structure (Figure 15.23). An interesting characteristic of the half-pincer compounds is the possibility of dimerization. In the absence of a suitable Lewis base (like, for example, pyridine), the Pd–Cl group bridges with another Pd–Cl group, forming either cis- or trans chloride bridged dimers. Upon addition of an excess of pyridine these dimers can be broken down to monometallic NC-half-pincer palladium units.

Three generations of these metallodendrimers functionalized with the corresponding cationic aqua complexes (as depicted in Figure 15.23, but with the chloride ion replaced by a water ligand), as well as the G0 and G1 dendrimers containing a shorter two carbon-atom linker per branching unit, were tested for their catalytic performance in the earlier mentioned aldol condensation reaction[3] and compared with their monomeric equivalents. All conversions were very high (96–99%) after 24 h, except for the G1-two carbon linker, which showed a significantly lower conversion of 55%. A slight decrease in total turnover numbers per palladium(II) atom was determined with increasing dendrimer generation. Therefore, it was concluded that the increased bulk around the palladium centers decreases the catalytic activity due to decreased accessibility. The less crowded, lower generation dendrimers are as active per palladium center as the monomeric species and thus might be interesting for application in catalysis in a continuous flow membrane reactor, provided that their retention is high enough [102].

15.3.6
Dendrimers Containing Functional Groups in the Vicinity of Palladacycles

Through the introduction of functional entities near the ECE-pincer metal center on a dendrimer, attempts have been made to influence the catalytic properties of the metal center. To the best of our knowledge, two reports on this subject have been published. These reports represent encapsulated pincer complexes or pincer complexes in the vicinity of functional groups that were expected to promote regio- and stereocontrol.

Rodríguez synthesized macrocyclic carbodiazasilane molecules containing NCN-pincer ligands [80]. In these complexes, the NCN-pincer palladium(II) sites are encapsulated by strategic placement of diphenylsilane moieties. By means of a *para*-hydroxy functional group on the pincer moiety of these encapsulated complexes, the pincers were connected to tricarboxy core molecules to yield new metallodendrimers. The pincer ligands could be palladated by addition of $Pd(dba)_2$, thereby forming *meso*-diastereoisomers in a selective way (Figure 15.24a). With these compounds, possible changes in catalytic properties by secondary interactions were studied. In aldol condensation reactions,[4] the catalytic activity per palladium site of these dendrimers was higher than for the mononuclear system (conversion >99% versus 89%): a small, yet significant positive dendritic effect. Other catalytic properties, like the diastereoselectivity, did not show a significant change.

Slagt prepared a pyrenoxy-based NCN-pincer palladium molecular tweezers that were coupled to inert carbosilane dendrimers [104]. This tweezer consists of three parts: (i) an NCN-pincer palladium complex, (ii) a pyrenoxy unit and, in between, (iii) a xylyl spacer. Crystal structures show a completely flattened conformation of the tweezer in the solid state, due to favorable intramolecular π-stacking interactions. The close proximity of the pyrenoxy unit and the NCN-pincer palladium

3) Reaction conditions: 0.94–1.19 mol.% Pd, 10 mol.% i-Pr_2NEt, CH_2Cl_2, rt.

4) Reaction conditions: used substrates are methylisocyanate and benzaldehyde, 1 mol.% Pd, 10 mol.% i-Pr_2NEt, CH_2Cl_2, rt.

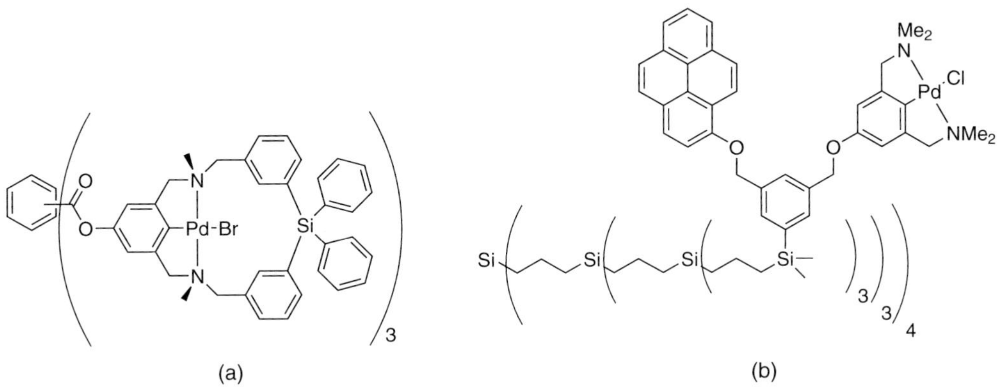

Figure 15.24 Multimetallic dendritic system with encapsulated catalytic sites synthesized in the group of Van Koten by Rodríguez (a) and Slagt (b).

unit, which is caused by cation–π interactions, leads to a small, but significant, rate enhancement of the aldol condensation reaction of aromatic aldehydes with methyl isocyanoacetate.

The tweezer was coupled to a carbosilane dendrimer by lithiation of the tweezer bromide with *t*BuLi, and subsequent quench with a G2 dendrimer, containing 36 peripheral chlorodimethylsilyl groups. After palladation, the molecule shown in Figure 15.24b was obtained. The monomeric tweezer and the dendrimer-decorated tweezers were both tested in the aldol condensation reaction of methyl isocyanoacetate and benzaldehyde.[5] Compared to the monomeric species, the dendritic compound showed a lower conversion, while at the same time, the cis : trans-ratio was not affected. This negative dendritic effect regarding conversion was attributed to the increase in steric crowding, making various catalytic sites inaccessible for substrate molecules.

15.3.7
ECE-Pincer Palladium Complexes on Polymers

Not only dendrimers have been used as support for palladacycles. The groups of Bergbreiter and Weck have performed extensive research on pincer palladium complexes attached to polymers to be used in carbon–carbon bond-formation reactions. Bergbreiter connected SCS-pincer palladium catalysts to poly(ethylene glycol) (PEG), poly(*N*-isopropylacrylamide) (PNIPAM) and poly(*N*-octadecylacrylamide) (PNODAM) polymers. These polymeric materials were selected because of their high thermal stability with respect to Pd(0) catalysts [105–107]. Heck and Suzuki reactions with aryl iodides and various acceptor substrates were

5) Reaction conditions: 1.0 mmol methyl isocyanoacetate, 1.2 mmol benzaldehyde, 1 mol.% catalyst, 10 mol.% *i*-Pr_2NEt, CH_2Cl_2, rt.

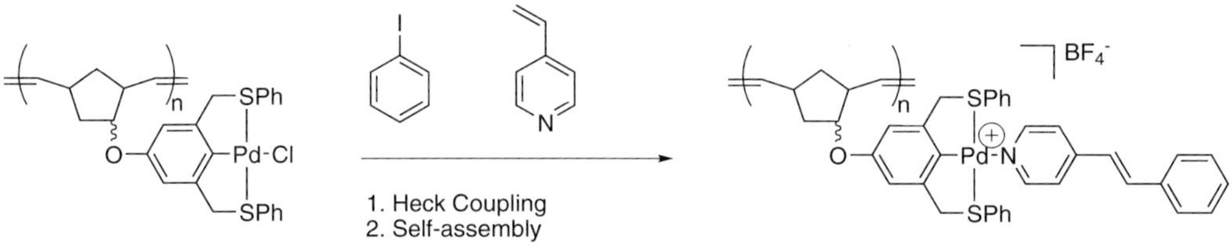

Scheme 15.6 Side-chain functionalized polymers possessing SCS-pincer Pd-complexes used for Heck catalysis and subsequent self-assembly of the formed species.

successfully performed in an *N*,*N*-dimethylacetamide/heptane mixture with triethylamine as the base in air at elevated temperatures.

In 2002, Weck and coworkers published reports on side-chain functionalized polymers containing palladated SCS-pincer complexes at each repeating unit [108, 109]. A palladated monomeric unit containing a bicyclic alkene was polymerized via ring-opening metathesis polymerization with a first generation Grubbs' catalyst to form these poly(norbornene) chains. These metallopolymers were used as catalysts in Heck coupling reactions and subsequently as coordinative centers for pyridine- or nitrile-containing substrates (Scheme 15.6). In another publication, cross-linked functionalized polymers were obtained by coordination via molecules like 1,2-(dipyridin-4-yl)-ethane, bearing two pyridine ligands in the same molecule, provide the possibility for cross-linking [110].

The groups of Bergbreiter and Weck independently found that both the homogeneous species and the tethered SCS-pincer palladium(II) complexes were not stable during the conditions used for Heck catalysis, while the conversion remained very high [111–113]. Furthermore, studies have shown that electronically different palladacycles showed a remarkable consistency in reactivity. After establishing mercury poisoning and kinetic studies, it was suggested that all pincer palladium(II) complexes act as pre-catalysts during Heck coupling reactions, and that the real catalytic species are poorly-defined, yet highly active palladium(0) nanoparticles.

Bergbreiter and coworkers obtained a further confirmation for this hypothesis by performing an unambiguous experiment that showed decomposition of palladacycles [111]. Here, in a thermomorphic system (i.e. a combination of liquids that in one temperature range is a homogeneous single phase but in another temperature range forms two immiscible layers) containing heptane and aqueous DMA, Heck chemistry was performed with PNODAM polymer-bound palladacycles. After the substrate was completely converted, the system was cooled down, and the formed layers were separated. The apolar PNODAM-bound palladacycles were, not surprisingly, found back in the heptane layer. The polar DMA-layer did not contain any measurable amounts of palladium [<0.1 ppm, value obtained by inductively coupled plasma (ICP) MS] yet remained catalytically active for freshly added substrates. Although its activity was slower than for the thermomorphic system, any activity in a phase that does not contain any polymers and thus palladacycles already shows that a palladacycle is not required for Heck catalysis.

Other studies on this and related supramolecular non-covalent assemblies were also performed by Weck *et al.* [114–116].

The group of Van Koten studied NCN-pincer palladium dendrimers bound to a "dendronized" polystyrene support [117]. The dendrons consist of propylamidobenzyl ether branching units with peripheral primary amine groups. After reaction with highly active succinimidyl ester-functionalized NCN-pincer palladium complexes, the first to third generation dendronized polymers were synthesized. Capping of the remaining unreacted free amines with a UV-active reagent (2,4-dinitrofluorobenzene) and measuring the UV absorbance revealed a loading of 91–93%, and thus an average number of 850 (G1), 1700 (G2) and 3400 (G3) NCN-pincer metal centers per molecule.

These so-called DenPol's (dendritic polymers) were tested in the earlier mentioned aldol condensation reaction of methyl isocyanate and benzaldehyde.[6] All generations showed a significantly lower activity than the parent propylamido-NCN palladium bromide, probably due to solubility problems and thus the more heterogeneous character of these DenPol's [117]. A more surprising observation, however, is that all DenPol catalysts, independent of the dendron generation, exhibit the same activity per palladium site for this reaction. Apparently, no interference between the various pincer metal sites takes place and therefore all the palladium centers seem to be kinetically equivalent. These results suggested that the pincer metal moieties are forced outwards, and no back-folding of groups occurs.

A collaboration between the groups of Van Koten and Frey resulted in the use of NCN-pincer palladium functionalized hyperbranched polymers for catalysis [118]. The main advantage of these polydisperse compounds compared to dendrimers is their easy synthesis. In a single-step, one-pot procedure a monomer can be converted into a hyperbranched polymer. Disadvantages are the fact that the reactive sites obtained after functionalization are randomly distributed throughout the whole molecule, and that, unlike dendrimers, the synthesis cannot be performed in a controllable way. Nanosized, hyperbranched polycarbosilane compounds functionalized with NCN-pincer palladium-complexes were synthesized. The polydispersity index (PDI) of these compounds is 1.8, making these polymers candidates for applications in continuous membrane reactor catalysis.

The catalytic behavior of a monomeric NCN-pincer palladium complex as a control was compared with the hyperbranched polycarbosilane compound in the aldol reaction of benzaldehyde and methyl isocyanoacetate.[7] It was shown that the total turnover numbers per palladium site of the two tested compounds are almost equal, although the initial TOF (after 1 h) of the monomer is about two times higher than the initial TOF of the hyperbranched polymer. Hajji has comprehensively reviewed hyperbranched polymers in catalysis [119].

SCS-pincer palladium complexes have also been attached to solid supports. Portnoy *et al.* synthesized aromatic polythioether dendrons which can serve as precursor to the SCS-pincer complex on Wang resin [120]. Metallation takes place

6) Reaction conditions: 2.5 mol.% Pd, 10 mol.% *i*-Pr_2NEt, CH_2Cl_2, rt.

7) Reaction conditions: 1 mol.% Pd, 10 mol.% *i*-Pr_2NEt, CH_2Cl_2, rt.

with $PdCl_2(PhCN)_2$. Preliminary catalytic results show that G2(CO_2Me) resins are efficient and recyclable pre-catalysts for the Heck reaction.

15.4
Concluding Remarks

In the broad field of supported palladium catalysis, the area of palladacycle-functionalized dendrimers is still in its infancy. For example, the only ligands supported on dendrimers are the ECE-pincer moieties (and its EC-half-pincer analogues) containing E = N, P or S donor atoms. Until now, organometallic palladadendrimers have mainly been used in two different domains by, coincidently, two Dutch research groups: (i) as (recyclable) catalysts (by Van Koten and coworkers) and (ii) for the synthesis of self-assembled nanostructures (in the group of Reinhoudt).

In catalysis, the reason palladacycle-functionalized dendrimers are used is that the covalent Pd–C bond creates robust dendritic catalysts, which therefore increases the sustainability of the nanosized catalysts. However, so far only a fraction of the possibilities have been investigated. Ironically, most research has focused on Heck catalysis, for which it was shown recently that the catalytic species are Pd-nanoparticles rather than the palladacycles themselves. Aldol-condensations are another class of reactions that has been studied to some extent. These investigations have shown that the application of palladacyclic dendrimers in continuous flow membrane reactors is a promising approach for the continuous production of, for example, pharmaceutically active compounds.

The coordination-based dendrimers of Reinhoudt's group have been used to visualize individual nanoscopic assemblies, and to introduce redox- and optical active fragments. An interesting development along this line of research is the synthesis of water-soluble palladadendrimers.

Considering the application of palladadendrimers in catalysis, several developments may be foreseen. Whereas to date mostly single catalytic transformations have been studied, a combination of dendritic palladacycles could lead to new and sustainable protocols for multi-step synthesis of fine chemicals. Such catalytic cascade or tandem reactions could either be accomplished by multiple-different palladacycles on a single dendritic object or otherwise by means of multiple dendritic objects carrying a single, yet different, type of palladacycle each. One of the constituents of palladadendrimers in particular and of dendritic catalyst in general, of which the overall effect on the catalytic properties is least well understood, is actually the dendritic backbone itself. It would therefore be of interest to study the effect of the nature and properties of the dendrimer backbone itself on catalysis in a more consistent manner.

To conclude, this chapter highlights the interesting properties of palladacycle-functionalized dendrimers, but also shows that a coming-of-age of these organometallic macromolecules and their application requires much more research effort.

References

1 Buhleier, E., Wehner, W. and Vögtle, F. (1978) *Synthesis*, 155–8.

2 Tomalia, D.A., Baker, H., Dewald, J., *et al.* (1985) *Polymer Journal (Tokyo)*, **17**, 117–32.

3 Tomalia, D.A., Naylor, A.M. and Goddard, W.A., III. (1990) *Angewandte Chemie-International Edition*, **29**, 138–75.

4 Tomalia, D.A. and Durst, H.D. (1993) *Topics in Current Chemistry*, **165**, 193–313.

5 Newkome, G.R., Yao, Z.Q., Baker, G.R. and Gupta, V.K. (1985) *Journal of Organometallic Chemistry*, **50**, 2003–2004.

6 Newkome, G.R., Yao, Z.Q., Baker, G.R., *et al.* (1986) *Journal of the American Chemical Society*, **108**, 849–50.

7 Hawker, C.J. and Fréchet, J.M.J. (1990) *Journal of the American Chemical Society*, **112**, 7638–47.

8 Hawker, C.J. and Fréchet, J.M.J. (1990) *Journal of the Chemical Society D-Chemical Communications*, **15**, 1010–13.

9 Wooley, K.J., Hawker, C.J. and Frechet, J.M.J. (1991) *Journal of the American Chemical Society*, **113**, 4252–61.

10 Fréchet, J.M.J. (1994) *Science*, **263**, 1710–15.

11 Oosterom, G.E., Reek, J.N.H., Kamer, P.C.J. and van Leeuwen, P.W.N.M. (2001) *Angewandte Chemie-International Edition*, **40**, 1828–49.

12 Dijkstra, H.P., van Klink, G.P.M. and van Koten, G. (2002) *Accounts of Chemical Research*, **35**, 798–810.

13 Astruc, D. and Chardac, F. (2001) *Chemical Reviews*, **101**, 2991–3023.

14 Majoral, J.P. and Caminade, A.M. (1999) *Chemical Reviews*, **99**, 845–80.

15 Kreiter, R., Kleij, A.W., Klein Gebbink, R.J.M. and van Koten, G. (2001) *Topics in Current Chemistry: Dendrimers IV*, **217**, 163–99.

16 Dahan, A. and Portnoy, M. (2005) *Journal of Polymer Science Part A-Polymer Chemistry*, **43**, 235–62.

17 Chase, P.A., Klein Gebbink, R.J.M. and van Koten, G. (2004) *Journal of Organometallic Chemistry*, **689**, 4016–54.

18 Kassube, J.K. and Gade, L.H. (2006) *Topics in Organometallic Chemistry*, **20** (Dendrimer Catal), 39–59.

19 Méry, D. and Astruc, D. (2006) *Coordination Chemistry Reviews*, **250**, 1965–79.

20 Twyman, L.J., King, A.S.H. and Martin, I.K. (2002) *Chemical Society Reviews*, **31**, 69–82.

21 Newkome, G.R., He, E. and Moorefield, C.N. (1999) *Chemical Reviews*, **99**, 1689–746.

22 Gorman, C. (1998) *Advanced Materials*, **10**, 295–309.

23 Andres, R., de Jesús, E. and Flores, J.C. (2007) *New Journal of Chemistry*, **31**, 1161–91.

24 Berger, A., Klein Gebbink, R.J.M. and van Koten, G. (2006) *Topics in Organometallic Chemistry*, **20** (Dendrimer Catal), 1–38.

25 Knapen, J.W.J., van der Made, A.W., de Wilde, J.C., *et al.* (1994) *Nature*, **372**, 659–63.

26 Ong, W. and McCarley, R.L. (2005) *Organic Letters*, **7**, 1287–90.

27 Dahan, A. and Portnoy, M. (2003) *Organic Letters*, **5**, 1197–200.

28 Reetz, M.T., Lohmer, G. and Schwickardi, R. (1997) *Angewandte Chemie-International Edition*, **36**, 1526–9.

29 Hurley, A. and Mohler, D.L. (2000) *Organic Letters*, **2**, 2745–8.

30 Valerio, C., Fillaut, J.L., Ruiz, J., *et al.* (1997) *Journal of the American Chemical Society*, **119**, 2588–9.

31 Breinbauer, R. and Jacobsen, E.N. (2000) *Angewandte Chemie-International Edition*, **39**, 3604–7.

32 Dijkstra, H.P., Slagt, M.Q., McDonald, A., *et al.* (2003) *European Journal of Inorganic Chemistry*, **5**, 830–8.

33 Helms, B. and Frechet, J.M.J. (2006) *Advanced Synthesis Catalysis*, **348**, 1125–48.

34 Hecht, S. and Fréchet, J.M.J. (2001) *Angewandte Chemie-International Edition*, **40**, 74–91.

35 Oosterom, G.E., van Haaren, R.J., Reek, J.N.H., *et al.* (1999) *Chemical Communications*, 1119–20.

36 de Groot, D., Eggeling, E.B., de Wilde, J.C., *et al.* (1999) *Chemical Communications*, 1623–4.

37 de Groot, D., Reek, J.N.H., Kamer, P.C.J. and van Leeuwen, P.W.N.M. (2002) *European Journal of Organic Chemistry*, **6**, 1085–95.

38 Heuzé, K., Mery, D., Gauss, D., *et al.* (2004) *Chemistry – A European Journal*, **10**, 3936–44.

39 Mizugaki, T., Murata, M., Ooe, M., *et al.* (2002) *Chemical Communications*, 52–3.

40 Lu, S.M. and Alper, H. (2005) *Journal of the American Chemical Society*, **127**, 14776–84.

41 Alper, H., Arya, P., Bourque, C., *et al.* (2000) *Canadian Journal of Chemistry*, **78**, 920–4.

42 Zweni, P.P. and Alper, H. (2006) *Advanced Synthesis Catalysis*, **348**, 725–31.

43 Antebi, S., Arya, P., Manzer, L.E. and Alper, H. (2002) *Journal of Organic Chemistry*, **67**, 6623–31.

44 Reynhardt, J.P.K. and Alper, H. (2003) *Journal of Organic Chemistry*, **68**, 8353–60.

45 Zweni, P.P. and Alper, H. (2004) *Advanced Synthesis Catalysis*, **346**, 849–54.

46 Smith, G., Chen, R. and Mapolie, S. (2003) *Journal of Organometallic Chemistry*, **673**, 111–15.

47 Smith, G.S. and Mapolie, S.F. (2004) *Journal of Molecular Catalysis A – Chemical*, **213**, 187–92.

48 Meder, M., Galka, C.H. and Gade, L.H. (2005) *Monatshefte Fur Chemie*, **136**, 1693–706.

49 Siggelkow, B.A. and Gade, L.H. (2005) *Zeitschrift Fur Anorganische Und Allgemeine Chemie*, **631**, 2575–84.

50 Huang, C.C., Lin, Y.C., Lin, P.Y. and Chen, Y.J. (2006) *European Journal of Organic Chemistry*, 4510–18.

51 Hoare, J.L., Lorenz, K., Hovestad, N.J., *et al.* (1997) *Organometallics*, **16**, 4167–73.

52 Hovestad, N.J., Hoare, J.L., Jastrzebski, J.T.B.H., *et al.* (1999) *Organometallics*, **18**, 2970–80.

53 Astruc, D., Lu, F. and Aranzaes, J.R. (2005) *Angewandte Chemie – International Edition*, **44**, 7852–72.

54 Scott, R.W.J., Wilson, O.M. and Crooks, R.M. (2005) *The Journal of Physical Chemistry B*, **109**, 692–704.

55 Jansen, J.F.G.A., De Brabander-van den Berg, E.M.M. and Meijer, E.W. (1994) *Science*, **266**, 1226–9.

56 Niu, Y.H., Yeung, L.K. and Crooks, R.M. (2001) *Journal of the American Chemical Society*, **123**, 6840–6.

57 Ooe, M., Mizugaki, T., Ebitani, K. and Kaneda, K. (2002) *Abstracts of Papers of the American Chemical Society*, **223**, U416–17.

58 Chung, Y.M. and Rhee, H.K. (2003) *Catalysis Letters*, **85**, 159–64.

59 Chandler, B.D. and Gilbertson, J.D. (2006) *Topics in Organometallic Chemistry*, **20** (Dendrimer Catal), 97–120.

60 Yeung, L.K. and Crooks, R.M. (2001) *Nano Letters*, **1**, 14–17.

61 Yeung, L.K., Lee, C.T., Johnston, K.P. and Crooks, R.M. (2001) *Chemical Communications*, 2290–1.

62 Liu, G.X. and Puddephatt, R.J. (1996) *Organometallics*, **15**, 5257–9.

63 Chessa, G., Canovese, L., Gemelli, L., Visentin, F. and Seraglia, R. (2001) *Tetrahedron*, **57**, 8875–82.

64 Angurell, I., Rossell, O., Seco, M. and Ruiz, E. (2005) *Organometallics*, **24**, 6365–73.

65 Vinogradov, S.A., Lo, L.W. and Wilson, D.F. (1999) *Chemistry – A European Journal*, **5**, 1338–47.

66 Rozhkov, V., Wilson, D. and Vinogradov, S. (2002) *Macromolecules*, **35**, 1991–3.

67 Albrecht, M. and van Koten, G. (2001) *Angewandte Chemie – International Edition*, **30**, 3750–81.

68 Moulton, C.J. and Shaw, B.L. (1976) *Journal of the Chemical Society – Dalton Transactions*, **11**, 1020–4.

69 van Koten, G., Timmer, K., Noltes, J.G. and Spek, A.L. (1978) *Journal of the Chemical Society D – Chemical Communications*, **6**, 250–2.

70 van Koten, G., Jastrzebski, J.T.B.H., Noltes, J.G., *et al.* (1978) *Journal of Organometallic Chemistry*, **148** , 233–45.

71 Creaserm, C.S. and Kaska, W.C. (1978) *Inorganica Chimica Acta*, **30**, L325–6.

72 Dijkstra, H.P., Ronde, N., van Klink, G.P.M., *et al.* (2003) *Advanced Synthesis Catalysis*, **345**, 364–9.

73 Beletskaya, I.P., Chuchurjukin, A.V., Dijkstra, H.P., *et al.* (2000) *Tetrahedron Letters*, **41**, 1081–5.

74 Huck, W.T.S., van Veggel, F.C.J.M., Kropman, B.L., Blank, *et al.* (1995) *Journal of the American Chemical Society*, **117**, 8293–4.

75 Dijkstra, H.P., Steenwinkel, P., Grove, D.M., *et al.* (1999) *Angewandte Chemie–International Edition*, **38**, 2185–8.

76 Huck, W.T.S., Snellink-Ruel, B., van Veggel, F.C.J.M. and Reinhoudt, D.N. (1997) *Organometallics*, **16**, 4287–91.

77 Beletskaya, I.P., Chuchuryukin, A.V., van Koten, G., *et al.* (2003) *Russian Journal of Organic Chemistry*, **39**, 1268–81.

78 Chanthateyanonth, R. and Alper, H. (2004) *Advanced Synthesis Catalysis*, **346**, 1375–84.

79 Dijkstra, H.P., Meijer, M.D., Patel, J., *et al.* (2001) *Organometallics*, **20**, 3159–68.

80 Rodríguez, G., Lutz, M., Spek, A.L. and van Koten, G. (2002) *Chemistry–A European Journal*, **8**, 45–57.

81 van de Coevering, R., Kuil, M., Alfers, A.P., *et al.* (2005) *Organometallics*, **24**, 6147–58.

82 Dijkstra, H.P., Chuchuryukin, A., Suijkerbuijk, B.M.J.M., *et al.* (2002) *Advanced Synthesis Catalysis*, **344**, 771–80.

83 Dijkstra, H.P., Kruithof, C.A., Ronde, N., *et al.* (2003) *Journal of Organic Chemistry*, **68**, 675–85.

84 Suijkerbuijk, B.M.J.M., Lutz, M., Spek, A.L., *et al.* (2004) *Organic Letters*, **6**, 3023–6.

85 Suijkerbuijk, B.M.J.M., Herreras Martínez, S.D., Van Koten, G. and Klein Gebbink, R.J.M. (2008) *Organometallics* **27**, 534–42.

86 van de Coevering, R., Kuil, M., Klein Gebbink, R.J.M. and van Koten, G. (2002) *Chemical Communications*, **15**, 1636–7.

87 van de Coevering, R., Alfers, A.P., Meeldijk, J.D., *et al.* (2006) *Journal of the American Chemical Society*, **128**, 12700–13.

88 Kleij, A.W., van de Coevering, R., Klein Gebbink, R.J.M., *et al.* (2001) *Chemistry–A European Journal*, **7**, 181–92.

89 Ribaudo, F., van Leeuwen, P.W.N.M. and Reek, J.N.H. (2006) *Topics in Organometallic Chemistry*, **20** (Dendrimer Catal), 39–59.

90 van de Coevering, R., Klein Gebbink, R.J.M. and van Koten, G. (2005) *Progress in Polymer Science*, **30**, 474–90.

91 van Manen, H.J., Nakashima, K., Shinkai, S., *et al.* (2000) *European Journal of Inorganic Chemistry*, 2533–40.

92 Huck, W.T.S., van Veggel, F.C.J.M. and Reinhoudt, D.N. (1996) *Angewandte Chemie–International Edition*, **35**, 1213–15.

93 Huck, W.T.S., Prins, L.J., Fokkens, R.H., *et al.* (1998) *Journal of the American Chemical Society*, **120**, 6240–6.

94 van Manen, H.J., Fokkens, R.H., Nibbering, N.M.M., *et al.* (2001) *Journal of Organic Chemistry*, **66**, 4643–50.

95 Huck, W.T.S., Rohrer, A., Anilkumar, A.T., *et al.* (1998) *New Journal of Chemistry*, **22**, 165–8.

96 Huisman, B.H., H., S., Huck, W.T.S., *et al.* (1999) *Angewandte Chemie–International Edition*, **38**, 2248–51.

97 van Manen, H.J., Auletto, T., Dordi, B., *et al.* (2002) *Advance Functional Materials*, **12**, 811–18.

98 Friggeri, A., van Manen, H.J., Auletta, T., *et al.* (2001) *Journal of the American Chemical Society*, **123**, 6388–95.

99 Huck, W.T.S., Hulst, R., Timmerman, P., *et al.* (1997) *Angewandte Chemie–International Edition*, **36**, 1006–8.

100 van Manen, H.J., Fokkens, R.H., van Veggel, F.C.J.M. and Reinhoudt, D.N. *European Journal of Organic Chemistry*, (2002), 18.

101 van Manen, H.J., van Veggel, F.C.J.M. and Reinhoudt, D.N. (2001) *Topics in Current Chemistry: Dendrimers IV*, **217**, 121–62.

102 Kleij, A.W., Klein Gebbink, R.J.M., Nieuwenhuijzen, P.A.J., *et al.* (2001) *Organometallics*, **20**, 634–47.

103 Albert, J., Cadena, J.M. and Granell, J. (1997) *Tetrahedron Asymmetry*, **8**, 991–4.

104 Slagt, M.Q., Jastrzebski, J.T.B.H., Klein Gebbink, R.J.M., *et al.* *European Journal of Organic Chemistry*, (2003), 1692–703.

105 Bergbreiter, D.E., Osburn, P.L., Wilson, A. and Sink, E.M. (2000) *Journal of the*

American Chemical Society, **122**, 9058–64.

106 Bergbreiter, D.E., Osburn, P.L. and Frels, J.D. (2001) *Journal of the American Chemical Society*, **123**, 11105–6.

107 Bergbreiter, D.E., Sung, S.D., Li, J., *et al.* (2004) *Organic Process Research and Development*, **8**, 461–8.

108 Pollino, J.M. and Weck, M. (2002) *Organic Letters*, **4**, 753–6.

109 Pollino, J.M. and Weck, M. (2002) *Synthesis*, **9**, 1277–85.

110 Pollino, J.M., Nair, K.P., Stubbs, L.P., *et al.* (2004) *Tetrahedron*, **60**, 7205–15.

111 Bergbreiter, D.E., Osburn, P.L. and Frels, J.D. (2005) *Advanced Synthesis Catalysis*, **347**, 172–84.

112 Yu, K.Q., Sommer, W., Richardson, J.M., *et al.* (2005) *Advanced Synthesis Catalysis*, **347**, 161–71.

113 Sommer, W.J., Yu, K.Q., Sears, J.S., *et al.* (2005) *Organometallics*, **24**, 4351–61.

114 Yu, K.Q., Sommer, W., Weck, M. and Jones, C.W. (2004) *Journal of Catalysis*, **226**, 101–10.

115 Gerhardt, W.W., Zucchero, A.J., Wilson, J.N., *et al.* (2006) *Chemical Communications*, 2141–3.

116 South, C.R., Higley, M.N., Leung, K.C.F., *et al.* (2006) *Chemistry–A European Journal*, **12**, 3789–97.

117 Suijkerbuijk, B.M.J.M., Shu, L., Gebbink, K., *et al.* (2003) *Organometallics*, **22**, 4175–7.

118 Schlenk, C., Kleij, A.W., Frey, H. and van Koten, G. (2000) *Angewandte Chemie–International Edition*, **39**, 3736.

119 Hajji, C. and Haag, R. (2006) *Topics in Organometallic Chemistry*, **20** (Dendrimer Catal), 149–76.

120 Dahan, A., Weissberg, A. and Portnoy, M. (2003) *Chemical Communications*, 1206–7.

Index

Palladacycles. Dupont, Pfeffer

ISBN: 978-3-527-31781-3

e

m

n

o

q

r